AF360699

Optimizing Plant Water Use Efficiency for a Sustainable Environment

Optimizing Plant Water Use Efficiency for a Sustainable Environment

Editors

Ivan Francisco Garcia Tejero
Victor Hugo Durán-Zuazo

MDPI • Basel • Beijing • Wuhan • Barcelona • Belgrade • Manchester • Tokyo • Cluj • Tianjin

Editors

Ivan Francisco Garcia Tejero
Forest and Natural Resources
Andalusian Institute of
Teaching and Agricultural
Research (IFAPA)
Alcalá del Río (Sevilla)
Spain

Victor Hugo Durán-Zuazo
Forest and Natural Resources
Andalusian Institute of
Teaching and Agricultural
Reserach (IFAPA)
Granada
Spain

Editorial Office
MDPI
St. Alban-Anlage 66
4052 Basel, Switzerland

This is a reprint of articles from the Special Issue published online in the open access journal *Agronomy* (ISSN 2073-4395) (available at: www.mdpi.com/journal/agronomy/special_issues/plant_water).

For citation purposes, cite each article independently as indicated on the article page online and as indicated below:

LastName, A.A.; LastName, B.B.; LastName, C.C. Article Title. *Journal Name* **Year**, *Volume Number*, Page Range.

ISBN 978-3-0365-5136-4 (Hbk)
ISBN 978-3-0365-5135-7 (PDF)

Cover image courtesy of Ivan Francisco Garcia Tejero

Contents

About the Editors

Ivan Francisco Garcia Tejero

Iván Francisco García-Tejero (Ph.D., 2010) started his research activity in 2005, in the Andalusian Institute of Teaching and Agricultural research, obtaining a PhD degree in 2010. He was then employed in this same institution for different projects related to the irrigation water management, achieving a Post-Doc contract in 2012 for 5 years. Once finished, he continued working on other research projects for the same institution, and finally, got the titular research position in 2019, within the "Agriculture and Environment"department and the research line "Water Use Efficiency". During his career as a researcher, he has had much experience about the water-use efficiency in Mediterranean woody crops, such as citrus, almonds or olives. Moreover, he has developed significant works related to conservation agriculture practices in rainfed systems by means of direct drill strategies or plant cover management. He has participated in different research projects at national, regional and EU levels on soil and water conservation, water-use efficiency, crop responses to drought conditions and irrigation scheduling under water scarcity scenarios. The results of these projects have led to the publication of more than 100 articles in SCI-JCR and non-SCI-JCR journals, 6 books and more than 20 book chapters, in addition to a series of technical reports and contributions to national and international conferences. Dr. García-Tejero is an Editorial Board member of *Agronomy* and *Agriculture* (MDPI, Switzerland) and Editor Reviewer of *Frontiers in Plant Science* (Frontiers Media SA, Switzerland), and collaborates as a reviewer with many other peer-reviewed scientific journals. He has carried out post-doctoral stays at Institute of Chemistry and Biological Technology (Lisbon, Portugal) and have directed 3 Ph.D. Theses and more than 20 Master Degree Theses and End-of-Degree Projects.

Victor Hugo Durán-Zuazo

Víctor Hugo Durán Zuazo (Ph.D., 2001) has expertise in soil management systems with rainfed woody crops (olive and almond trees) in areas with high erosion risk, in water stress management strategies for irrigated fruit trees (avocado, mango, citrus, and almond trees), and in woody crops for bioenergy. He has been involved in different research projects at national, regional and EU levels, on land use changes, soil and water conservation, water-use efficiency, soil quality, and biomass. The results of these projects have led to the publication of more than 69 articles in SCI-JCR journals, 72 articles in non-SCI-JCR journals, 12 books and 35 book chapters, in addition to a series of technical reports and contributions to national and international conferences. Dr. Durán Zuazo is an Associate Editor of *Agronomy for Sustainable Development* (INRAE, France) and *Comunicata Scientiae* (FUP-Brazil) and collaborates as a reviewer with many others peer-reviewed scientific journals. He has carried out post-doctoral stays at Wageningen University, Netherlands (Environmental Sciences-Soil & Water Conservation Department) and at the Agricultural Research Service USDA (National Soil Erosion Research Laboratory-NSERL, West Lafayette, Indiana, USA). Dr. Durán Zuazo is the head of the PAIDI non-university research group AGR-144 of Andalusian Institute of Agricultural and Fisheries Research and Training (IFAPA) "Conservation and Sustainable Use of Soil, Water and Biodiversity in Agricultural Systems".

Preface to "Optimizing Plant Water Use Efficiency for a Sustainable Environment"

The rising shortage of water resources in crop-producing regions worldwide and the need for irrigation optimisation call for sustainable water savings. The allocation of irrigation water will be an ever-increasing source of pressure because of vast agricultural demands under changing climatic conditions. Consequently, irrigation has to be closely linked with water-use efficiency with the aim of boosting productivity and improving food quality, singularly in those regions where problems of water shortages or collection and delivery are widespread. The present Special Issue (SI) showcases 19 original contributions, addressing water-use efficiency in the context of sustainable irrigation management to meet water scarcity conditions. These papers cover a wide range of subjects including (i) interaction mineral nutrition and irrigation in horticultural crops, (ii) sustainable irrigation in woody fruit crops, (iii) medicinal plants, (iv) industrial crops, and (v) other topics devoted to remote sensing techniques and crop water requirements, genotypes for drought tolerance, and agricultural management. The studies were carried out in both field and laboratory surveys, with modelling studies also being conducted, and a wide range of geographic regions are also covered. The collection of these manuscripts presented in this SI updates on and provides a relevant contribution for efficient saving water resources.

To Elías and Noel, who struggled for living, and now they are happily with us.

Ivan Francisco Garcia Tejero and Victor Hugo Durán-Zuazo
Editors

Editorial

Plant Water Use Efficiency for a Sustainable Agricultural Development

Iván Francisco García-Tejero [1] and Víctor Hugo Durán-Zuazo [2,*]

1   IFAPA Centro "Las Torres", Carretera Sevilla-Alcalá del Río km 12,2, 41200 Alcalá del Río, Spain; ivanf.garcia@juntadeandalucia.es
2   IFAPA Centro "Camino de Purchil", Camino de Purchil s/n, 18004 Granada, Spain
*   Correspondence: victorh.duran@juntadeandalucia.es; Tel.: +34-958-895-200

Abstract: The rising shortage of water resources worldwide in crop-producing regions and the need for irrigation optimisation call for sustainable water savings. That is, the allocation of irrigation water will be an ever-increasing source of pressure because of vast agricultural demands under changing climatic conditions. Consequently, irrigation has to be closely linked with water-use efficiency with the aim of boosting productivity and improving food quality, singularly in those regions where problems of water shortages or collection and delivery are widespread. The present Special Issue (SI) contains 19 original contributions addressing water-use efficiency under challenging topic of sustainable irrigation management to meet water scarcity conditions. These papers cover a wide range of subjects, including (i) interaction mineral nutrition and irrigation in horticultural crops, (ii) sustainable irrigation in woody fruit crops, (iii) medicinal plants, (iv) industrial crops, and (v) others devoted to remote sensing techniques and crop water requirements, genotypes for drought tolerance, and agricultural management platform. The studies have been carried out in both field and laboratory surveys, as well as modelling studies, and a wide range of geographic regions are also covered. The collection of these manuscripts presented in this SI updates and provides a relevant knowledge contribution for efficient saving water resources.

Keywords: deficit irrigation; crop-water requirements; smart farming; crop-production functions; food quality; remote sensing; crop physiological response to drought scenarios

Citation: García-Tejero, I.F.; Durán-Zuazo, V.H. Plant Water Use Efficiency for a Sustainable Agricultural Development. *Agronomy* **2022**, *12*, 1806. https://doi.org/10.3390/agronomy12081806

Academic Editor: Peter Langridge

Received: 25 July 2022
Accepted: 26 July 2022
Published: 30 July 2022

Publisher's Note: MDPI stays neutral with regard to jurisdictional claims in published maps and institutional affiliations.

1. Introduction

Water is considered the most vital resource for agricultural development, and under scarcity conditions and climate change, substantial effort must be devoted to introducing measures in improving their sustainable use [1,2]. Various studies have reported the predictably effects of climate change on the agricultural systems [3–5]. Augmenting agricultural water through irrigation systems to complement soil moisture deficit has driven enhanced agricultural productivity in large areas worldwide. The consequences include unintended side effects such as exhaustion of river flow, river basin closure, groundwater reduction, and water pollution [6]. The Green revolution promoted food production but turned out not to be sustainable due to improper management by the agricultural sector.

In many dry climate countries, the dominant form of water is the contained in the soil, i.e., infiltrated rain, and boosting crop production is therefore an issue of water security, which can be achieved only by overcoming the difficulties of rainfall variability exacerbated by climate change. In this line, rainfed agriculture is the most common method of agriculture in developing regions, 80% of the land farmed around the world being rainfed and can cover more than 75% of the needed increase in food production by the year 2025 [7].

Agricultural water management should balance the need for crops with the preservation of a sustainable environment. In this line, water-use efficiency is the main challenge in worldwide farming practices, where water shortages are becoming more frequent. Today, the agricultural sphere is undergoing significant changes regarding irrigation, the implementation of adaptive and water-saving strategies being urgent [8]. Moreover, under climate change conditions, traditional and calculated irrigation at full-water requirements based on crop evapotranspiration (ETC) will probably not be practicable in the medium-long term, requiring alternative strategies to face the current climatic circumstances [9]. In this line, deficit-irrigation could be considered a sustainable option to achieve a balance in irrigated crops in saving water and producing assumable yield losses [10]. That is, deficit-irrigation practices are a tool to mitigate climate change effects, attaining environmental, social, and economic benefits.

In this SI, we tried to collect studies regarding water scarcity as the most limiting factor in agriculture, together with the climate change scenario that promotes a framework of uncertainty and great challenges regarding the sustainability and viability of current agroecosystems. Likewise, this SI provides updates and recent developments in physiological and biochemical perspectives on the response to water deficit in field and greenhouse crops, as well as new tools to assess the crop water status, monitor the continuous soil–plant–atmosphere system, or integrate information systems based on big-data and smart farming tools that reinforce our knowledge to offer an appropriate response to the current challenges for achieve a higher water-use efficiency.

2. Overview of This SI

This SI presents 19 original contributions focused on plant water-use efficiency and aimed to address the challenging topic of sustainable irrigation management to face water scarcity conditions. From a methodological perspective, the contributions involve both field [11–25] and laboratory [26,27] experiments, and modelling [18,28,29] studies. The present SI contains studies at different spatial scales, from the field- to regional-scales. Particularly, the contributions stress five main subjects, including: (i) the interaction of mineral nutrition and irrigation in horticultural crops [11,12,21,27]; (ii) sustainable irrigation in woody fruit crops [15,17,19,22,25,26]; (iii) medicinal plants [14,24]; (iv) industrial crops [23]; and (v) remote sensing techniques and crop water requirements, genotypes for drought tolerance, and agricultural management platform [18,28,29]. In addition, a wide range of geographic regions worldwide are also covered, including Asia [12–14,16,20,24], Australia [23], Brazil [18], the Mediterranean basin [15,17,19,21,22,25,28,29], and Central [11,26] and Western [27] Europe.

Concretely, subject (i) comprises four papers. In general, these works display the impact of interaction fertilization and irrigation on the yield and growth of artichoke (*Helianthus tuberosus* L.) [11], tomato (*Solanum lycopersicum* L.) [12], maize (*Zea mays* L.) [21], and cassava (*Manihot esculenta* Crantz) [27]. These findings illustrate the importance of assessing the effects on crop productivity under various interaction levels of plant nutrients and water doses for proper management strategies, with the aim to reach a sustainable cultivation for a particular environment. According to Bogucka et al. [11], potassium fertilizer applied at rate of 150 kg K_2O ha^{-1} contributed to the greatest increase in the above-ground biomass yield of artichoke, while the irrigation had a significant effect on total tuber and above-ground biomass yields, which increased by 59 and 42%, respectively. On the other hand, the irrigation level based on 80% ET_C and nitrogen (N) incorporation of 15 mM (N_{100}) for soilless tomato production in greenhouses was found to be optimal regarding yield and irrigation water-use efficiency. In the study by Ibrahim et al. [21], it was determined that the potassium silicate (K_2SiO_3) supply, particularly at 2 mM as a foliar spray, may have benefits on maize under limited irrigation supply. These effects were associated with changes at physiological and biochemical levels, including the adjustment of relative water content and osmolytes, the alleviation of oxidative damage and reduction in cell membrane dysfunction, as well as the enhancement of nutrient uptake and regulation of nonenzymatic and enzymatic antioxidant systems. Finally, Wasonga et al. [27] found significant interactions between deficit irrigation and K nutrition, whereby decreasing the irrigation dose to 60% together with 16 mM K resulted in the least reduction in growth and cassava yield. Thus, it seems that deficit irrigation strategies could be used as a tool to develop management practices to improve cassava productivity by means of K fertigation under low moisture field conditions.

Subject (ii), regarding woody fruit trees and water stress induced by deficit irrigation, includes six papers. Millan et al. [15], in a drip-irrigated orchard with early-maturing Japanese plumb (*Prunus salicina* L.) cv. Red Beaut, reported that Ψ_{stem} was found to be the physiological parameter that detected water stress the earliest. Regarding the capacitance probes located closest to the drippers, a drop in the relative soil water content (RSWC) below 0.2 would not be advisable for "non-stress" scheduling in the preharvest period. However, the probes located between the dripper at 0.15 and 0.30 m depth provide information on moderate water stress if the RSWC values falls below 0.2. Severe tree water stress was detected below 0.1 RSWC in capacitance probes located at 60 cm depth from this same position. In vitro culture experiment by Kovalikova et al. [26] revealed that drought stress negatively affects the water content, leaf areas, and chlorophyll content in cherry (*Prunus avium*) and apple (*Malus* × *domestica*) plants. The oxidative status and membrane damage of plants under water deficiency conditions were observed to be important indicators of the stress tolerance mechanism. However, cherries exhibited higher hydrogen peroxide levels compared to apples, whereas their malondialdehyde values were generally lower. The overall findings indicated a wide tolerance range to water deficit among apple and cherry, as well as among cultivars within single plant species. In relation to irrigation strategies, the impact of sustained deficit irrigation (SDI) on parameters related to almond (*Prunus dulcis* L.), including functionality, aroma, and sensory profile of three almond cultivars (Marta, Guara, and Lauranne), was studied by Garcia-Tejero et al. [17]. These authors determined that the SDI strategies allowed the improvement in physical parameters such as unit weight, kernel length, kernel thickness, or color. Higher total phenolic compounds, organic acids, and sugars were found in SDI almonds. Likewise, the highest concentrations of volatile compounds were obtained under SDI, this being a clear advantage in relation to almond flavor. Consequently, a moderate SDI strategy offered relevant improvements in parameters regarding marketability by enhancing the final added value of hydroSOStainable almonds with respect to those cultivated under full irrigation conditions. Similarly, Lipan et al. [19] highlighted that key quality parameters can be used as makers of hydroSOStainable almonds. In addition, these researchers claimed that controlling water stress in almond trees by using deficit irrigation strategies can lead to appropriate yields, improve the product quality, and, accordingly, lead to a final added value. In this context, Gutierrez-Gordillo et al. [22] reported the cultivar effect when a deficit irrigation strategy is being applied because different physiological behaviors with different responses in terms of yield and its components are possible. In this sense, cv. Guara registered the lesser promising results, with significant yield reductions (~14%) when water restrictions around 35% of irrigation requirements (IR) were applied; these are particularly promoted by depletions in the fruit number per tree. Therefore, sustained deficit irrigation at 65% IR (SDI_{65}) was a suitable strategy for cvs. Lauranne and Marta, whereas, for the case of cv. Guara, a more moderate SDI strategy should be selected (such as SDI_{75}). Lastly, Martín-Palomo et al. [25], in a deficit irrigation experiment with table olive (*Olea europea* L.) trees, applied restrictions at 4 and 2 weeks before harvest, irrigation being controlled using the Ψ_{stem}, with a threshold value of -2 MPa, and compared with fully irrigated trees. This water stress did not reduce gas exchange during the deficit period, and the effect on yield was not significant in any of the three-monitoring seasons. In addition, in the high-fruit load season, fruit volume was slightly affected (~10%), but this was not significant at harvest. These findings suggest an early affection of fruit growth with water stress, but with a slow rate of decrease. Thus, a moderate water stress could be useful for the management of deficit irrigation in table olive trees.

The development of aromatic plants under water stress has been described in subject (iii), covering two contributions. In this sense, Kiani et al. [14] evaluated the morphological, phenological, and physiological responses of six *Linum album* accessions under different levels of water deficit treatments (100%, 75%, 50%, and 25% available water) in pot conditions. Accessions UTLA7, UTLA9, and UTLA10 showed a higher seed yield and dry weight of the vegetative part. The maturation process was accelerated in plants under stress conditions, and accession UTLA9 completed its complete growth cycle faster than the other accessions. In addition, the physiological responses of the different accessions did not show the same pattern based on the characteristics studied, and significant differences were observed depending on the trait and accession. Therefore, based on these results, the morphological features (seed yield per plant, plant height, number of inflorescences per plant, shoot, and root dry weight) could be used to select tolerant accessions. On the other hand, Park et al. [24] confirmed that the growth and bioactive compounds of *Crepidiastrum denticulatum*, which is used as a plant-derived raw material for functional food, can be influenced by the water content of the substrate. This study reported that a water content of 45% in the substrate increased the biomass of the shoot and root and increased the phenolic content, antioxidant capacity, and hydroxycinnamic acids content per shoot. These findings are useful for the stable mass production of high-quality *C. denticulatum* in greenhouses or plant factories capable of controlling water content in the root zone.

The subject (iv) includes the industrial crop cultivation. In this line, Braunack et al. [23] with cotton (*Gossypium hirsutum* L.) plantations studied the preformed biodegradable and next-generation sprayable biodegradable polymer membrane (SBPM) formulations, which biodegrade to non-harmful products (water, carbon dioxide, and microbial biomass) and have been introduced as an alternative to plastic mulch films to mitigate plastic pollution. The results showed a higher crop water productivity and crop yield, and increased soil water content with SBPM cover. In addition, this experiment showed that SBPM technology could perform at similar level as oxo-degradable plastic or comparable films under field conditions and, at the same time, provide environmentally sustainable agricultural cropping practices. That is, this innovative technology has shown a high potential even at this early stage of development, indicating that advances in formulation and further testing can lead to significant improvements, and thus increased use in crop production systems.

Lastly, subject (v) includes three contributions in relation to remote sensing and crop water requirements, genotypes for drought tolerance, and agricultural management platform. In this sense, Elnashar et al. [28] reported that the ALESarid-GIS facilitates the selection of suitable crops to improve the estimation of irrigation crop water requirements (CWR) based on crop suitability. In addition, the remote sensing technique and the Surface Energy Balance Algorithms for Land Model (SEBAL) model offer a tool that can be used for estimating the ETa and to support land and water management. The mean daily satellite-based CWR was based on SEBAL ranges between 4.79 and 3.62 mm in Toshka and Abu Simbel areas (Egypt), respectively. This study provides a new approach for coupling Agriculture Land Evaluation System for Arid and Semi-arid regions (ALES-Arid), Ref-ET, and SEBAL models to facilitate the selection of suitable crops and offers an excellent source for predicting CWR in arid environments. In addition, Kamphorst et al. [18] evaluated maize (*Zea mays* L.) genotypes, specifically popcorn inbred lines, under water-stressed (WS) and well-watered (WW) conditions regarding agronomic attributes, root morphology, and leaf "greenness" index (SPAD index), in addition to investigating the viability of indirect selection by canonical correlations (CC) of grain yield (GY) and popping expansion (PE). The WS (-29% less water than WW) significantly affected the GY (-55%) and PE (-28%), increased the brace and crown root density, and more vertically oriented the brace and crown angles. The higher SPAD index was associated with a higher yield, and these measurements were the only ones with no significant genotype $\times$ water condition interaction, which may render concomitant selection for WS and WW easier. For associating the corrections of the different traits, CC proved to have better potential than simple correlations. Thus, the evaluation of SPAD index at 29 days after the anthesis showed the best CC, and based on the previous results of SPAD index, may be used regardless of the water condition. Finally, Aguilar Morales et al. [29] developed an agricultural management platform, based on Enterprise Resource Planning (ERP) principles and with the ability to collect geolocated information from different plots related to Protected Designation of Origin (PDO) and Protected Geographical Indication (PGI) wine (*Vitis vinifera* L.) production. The results showed that the end user completes information database, complies with the legal requirements, and obtains benefits derived from the data analysis. Therefore, the platform: (1) solves the lack of agricultural data problem; (2) provides the user with management tools for its agricultural operations; (3) allows the decision maker to obtain geolocated information in real time; and (4) sets out the bases for the future development of agricultural systems based on Big Data.

3. Conclusions

The research studies contained in this SI described several specific processes and their links with environmental irrigation, balancing environmental protection with improved agricultural production. That is, sustainable irrigation must be based on applying uniform and precise amounts of water, based on rational agricultural knowledge of the plant's water needs. It is clear that sustainable water-management practices are essential to boost productivity, promote regional growth, and protect the environment. This entails addressing the potentially negative impacts of water scarcity on food security. Thus, improving water management in agriculture must be based on implementing sustainable irrigation strategies, developing crop modifications that help tolerate water stress, and promoting cooperation among multidisciplinary researchers.

Author Contributions: Both authors had an equal contribution during the Editorial development.

Funding: This research received no external funding.

Acknowledgments: The Editors would like to thank all the contributions to the present SI, the time invested by each author, as well as to the anonymous reviewers and editorial managers who have contributed to the development of the articles in this SI.

Conflicts of Interest: The authors declare no conflict of interest.

References

1. García-Tejero, I.F.; Durán, Z.V.H. *Water Scarcity and Sustainable Agriculture in Semiarid Environment: Tools, Strategies and Challenges for Woody Crops*; Academic Press: Cambridge, MA, USA; Elsevier: London, UK, 2018; p. 624.
2. García-Tejero, I.F.; Durán, Z.V.H.; Muriel, F.J.L.; Rodríguez, P.C.R. Water and Sustainable Agriculture. In *Springerbriefs in Agriculture*; Springer: Dordrecht, The Netherlands, 2011; p. 94. [CrossRef]
3. Gornall, J.; Betts, R.; Burke, E.; Clark, R.; Camp, J.; Willett, K.; Wiltshire, A. Implications of climate change for agricultural productivity in the early twenty-first century. *Philos. Trans. R. Soc. Ser. B Biol. Sci.* **2010**, *365*, 2973–2989. [CrossRef] [PubMed]
4. Souissi, I.; Temani, N.; Belhouchette, H. Vulnerability of Mediterranean agricultural systems to climate: From regional to field scale analysis. In *Climate Vulnerability: Understanding and Addressing Threats to Essential Resources*; Pielke, R., Ed.; Academic Press: Cambridge, MA, USA; Elsevier: London, UK, 2013; pp. 89–103. [CrossRef]
5. Lamboll, R.; Stathers, T.; Morton, J. Climate Change and Agricultural Systems. In *Agricultural Systems: Agroecology and Rural Innovation for Development*, 2nd ed.; Snapp, S., Pound, B., Eds.; Academic Press: Cambridge, MA, USA; Elsevier: London, UK, 2017; pp. 441–490. [CrossRef]
6. Trout, T.J. Environmental effects of irrigated agriculture. *Acta Hortic.* **2000**, *537*, 605–610. [CrossRef]
7. Siderius, C.; Van Walsum, P.E.V.; Roest, C.W.J.; Smit, A.A.M.F.R.; Hellegers, P.J.G.J.; Kabat, P.; Van Ierland, E.C. The role of rainfed agriculture in securing food production in the Nile Basin. *Environ. Sci. Policy* **2016**, *61*, 14–23. [CrossRef]
8. Iglesias, A.; Garrote, L. Adaptation strategies for agricultural water management under climate change in Europe. *Agric. Water Manag.* **2015**, *155*, 113–124. [CrossRef]
9. Durán, Z.V.H.; Cárceles, R.B.; Gutiérrez, G.S.; Bilbao, B.M.; Cermeño, S.P.; Pérez, P.J.; García-Tejero, I.F. Rethinking irrigated almond and pistachio intensification: A shift towards a more sustainable water management paradigm. *Rev. Ciênc. Agrár.* **2021**, *43*, 24–49. [CrossRef]
10. Ünlü, M.; Kanber, R.; Levent Koç, D.; Tekin, S.; Kapur, B. Effects of deficit irrigation on the yield and yield components of drip irrigated cotton in a Mediterranean environment. *Agric. Water Manag.* **2011**, *98*, 597–605. [CrossRef]
11. Bogucka, B.; Pszczółkowska, A.; Okorski, A.; Jankowski, K. The Effects of Potassium Fertilization and Irrigation on the Yield and Health Status of Jerusalem Artichoke (Helianthus tuberosus L.). *Agronomy* **2021**, *11*, 234. [CrossRef]
12. Ullah, I.; Mao, H.; Rasool, G.; Gao, H.; Javed, Q.; Sarwar, A.; Khan, M.I. Effect of Deficit Irrigation and Reduced N Fertilization on Plant Growth, Root Morphology and Water Use Efficiency of Tomato Grown in Soilless Culture. *Agronomy* **2021**, *11*, 228. [CrossRef]

13. Shabbir, A.; Mao, H.; Ullah, I.; Buttar, N.A.; Ajmal, M.; Solangi, K.A. Improving Water Use Efficiency by Optimizing the Root Distribution Patterns under Varying Drip Emitter Density and Drought Stress for Cherry Tomato. *Agronomy* **2021**, *11*, 3. [CrossRef]
14. Kiani, R.; Nazeri, V.; Shokrpour, M.; Hano, C. Morphological, Physiological, and Biochemical Impacts of Different Levels of Long-Term Water Deficit Stress on Linum album Ky. ex Boiss. Accessions. *Agronomy* **2020**, *10*, 1966. [CrossRef]
15. Millán, S.; Campillo, C.; Vivas, A.; Moñino, M.J.; Prieto, M.H. Evaluation of Soil Water Content Measurements with Capacitance Probes to Support Irrigation Scheduling in a "Red Beaut" Japanese Plum Orchard. *Agronomy* **2020**, *10*, 1757. [CrossRef]
16. Shabbir, A.; Mao, H.; Ullah, I.; Buttar, N.A.; Ajmal, M.; Lakhiar, I.A. Effects of Drip Irrigation Emitter Density with Various Irrigation Levels on Physiological Parameters, Root, Yield, and Quality of Cherry Tomato. *Agronomy* **2020**, *10*, 1685. [CrossRef]
17. García-Tejero, I.F.; Lipan, L.; Gutiérrez-Gordillo, S.; Durán Zuazo, V.H.; Jančo, I.; Hernández, F.; Cárceles Rodríguez, B.; Carbonell-Barrachina, A.A. Deficit Irrigation and Its Implications for HydroSOStainable Almond Production. *Agronomy* **2020**, *10*, 1632. [CrossRef]
18. Kamphorst, S.H.; Gonçalves, G.M.B.; do Amaral Júnior, A.T.; de Lima, V.J.; Leite, J.T.; Schmitt, K.F.M.; dos Santos Junior, D.R.; Santos, J.S.; de Oliveira, F.T.; Corrêa, C.C.G.; et al. Screening of Popcorn Genotypes for Drought Tolerance Using Canonical Correlations. *Agronomy* **2020**, *10*, 1519. [CrossRef]
19. Lipan, L.; Cano-Lamadrid, M.; Hernández, F.; Sendra, E.; Corell, M.; Vázquez-Araújo, L.; Moriana, A.; Carbonell-Barrachina, A.A. Long-Term Correlation between Water Deficit and Quality Markers in HydroSOStainable Almonds. *Agronomy* **2020**, *10*, 1470. [CrossRef]
20. Ao, S.; Russelle, M.P.; Feyereisen, G.W.; Varga, T.; Coulter, J.A. Maize Hybrid Response to Sustained Moderate Drought Stress Reveals Clues for Improved Management. *Agronomy* **2020**, *10*, 1374. [CrossRef]
21. Ibrahim, M.F.M.; Abd El-Samad, G.; Ashour, H.; El-Sawy, A.M.; Hikal, M.; Elkelish, A.; El-Gawad, H.A.; El-Yazied, A.A.; Hozzein, W.N.; Farag, R. Regulation of Agronomic Traits, Nutrient Uptake, Osmolytes and Antioxidants of Maize as Influenced by Exogenous Potassium Silicate under Deficit Irrigation and Semiarid Conditions. *Agronomy* **2020**, *10*, 1212. [CrossRef]
22. Gutiérrez-Gordillo, S.; Durán Zuazo, V.H.; Hernández-Santana, V.; Ferrera Gil, F.; García Escalera, A.; Amores-Agüera, J.J.; García-Tejero, I.F. Cultivar Dependent Impact on Yield and Its Components of Young Almond Trees under Sustained-Deficit Irrigation in Semi-Arid Environments. *Agronomy* **2020**, *10*, 733. [CrossRef]
23. Braunack, M.V.; Adhikari, R.; Freischmidt, G.; Johnston, P.; Casey, P.S.; Wang, Y.; Bristow, K.L.; Filipović, L.; Filipović, V. Initial Experimental Experience with a Sprayable Biodegradable Polymer Membrane (SBPM) Technology in Cotton. *Agronomy* **2020**, *10*, 584. [CrossRef]
24. Park, S.-Y.; Kim, J.; Oh, M.-M. Determination of Adequate Substrate Water Content for Mass Production of a High Value-Added Medicinal Plant, Crepidiastrum denticulatum (Houtt.) Pak & Kawano. *Agronomy* **2020**, *10*, 388. [CrossRef]
25. Martín-Palomo, M.J.; Corell, M.; Girón, I.; Andreu, L.; Galindo, A.; Centeno, A.; Pérez-López, D.; Moriana, A. Absence of Yield Reduction after Controlled Water Stress during Prehaverst Period in Table Olive Trees. *Agronomy* **2020**, *10*, 258. [CrossRef]
26. Kovalikova, Z.; Jiroutova, P.; Toman, J.; Dobrovolna, D.; Drbohlavova, L. Physiological Responses of Apple and Cherry In Vitro Culture under Different Levels of Drought Stress. *Agronomy* **2020**, *10*, 1689. [CrossRef]
27. Wasonga, D.O.; Kleemola, J.; Alakukku, L.; Mäkelä, P.S.A. Growth Response of Cassava to Deficit Irrigation and Potassium Fertigation during the Early Growth Phase. *Agronomy* **2020**, *10*, 321. [CrossRef]
28. Elnashar, A.; Abbas, M.; Sobhy, H.; Shahba, M. Crop Water Requirements and Suitability Assessment in Arid Environments: A New Approach. *Agronomy* **2021**, *11*, 260. [CrossRef]
29. Aguilar Morales, D.; Sánchez-Bravo, P.; Lipan, L.; Cano-Lamadrid, M.; Issa-Issa, H.; del Campo-Gomis, F.J.; Lluch, D.B.L. Designing of an Enterprise Resource Planning for the Optimal Management of Agricultural Plots Regarding Quality and Environmental Requirements. *Agronomy* **2020**, *10*, 1352. [CrossRef]

 *agronomy*

Article

Crop Water Requirements and Suitability Assessment in Arid Environments: A New Approach

Abdelrazek Elnashar, **Mohamed Abbas**, **Hassan Sobhy and Mohamed Shahba** *

Natural Resources Department, Faculty of African Postgraduate Studies, Cairo University, 12613 Giza, Egypt; abdelrazek.elnashar@cu.edu.eg (A.E.); msaelsarawy@cu.edu.eg (M.A.); hassansobhy20@yahoo.com (H.S.)
* Correspondence: shahbam@cu.edu.eg; Tel.: +20-1093162071

Abstract: Efficient land and water management require the accurate selection of suitable crops that are compatible with soil and crop water requirements (CWR) in a given area. In this study, twenty soil profiles are collected to represent the soils of the study area. Physical and chemical properties of soil, in addition to irrigation water quality, provided data are utilized by the Agriculture Land Evaluation System for Arid and semi-arid regions (ALES-Arid) to determine crop suitability. University of Idaho Ref-ET software is used to calculate CWR from weather data while the Surface Energy Balance Algorithms for Land Model (SEBAL) is utilized to estimate CWR from remote sensing data. The obtained results show that seasonal weather-based CWR of the most suitable field crops (S1 and S2 classes) ranges from 804 to 1625 mm for wheat and berssem, respectively, and ranges from 778 to 993 mm in the vegetable crops potato and watermelon, respectively, under surface irrigation. Mean daily satellite-based CWR are predicted based on SEBAL ranges between 4.79 and 3.62 mm in Toshka and Abu Simbel areas respectively. This study provides a new approach for coupling ALES-Arid, Ref-ET and SEBAL models to facilitate the selection of suitable crops and offers an excellent source for predicting CWR in arid environments. The findings of this research will help in managing the future marginal land reclamation projects in arid and semi-arid areas of the world.

Keywords: crop suitability; remote sensing; ALES-Arid; SEBAL; landsat

Citation: Elnashar, A.; Abbas, M.; Sobhy, H.; Shahba, M. Crop Water Requirements and Suitability Assessment in Arid Environments: A New Approach. *Agronomy* **2021**, *11*, 260. https://doi.org/10.3390/agronomy11020260

Academic Editors: Iván Fco. García-Tejero and Víctor H. Durán-Zuazo
Received: 1 December 2020
Accepted: 28 January 2021
Published: 30 January 2021

Publisher's Note: MDPI stays neutral with regard to jurisdictional claims in published maps and institutional affiliations.

1. Introduction

Arid and semi-arid zones represent more than one-third of the land area of the world [1], and are characterized by a long dry season as well as sporadic precipitation [2]. Generally, drylands have been used for livestock production, but recently they are increasingly being used for crop production [3–5]. Egypt lies primarily in arid and semi-arid regions and faces increasing food and water demand. As a result, it struggles to meet its basic food and water needs, due to the continuous increase in population. Increasing crop production without depleting water and land resources in addition to efficient management are significant challenges. The Lake Nasser area in the Aswan governorate of Egypt (22°–24′ N and 31°–33.5′ E) is a good representative for arid and semi-arid environments (Figure 1).

Land suitability is defined as the fitness of a given type of land for specified use, and such suitability can be determined through analytical methods [6–8]. Selecting of a suitable crop is considered an important factor of sustainable agriculture relying on land suitability assessment and also involves assessment of water requirement [9]. Selecting suitable crops for a given area also plays a vital role in efficient water management of time [10,11]. The broad objective of sustainable agriculture is to balance the available land resources with crop requirements, paying particular attention to the optimization of resources used to achieve sustained productivity over a long period [12,13]. Under good management policies in arid regions, the deciding real and exact land resources suitability for specific crop production could likely be more effective and suitable [14].

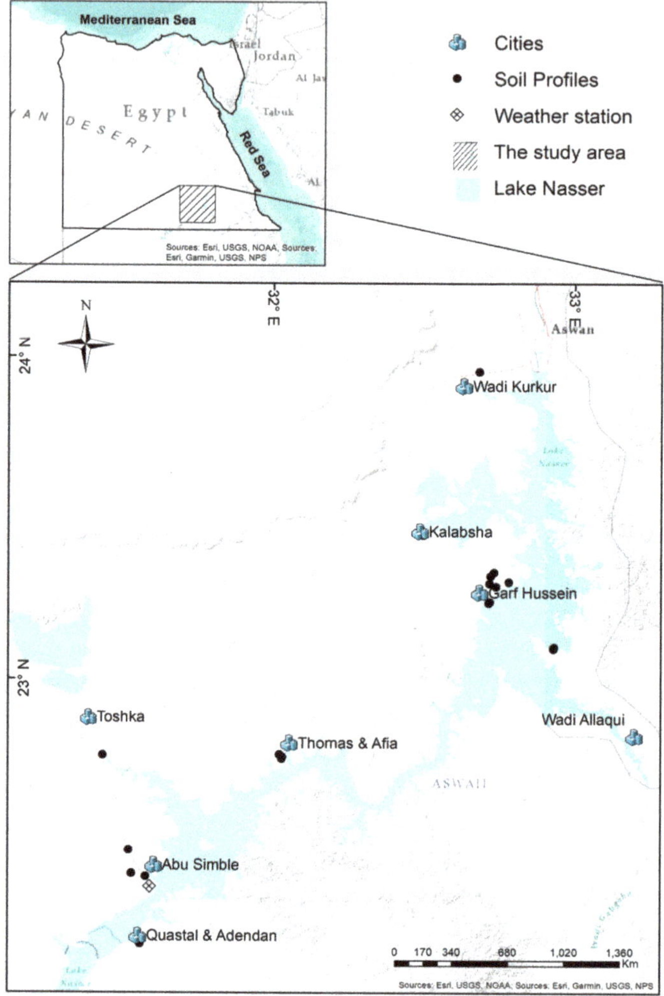

Figure 1. Lake Nasser area, Aswan governorate, Egypt.

Several land evaluation models have been developed to provide a quantified procedure to match land with various actual and proposed uses. For instance, Automated Land Evaluation System (ALES) [6]), Microcomputer-based Mediterranean Land Evaluation Information System (MicroLEIS [15]), Land Evaluation system for Central Ethiopia (LEV-CET [16,17]), Applied System of Land Evaluation and Agricultural Land Evaluation System for arid and semi-arid regions (ASEL/ALES-Arid: [18]), and Agriculture Land Suitability Evaluator (ALSE [19]). However, there is no single or unified land evaluation modelling approach [20,21]. ALESarid-GIS is the updated version of ALES-Arid developed to assess the agricultural land capability and crop suitability in the Geographic Information System (GIS) environment [22]. ALESarid-GIS provides a reasonable solution balancing accuracy, ease of application, and moderate data demand, so its usage has been preferred in evaluating soils for specific crop production in several studies: for instance, in Wahab, et al. [23], Darwish and Abdel Kawy [24], Abd El-Kawy, et al. [25], and Mahmoud, et al. [26]. However, little attention has been paid to estimate the CWR of suitable crops, which is defined by land evaluation for a given area.

Actual evapotranspiration (ETa) is a crucial input to calculate CWR. It can be estimated quite accurately using the aid of weighing lysimeters [27], Eddy correlation [28], and the Bowen ratio [29]. These methods offer potent alternatives for measuring land surface evapotranspiration with high accuracy for a homogeneous area. However, their practical use over large areas is limited due to the number of sites needed to provide point values of evapotranspiration for a specific location. Moreover, it cannot be easily extrapolated to produce accurate maps over a landscape or region. Traditionally, ETa has been estimated by multiplying weather-based reference evapotranspiration (ETr) with crop coefficients (Kc). This method is commonly flawed for multiple reasons. For instance: ETr is a function of weather data alone. Kc values for the same crop showed a significant variation among locations due to differences in crop growth stage, crop variety, soil properties, irrigation method and frequency, climate, and crop management practices. It also does not consider the soil moisture stress level. Furthermore, ETa estimated using this procedure is relatively accurate with an error of $\pm 20\%$ if done well, compared to lysimeters data. Moreover, the accuracy of this methodology is restricted to climatic data, which are not always reliable in many parts of the world [30–32]. However, the role of this method cannot be denied for management and planning purposes—for example, in estimating CWR of the proposed suitable crops for current or newly developed areas.

Therefore, these limitations have encouraged using remotely sensed data to estimate ETa over huge areas. Nowadays, satellite images provide an excellent method for mapping spatial and temporal ETa above the canopy for an entire satellite image. Hence, the estimation of ETa based on remotely sensed data has become a desirable and adequate tool in water resources planning and management [33–36]. Several remote sensing models have been developed to estimate ETa from satellite images particularly at the field/human scale: for instance, the Surface Energy Balance Algorithms for Land Model (SEBAL [37]), Surface Energy Balance System (SEBS [38]), Mapping EvapoTranspiration at High Resolution with Internalized Calibration (METRIC: [30], operational Simplified Surface Energy Balance (SSEBop [39]), and The Atmosphere-Land Exchange Inverse (ALEXI [40]); for more models of remotely sensed ETa see [41–44]. Among these models, SEBAL requires the least amount of inputs with acceptable accuracy. Thus, it has excellent potential for use in developing countries where water management policies are generally inadequate, and ground information is scarce. Moreover, SEBAL has been tested in many countries, especially in arid–semi-arid regions under several different irrigation conditions [45–50].

It is for the abovementioned reasons; this study aims to combine ALESarid, Ref-ET, and SEBAL models as a new and comprehensive approach to improve the selection of suitable crops for available land and water resources, which could be considered the novelty of the current work. This study could be used as a rapid assessment tool to help decision-makers and land managers to prioritize suitable crops based on land and water resources. Section 2 describes the materials and methods. Section 3 presents and discusses the results using data for the area around Lake Naser, Upper Egypt. Conclusions are provided in Section 4.

2. Materials and Methods

2.1. Soil and Water Sampling and Analyses

Twenty representative soil profiles were selected and geo-referenced using the Global Positioning System (GPS) in the study area (Figure 1) around Lake Nasser, Aswan governorate, Egypt ($22°–24'$ N and $31°–33.5'$ E). Soil samples were collected and analyzed in the Laboratories of the Natural Resources Department, Faculty of African Postgraduate Studies, Cairo University in Giza, Egypt, during 2014–2017. Soil physical, chemical, and fertility properties were assessed. Moreover, irrigation water samples representing different soil profiles at 10 cm below the water surface were collected to determine the irrigation water properties. Soil samples were air-dried, ground gently, and sieved through a 2 mm sieve to obtain the fine soil particles. Data of water and soil samples were compiled in ALESarid-GIS system. Physical soil properties (including clay (%), available water (%), hydraulic conductivity (Ks, m/hr), soil depth (cm) and groundwater depth), and chemical soil properties (including soil pH, electrical conductivity (EC, dS/m), cations exchange capacity (CEC, meq/100 g soil), exchangeable sodium percentage (ESP, %), total carbonate (%) and gypsum content (%)) were assessed following USDA [51]. Soil fertility properties (including organic matter (OM, %) and available NPK (ppm)) in addition to irrigation water quality parameters (pH, EC (dS/m), sodium adsorption ratio (SAR), sodium and chloride (meq/L) and boron (B, ppm) were also measured.

2.2. Crop Suitability Using ALESarid-GIS

Soil and water data have been used in the ALESarid-GIS system to assess crop suitability [22]. The evaluation is based on crop suitability affected by the environmental characteristics at the site, such as physical, chemical, and fertility characteristics of the soil, irrigation water quality, and climatic conditions that represent the main factors affecting agricultural soil suitability and productivity in arid and semi-arid regions. Input data of this model are soil physical properties (e.g., soil texture, soil depth, available water and soil permeability), soil chemical properties (e.g., soil salinity, soil alkalinity, calcium carbonate content, gypsum content, cation exchange capacity, and soil reaction), soil fertility properties (e.g., organic matter, available forms of N, P and K), irrigation water characteristics and qualities (e.g., water salinity and toxicity), and finally climate data (e.g., mean summer and winter temperature). Firstly, the model calculates the weighted average value (AV) for each soil property related to a particular soil profile, Equation (1).

$$\mathrm{AV} = \frac{\sum_{i=1}^{n}(v_i \times t_i)}{T} \tag{1}$$

where: v_i is the soil property value relating to soil horizon i; t is the soil horizon thickness (cm), n is the number of horizons within a soil profile, and T is the total soil profile depth (cm). Then, based on the match between the weighted average values of soil parameters and suggested ratings that coded within the model, the land suitability indices and classes for crops were calculated according to the match between the standard crop requirements, which are internally coded data within the model, and various soil parameter levels in the studied area. Finally, the land suitability class was determined by assigning each land suitability index to the confined categories (Table 1). Ismail, Bahnassy and Abd El-Kawy [18] and Abd El-Kawy, Ismail, Rod and Suliman [22] have provided a more detailed description of this model. It is worth noting that ALES-Arid was designed for the arid and semi-arid area. However, for studies in different areas, other land evaluation models can be used (e.g., ALES, MicroLEIS, LEV-CET, and ALSE).

Table 1. Land suitability classes, description and ranges used by ALESarid-GIS.

Class	Description	Rating (%)
S1	Highly suitable	80–100
S2	Moderately suitable	60–80
S3	Marginally suitable	40–60
S4	Conditionally suitable	20–40
NS1	Potentially suitable	10–20
NS2	Actually unsuitable	<10

2.3. Climatic and Remote Sensing Data

Weather data for 2014 were obtained from Abu Simbel weather station located in 22°21′36″ N, 31°36′36″ E with an elevation of 192 m. Data collected were daily minimum and maximum air temperatures, relative humidity, and wind speed. Multi-temporal Landsat-8 images (path 175, row 44) were acquired from earthexplorer.usgs.gov between 20 February and 21 December 2014. Landsat-8 data was provided at the 16-day temporal resolution, 16-bit radiometric resolution, 30 m spatial resolution, LIT processing level (geometric and terrain correction) and free cloud. Satellite image processing was implemented using the geospatial data abstraction library, gdal, [52] in Python programming language.

2.4. Weather-Based CWR Using Ref-ET

Daily reference evapotranspiration (ET_r) was calculated using the University of Idaho Ref-ET software [53,54] as Equation (2).

$$ET_r = \frac{0.408(R_n - G) + \gamma \frac{C_n}{T_a + 273.15} + u_2(e_s - e_a)}{\Delta + \gamma(1 + C_d u_2)} \tag{2}$$

where ET_r is the alfalfa reference evapotranspiration [mm/day]; R_n is the net radiation at the crop surface [MJ/m^2 day]; G is the soil heat flux density at the soil surface [MJ/m^2 day]; T_a is the mean daily or hourly air temperature at 1.5–2.5 m height [C]; u_2 is the mean daily wind speed at 2 m height [m/s]; e_s is the saturation vapor pressure at 1.5–2.5 m height [KPa]; e_a is the actual vapour pressure at 1.5–2.5 m height [KPa]; Δ is the slope of the saturation vapor pressure-temperature curve [KPa/C]; γ is the psychometric constant [KPa/C]; C_n is the numerator constant that changes with reference type and calculation time step; C_d is the denominator constant that changes with reference type and calculation time step; 0.408 coefficient [m^2 mm/MJ]. Cumulative ETa and CWR [55] were estimated by Equations (3) and (4) respectively.

$$ET_{a\ Cumulative-WB} = \sum_{i=1}^{n} ET_r\ K_{cr} \tag{3}$$

$$CWR_{WB} = ET_{a\ Cumulative-WB}\ /\ \text{Irrigation efficiency} \tag{4}$$

where $ET_{a\ Cumulative}$ is the weather-based cumulative ET_a [mm] from the day i through the day n; ET_r is the reference ET [mm] for the day i from Equation (2); K_{cr} is the alfalfa-based single crop coefficient [dimensionlessfor the day i, irrigation efficiency ranging between 0 and 1, and CWR_{WB} is the weather-based crop water requirement [mm].

2.5. Satellite-Based CWR Using SEBAL

Extensive SEBAL formulation is available in its original literature [37,56–58], so here we introduce a short description of the SEBAL model. Landsat-8 data converted from digital numbers to reflectance and radiance to calculate vegetation indices, surface albedo, and surface temperatures following [59]. It is worth noting that the SEBAL Calibrated using Inverse Modeling of Extreme Conditions (CIMIC) approach is used to generate image-date specific sensible heat flux (H) map where CIMIC effectively minimizes systematic biases in Rn, G, Ts, and Z_{0m} [37]. ET_a is predicted from the residual amount of energy remaining from the energy balance that includes all major sources (R_n) and consumers (G, H and LE) of energy as Equation (5):

$$R_n - G - H - LE = 0 \tag{5}$$

where R_n is the net radiation, H is the sensible heat, G is the soil heat flux, LE is the latent heat flux. All are instantaneous values in [W/m^2]. Net radiation was calculated as Equation (6):

$$R_n = (1 - \alpha)R_{S\downarrow} + R_{L\downarrow} - R_{L\uparrow} - (1 - \varepsilon_0)R_{L\downarrow} \tag{6}$$

where α is the surface albedo [dimensionless]; $R_{S\downarrow}$ is the incoming short-wave radiation [W/m^2]; $R_{L\downarrow}$ is the incoming longwave radiation [W/m^2]; $R_{L\uparrow}$ is the outgoing longwave radiation [W/m^2]; ε_0 is the broad-band surface emissivity [dimensionless]. Soil heat flux calculated as Equation (7):

$$G = \left(((Ts - 273.15)/\alpha)\left(0.0038\alpha + 0.0074\alpha^2\right)\left(1 - 0.98NDVI^4\right)\right)R_n \tag{7}$$

where Ts is the surface temperature [K]; NDVI is the Normalized Differences Vegetation Index [dimensionless].

Momentum roughness length was calculated as Equation (8):

$$Z_{0m} = \exp[(a\ NDVI/\alpha) + b] \tag{8}$$

where Z_{0m} is the momentum roughness length [m]; a and b are regression constants derived from a plot of initial $\ln(Z_{0m})$ vs NDVI/α [56]. These two parameters should be defined by the SEBAL operator, thus, they play an important role in the model performance. Sensible heat flux calculated as Equation (9):

$$H = \rho_a\ C_P(dT/r_{ah}) \tag{9}$$

where ρ_a is the air density [Kg/m^3]; C_P is the specific heat [J/Kg $\times$ K]; r_{ah} is the aerodynamic resistance for heat transport [s/m]. The relationship between the temperature differences and remotely sensed surface temperature is very close as Equation (10):

$$dT = a\ T_s + b \tag{10}$$

where dT is the temperature differences between two heights at 0.1 m and 2 m above the canopy [K]; a $[-]$, b[K] are the calibration coefficients derived using the cold and hot pixels site and time-specific candidates. It should be highlighted that cold and hot pixels location are operator-specific, which means a SEBAL operator has to define these two locations for each image carefully as described, in detail, in SEBAL literature.

Once the instantaneous net radiation, soil heat flux, and sensible heat flux were determined, the instantaneous latent heat flux was estimated at the moment of satellite overpass on a pixel-by-pixel level, then converted to an equivalent amount of water depth. The instantaneous evaporative fraction was calculated as Equation (11):

$$\Lambda = LE/R_n - G \tag{11}$$

Evaporative fraction expresses the ratio of actual to crop evaporative demand when atmospheric moisture conditions are in equilibrium with soil moisture conditions [60]. Studies have shown that the evaporative fraction remains constant throughout the day [61,62]. Therefore, daily ET_a was calculated from the energy balance equation as Equation (12):

$$ET_{a24} = 86400\ \Lambda\ (R_{n24} - G_{24})/\lambda \tag{12}$$

where: Λ is the evaporative fraction [dimensionless]; R_{n24} is the daily net radiation calculated on a daily time step [W/m^2]; G_{24} is the daily soil heat flux [W/m^2]; λ is the latent heat of vaporization [J/kg]; 86400 is a time conversion from seconds to days. The daily ET_a for the entire image area changes in proportion to the change in the daily ET_r on the index weather site [30,63]. Thereby, Cumulative ET_a calculated as Equation (13):

$$ET_{a\ Cumulative-RS} = \sum_{i=1}^{n}(ET_{a24})_i \times (K_m)_i \tag{13}$$

$$K_m = (ET_{r\ cumulative}/ETr_i) \tag{14}$$

where $ET_{a\ Cumulative-RS}$ is the remotely sensed cumulative ET_a [mm] from the day i through the day n; ET_{a24} is the daily ET_a [mm] for day i; $ETrF_{24}$ is the daily ET_r fraction [mm] for day i; K_m is multiplier [dimensionless] for each period to convert ET_a for the day of the image into ET_a for the period; $ET_{r\ Cumulative}$ is the cumulative reference ET [mm] for the period; ETr_i is the reference ET [mm] for day i. Finally, remote sensing CWR can be estimated as Equation (15):

$$CWR_{RS} = ET_{a\ Cumulative-RS} \tag{15}$$

where $ET_{a\,Cumulative-RS}$ is the remotely sensed cumulative ET_a [mm] from the day i through the day n; and CWR_{RS} is the remote sensing CWR [mm].

3. Results and Discussion

3.1. Soil and Irrigation Water Properties

Soil analysis indicated low clay content, low water availability, and high hydraulic conductivity (Table 2). Most of the investigated soil could be considered as alkaline and non-saline with low CEC. These results are in agreement with those obtained by previous studies [64–66]. In accordance with Khalifa [64], and Abbas, El-Husseiny, Mohamed and Abuzaid [65], soil OM content was very low, and the available NPK values were not sufficient. The difference in soil properties may be due to the variability of topography and parent rocks. Taghizadeh-Mehrjardi, et al. [67] assessed land suitability in Kurdistan province in Iran for crop production and conclude that the differences in soil characteristics were due to variability in topography, climate, and parent material. Additionally, they considered topography and climate data as the essential auxiliary data for predicting land suitability class.

Table 2. Soil depth (SD), clay content average (%), available water (AW, %), hydraulic conductivity (Ks, m/hr), total carbonates (TC, %), gypsum content (GC, %), exchangeable sodium percentage (ESP, %), soil pH, cations exchangeable capacity (CEC, meq/100 g soil) electrical conductivity (EC, dS/m), organic matter (OM, %) and available nitrogen (N, ppm), phosphorous (P, ppm) and potassium (K, ppm).

ID	SD	Clay	AW	Ks	TC	GC	ESP	pH	CEC	EC	OM	N	P	K
1	85	0.72	2.48	0.63	2.21	0.08	13.96	7.82	3.32	2.09	0.04	0.11	0.28	2.12
2	90	8.10	2.80	0.22	1.70	0.07	12.20	8.11	6.30	1.20	0.03	0.13	0.23	1.90
3	90	8.12	2.64	0.22	1.86	0.06	11.63	8.12	6.41	1.25	0.04	0.11	0.33	1.88
4	95	7.93	2.80	0.23	1.83	0.05	14.43	8.05	6.17	1.27	0.04	0.10	0.40	2.80
5	95	5.55	2.61	0.37	1.05	0.06	5.34	7.74	5.64	0.90	0.05	0.14	0.52	2.89
6	90	1.00	2.63	0.62	2.50	0.07	14.07	7.76	3.80	2.32	0.03	0.07	0.23	1.47
7	70	7.93	3.03	0.23	1.63	0.05	11.81	7.63	6.39	2.45	0.01	0.04	0.13	0.77
8	90	5.95	2.50	0.34	1.10	0.08	5.25	7.67	5.15	0.79	0.04	0.05	0.30	2.05
9	95	5.64	2.11	0.36	0.99	0.07	4.88	7.72	5.06	0.94	0.04	0.04	0.31	1.60
10	90	5.05	1.90	0.39	1.05	0.07	4.80	7.60	4.95	0.89	0.05	0.06	0.25	1.90
11	90	1.50	2.70	0.59	3.80	0.07	20.10	7.95	3.05	2.87	0.05	0.15	0.65	3.05
12	85	6.94	2.94	0.29	1.75	0.06	14.06	8.08	5.44	0.64	0.04	0.12	0.48	2.35
13	90	5.60	2.00	0.36	1.10	0.06	4.80	7.61	4.55	3.45	0.05	0.15	0.55	2.60
14	95	1.64	2.54	0.58	3.81	0.07	4.99	7.86	2.99	2.81	0.06	0.15	0.66	2.21
15	80	0.78	1.51	0.63	2.29	0.07	12.28	7.88	2.60	1.27	0.03	0.04	0.19	2.18
16	50	1.72	2.54	0.58	3.44	0.06	11.14	8.66	2.82	1.97	0.04	0.14	0.36	3.04
17	85	6.44	3.32	0.32	1.71	0.06	12.82	8.08	5.51	0.62	0.04	0.10	0.37	1.56
18	90	6.80	3.24	0.30	1.71	0.07	13.83	7.68	5.59	3.06	0.06	0.18	0.62	4.00
19	95	6.94	3.37	0.29	1.58	0.07	13.76	7.59	5.85	3.47	0.04	0.05	0.14	1.37
20	60	11.00	2.95	0.06	4.60	0.07	12.05	7.89	7.00	4.72	0.06	0.10	0.55	2.45
Min	50.00	0.72	1.51	0.06	0.99	0.05	4.80	7.59	2.60	0.62	0.01	0.04	0.13	0.77
Max	95.00	11.00	3.37	0.63	4.60	0.08	20.10	8.66	7.00	4.72	0.06	0.18	0.66	4.00
Mean	85.50	5.27	2.63	0.38	2.09	0.07	10.91	7.88	4.93	1.95	0.04	0.10	0.38	2.21
SD	11.82	2.94	0.47	0.16	1.02	0.01	4.25	0.25	1.34	1.13	0.01	0.04	0.16	0.71
CV (%)	13.83	55.73	17.68	42.95	48.79	12.12	38.98	3.21	27.09	57.95	27.77	42.29	43.49	32.16

Irrigation water properties for all collected samples were similar among different sectors (Table 3). This result was expected as irrigation water came from the same source (Lake Nasser), which has high-quality irrigation water for the proposed crops according to FAO [68] and El-Mahdy, et al. [69], who indicated the suitability of Lake Naser water for drinking and irrigation. These findings also are found to be in agreement with previous work of Fayed, et al. [70]. They tested the chemical properties of Lake Nasser water and found that the concentration of elements in Lake Nasser water was within the permissible limits.

Table 3. Irrigation water properties in the study area.

Samples	EC (dS/m)	pH	SAR	Na$^+$ (meq/L)	Cl^{-1} (meq/L)	B^{-1} (ppm)
1	0.20	8.38	3.92	3.30	1.20	0.02
2	0.20	8.53	4.28	3.37	1.00	0.13
3	0.24	7.79	3.31	3.13	1.20	0.08
4	0.24	7.32	3.54	3.19	1.20	0.04
5	0.21	7.37	3.67	3.13	1.00	0.11
6	0.19	7.67	3.16	2.85	1.20	0.11
7	0.22	7.67	3.16	2.92	2.20	0.07
8	0.71	6.85	2.99	4.31	1.80	0.03
Min	0.19	6.85	2.99	2.85	1.00	0.02
Max	0.71	8.53	4.28	4.31	2.20	0.13
Mean	0.27	7.70	3.50	3.27	1.35	0.08
SD	0.16	0.52	0.41	0.43	0.40	0.04
CV (%)	59.48	6.71	11.70	13.00	29.40	53.04

3.2. Crop Suitability Assessment Using ALESarid-GIS

Crop suitability is divided into five classes: S1, S2, S3, S4, and NS2, indicating highly suitable, moderately suitable, marginally suitable, conditionally suitable and unsuitable, respectively. Table 4 previews land suitability for 28 field crops in the study area. Since the total number of soil profiles are 20 profiles and each soil profile covers a different area, crop suitability class (%) is calculated as n of soil profiles in each class divided by the total number of soil profiles. For instance, wheat crop classified as S1 (highly suitable) for four soil profiles (2, 3, 4, and 20), thus, wheat is highly suitable for 20% of the study area. Based on S1 and S2 classes of suitability, alfalfa and sorghum were the highest suitable crops (95%), followed by onion, wheat and barley (90%), sugar beet (80%), sugarcane, peppers, and watermelons (70%), and pear (50%). Some crops were found to be completely unsuitable such as date palm, fig, olives, grapes, citrus, tomatoes, cabbage, peas, peanuts, and rice (Table 4). According to Aswan governorate statistical guide [71], most of these crops are actually planted in the study area indicating the validity of ALESarid estimates. At the same time, there are other crops not included in ALESarid database but cultivated in the study area (i.e., eggplant, courgettes, garlic, okra, spinach, corchorus, hibiscus, henna, sesame and fenugreek). Similar findings were reported by Hassan, et al. [72], who studies land suitability for wheat, maize, potatoes, sugar beet, alfalfa, peach, citrus, and olive in Hala'ib and Shalateen regions, South-Eastern of the study area.

3.3. Weather-Based CWR

Monthly reference evapotranspiration (ET$_r$) increased from January to July, then gradually decreased to reach its minimum in December (Figure 2). Monthly ET$_r$ was 5.79, 10.94, and 4.80 mm/day in January, July, and December, respectively. There was a positive association between the change in ET$_r$ and the change in air temperature. The difference in ET$_r$ was negatively associated with the change in humidity. Data collected from the nearest weather station agreed with our findings. Crop water requirements (CWR) were calculated based on 60%, 75%, and 85% efficiency for surface, sprinkler and drip irrigation respectively [73]. Crop coefficient (Kc) values, planting date and harvesting date were obtained from the previous studies [74–77].

Table 4. Land suitability for 28 field crops around Lake Nasser, Aswan, Egypt, determined during 2014–2017.

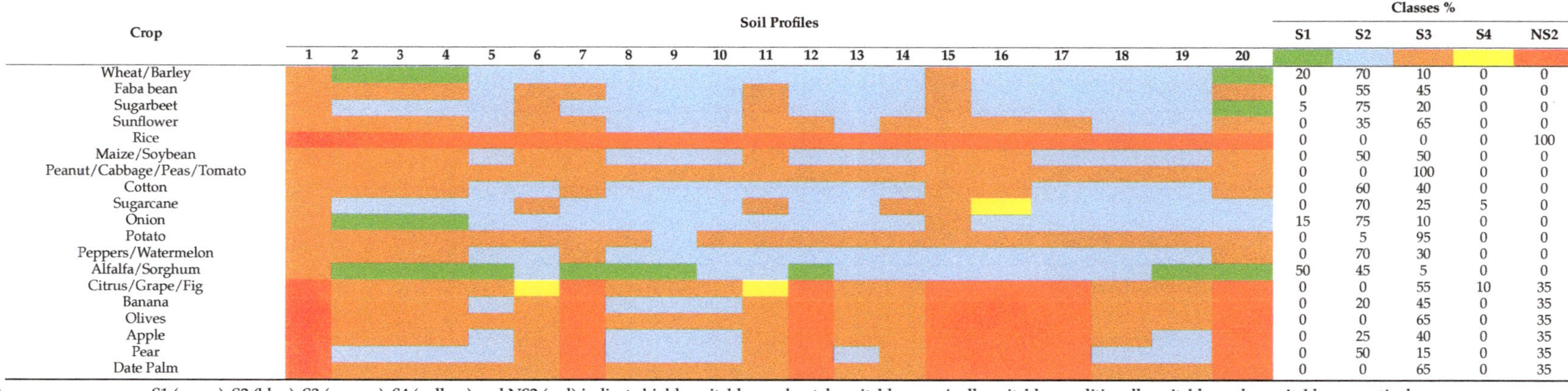

Crop	Soil Profiles																				Classes %				
	1	2	3	4	5	6	7	8	9	10	11	12	13	14	15	16	17	18	19	20	S1	S2	S3	S4	NS2
Wheat/Barley																					20	70	10	0	0
Faba bean																					0	55	45	0	0
Sugarbeet																					5	75	20	0	0
Sunflower																					0	35	65	0	0
Rice																					0	0	0	0	100
Maize/Soybean																					0	50	50	0	0
Peanut/Cabbage/Peas/Tomato																					0	0	100	0	0
Cotton																					0	60	40	0	0
Sugarcane																					0	70	25	5	0
Onion																					15	75	10	0	0
Potato																					0	5	95	0	0
Peppers/Watermelon																					0	70	30	0	0
Alfalfa/Sorghum																					50	45	5	0	0
Citrus/Grape/Fig																					0	0	55	10	35
Banana																					0	20	45	0	35
Olives																					0	0	65	0	35
Apple																					0	25	40	0	35
Pear																					0	50	15	0	35
Date Palm																					0	0	65	0	35

S1 (green), S2 (blue), S3 (orange), S4 (yellow) and NS2 (red) indicate highly suitable, moderately suitable, marginally suitable, conditionally suitable, and unsuitable, respectively.

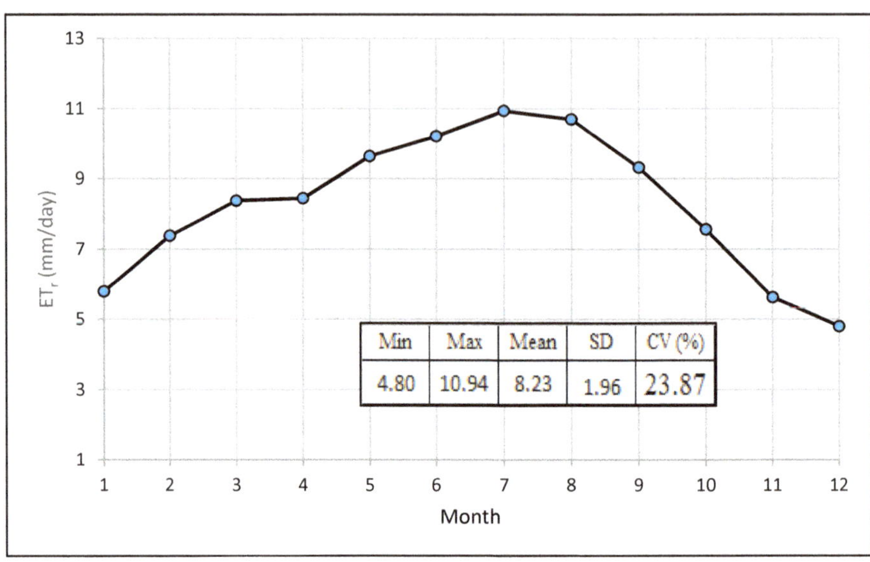

Figure 2. Monthly reference evapotranspiration (ET_r; mm/day) based on daily time step climatic data from Abu Simbel weather station (22°21′36″ N, 31°36′36″ E) for 2014.

Land suitability level for 28 field crops around Lake Nasser in Aswan, Egypt, determined during 2014–2017 was graphically presented in Table 4. Crop water requirements for summer field crops ranged from 820 to 3406 mm for sunflower and sugarcane, while it ranged for winter crops from 658 to 1625 mm for faba bean and berssem (5 cuts), respectively (Table 5).

Table 5. Crop water requirements for field crops, vegetable crops, and fruit trees under different irrigation systems.

					Surface	Sprinkler	Drip
Crop	Days	Planting Date	Harvesting Date	ETa (mm)		CWR (mm)	
Summer field crops							
Sunflower	90	01/05/2014	30/07/2014	492	820	656	579
Sorghum	120	15/05/2014	12/09/2014	675	1126	900	
Maize	120	15/04/2014	13/08/2014	680	1133	906	799
Peanut	120	15/04/2014	13/08/2014	697	1162	930	820
Sugarcane	365	01/02/2014	01/02/2015	2044	3406	2725	2405
Soybean	123	01/05/2014	01/09/2014	641	1069	855	755
Winter field crops							
Wheat	165	01/11/2014	15/04/2015	482	804	643	
Barley	150	15/10/2014	14/03/2015	482	803	643	
Berssem	240	15/09/2014	13/05/2015	975	1625	1300	
Faba bean	122	01/11/2014	03/03/2015	395	658	527	465
Onion	151	01/10/2014	01/03/2015	485	808	646	570
Annual field crops							
Alfalfa	365	01/01/2014	01/01/2015	2025	3374	2699	2382

Table 5. *Cont.*

Crop	Days	Planting Date	Harvesting Date	ETa (mm)	Surface	Sprinkler	Drip
						CWR (mm)	
Summer vegetable crops							
Watermelon	122	01/03/2014	01/07/2014	596	993	794	701
Peppers	153	01/04/2014	01/09/2014	793	1321	1057	933
Cabbage	153	15/04/2014	15/09/2014	783	1305	1044	921
Tomato	150	15/01/2014	14/06/2014	678	1130	904	797
Potato	120	01/02/2014	01/06/2014	544	907	726	640
Winter vegetable crops							
Cabbage	151	15/10/2014	15/03/2015	483	806	644	569
Tomato	151	15/09/2014	13/02/2015	529	882	705	622
Potato	123	01/10/2014	01/02/2015	389	648	518	457
Peppers	150	01/10/2014	28/02/2015	481	801	641	566
Peas	150	15/09/2014	12/02/2015	490	816	653	576
Deciduous fruit trees							
Grape	275	01/3/2014	01/12/2014	933	1555	1244	1098
Fig	275	01/3/2014	01/12/2014	948	1579	1263	1115
Evergreen fruit trees							
Date Palm	365	01/01/2014	01/01/2015	1119	1865	1492	1316
Olives	365	01/01/2014	01/01/2015	1119	1865	1492	1316
Citrus	365	01/01/2014	01/01/2015	1548	2581	2065	1822
Banana	365	01/01/2014	01/01/2015	2022	3369	2695	2378

Summer and winter vegetable crop harvests varied significantly for the same crop. For a summer harvest, CWR ranged from 907 to 1321 mm, and for winter harvest ranged from 648 to 882 mm in potato and tomato, respectively (Table 5). Crop water requirements for deciduous fruit trees varied from 1555 to 1579 mm for grape and fig, respectively, and ranged from 1865 to 3369 mm in the evergreen fruit trees date palm and banana, respectively. These findings can be confirmed by the study of Mahmoud and El-Bably [78]. Precise predictions of CWR depend on accurate crop ET assessment, accessible satellite images source and precise forecasting of meteorological data [79].

3.4. Weather-Based CWR of Suitable Crops

Crop suitability that represented by S1 and S2 classes along with their CWR (Table 6) indicated that the range of CWR for the most suitable field crops is between 804 and 1625 mm for wheat and berssem (5 cuts), respectively. Vegetable crops CWR ranged from 778 to 993 mm for potato and watermelon, respectively. For banana trees, CWR was 3369 mm under surface irrigation. ALESarid-GIS output based on soil and water properties indicated that sugar beet, cotton, apple, and pear are the most suitable crops. However, based on the physiological demand of these crops, they cannot grow in the study area because of other factors, such as climatic conditions. At the same time, date palm that was proven as unsuitable (S3) is successfully cultivated in the study area. In arid regions, a suitable cropping pattern for an area could be decided based on both the actual and potential status of the area defined by land suitability indices for different crops [14] while Abd El-Hady and Abdelaty [80] indicated that crops soil suitability is mainly determined by soil properties, crop rooting depth, and crops salinity tolerance. However, this study highly recommends integrating CWR of the most suitable crops for a region to ensure a real match between these crops and water availability for irrigation.

Table 6. Crop water requirements (CWR) of the most suitable crops under the surface, sprinkler, and drip irrigation systems.

			Surface	Sprinkler	Drip
Crop	**S1%**	**S2%**	**CWR [mm]**		
Field crops					
Faba bean		55	658	527	465
Wheat	20	70	804	643	
Barley	20	70	803	643	
Sunflower		35	820	656	579
Maize		50	1133	906	799
Sugarbeet	5	75			
Soybean		50	1069	855	755
Onion	15	75	808	646	570
Berssem	50	45	1625	1300	
Alfalfa	50	45	3374	2699	
Cotton		60	-		
Vegetable crops					
Potato		5	778	622	549
Watermelon		70	993	794	701
Fruit trees					
Apple		25			
Pear		50			
Banana		20	3369	2695	2378

3.5. Actual CWR Using SEBAL

Calculations of ET_a based on remotely sensed data and SEBAL approach were done with sprinkler and surface irrigation systems in Toshka and Abu Simbel locations, respectively (Figure 3). Those two locations were selected to investigate the applicability of remote sensing data with the SEBAL model in CWR estimation, given that they represent two different irrigation and management systems and cover most of the study area. The essential elements in SEBAL are the sensible heat flux and the momentum roughness length calculation, which depend upon the operator, time, and site-specific parameters; coefficients a and b in Equations (8) and (10). These coefficients are defined for each day-image and presented in Table A1. Paula, et al. [81] assured that the atmospheric stability conditions ensure reasonable estimates of ETa.

From Figure 3, ETa spatial variations between Toshka and Abu Simbel locations can be attributed to the differences in the land and water management in each location where more water is consumed at Toshka location because of the well-managed agriculture system (e.g., sprinkler irrigation) compared with that at Abu Simbel location (flood irrigation). Figure 4 presents daily ET_a at cold pixels, mean daily ETa at Toshka and Abu Simbel locations, as well as weather-based ET_r calculated based on weather data from the Abu Simbel weather station. Daily ET_a at cold pixels represents a well-watered vegetation condition that has a minimum surface temperature (Ts) above the canopy with maximum vegetation cover (NDVI) and surface albedo (α). In this situation, the temperature difference (dT) is minimal or zero and this leads to sensible heat flux (H) that has become minimal or zero too. Latent heat flux (LE) and the evaporative fraction (Λ) becomes a maximal rate due to all the available energy consumed in the latent heat flux [30,37]. Thus, these cold pixel values refer to well-managed fields. Compared to the temporal change in daily ET_a at cold pixels versus mean daily ETa at Toshka and Abu Simbel locations: (1) The mean daily ETa at Abu Simbel location is always lower than at Toshka location, and (2) the mean daily ETa at Toshka location is very close to daily ET_a at cold pixels confirming the results that obtained in Figure 3 and Table 7. Remotely sensed CWR of each cultivated crop could be achieved by using a crop type map. Unfortunately, this map is not available for this study to precisely compare between weather-based and remote sensing-based CWR, which is highly recommended in future studies. However, daily ETr from Figure 4 and Table 7 is higher than ETa by about 50% with SD and CV reaching 2.4 and 26.92% respectively, thus indicating, in general, a higher estimation of weather-based CWR (Table 4; Table 5). Therefore, the calculation of ETa using satellite data and SEBAL model is useful for guiding the daily operation of water management in the arid region [82]. Moreover, Sun, et al. [83] demonstrated the considerable potential of the SEBAL model for estimation of spatial ETa with little ground-based weather data over large areas at the field scale. These findings also can be confirmed by the mean NDVI spatial variation maps (Figure 5). The maximum NDVI values were clustered over Toshka at 0.80 (mean = 0.33; CV = 40%) while at Abu Simbel it was at 0.73 (mean = 0.27; CV = 42%). Both ETa and NDVI spatial variation maps are completely agreed with each other where lower ETa (NDVI) with higher CV value mapped over Abo Simbel and higher ETa (NDVI) with lower CV value clustered over Toshka.

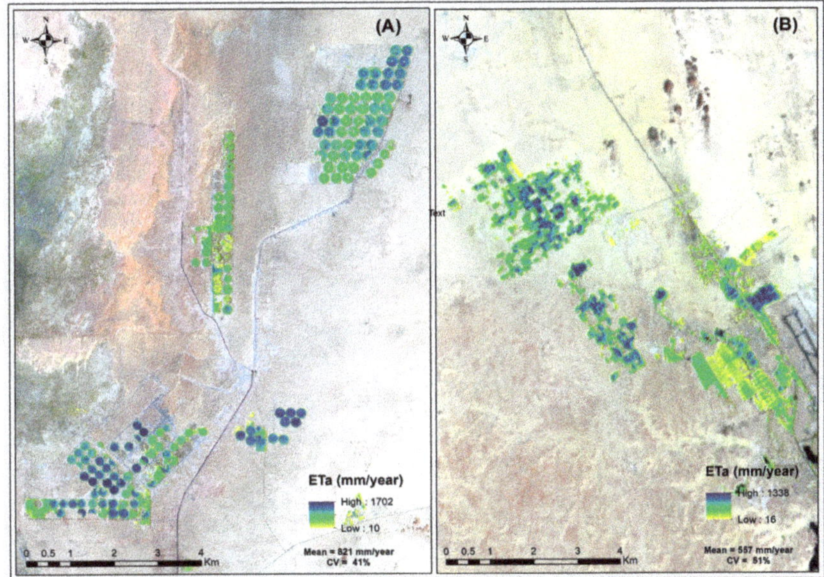

Figure 3. Annual actual evapotranspiration (mm) at Toshka (**A**) and Abu Simbel (**B**) for 2014.

Table 7. Minimum, maximum, mean, standard deviation (SD) and coefficient of variation (CV) of daily ETa at cold pixels, mean daily ETa at Toshka and Abu Simbel locations and weather-based ET_r.

	ETa (mm)			ETr
	Cold Pixels	**Toshka**	**Abu Simbel**	
Minimum	2.81	2.40	2.74	4.49
Maximum	5.74	6.56	4.77	13.40
Mean	4.73	4.79	3.62	8.90
SD	0.88	1.08	0.69	2.40
CV (%)	18.53	22.67	19.16	26.92

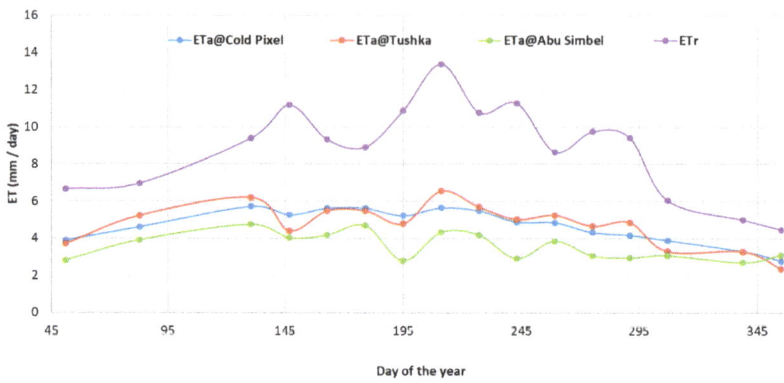

Figure 4. Daily ET_a at cold pixels, mean daily ETa at Toshka and Abu Simbel locations and weather-based ET_r for 2014.

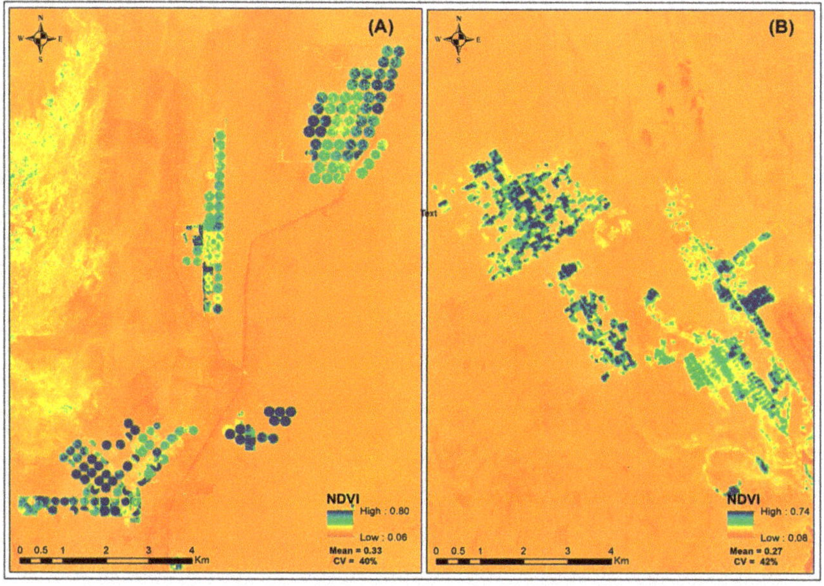

Figure 5. Mean NDVI at Toshka (**A**) and Abu Simbel (**B**) for the year 2014.

3.6. Study Llimitations and Innovation

The study area has only one weather station used for calculating weather-based CWR and in SEBAL calibration. Thus, it is considered one of the limitations of this study. In addition, a crop map was not available for this study, which plays an important role in linking the proposed CWR using climate data and the actual CWR using remote sensing data. Therefore, we highly recommend this point in future studies. Despite that, the innovation of the study is integrating ALESarid-GIS, Ref-ET, and SEBAL models for selecting crop suitability and assessing its water requirements using weather and remote sensing data in a given area. Besides, we highly encourage to add some crops which are planted in the study area, but not included in ALESarid database (i.e., eggplant, courgettes, garlic, okra, spinach, corchorus, hibiscus, henna, sesame and fenugreek).

4. Conclusions

Crop type and water management must be compatible with land and water resources. When selecting cropping systems, several factors related to soil properties and water quality have to be considered, along with other climatic factors that may affect the physiological performance of each individual crop differently. ALESarid-GIS facilitates the selection of suitable crops to improve the estimation of irrigation crop water requirements based on crop suitability. Remote sensing techniques and the SEBAL model offer a great tool that can be used for estimating the ETa and support land and water management, especially in arid and semiarid regions of the world. Our results reveal that: (1) The highly suitable crops are alfalfa and sorghum (95%) followed by onion, wheat and barley (90%), sugar beet (80%), sugarcane, peppers and watermelons (70%), and pear (50%); (2) their weather-based CWR ranges from 804 to 1625 mm for wheat and berssem (5 cuts), respectively; and (3) satellite-based CWR spatial distribution for Toshka pivots irrigation system ranges between 10 and 1702 mm/year (mean = 821 mm/year), while this finding for Abu Simbel flood irrigation system it ranges from 16 to 1338 mm/year (mean = 557 mm/year). The findings of the present research may help decision-makers to plan and manage the future marginal land reclamation projects in Egypt and arid and semi-arid areas of the world. The concept of the current study can be applied to other sites of a similar subject.

Author Contributions: Conceptualization, H.S. and M.S., Data acquisition, A.E. Design of methodology H.S., M.A., and A.E. Writing and editing, M.S. and M.A. All authors have read and agreed to the published version of the manuscript.

Funding: This research was partially funded by Cairo University, Egypt.

Conflicts of Interest: The authors declare no conflict of interest.

Appendix A

Table A1. Coefficient parameters (a and b) of momentum roughness length (Z0m) and temperature differences (dT).

DOY	Z0m=exp[(a × NDVI/α)+b]		dT = (a × T_s) + b	
	a	b	a	b
51	5.02	−6.45	0.36	−107.5
83	4.99	−6.44	0.30	−88.91
131	5.13	−6.47	0.17	−50.55
147	5.11	−6.44	0.15	−45.36
163	5.05	−6.43	0.16	−48.33
179	4.88	−6.38	0.13	−40.58
195	5.06	−6.42	0.26	−79.56
211	5.07	−6.45	0.17	−51.52
227	4.88	−6.38	0.18	−55.86
243	5.02	−6.42	0.25	−76.85
259	4.99	−6.42	0.20	−62.17
275	4.94	−6.41	0.25	−73.90
291	4.78	−6.35	0.22	−66.47
307	4.98	−6.42	0.25	−76.29
339	5.25	−6.52	0.31	−92.15
355	4.96	−6.43	0.49	−142.51
Min.	4.78	−6.52	0.13	−142.51
Max.	5.25	−6.35	0.49	−40.58
Mean	5.01	−6.43	0.24	−72.41
SD	0.11	0.04	0.09	25.68
CV (%)	2.14	−0.57	37.21	−35.46

Note DOY, day of the year; Min, minimum; Max, maximum; SD, standard deviation; CV, coefficient of determination.

References

1. FAO. *Arid Zone Forestry: A Guide for Field Technicians*; Food and Agriculture Organization: Rome, Italy, 1989.
2. D'Odorico, P.; Bhattachan, A. Hydrologic variability in dryland regions: Impacts on ecosystem dynamics and food security. *Philos. Trans. R. Soc. B Biol. Sci.* **2012**, *367*, 3145–3157. [CrossRef] [PubMed]
3. Wheater, H.; Sorooshian, S.; Sharma, K.D. *Hydrological Modelling in Arid and Semi-Arid Areas*; Cambridge University Press: London, UK, 2007.
4. Camarasa-Belmonte, A.M.; Soriano, J. Empirical study of extreme rainfall intensity in a semi-arid environment at different time scales. *J. Arid Environ.* **2014**, *100–101*, 63–71. [CrossRef]
5. Williams, M. *Climate Change in Deserts: Past, Present and Future*; Cambridge University Press: London, UK, 2014.
6. Rossiter, D.G. ALES: A framework for land evaluation using a microcomputer. *Soil Use Manag.* **1990**, *6*, 7–20. [CrossRef]
7. De la Rosa, D.; Mayol, F.; Diaz-Pereira, E.; Fernandez, M.; de la Rosa, D., Jr. A land evaluation decision support system (MicroLEIS DSS) for agricultural soil protection: With special reference to the Mediterranean region. *Environ. Model. AMP Softw.* **2004**, *19*, 929–942. [CrossRef]
8. Akıncı, H.; Özalp, A.Y.; Turgut, B. Agricultural land use suitability analysis using GIS and AHP technique. *Comput. Electron. Agric.* **2013**, *97*, 71–82. [CrossRef]
9. Ali, R.R.; Shalaby, A. Sustainable agriculture in the arid desert west of the Nile Delta: A crop suitability and water requirements perspective. *Int. J. Soil Sci.* **2012**, *7*, 116–131. [CrossRef]
10. Abdel Kawy, W.A.; Abou El-Magd, I.H. Assessing crop water requirements on the bases of land suitability of the soils South El Farafra Oasis, Western Desert, Egypt. *Arab. J. Geosci.* **2013**, *6*, 2313–2328. [CrossRef]
11. Elnashar, A.F. Assessing Crop Suitability and Water Requirements of the Common Land Use in Egypt, Sudan and Ethiopia. Institute of African Research and Studies. Master's Thesis, Cairo University, Giza, Egypt, 2016.
12. Joshua, J.K.; Anyanwu, N.C.; Ahmed, A.J. Land suitability analysis for agricultural planning using GIS and multi criteria decision analysis approach in Greater Karu Urban Area, Nasarawa State, Nigeria. *Afr. J. Agric. Sci. Technol.* **2013**, *1*, 14–23.
13. Bozdag, A.; Yavuz, F.; Gunay, A.S. AHP and GIS based land suitability analysis for Cihanbeyli (Turkey) County. *Environ. Earth Sci.* **2016**, *75*, 813–827. [CrossRef]
14. Ghabour, T.K.; Ali, R.R.; Wahba, M.M.; El-Naka, E.A.; Selim, S.A. Spatial decision support system for land use management of newly reclaimed areas in arid regions. *Egypt. J. Remote Sens. Space Sci.* **2019**, *22*, 219–225. [CrossRef]
15. De la Rosa, D.; Moreno, J.A.; Garcia, L.V.; Almorza, J. MicroLEIS: A microcomputer-based Mediterranean land evaluation information system. *Soil Use Manag.* **1992**, *8*, 89–96. [CrossRef]
16. Yizengaw, S.; Verheye, W. Computer aided decision support system in land evaluation—A case study. *Agropedology* **1994**, *4*, 1–18.
17. Yizengaw, T.; Verheye, W. Application of computer captured knowledge in land evaluation, using ALES in central Ethiopia. *Geoderma* **1995**, *66*, 297–311. [CrossRef]
18. Ismail, H.A.; Bahnassy, M.H.; Abd El-Kawy, O.R. Integrating GIS and modelling for agricultural land suitability evaluation at East Wadi El-Natrun, Egypt. *Egypt. J. Soil Sci.* **2005**, *45*, 297–322.
19. Elsheikh, R.; Mohamed Shariff, A.R.B.; Amiri, F.; Ahmad, N.B.; Balasundram, S.K.; Soom, M.A.M. Agriculture Land Suitability Evaluator (ALSE): A decision and planning support tool for tropical and subtropical crops. *Comput. Electron. Agric.* **2013**, *93*, 98–110. [CrossRef]
20. Rossiter, D.G. A theoretical framework for land evaluation. *Geoderma* **1996**, *72*, 165–190. [CrossRef]
21. Rossiter, D.G. Biophysical models in land evaluation. In *Encyclopedia of Life Support Systems (EOLSS), Section 1.5 "Land Use and Land Cover"*; Verheye, W.H., Ed.; EOLSS Publishers Co. Ltd.: Oxford, UK, 2003; pp. 1–16.
22. El-Kawy, A.O.R.; Ismail, H.A.; Rod, J.K.; Suliman, A.S. A Developed GIS-based land evaluation model for agricultural land suitability assessments in arid and semi-arid regions. *Res. J. Agric. Biol. Sci.* **2010**, *6*, 589–599.
23. Wahab, M.A.; El Semary, M.A.; Ali, R.R.; Darwish, K.M. Land resources assessment for agricultural use in some areas west of Nile valley, Egypt. *J. Appl. Sci. Res.* **2013**, *9*, 4288–4298.
24. Darwish, K.M.; Abdel Kawy, W.A. Land suitability decision support for assessing land use changes in areas west of Nile Delta, Egypt. *Arab. J. Geosci.* **2014**, *7*, 865–875. [CrossRef]
25. Abd El-Kawy, O.R.; Flous, G.M.; Abdel-Kader, F.H.; Suliman, A.S. Land suitability analysis for crop cultivation in a newly developed area in Wadi Al-Natrun, Egypt. *Alex. Sci. Exch. J.* **2019**, *40*, 683–693. [CrossRef]
26. Mahmoud, H.; Binmiskeen, A.; Saad Moghanm, F. Land evaluation for crop production in the Banger El-Sokkar Region of Egypt using a geographic information system and ALES-arid Model. *Egypt. J. Soil Sci.* **2020**, *60*, 129–143. [CrossRef]
27. Bryla, D.R.; Trout, T.J.; Ayars, J.E. weighing lysimeters for developing crop coefficients and efficient irrigation practices for vegetable crops. *HortScience* **2010**, *45*, 1597–1604. [CrossRef]
28. Jia, X.; Dukes, M.D.; Jacobs, J.M. Bahiagrass crop coefficients from eddy correlation measurements in central Florida. *Irrig. Sci.* **2009**, *28*, 5–15. [CrossRef]
29. Pivec, J.; Brant, V.; Hamouzová, K. *Evapotranspiration and Transpiration Measurements in Crops and Weed Species by the Bowen Ratio and Sapflow Methods under the Rainless Region Conditions*; Gerosa, G., Ed.; InTech: London, UK, 2011. [CrossRef]
30. Allen, R.; Tasumi, M.; Trezza, R. Satellite-based energy balance for mapping evapotranspiration with internalized calibration (METRIC)-Model. *J. Irrig. Drain. Eng.* **2007**, *133*, 380–394. [CrossRef]

31. Morton, C.G.; Huntington, J.L.; Pohll, G.M.; Allen, R.G.; McGwire, K.C.; Bassett, S.D. Assessing calibration uncertainty and automation for estimating evapotranspiration from agricultural areas using METRIC. *J. Am. Water Resour. Assoc.* **2013**, *49*, 549–562. [CrossRef]

32. Tasumi, M.; Allen, R.G.; Trezza, R.; Wright, L. Satellite-based energy balance to assess within-population variance of crop coefficient curves. *J. Irrig. Drain. Eng.* **2005**, *131*, 94–109. [CrossRef]

33. El-Magd, I.A.; Tanton, T. Remote sensing and GIS for estimation of irrigation crop water demand. *Int. J. Remote Sens.* **2005**, *26*, 2359–2370. [CrossRef]

34. Allen, R.; Irmak, A.; Trezza, R.; Hendrickx, J.M.H.; Bastiaanssen, W.; Kjaersgaard, J. Satellite-based ET estimation in agriculture using SEBAL and METRIC. *Hydrol. Process.* **2011**, *25*, 4011–4027. [CrossRef]

35. Yang, Y.; Shang, S.; Jiang, L. Remote sensing temporal and spatial patterns of evapotranspiration and the responses to water management in a large irrigation district of North China. *Agric. For. Meteorol.* **2012**, *164*, 112–122. [CrossRef]

36. Fisher, J.B.; Melton, F.; Middleton, E.; Hain, C.; Anderson, M.; Allen, R.; McCabe, M.F.; Hook, S.; Baldocchi, D.; Townsend, P.A.; et al. The future of evapotranspiration: Global requirements for ecosystem functioning, carbon and climate feedbacks, agricultural management, and water resources. *Water Resour. Res.* **2017**, *53*, 2618–2626. [CrossRef]

37. Bastiaanssen, W.; Menenti, M.; Feddes, R.; Holtslag, A. A remote sensing surface energy balance algorithm for land (SEBAL): 1. Formulation. *J. Hydrol.* **1998**, *212–213*, 198–212. [CrossRef]

38. Su, Z. The Surface Energy Balance System (SEBS) for estimation of turbulent heat fluxes. *Hydrol. Earth Syst. Sci.* **2002**, *6*, 85–99. [CrossRef]

39. Senay, G.B. Satellite psychrometric formulation of the operational Simplified Surface Energy Balance (SSEBop) model for quantifying and mapping evapotranspiration. *Appl. Eng. Agric.* **2018**, *34*, 555–566. [CrossRef]

40. Anderson, M.C.; Norman, J.M.; Mecikalski, J.R.; Otkin, J.A.; Kustas, W.P. A climatological study of evapotranspiration and moisture stress across the continental United States based on thermal remote sensing: 1. Model formulation. *J. Geophys. Res. Atmos.* **2007**, *112*, D10117. [CrossRef]

41. Courault, D.; Seguin, B.; Olioso, A. Review on estimation of evapotranspiration from remote sensing data: From empirical to numerical modeling approaches. *Irrig. Drain. Syst.* **2005**, *19*, 223–249. [CrossRef]

42. Liou, Y.; Kar, S.K. Evapotranspiration estimation with remote sensing and various surface energy balance algorithms—A review. *Energies* **2014**, *7*, 2821–2849. [CrossRef]

43. Subedi, A.; Chávez, J.L. Crop evapotranspiration (ET) estimation models: A review and discussion of the applicability and limitations of ET methods. *J. Agric. Sci.* **2015**, *7*, 50–68. [CrossRef]

44. Zhang, K.; Kimball, J.S.; Running, S.W. A review of remote sensing based actual evapotranspiration estimation. *Wiley Interdiscip. Rev. Water* **2016**, *3*, 834–853. [CrossRef]

45. Bastiaanssen, W.; Pelgrum, H.; Wang, J.; Ma, Y.; Moreno, J.; Roerink, G.; van der Wal, T. A remote sensing surface energy balance algorithm for land (SEBAL): 2. Validation. *J. Hydrol.* **1998**, *212–213*, 213–229. [CrossRef]

46. Zamani Losgedaragh, S.; Rahimzadegan, M. Evaluation of SEBS, SEBAL, and METRIC models in estimation of the evaporation from the freshwater lakes (Case study: Amirkabir dam, Iran). *J. Hydrol.* **2018**, *561*, 523–531. [CrossRef]

47. Papadavid, G.; Perdikou, S.; Hadjimitsis, M.; Hadjimitsis, D. Remote sensing applications for planning irrigation management. The use of SEBAL methodology for estimating crop evapotranspiration in Cyprus. *Environ. Clim. Technol.* **2012**, *9*, 17–21. [CrossRef]

48. Bhattarai, N.; Dougherty, M.; Marzen, L.J.; Kalin, L. Validation of evaporation estimates from a modified surface energy balance algorithm for land (SEBAL) model in the south-eastern United States. *Remote Sens. Lett.* **2012**, *3*, 511–519. [CrossRef]

49. Jaafar, H.H.; Ahmad, F.A. Time series trends of Landsat-based ET using automated calibration in METRIC and SEBAL: The Bekaa Valley, Lebanon. *Remote Sens. Environ.* **2019**. [CrossRef]

50. Laipelt, L.; Ruhoff, A.L.; Fleischmann, A.S.; Kayser, R.H.B.; Kich, E.d.M.; da Rocha, H.R.; Neale, C.M.U. Assessment of an automated calibration of the SEBAL algorithm to estimate dry-season surface-energy partitioning in a forest–savanna transition in Brazil. *Remote Sens.* **2020**, *12*, 1108. [CrossRef]

51. USDA. *Soil Survey Field and Laboratory Methods Manual*; National Soil Survey Center, Natural Resources Conservation Service; U.S. Department of Agriculture: Lincoln, NE, USA, 2009.

52. Appel, M.; Lahn, F.; Buytaert, W.; Pebesma, E. Open and scalable analytics of large Earth observation datasets: From scenes to multidimensional arrays using SciDB and GDAL. *ISPRS J. Photogramm. Remote Sens.* **2018**, *138*, 47–56. [CrossRef]

53. ASCE-EWRI. *The ASCE Standardized Reference Evapotranspiration Equation*; ASCE-EWRI Standardization of Reference Evapotranspiration Task Committee Report; The American Society of Civil Engineers (ASCE): Washington, DC, USA, 2005; p. 216.

54. Allen, R. *REF-ET: Reference Evapotranspiration Calculation Software for FAO and ASCE Standardized Equations Version 4.1. for Windows*; University of Idaho: Moscow, ID, USA, 2015.

55. Allen, R.; Pereira, L.S.; Raes, D.; Smith, M. *Crop Evapotranspiration: Guidelines for Computing Crop Water Requirements*; Food and Agriculture Organization: Rome, Italy, 1998.

56. Waters, R.; Allen, R.; Tasumi, M.; Trezza, R.; Bastiaanssen, W. *SEBAL: Surface Energy Balance Algorithms for Land—Advanced Training and Users Manual Version 1.0*; The Idaho Department of Water Resources: Boise, ID, USA, 2002; p. 98.

57. Bastiaanssen, W.; Ahmad, M.-u.-D.; Chemin, Y. Satellite surveillance of evaporative depletion across the Indus basin. *Water Resour. Res.* **2002**, *38*, 1273. [CrossRef]

58. Bastiaanssen, W.; Noordman, E.; Pelgrum, H.; Davids, G.; Thoreson, B.; Allen, R. SEBAL model with remotely sensed data to improve water-resources management under actual field conditions. *J. Irrig. Drain. Eng.* **2005**, *131*, 85–93. [CrossRef]

59. Beg, A.A.F.; Al-Sulttani, A.H.; Ochtyra, A.; Jarocińska, A.; Marcinkowska, A. Estimation of evapotranspiration using SEBAL algorithm and Landsat-8 data-A case study: Tatra mountains region. *J. Geol. Resour. Eng.* **2016**, *6*, 257–270. [CrossRef]

60. Farah, H.; Bastiaanssen, W. Impact of spatial variations of land surface parameters on regional evaporation: A case study with remote sensing data. *Hydrol. Process.* **2001**, *15*, 1585–1607. [CrossRef]

61. Brutsaert, W.; Sugita, M. Application of self-preservation in the diurnal evolution of the surface energy budget to determine daily evaporation. *J. Geophys. Res.* **1992**, *97*, 18377–18382. [CrossRef]

62. Ahmad, M.D.; Biggs, T.; Turral, H.; Scott, C.A. Application of SEBAL approach and MODIS time-series to map vegetation water use patterns in the data scarce Krishna river basin of India. *Water Sci. Technol.* **2006**, *53*, 83–90. [CrossRef] [PubMed]

63. Bastiaanssen, W. SEBAL-based sensible and latent heat fluxes in the irrigated Gediz basin, Turkey. *J. Hydrol.* **2000**, *229*, 87–100. [CrossRef]

64. Khalifa, A.M. Mineralogical and Chemical Properties of Toshka Soils. Institute of African Research and Studies. Master's Thesis, Cairo University, Giza, Egypt, 2001.

65. Abbas, H.H.; El-Husseiny, O.H.; Mohamed, M.K.; Abuzaid, A.S. Land capability and suitability of some soils in Toshka area, Southwestern Egypt. *Ann. Agric. Sci. MoshtohorEgypt* **2010**, *48*, 1–12.

66. Hamzawy, M.M.H. Soil Studies in Lake Nasser Region Using Remote Sensing and GIS Capabilities. Ph.D. Thesis, Al-Azhar University, Cairo, Egypt, 2014.

67. Taghizadeh-Mehrjardi, R.; Nabiollahi, K.; Rasoli, L.; Kerry, R.; Scholten, T. Land suitability assessment and agricultural production sustainability using machine learning models. *Agronomy* **2020**, *10*, 573. [CrossRef]

68. FAO. *Water Quality for Agriculture*; Food and Agriculture Organization of the United Nations: Rome, Italy, 1985.

69. El-Mahdy, M.E.; Abbas, M.S.; Sobhy, H.M. Investigating the water quality of the water resources bank of egypt: Lake nasser. In *Conventional Water Resources and Agriculture in Egypt*; Negm, A.M., Ed.; Springer International Publishing: Cham, Swetzerland, 2019; pp. 639–655. [CrossRef]

70. Fayed, R.M.; Hussin, M.A.; Rizk, A.H.; Tawfik, T.A.; M Shreif, M.M. Effect of wells water quality on some soil properties and productivity in Toshka area. *J. Soil Sci. Agric. Eng.* **2010**, *1*, 873–881. [CrossRef]

71. IDMC. *Aswan Governorate Statistical Guide*; Information and Decision Making Center: Aswan, Aswan Governorate, Egypt, 2014. (In Arabic)

72. Hassan, F.O.; Salam, A.A.A.; Rashed, H.S.; Faid, A.M. Land evaluation and suitability of Hala'ib and Shalateen region, Egypt, by integrated use of GIS and remote sensing techniques. *Ann. Agric. Sci. Moshtohor.* **2017**, *55*, 151–162.

73. Brouwer, C.; Prins, K.; Heibloem, M. *Irrigation Water Management: Irrigation Scheduling*; Food and Agriculture Organization: Rome, Italy, 1989.

74. El-Marsafawy, S.M.; Eid, H.M. Estimation of water consumptive use for Egypt. In Proceedings of the Third Conference of On-Farm Irrigation and Agroclimatology, Cairo, Egypt, 25–27 January 1999.

75. Eid, H.M.; Ainer, N.G.; El-Marsafawy, S.M.; Khater, A.N. Crop water needs under different irrigation system in new land. In Proceedings of the Third Conference of On-Farm Irrigation and Agroclimatology, Cairo, Egypt, 25–27 January 1999.

76. Eid, H.M.; El-Marsafawy, S.M.; Abbas, F.A.; Ali, M.A.; Khater, I.N.; Eissa, M.M. Estimation of water needs for vegetable crops in the new land. *Meteorol. Res. Bull.* **2002**, *16*, 156–179.

77. Eid, H.M.; El-Marsafawy, S.M.; Ibrahim, M.M.; Eissa, M.M. Estimation of water needs for Orchard trees in the old land. *Meteorol. Res. Bull.* **2002**, *17*, 131–139.

78. Mahmoud, M.A.; El-Bably, A.Z. Crop water requirements and irrigation efficiencies in Egypt. In *Conventional Water Resources and Agriculture in Egypt*; Negm, A.M., Ed.; Springer International Publishing: Cham, Swetzerland, 2019; pp. 471–487. [CrossRef]

79. Calera, A.; Campos, I.; Osann, A.; D'Urso, G.; Menenti, M. Remote sensing for crop water management: From ET modelling to services for the end users. *Sensors* **2017**, *17*, 1104. [CrossRef]

80. Abd El-Hady, A.M.; Abdelaty, E.F. GIS—Comprehensive analytical approach for soil use by linking crop soil suitability to soil management and reclamation. *Alex. Sci. Exch. J.* **2019**, *40*, 60–81. [CrossRef]

81. Paula, A.C.P.d.; Silva, C.L.d.; Rodrigues, L.N.; Scherer-Warren, M. Performance of the SSEBop model in the estimation of the actual evapotranspiration of soybean and bean crops. *Pesqui. Agropecuária Bras.* **2019**, *54*. [CrossRef]

82. Biro, K.; Zeineldin, F.; Al-Hajhoj, M.R.; Dinar, H.A. Estimating irrigation water use for date palm using remote sensing over an Oasis in arid region. *Iraqi J. Agric. Sci.* **2020**, *51*, 1173–1187. [CrossRef]

83. Sun, Z.; Wei, B.; Su, W.; Shen, W.; Wang, C.; You, D.; Liu, Z. Evapotranspiration estimation based on the SEBAL model in the Nansi Lake Wetland of China. *Math. Comput. Model.* **2011**, *54*, 1086–1092. [CrossRef]

 *agronomy*

Article

The Effects of Potassium Fertilization and Irrigation on the Yield and Health Status of Jerusalem Artichoke (*Helianthus tuberosus* L.)

Bożena Bogucka [1,*], Agnieszka Pszczółkowska [2], Adam Okorski [2] and Krzysztof Jankowski [1]

1   Department of Agrotechnology and Agribusiness, University of Warmia and Mazury in Olsztyn, Oczapowskiego 8, 10-719 Olsztyn, Poland; krzysztof.jankowski@uwm.edu.pl
2   Department of Entomology, Phytopathology and Molecular Diagnostics, University of Warmia and Mazury in Olsztyn, Plac Łódzki 5, 10-727 Olsztyn, Poland; agnieszka.pszczolkowska@uwm.edu.pl (A.P.); adam.okorski@uwm.edu.pl (A.O.)
*   Correspondence: bozena.bogucka@uwm.edu.pl

Abstract: The objective of this study was to determine the effects of potassium fertilization (applied to soil at 150, 250, and 350 kg K_2O ha^{-1}) and irrigation on the yield (fresh matter yield and dry matter yield of above-ground biomass and tubers) and the health status of tubers and leaves of three Jerusalem artichoke—JA (*Helianthus tuberosus* L.) cultivars (Topstar, Violette de Rennes, Waldspindel). The Topstar cultivar was characterized by the highest total tuber yield (60.53 Mg FM ha^{-1}) and the highest above-ground biomass yield (65.74 Mg FM ha^{-1}). An increase in the rate of potassium fertilizer to 350 kg K_2O ha^{-1} did not affect total tuber yields. The greatest increase in above-ground biomass yields was observed in response to the potassium fertilizer rate of 150 kg K_2O ha^{-1} (64.40 Mg FM ha^{-1}). Irrigation increased tuber yields by 59% and above-ground biomass yields by 42% on average. Phytopathological analyses revealed that JA leaves were most frequently colonized by fungi of the genera *Alternaria*, *Fusarium*, and *Epicoccum*. *Alternaria* and *Fusarium* fungi were more prevalent in non-irrigated than in irrigated plots. A higher number of fungal pathogens was isolated from the leaves of cv. Violette de Rennes grown in a non-irrigated plot fertilized with 250 kg K_2O ha^{-1}. Tubers were most heavily colonized by fungi of the genera *Penicillium*, *Fusarium*, *Alternaria*, *Botrytis*, and *Rhizopus*. Fungal species of the genus *Fusarium* were isolated from tubers in all irrigated treatments, and they were less frequently identified in non-irrigated plots. Only the tubers of cv. Topstar grown in non-irrigated plots and supplied with 150 kg K_2O ha^{-1} were free of *Fusarium* fungi. The number of cultures of pathogenic species isolated from Jerusalem artichoke tubers had a minor negative impact on fresh and dry matter yield.

Keywords: Jerusalem artichoke; mineral fertilization; irrigation; yield; diseases; fungi

Citation: Bogucka, B.; Pszczółkowska, A.; Okorski, A.; Jankowski, K. The Effects of Potassium Fertilization and Irrigation on the Yield and Health Status of Jerusalem Artichoke (*Helianthus tuberosus* L.). *Agronomy* **2021**, *11*, 234. https://doi.org/10.3390/agronomy11020234

Academic Editor: Ivan Francisco Garcia Tejero and Victor Hugo Durán-Zuazo
Received: 22 December 2020
Accepted: 24 January 2021
Published: 27 January 2021

1. Introduction

Jerusalem artichoke (JA) has been long grown as a source of animal feed, but in recent years, its popularity increased in the food processing industry, mainly in the production of functional ingredients such as inulin, oligofructose, and fructose [1–4]. The leaves and stems of JA are also a source of bioactive compounds that are used in the treatment of wounds and swelling [5–7].

In recent years, JA tubers have also been recognized as a valuable source of sugars for bioethanol production. Its tubers are more abundant in ethanol (1500–11,000 L ha^{-1}) than sugar beetroots (5000–6000 L ha^{-1}) and maize (2000–6698 L ha^{-1}) [8–15]. Jerusalem artichoke is a promising source of agricultural biomass due to its wide range of applications and low production costs.

Jerusalem artichoke has numerous advantages over other agricultural crops, including a rapid growth rate, tolerance to low temperatures, and high resistance to pests. New cultivars are characterized by high yields. However, it should be stressed that JA thrives in moist soils and has high fertilization requirements. The goal of every agricultural production system is to maximize yields per unit area. Fecundity is determined by the optimal combination of genetic and agronomic factors. Macronutrient deficiencies, including nitrogen, potassium, and phosphorus, as well as drought compromise yields [8,14,16–20]. According to Rossini [14], JA thrives in regions where annual precipitation exceeds 500 mm. Soja et al. [8] observed that potassium deficiency is more likely to slow down the growth of tubers than aerial plant parts. However, nitrogen has a stronger influence on yields than potassium because it determines the photosynthetic potential of plants and increases their water use efficiency [21]. In turn, potassium speeds up the translocation of sugars from leaves to tubers. Potassium and nitrogen fertilization increases the yield of tubers and improves their quality [22]. According to some authors [16–18], water supply is the key limiting factor in the production of JA. Research indicates that early-maturing varieties are more sensitive to drought that late-maturing varieties. For this reason, drought as well as nitrogen, phosphorus, and potassium (NPK) fertilizers strongly influence the accumulation of dry matter [23,24].

Jerusalem artichoke is colonized by various herbivorous insects and microorganisms, but not all of them contribute to a decrease in yields [14,16]. The threats associated with pathogens are determined by agronomic conditions [25]. According to Doneroy [16], aboveground plant parts are less susceptible to disease than tubers, in particular in the last stages of growth and during storage. The most dangerous pathogens of JA are *Sclerotinia sclerotiorum*, which causes sclerotinia wilt/rot, and *Sclerotium rolfsii* which causes southern wilt, also known as southern blight or collar rot [16,26]. *Sclerotium rolfsii* can decrease JA yields by as much as 60% [27]. Jerusalem artichoke is also susceptible to rust caused by *Puccinia helianthin* and powdery mildew caused by *Erisyphe chicoracearum*, but these pathogens does not compromise yields [16,28]. Various pathogens are responsible for leaf spot diseases in JA. *Alternaria helianthi* causes small yellow spots on leaves, followed by leaf damage and defoliation; it reduces the photosynthetic capacity of plants [29,30] and can decrease sunflower yields by up to 80% [29]. The disease is most severe in tropical regions. In China, JA is susceptible to *Bipolaris zeae* which causes brown spot disease [31]. Pathogens also affect JA tubers. Symptoms of disease are also observed during storage, in particular in tubers that were damaged during harvest. Tuber pathogens include *Botrytis cinerea*, *Rhizopus nigricans* as well as species of the genera *Fusarium* and *Pennicillum* [32,33].

The aim of this study was to determine the optimal rates of potassium fertilizer applied to soil, and to evaluate the effect of irrigation on the yield and health status of JA tubers.

2. Materials and Methods

2.1. Field Experiment

Jerusalem artichoke was grown in 2018 during a field experiment conducted in the Agricultural Experiment Station in Tomaszkowo (53°42′ N, 20°26′ E, NE Poland). The experiment had a three-factorial split-split-plot design with three replications. The analyzed variables were: (i) cultivar: Topstar (early edible cultivar with yellow-brown tubers), Violette de Rennes (mid-late edible cultivar with red tubers), Waldspindel (mid-late cultivar with red tubers which is processed in herbal and distilling industries); (ii) rate of potassium fertilizer applied to soil (kg K_2O ha^{-1}): 150, 250, 350; and (iii) irrigation: treatments that were and were not irrigation.

The examined cultivars were acquired from an organic farm (Die Topinambur Manufaktur, Heimenkirch, Bavaria, Germany). The tubers were planted in mid-April at a depth of 6–8 cm, with a spacing of 75 × 30 cm. Potassium fertilizer in the form of potassium sulfate (50%), 80 kg N ha^{-1} (urea, 46%), 70 kg P_2O_5 ha^{-1} (enriched superphosphate, 40%), and 90 CaO kg ha^{-1} (ground dolomite, 52% CaO, 37% $MgCO_3$, 48% $CaCO_3$) was applied before planting based on the experimental design. Organic fertilizer was not applied. Fertilizer rates were determined based on the results of an experiment conducted in Germany during 1994–2001 [34].

The moisture content of soil was monitored from the beginning of tuber formation, during plant growth, until leaf ageing and the transfer of sugars to tubers (from mid-June to mid-October). The optimal soil moisture content was established at 14.3–16.5%, i.e., 65–75% of field water capacity at a depth of 30 cm. The soil moisture was measured twice a week. Jerusalem artichoke was irrigated every 5–7 days at 20 dm^3 m^{-2} when field water capacity fell below 60% ($\leq$13.2% soil moisture content). The irrigation schedule was based on the irrigation regime for late potato cultivars (*Solanum tuberosum L.*) and field water capacity for various types of soil [35]. Soil moisture was measured with the SM 150-KIT probe (Geomor-Technik Ltd., Szczecin, Poland). Each irrigation treatment involved 220 mm of water per m^2 of plot area, and 11 treatments were applied during the growing season of *H. tuberosus* (on 13 and 19 June; 3 and 9 July; 10 and 23 August; 3, 11, 20 and 28 September; 6 October).

The tuber fresh matter yield was determined at harvest in each plot. Tuber samples of 0.5 kg each were collected from each plot, and their dry matter was determined by the gravimetric method. Ground analytical samples of 5 g each were dried to constant weight at a temperature of 105 °C for 5–8 h (based on the results of three consecutive measurements) and were left in the desiccator until the achievement of room temperature (SUP 100 W laboratory drier, WAMED Warszawa, Poland). The weight of the harvested tubers was expressed per 1 ha.

Fresh matter yield was also determined in the above-ground biomass harvested from each plot. Dry matter content was determined in approximately 1 kg samples of above-ground biomass (stems, leaves, inflorescence) collected from each plot. The samples were cut into segments with a length of 1 cm, dried at a temperature of 65 °C for 10 h (BINDER GmBH, FED 720 drying oven, Binder Ltd., Tuttlingen, Germany), and weighed. The weight of the harvested aerial plant parts was expressed per 1 ha.

Jerusalem artichoke was grown on Haplic Luvisol loamy sand [36] in plots with an area of 2.7 m^2 each. The preceding crop was oat (*Avena sativa* L.). Composite soil samples were obtained at a depth of 20 cm from each plot to for analyses of the chemical properties of soil. Soil pH was determined at 5.4 with a digital pH meter. Nutrient levels in soil samples were determined at 74 mg P kg^{-1} (Egner-Riehm method), 145 mg K kg^{-1} (Egner-Riehm method) and 69 mg Mg kg^{-1} (AAS) [37]. Tubers were ridged once after planting. Crops were harvested in early November.

2.2. Mycological Analyses of Jerusalem Artichoke Leaves and Tubers

Mycological analyses were carried out in three JA cultivars: Topstar, Violette de Rennes and Waldspindel. Jerusalem artichoke was grown in non-irrigated or irrigated plots with various rates of potassium fertilizer. Plant health was evaluated once, in the last ten days of August in all plots.

To evaluate the health status of plants, three leaves were collected from the middle segment of randomly selected plants in each replication in each treatment. In the laboratory, the collected plant materials were pooled (samples of 9 leaves each), and next six leaves were collected randomly from each treatment. The leaves were cut into 5 mm × 5 mm segments. Leaf segments were rinsed under running water, disinfected in 1% sodium hypochlorite solution for 5 min and in 70% ethyl alcohol for 5 min, and rinsed with sterile distilled water. The prepared specimens were plated on PDA (5 specimens per plate) and incubated at 23 °C in a 12-h dark and 12-h UV light cycle. After 10 days, fungal cultures were transferred onto PDA in sterile Petri plates. After 14 days, fungal colonies were identified to genus and species level based on the literature [37–39].

The health status of the harvested tubers was evaluated in each JA cultivar. Ten tubers were sampled from each cultivar in the first ten days of December. The tubers were washed under running water, disinfected in 1% sodium hypochlorite solution for 5 min and 70% ethyl alcohol for 5 min, rinsed with sterile distilled water, and cut into segments measuring 5 mm × 5 mm × 3 mm. The prepared specimens representing each cultivar and each treatment were plated on PDA (5 specimens per plate, a total of 10 plates) and incubated at

23 °C in a 12-h dark and 12-h UV light cycle. After 10 days, fungal colonies were transferred onto PDA in sterile Petri plates. After 14 days, fungal colonies were identified to genus and species level based on the literature [38–40].

2.3. Analysis of Pathogenic and Saprotrophic Fungi

Relative frequency [*RF*] was calculated with the use of Equation (1), dominance [*Y*]—with Equation (2), species richness [*S*] (number of species colonizing the leaves and tubers of each JA cultivar), and Margalef index [*D′*]—with Equation (3), Shannon–Wiener index [*H′*]—with Equation (4), and the dominance index [λ]—with Equation (5). The calculated indices were used in quantitative analyses of the abundance, distribution preference, and composition of pathogenic and saprotrophic fungal species colonizing Jerusalem artichoke leaves and tubers.

$$RF\ (\%) = \left(\frac{n_i}{N_i}\right) x\ 100\% \tag{1}$$

$$Y = \left(\frac{n_i}{N_i}\right) x f_i \tag{2}$$

$$D' = (S - 1)/lnN_t \tag{3}$$

$$H' = -\sum_{i=1}^{s} P_i ln P_i,\ P_i = N_i/N_t \tag{4}$$

$$\lambda = \sum_{i}^{s} p_i^2, \tag{5}$$

where N_t is the number of isolated cultures, N_i is the number of isolates belonging to the *i*-th species, and f_i is the frequency of taxa belonging to a given genus [41].

2.4. Weather Conditions

In 2018, the growing season had 201 days. Jerusalem artichokes were harvested in the first week of November. Weather conditions during the growing season of 2018 are presented in Table 1. The mean monthly air temperatures were similar to the long-term average for 1981–2010. The rapid growth of aerial plant parts and tuber formation began in mid-June. During the growing season, the total rainfall was determined at 418.8 mm, and it was 7% lower than the long-term average (450.1 mm). Rainfall was not evenly distributed across months, and dry spells were observed in May and September. May and September were the driest months (when precipitation was 57% and 64% lower than the long-term average, respectively), and irrigation was required (3, 11, 20, and 28 September). The optimal moisture content of soil was set based on the irrigation regime for late-maturing potato varieties. The precipitation levels in August also failed to meet JA's water needs. July was the only month when rainfall exceeded the long-term average by 90% and the crops' water requirements by 45%. However, the field water capacity fell below 60% (to ≤13.2%), and the plots had to be irrigated.

During crop production, atmospheric drought can be determined by calculating Selyaninov's hydrothermal coefficient (K) with the use of the following formula (Equation (6)):

$$K = \frac{p}{0.1 \sum t} \tag{6}$$

where:

P—total monthly precipitation

t—sum of monthly temperatures divided by 10

Values below 1.0 indicate drought, and values below 0.5 denote severe drought. The analyzed growing period was characterized mostly by drought (April, June, and August) and severe drought (May and September). The rapid growth of aerial plant plants and tuber setting begins in the second half of June and ends in late October [35]. According to Denoroy [16], JA is particularly susceptible to drought during seedling emergence, flowering, and late stages of tuber growth. However, water deficit during seedling emergence is less detrimental to final yields than drought in the remaining two stages which can decrease yields by as much as 20%.

Table 1. Meteorological data for the growing season in 2018 and the long-term average for 1981–2010.

Specification	Month							
	April	May	June	July	August	September	October	November
Mean air temperature (°C)	10.8	15.7	17.2	19.7	19.2	14.5	8.7	3.3
30-yr mean	7.7	13.5	16.1	18.7	17.9	12.8	8.0	2.9
Total rainfall (mm)	33.5	25.0	53.7	141.0	44.6	20.3	84.7	16.0
30-yr mean	33.3	58.5	80.4	74.2	59.4	56.9	42.6	44.8
Water requirements of late-maturing potato varieties [35]	-	62	74	97	79	50	-	-
Selyaninov's hydrothermal coefficient (K) * [41,42]	1.03	0.51	1.04	2.30	0.75	0.47	3.14	1.61

* K: 0–0.5—severe drought, 0.6–1.0—drought, 1.0–2.0—moist, >2.1—wet.

2.5. Statistical Analysis

The results were analyzed statistically with the use of one-way analysis of variance (ANOVA) in the Statistica 13.3 program [42,43]. Significant differences ($p < 0.05$) between means were determined in Tukey's (honestly significant difference (HSD) test for multiple comparisons to assess significant differences between means. The results of the F-test for fixed effects in ANOVA are presented in Table 2.

Table 2. F-test statistics in ANOVA.

Parameter	Cv	K	Irrigation (IR)	Cv × K	Cv × IR	K × IR	Cv × K × IR
Total tuber yield (FM Mg ha^{-1})	9.826 ***	2.696*ns*	59.120 ***	3.105 **	0.601*ns*	3.313 **	1.112*ns*
Total tuber yield (DM Mg ha^{-1})	14.880 ***	2.244*ns*	50.849 ***	3.271 **	0.677*ns*	3.644 **	0.986*ns*
Above-ground biomass yield (FM Mg ha^{-1})	10.896 ***	7.714 **	77.245 ***	6.702 ***	1.403*ns*	3.504 **	2.137*ns*
Above-ground biomass yield (DM Mg ha^{-1})	8.051 ***	5.051 **	54.275 ***	4.158 ***	0.799*ns*	2.098*ns*	1.498*ns*

** significant $p < 0.01$, *** significant $p < 0.001$, *ns*—not significant.

The relationships between above-ground biomass yield, total tuber yield (fresh and dry matter yield), and the number of pathogens and saprotrophic fungi isolated from Jerusalem artichokes were determined with the use of linear regression methods and Pearson's linear correlation coefficients.

3. Results and Discussion

3.1. Total Tuber Yields and Above-Ground Biomass Yields

The total tuber yield was highest in cv. Topstar and similar to that reported by Rodrigues et al. [44] at 65.6 Mg ha^{-1} (Table 3). Topstar is an early cultivar, and its yields exceeded the values noted in mid-late cultivars: by 13.6 Mg ha^{-1} (FM) in comparison with cv. Waldspindel and by 17.5 Mg ha^{-1} (FM) in comparison with cv. Violette de Rennes. The fresh matter yield of JA tubers ranged from 55.5 to 90 Mg ha^{-1} in the work of Conde et al. [45], Baldini et al. [46], and Kim and Kim [47]. The fresh matter yield of JA tubers grown in a high-input system was determined at 30–80 Mg ha^{-1} by Denoroy [16], and Izsaki and Kadi [19]. The lowest tuber yields in the range of 3 to 46 Mg ha^{-1} were reported by Swanton et al. [9] and Pimsaen [48].

Table 3. The effect of the experimental factors on the yields of Jerusalem artichoke tubers and above-ground biomass ($Mg\ ha^{-1}$).

Parameter	Total Tuber Yield FM	Total Tuber Yield DM	Above-Ground Biomass Yield FM	Above-Ground Biomass Yield DM
Cultivar				
Violette de Rennes	43.17 [b]	8.84 [b]	55.97 [b]	20.69 [b]
Waldspindel	46.95 [b]	12.23 [a]	53.39 [b]	19.59 [b]
Topstar	60.53 [a]	14.18 [a]	65.74 [a]	24.42 [a]
Potassium fertilizer (kg $K_2O\ ha^{-1}$)				
150	55.73 [a]	12.77 [a]	64.40 [a]	23.82 [a]
250	47.36 [a]	11.81 [a]	53.70 [b]	19.97 [b]
350	47.55 [a]	10.67 [a]	57.00 [b]	20.93 [ab]
Irrigation				
Irrigated	63.14 [a]	14.64 [a]	68.38 [a]	25.36 [a]
Not irrigated	37.29 [b]	8.86 [b]	48.35 [b]	17.78 [b]

Means with the same letters do not differ significantly at $p \leq 0.05$ in Tukey's test.

The difference in the tuber dry matter yield between cvs. Topstar and Waldspindel reached only 1.95 Mg ha^{-1}, and it was not significant (Table 3). The tubers of cv. Violette de Rennes were characterized by the lowest dry matter yield which was 5.34 Mg ha^{-1} lower than in cv. Topstar and 3.4 Mg ha^{-1} lower in than in cv. Waldspindel. In a study by Rodrigues et al. [44], the dry matter yield of JA tubers was much higher at 18.4 Mg ha^{-1}.

An increase in the mineral fertilizer rate to 350 kg K$_2$O ha^{-1} did not influence the fresh matter yield or the dry matter yield of JA tubers (Table 3). Similar observations were made by Matias et al. [10]; in their study, the total tuber yields were not significantly affected by the applied rate of NPK fertilizer, which could be attributed to high soil fertility resulting from the choice of an adequate preceding crop. Izsaki and Kadi [11] demonstrated that tuber yields peaked at 10 Mg ha^{-1} in response to a potassium rate of 120 Mg ha^{-1}, which was found to be optimal. Izsaki and Kadi [11] and Raso [49] reported no interaction between potassium in the form of potassium sulfate (120 and 240 kg K$_2$O ha^{-1}) and nitrogen fertilizers for JA tuber yields (the yield reached 34 Mg ha^{-1} in the treatment with 50 kg N ha^{-1}).

The analyzed JA cultivars also differed in their responses to higher potassium rates—the tuber fresh matter yield was highest in cv. Topstar in plots fertilized with 150 kg K$_2$O ha^{-1} and the lowes in cvs. Violette de Rennes and Waldspindel in plots fertilized with 250 kg and 350 kg K$_2$O ha^{-1} and (Figure 1). Similar observations were made in an analysis of the tuber dry matter yield (Figure 2). In the early cultivar Topstar, the tuber dry matter yield was highest in response to 250 kg K$_2$O ha^{-1} (Figure 2).

The application of 350 kg K$_2$O ha^{-1} increased both tuber fresh matter and dry matter yields only in cv. Waldspindel (Figures 1 and 2).

The lowest tuber dry matter yields were noted in cv. Violette de Rennes fertilized with 350 kg K$_2$O ha^{-1}. The difference between the highest yielding cv. Topstar and the lowest yielding cv. Violette de Rennes reached 17.36 Mg FM ha^{-1} and 5.34 Mg DM ha^{-1} (Tables S1 and S2).

Jerusalem artichoke responded strongly to irrigation. Tuber fresh matter and dry matter yields increased by 69.3% and 65.2%, respectively, in response to irrigation (Tables S1 and S2). Irrigation also induced differences in the fresh matter yield of JA tubers in response to an increase in fertilization levels. The highest tuber fresh matter and dry yield peaked in response to 150 kg K$_2$O ha^{-1} in irrigated plots, whereas an the lowest was noted in non irrigated plots with 250 kg K$_2$O ha^{-1} fertilization (Figures 3 and 4, Tables S1 and S2). In the study, no significant effect of potassium fertilization on increase the tuber fresh matter and dry matter yield of Jerusalem artichoke was shown (Tables S1 and S2).

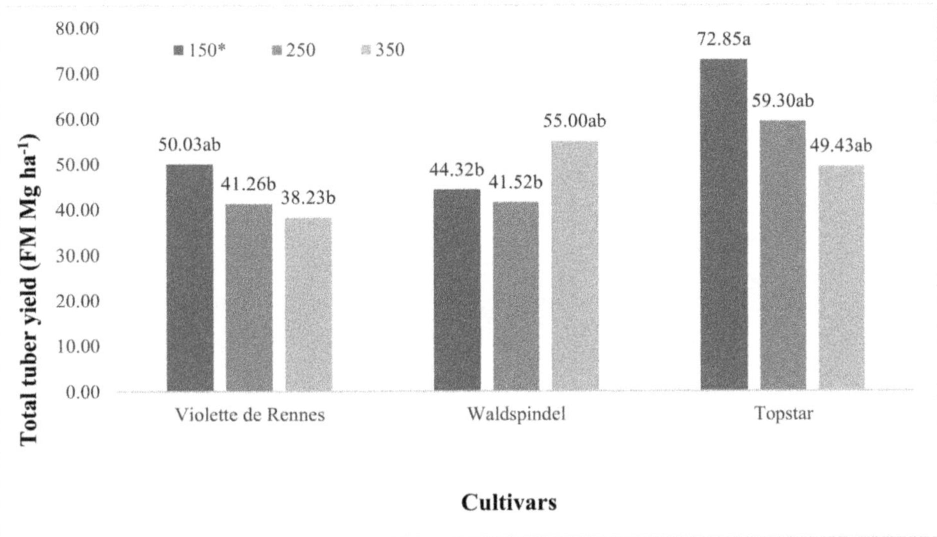

Figure 1. The effect of potassium fertilization on the total tuber yield (fresh matter basis) of the analyzed Jerusalem artichoke cultivars. Means with the same letters do not differ significantly at $p \leq 0.05$ in Tukey's test. * 150, 250, 350 kg K_2O ha^{-1}—potassium fertilization.

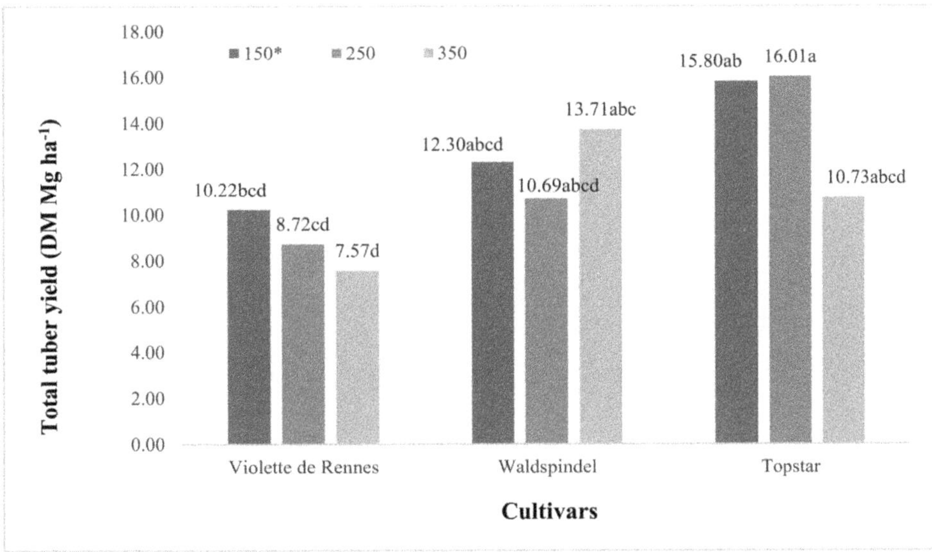

Figure 2. The effect of potassium fertilization on the total tuber yield (dry matter basis) of the analyzed Jerusalem artichoke cultivars. Means with the same letters do not differ significantly at $p \leq 0.05$ in Tukey's test. * 150, 250, 350 kg K_2O ha^{-1}—potassium fertilization.

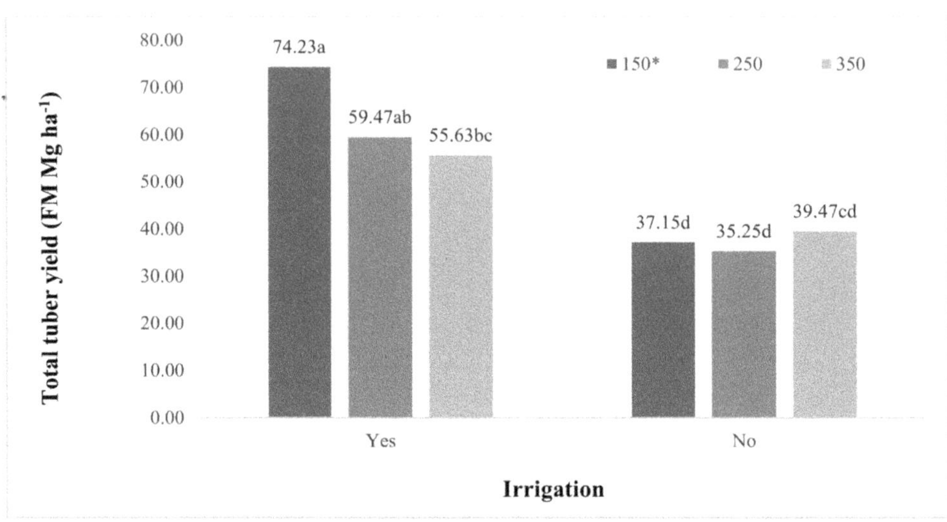

Figure 3. The effect of irrigation and potassium fertilization on the total tuber yield (fresh matter basis) of Jerusalem artichoke. Means with the same letters do not differ significantly at $p \leq 0.05$ in Tukey's test. * 150, 250, 350 kg K_2O ha^{-1}—potassium fertilization.

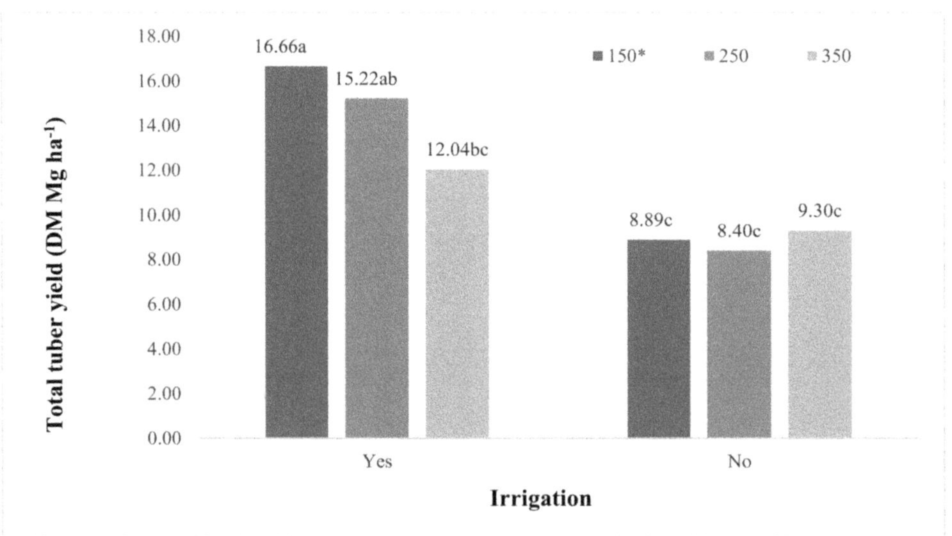

Figure 4. The effect of irrigation and potassium fertilization on the total tuber yield (dry matter basis) of Jerusalem artichoke. Means with the same letters do not differ significantly at $p \leq 0.05$ in Tukey's test. * 150, 250, 350 kg K_2O ha^{-1}—potassium fertilization.

In a study by Baldini et al. [46], irrigation increased the tuber dry matter yield by 24.2%. Schittenhelm [50] reported a 12 Mg FM ha^{-1} and 5.1 Mg DM ha^{-1} under water stress conditions and concluded that JA was more sensitive to drought than sugar beetroots and chicory.

In the present study, cv. Topstar was characterized by the highest fresh matter and dry matter yields of aerial plant parts. The yields of the early cv. Topstar exceeded the values noted in the mid-late cultivars by 9.8 Mg FM ha^{-1} and 3.7 Mg DM ha^{-1} (cv. Violette de Rennes) and by 12.3 Mg FM ha^{-1} and 4.8 Mg DM ha^{-1} (cv. Waldspindel) (Table 3).

In the work of Baldini et al. [46], the fresh matter yield of above-ground biomass ranged from 29.5 to 58.7 Mg ha^{-1} and was similar to that noted in this study (58.4 Mg ha^{-1} on average). Izsaki and Kadi [19] and Monti et al. [23] observed that plants grown in irrigated fields adapted to water stress with the growth and development of the root system. Under favorable water conditions, tubers compete with aerial plant parts for water [16].

An increase in the potassium fertilizer rate to 250 kg K$_2$O ha^{-1} decreased the fresh matter and the dry matter yields of above-ground biomass by around 10.7 and 3.8 Mg ha^{-1}, respectively, relative to plots fertilized with 150 kg K$_2$O ha^{-1}. When the potassium rate was increased to 350 kg K$_2$O ha^{-1}, a minor and non-significant increase was noted in fresh matter yield (3.3 Mg ha^{-1}) and dry matter yield (1.0 Mg ha^{-1}). The lowest potassium rate (150 kg K$_2$O ha^{-1}) exerted the greatest yield-forming effect (Table 3).

The influence of higher fertilizer rates on the fresh matter and dry matter yield of aerial plant parts also differed among the analyzed JA cultivars. In cv. Topstar fresh matter yield peaked in response to 150 kg K$_2$O ha^{-1} (Figures 5 and 6, Tables S3 and S4). The lowest fresh matter yields were observed for the cultivar Violette de Rennes with with application of 250 and 350 kg K$_2$O ha^{-1} and cv. Waldspindel with 150 and 250 kg K$_2$O ha^{-1} doses (Figure 5). Similar relationships were observed in the dry matter yield of above-ground biomass (Figure 6).

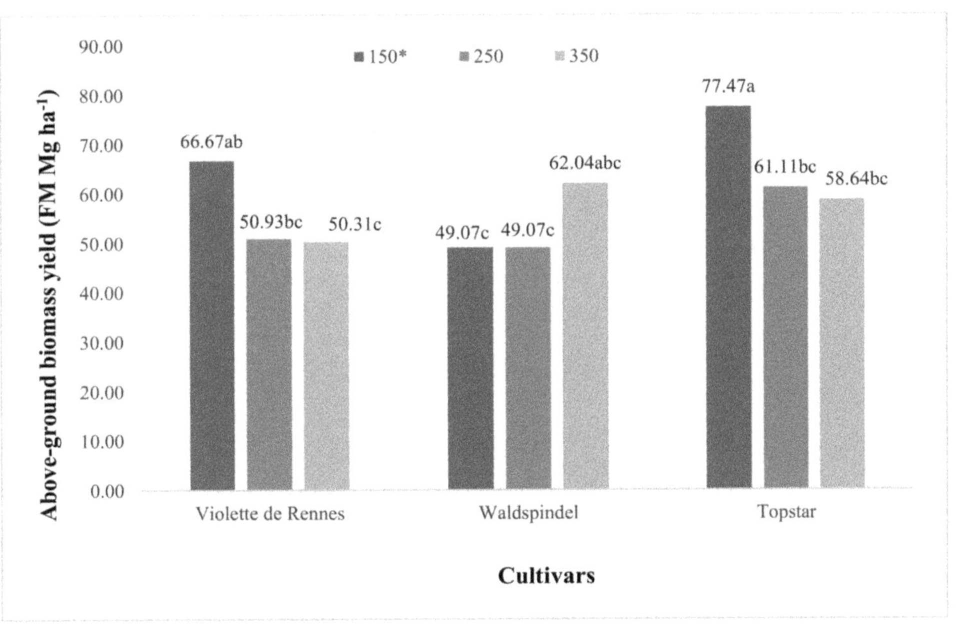

Figure 5. The effect of potassium fertilization on the above-ground biomass yield (fresh matter basis) of the analyzed Jerusalem artichoke cultivars. Means with the same letters do not differ significantly at $p \leq 0.05$ in Tukey's test. * 150, 250, 350 kg K$_2$O ha^{-1}—potassium fertilization.

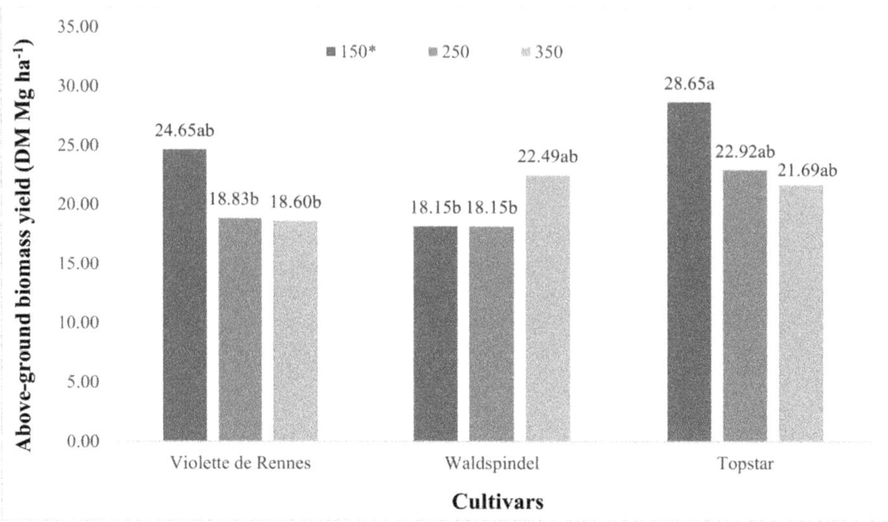

Figure 6. The effect of potassium fertilization on the above-ground biomass yield (dry matter basis) of the analyzed Jerusalem artichoke cultivars. Means with the same letters do not differ significantly at $p \leq 0.05$ in Tukey's test. * 150, 250, 350 kg K_2O ha^{-1}—potassium fertilization.

Jerusalem artichoke above-ground fresh and dry matter basis biomass yield responded strongly to irrigation. Irrigation increased the fresh matter and dry matter yields of aerial plant parts by 42% and 43%, respectively (Table 3).

Irrigation also induced differences in the fresh matter yield of above-ground biomass in response to higher fertilizer rates. The fresh matter yield of aerial plant parts was the highest after the application of 150 kg K_2O ha^{-1}, but it significantly decreased by 14.4 Mg ha^{-1} when the potassium rate was increased to 350 kg K_2O ha^{-1} and 16.3 Mg ha^{-1} with a dose of 250 kg K_2O ha^{-1} on irrigated plots (Figure 7). The lowest fresh matter yields were recorded in plots without irrigation (Figure 7).

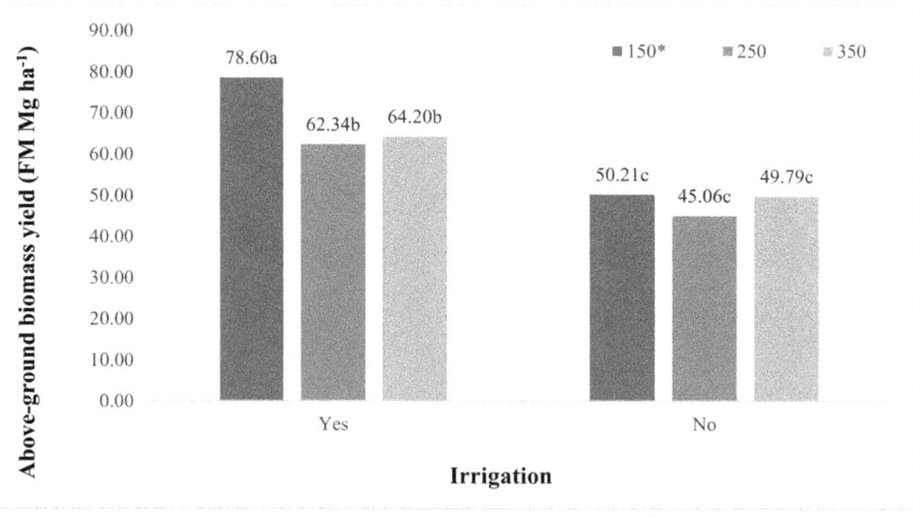

Figure 7. The effect of irrigation and potassium fertilization on the above-ground biomass yield (fresh matter basis) of Jerusalem artichoke. Means with the same letters do not differ significantly at $p \leq 0.05$ in Tukey's test. * 150, 250, 350 kg K_2O ha^{-1}—potassium fertilization.

According to Gao [51], irrigation significantly improves the yields of aerial plant parts as well as tubers.

3.2. Mycological Analyses of Tubers and Aerial Plant Parts

Jerusalem artichoke is resistant to biotic stresses such as pests and disease [52] and potentially resistant to abiotic stresses, including drought, frost, and high temperature [53]. However, there is considerable evidence [14,16,25,26,33,54] to indicate that similar to other crops, JA is susceptible to various pathogens, in particular fungi. Phytopathogens exert a negative influence on plant growth and development, and they compromise the quality of crops. In the current study, various fungal species and genera were identified on aerial plant parts (leaves) and tubers of three JA cultivars. A total of 355 fungal isolates were obtained from 900 cultured leaf segments. Of those, 164 isolates were obtained from leaves grown in irrigated treatments, and 191 from leaves grown in non-irrigated treatments (Table 4). Phytopathological analyses revealed that JA leaves were colonized mostly by fungi of the genera *Alternaria*, *Fusarium*, and *Epicoccum*. The most prevalent fungal species of the genus *Fusarium* were *Fusarium avenaceum* and *F. sporitrichioides*. Jerusalem artichoke leaves were also colonized by *Botrytis cinerea*, *Epicoccum nigrum*, *Nigrospora sphaerica*, *Didymella pinodella* as well as *Mucor* spp., *Penicillium* spp., and *Chaetomium* spp. (Table 5). The analysis of the composition of saprotrophs and pathogens revealed that pathogenic species were predominant on Jerusalem artichoke leaves, which was confirmed by nearly all biodiversity indicators (Table 5). The number of fungal isolates obtained from the analyzed cultivars was similar in irrigated plots (54–56 isolates). In non-irrigated treatments, fungal pathogens were most frequently isolated from the leaves of cv. Violette de Rennes and were least prevalent on the leaves of cv. Waldspindel (Table 4, Figure 8A,B). Fungi of the genera *Alternaria* and *Fusarium* were more prevalent in non-irrigated than in irrigated plots (Table 4).

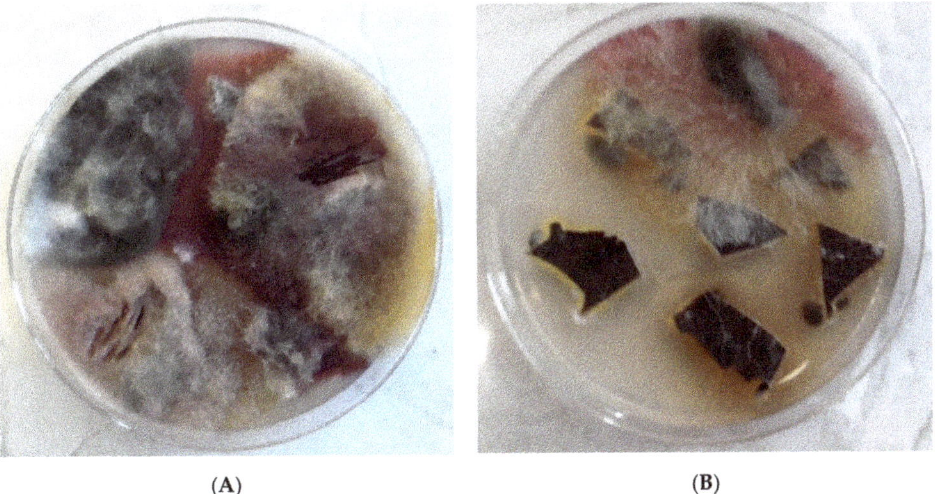

(A)   **(B)**

Figure 8. (**A**,**B**) Jerusalem artichoke leaves (segments) infected by fungi.

The values of Rf, Y, S, D', Shannon–Wiener and dominance indicators were higher in the non-irrigated plots than in the irrigated ones (Table 5).

According to the literature, leaf spot diseases caused by the fungi of the genus *Alternaria* can reduce the photosynthetic capacity of leaves and decrease yields [30,54]. Similar observations were made by Lagopidi and Thanassoulopoulos [55] who investigated sunflower leaf spot caused by *A. alternata* in Greece. In recent years, Alternaria leaf blight caused by *Alternaria alternata* emerged as the predominant disease of sunflowers in the

Republic of South Africa [56,57]. In severely infected plants, pathogenic changes can lead to defoliation and plant death [58]. Pathogens can also infect seeds, compromise seed germination, and decrease yields [59]. In a study by Lagopidi and Thanassoulopoulos [55], sunflower leaf spot caused by *A. alternata* decreased the number of seeds per head by 16–65% and reduced seed weight by 15–79%. According to Viriyasuthee et al. [54], relative humidity can be the main determinant of conidial development, leaf penetration by pathogens, and the progression of infections caused by *Alternaria* spp. In the cited study, relative humidity was higher in the early (72–97%) than late rainy season, which increased the area under the disease-progress curve (AUDPC) and the disease severity index (DSI). Maldaner et al. [60] examined the effect of irrigation and fungicides on the prevalence of infections and yields in two sunflower genotypes. They found that irrigation increased sunflower yields, but only in periods when weather conditions hindered the development of diseases caused by *Alternaria* spp. and *Sclerotinia* spp. Viriyasuthee et al. [30] evaluated the effectiveness of *Trichoderma harzianum* isolate T9 in protecting resistant JA genotypes against Alternaria leaf spot in treatments supplied with two different fertilizer rates. Most disease parameters were more severe in treatments with a lower fertilizer rate. The applied *T. harzianum* isolate was not effective against *Alternaria* leaf spot. According to Wright [61], plant nutrition can also play a role in combatting Alternaria leaf spot. In turn, Blachinski et al. [62] found that foliar urea ($CO(NH_2)_2$) and potassium nitrate (KNO_3) fertilizers did not reduce the severity of infections caused by *Alternaria* spp. in field-grown potatoes and cotton relative to the control treatment. In the present study, *Alternaria* spp. were most frequently isolated from JA leaves. Alternaria leaf spot was more severe in non-irrigated than in irrigated plots. Potassium (K_2O) fertilizer did not exert a clear influence on leaf colonization by *Alternaria* spp. Different potassium rates did not affect the severity of fungal infections in cvs. Topstar and Waldspindel, but in cv. Violette de Rennes, the number of fungal isolates increased in non-irrigated plants supplied with 250 kg K_2O ha^{-1} (Table 4). According to Denoroy [16], Koike [26], and Jansopa et al. [63], the above-ground parts of JA are most susceptible to southern wilt (also known as southern blight or collar rot) caused by *Sclerotium rolfsii*, as well as sclerotinia wilt/rot caused by *Sclerotinia sclerotiorum*. These pathogens can lead to the death of whole plants. Excessive nitrogen fertilization and low soil pH contribute to the development of *S. sclerotiorum*, whereas *S. rolfsii* thrives in moist and warm environments [16]. Avad and Ahmed [64] analyzed the influence of three types of fermented organic fertilizers, including farmyard, poultry and pigeon manure, and their combinations on the incidence of southern blight (caused by *Sclerotium rolfsii*), vegetative growth parameters, chemical composition, yield, and yield components in *Helianthus tuberosus* L., and found that all fertilizer combinations minimized the severity of infections.

The linear regression analysis did not reveal significant relationships between above-ground biomass yield (fresh and dry matter yield) and the number of cultures of pathogenic and saprotrophic fungal species (Figure 9). These findings indicate that despite a high number of pathogenic species (confirmed by the values of biodiversity indicators), their effect on biomass yield was not significant.

Table 4. Fungi colonizing Jerusalem artichoke leaves grown in treatments with different potassium fertilizer rates and different soil moisture levels.

Fungal Genus/Species	Irrigated — Topstar 150*	Irrigated — Topstar 250	Irrigated — Topstar 350	Irrigated — Waldspindel 150	Irrigated — Waldspindel 250	Irrigated — Waldspindel 350	Irrigated — Violette de Rennes 150	Irrigated — Violette de Rennes 250	Irrigated — Violette de Rennes 350	Total Σ	Not Irrigated — Top Star 150	Not Irrigated — Top Star 250	Not Irrigated — Top Star 350	Not Irrigated — Waldspindel 150	Not Irrigated — Waldspindel 250	Not Irrigated — Waldspindel 350	Not Irrigated — Violette de Rennes 150	Not Irrigated — Violette de Rennes 250	Not Irrigated — Violette de Rennes 350	Total Σ
										Number of fungal isolates										
Alternaria alternata (Fr.) Keissl.	9	13	16	12	15	14	8	13	11	111	14	14	19	10	9	7	12	22	13	120
Alternaria spp.		3	2		1		1	4	2	13	2	5	3	1			1	4		16
Botrytis cinerea Pers.										0				1						1
Chaetomium spp.										0							1			1
Cladosporium cladosporioides (Fresen.) G.A. de Vries		2	1						1	4				2		1		4		7
Epicoccum nigrum Link		3		3	1	1	5	3	2	18	4	2		2			4	1	2	15
Fusarium avenaceum (Fr.) Sacc.	1				1		1	1	1	5		1			3			3	4	11
Fusarium culmorum (Wm. G. SM.) Sacc.	2									2					1	2		1		4
Fusarium equiseti (Corda) Sacc.	1									1										0
Fusarium poae (Pack) Wollenw.					1			1		2										0
Fusarium tricinctum (Corda) Sacc.								1		1										0
Fusarium sporitrichioides Scherb.										0			1		2	1		6	1	11
Fusarium spp.	1	1			1					3										0
Mucor spp.							1			1										0
Nigrpospora sphaerica (Sacc.) E.W. Mason					1					1										0
Penicillium spp.				1						1	3		1					1		5
Didymella pinodella (L.K. Jones) Qian Chen&L.Cai				1						1										0
Total	14	20	20	18	21	15	16	23	17		23	22	24	16	15	11	18	42	20	
Total for cultivars		54			54			56				69			42			80		
Total for irrigated/non-irrigated plots					164										191					

* 150, 250, 350 kg K_2O ha^{-1}—potassium fertilization.

Table 5. Diversity analysis of pathogenic and saprotrophic fungi isolated from Jerusalem artichoke leaves grown in treatments with different potassium fertilizer rates and different soil moisture levels.

Diversity Index	Group of Fungi	Irrigated									Σ	Not Irrigated									Σ
		Topstar			Waldspindel			Violette de Rennes				Top Star			Waldspindel			Violette de Rennes			
		150	250	350	150	250	350	150	250	350		150	250	350	150	250	350	150	250	350	
Relative frequency [Rf]	pathogens	3.94	5.63	5.07	4.51	5.63	4.23	4.23	6.48	4.51	44.2	5.63	6.2	6.48	3.94	4.23	2.82	4.79	10.4	5.63	50.1
	saprotrophs	0	0	0.56	0.56	0.28	0	0.28	0	0.28	1.97	0.85	0	0.28	0.56	0	0.28	0.28	1.41	0	3.66
Dominance [Y]	pathogens	0.16	0.32	0.26	0.2	0.32	0.18	0.18	0.42	0.2	19.6	0.32	0.38	0.42	0.16	0.18	0.08	0.23	1.09	0.32	25.1
	saprotrophs	0	0	0	0	0	0	0	0	0	0.04	0.01	0	0	0	0	0	0	0.02	0	0.13
Species richness [S]	pathogens	5	4	2	3	6	2	4	6	4	9	3	4	3	4	4	3	3	6	4	7
	saprotrophs	1	1	1	2	1	1	1	1	1	4	1	1	1	1	1	1	1	2	1	3
Marglef index [D']	pathogens	0.68	0.51	0.17	0.34	0.85	0.17	0.51	0.85	0.51	1.36	0.34	0.51	0.34	0.51	0.51	0.34	0.34	0.85	0.51	1.02
	saprotrophs	0	0	0	0.17	0	0	0	0	0	0.51	0	0	0	0	0	0	0	0.17	0	0.34
Shannon-Wiener index [H']	pathogens	154	256	355	155	256	354	154	256	355	44.2	156	256	356	154	254	353	155	260	356	50.1
	saprotrophs	0.16	0.32	0.82	0.77	0.6	0.18	0.46	0.42	0.49	21.6	1.17	0.38	0.7	0.72	0.18	0.36	0.51	2.51	0.32	28.9
Dominance index [λ]	pathogens	154	256	356	155	256	354	155	257	355	65.8	157	257	357	155	254	353	155	263	356	78.9
	saprotrophs	6	5	3	5	7	3	5	7	5	13	4.01	5	4	5	5	4	4	8.02	5	10.1

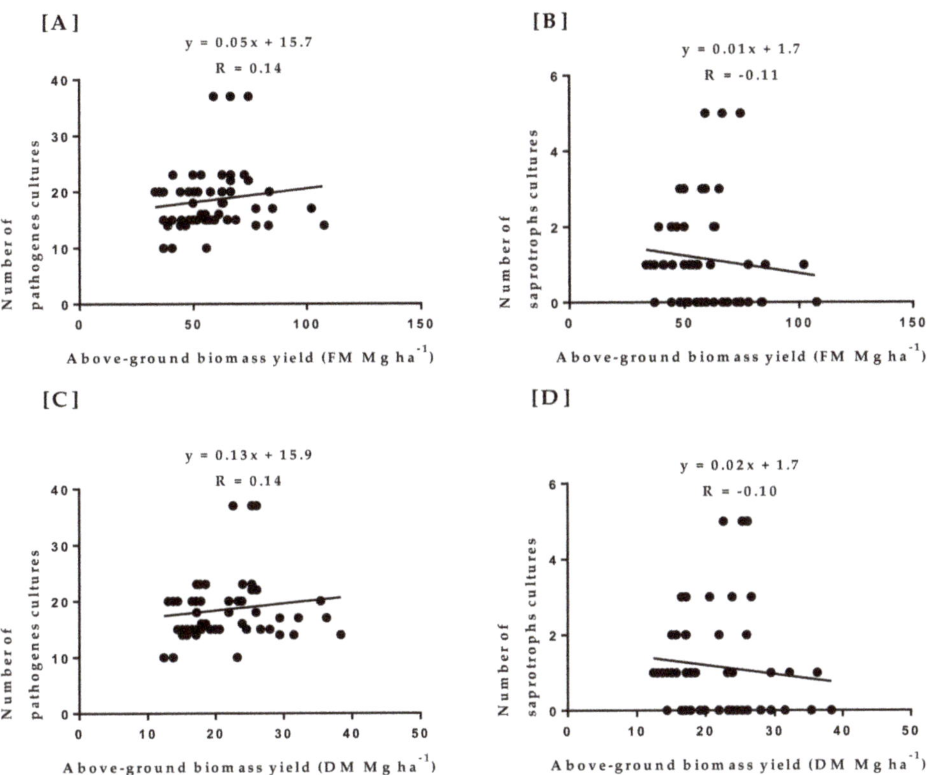

Figure 9. Analysis of linear regression between above-ground biomass yield (fresh matter yield) of Jerusalem artichokes and the number of cultures of pathogenic (**A**) and saprotrophic (**B**) fungal species; above-ground biomass yield (dry matter yield) and the number of cultures of pathogenic (**C**) and saprotrophic (**D**) fungal species.

An evaluation of the health status of JA tubers supported the identification of 10 fungal genera (total of 946 isolates) (Table 6, Figure 10A,B).

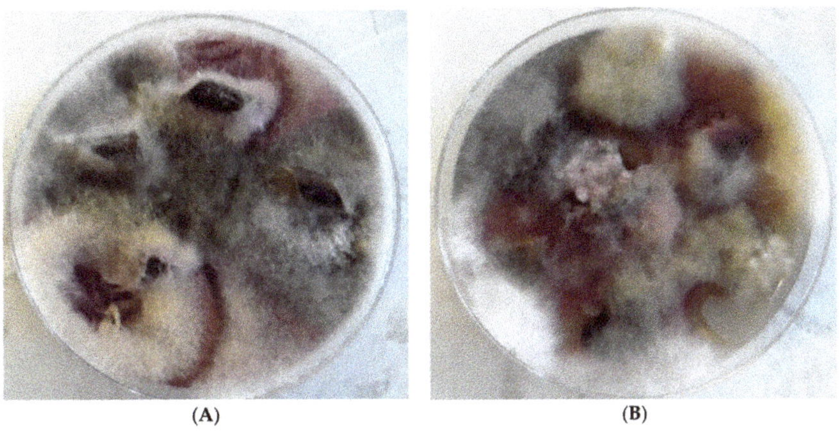

Figure 10. (**A,B**) Jerusalem artichoke tubers (segments) infected by fungi.

Agronomy **2021**, *11*, 234

Table 6. Fungi colonizing Jerusalem artichoke tubers grown in treatments with different potassium fertilizer rates and different soil moisture levels.

Number of Fungal Isolates

Fungal Genus/Species	Irrigated — Topstar 150*	250	350	Waldspindel 150	250	350	Violette de Rennes 150	250	350	Total Σ	Not Irrigated — Topstar 150	250	350	Waldspindel 150	250	350	Violette de Rennes 150	250	350	Total Σ
Alternaria alternata	7	7	8	3	4	5	2	6	2	44		2	2	5		5	2	1	1	18
Aspergillus spp.				2			1		1	4	1	1	1			1				4
Botrytis cinerea		2	1	6	12	1	10		2	34	14		1		20	5	1	11	1	43
Cladosporium cladosporioides		5		1			1	1	3	11			2					3		5
Cylindrocarpon spp.		5								5										0
Epicoccum nigrum	1					1		1		3										0
Fusarium avenaceum	5	1	2	3	2	2		1	1	17		1	1	10	2	1			2	18
Fusarium culmorum	6									6										0
Fusarium equiseti	1						3		2	6								3		3
Fusarium oxysporum	1							2	1	5					1					1
Fusarium tricinctum	3									3										0
Fusarium sambucunum				1		2	1		1	5						2		1		3
Fusarium solani	1	2	1	4	1	4	3	2	2	20			1	3			2	1		7
Fusarium sporitrichioides	2									2										0
Fusarium spp.	1	1		2	1	1		1		7		4		1	3		1		2	11
Mucor spp.	1			2	1	4	4	1		13		1	1	1			2		1	6
Penicillium spp.	25	21	25	24	22	30	22	33	32	234	27	31	35	23	31	18	25	28	22	240
Rhizopus nigricans Ehrenb.																				
Rhizopus spp.	11	4	13	4	8		4	6	8	58	11	13	10	13	12	12	10	7	13	101
Total	65	48	50	52	51	53	52	54	52		53	53	54	57	70	44	41	55	42	
Total for cultivars		163			156			158				160			171			138		
Total for irrigated/non-irrigated plots					477										469					

* 150, 250, 350 kg K$_2$O ha^{-1}—potassium fertilization.

The most prevalent fungal genera were *Penicillium, Fusarium, Alternaria, Botrytis,* and *Rhizopus.* A similar number of fungal isolates was obtained from JA tubers grown in irrigated and non-irrigated plots (477 and 469, respectively).

According to the literature, tuber rot diseases in JA are caused by *Sclerotium rolfsii, Sclerotinia sclerotiorum* [65], *Botrytis cinerea, Rhizopus stolonifera, Penicillium* spp., *Fusarium* spp. [25], as well as *Rhizoctonia solani* [66]. In a study by Ghoneem et al. [67], JA tubers were colonized by 17 fungal species belonging to 12 genera, including *S. rolfsii* (61.7%), *Fusarium incarnatum* (22%) and *Geotrichum candidum* (2.7%). AbdAl-Aziz et al. [33] isolated 24 species belonging to more than five fungal genera from rotten JA tubers. *Alternaria, Aspergillus, Fusarium, Pencillium,* and *Trichoderma* were the most prevalent inulinolytic genera that accounted for more than 90% of the isolated fungi. This study demonstrated that pathogenic species were characterized by higher species richness (S) and higher values of the Marglef index (D') (Table 7), but the values of the remaining biodiversity indicators revealed that the analyzed communities were dominated by saprotrophic species. In the current study, *Penicillium* spp. were isolated from irrigated and non-irrigated tubers (Table 4). Tubers from irrigated treatments were abundantly colonized by fungi of the genus *Fusarium,* including *F. solani, F. avenaceum, F. culmorum. F. equiseti, F. oxysporum, F. sambusinum, F. tricinctum* and *F. sporotrichioides,* as well as *Alternaria* spp. *Fusarium* spp. were isolated from tubers from all irrigated plots, and their prevalence ranged from 5% to 32.30%.

In pathogenic species, the values of species dominance indicators Y and λ were higher in irrigated than in non-irrigated plots (Table 7). In irrigated plots, the values of the remaining biodiversity indicators (species richness index, Marglef index and Shannon–Wiener index) were also higher for pathogenic species. *Fusarium* fungi were less prevalent in non-irrigated plots, and they were absent only in cv. Topstar fertilized with 150 kg K_2O ha^{-1} (Table 5). Ghoneem et al. [67], identified several *Fusarium* pathogens, including *F. incarnatum, F. vertocilioises, F. oxysporum,* and *F. solani,* as well as *Penicillium* spp. and *Alternaria alternata* from tubers. According to Kays and Nottingham [25], high temperature contributes to blue mold (*Penicillium* spp.) and Fusarium rot in JA tubers. In the present study, *Rhizopus* spp. was also a common pathogen, but it was less frequently isolated from irrigated plots (Table 5). Kays and Nottingham [25] observed that tubers stored in winter are susceptible to fungal infections and can develop symptoms of soft rot. The cited authors isolated *Rhizopus stolonifer* from chill-stored JA tubers. Yang et al. [68] demonstrated that *R. arrhizus* caused soft rot of JA tubers in China and that the disease resulted in the loss of 30–50% of stored tubers each year. In the current study, *Botrytis cinerea* was isolated from tubers grown in both irrigated and non-irrigated plots. In irrigated treatments, *B. cinerea* infections were identified in 0–23.53% tubers. This pathogen was not isolated from cv. Topstar fertilized with 150 kg K_2O ha^{-1} and from cv. Violette de Rennes supplied with 250 kg K_2O ha^{-1}. The prevalence of *B. cinerea* in non-irrigated treatments ranged from 0 to 27.78%. The pathogen was not isolated from cv. Topstar fertilized with 250 kg K_2O ha^{-1} and cv. Waldspindel supplied with 150 kg K_2O ha^{-1} (Table 5). According to Doehlemann et al. [69], *B. cinerea* is the most common cause of grey mold, a serious disease that leads to significant economic losses around the world. Grey mold is devastating for senescent and damaged tissues in dicotyledonous plants. The pathogen penetrates host tissues in early stages of plant development, and it may remain inactive until environmental conditions and the host's physiological status are favorable for colonization. This mechanism of action explains serious losses during the storage of seemingly healthy crops. In the present study, the prevalence of fungal infections in tubers was highest in cv. Waldspindel in non-irrigated plots and in cv. Topstar in irrigated treatments. Different potassium fertilizer rates exerted varied effects on fungal colonization.

Table 7. Diversity analysis of pathogenic and saprotrophic fungi isolated from Jerusalem artichoke tubers grown in treatments with different potassium fertilizer rates and different soil moisture levels.

| Diversity Index | Group of Fungi | Irrigated | | | | | | | | | | Not Irrigated | | | | | | | | | |
| | | Topstar | | | Waldspindel | | | Violette de Rennes | | | Σ | Top Star | | | Waldspindel | | | Violette de Rennes | | | Σ |
		150 *	250	350	150	250	350	150	250	350		150	250	350	150	250	350	150	250	350	
Relative frequency [Rf]	pathogens	2.960	1.903	1.268	2.008	2.114	1.903	2.114	1.163	1.163	11.95	1.480	0.740	0.529	2.114	2.854	1.374	0.423	1.797	0.634	10.99
	saprotrophs	3.911	3.700	4.017	3.488	3.277	3.700	3.383	4.545	4.334	34.36	4.123	4.863	5.180	3.911	4.545	3.277	3.911	4.017	3.805	37.63
Dominance [Y]	pathogens	0.088	0.036	0.016	0.040	0.045	0.036	0.045	0.014	0.014	1.43	0.022	0.005	0.003	0.045	0.081	0.019	0.002	0.032	0.004	1.21
	saprotrophs	0.153	0.117	0.161	0.122	0.107	0.137	0.114	0.207	0.188	11.62	0.170	0.236	0.268	0.153	0.207	0.107	0.153	0.161	0.072	6.33
Species richness [S]	pathogens	10	6	4	6	5	8	6	5	7	13	2	3	4	5	4	4	3	5	4	8
	saprotrophs	3	2	2	4	3	2	4	3	3	4	3	4	4	3	2	3	3	2	3	4
Marglef index [D′]	pathogens	1.313	0.730	0.438	0.730	0.584	1.022	0.730	0.584	0.876	1.75	0.146	0.292	0.438	0.584	0.438	0.438	0.292	0.584	0.438	1.02
	saprotrophs	0.292	0.146	0.146	0.438	0.292	0.146	0.438	0.292	0.292	0.44	0.292	0.438	0.438	0.292	0.146	0.292	0.292	0.146	0.292	0.44
Shannon–Wiener index [H′]	pathogens	0.164	0.105	0.068	0.112	0.106	0.112	0.112	0.067	0.074	0.65	0.062	0.043	0.035	0.109	0.126	0.076	0.028	0.092	0.041	0.42
	saprotrophs	0.155	0.135	0.155	0.150	0.135	0.140	0.148	0.175	0.162	0.65	0.161	0.185	0.198	0.157	0.167	0.138	0.157	0.159	0.154	0.67
Dominance index [λ]	pathogens	0.004	0.003	0.002	0.002	0.004	0.002	0.003	0.001	0.001	0.05	0.004	0.001	0.000	0.004	0.007	0.002	0.000	0.003	0.000	0.04
	saprotrophs	0.012	0.008	0.013	0.010	0.009	0.013	0.009	0.015	0.015	0.16	0.013	0.016	0.017	0.012	0.016	0.009	0.012	0.013	0.011	0.18

* 150, 250, 350 kg K_2O ha^{-1}—potassium fertilization.

In irrigated plots, disease severity increased only in cv. Topstar supplied with 150 kg K_2O ha^{-1}, whereas in non-irrigated treatments, infections were more frequently noted in cvs. Waldspindel and Violette de Rennes fertilized with 250 kg K_2O ha^{-1} (Table 5). According to the literature [25,26,65,67,70], tuber rot caused by *S. rolfsii* is a devastating disease of both field-grown and stored crops. The pathogen can decrease yields by as much as 60% [27]. *Sclerotium rolfsii* was not isolated from JA in the current study. According to Kosaric et al. [32], frozen storage can minimize the spread of pathogenic fungi on tubers.

In this study, the number of pathogenic species isolated from Jerusalem artichoke tubers had a minor negative impact on fresh (R = −0.37) and dry matter (R = −0.41) yield (Figure 11). These findings suggest that despite the low abundance and frequency of pathogenic species relative to saprotophicspecies, as well as the dominance of saprotrophic species in all plots (indicators Y and λ), pathogenic species exerted a considerable influence on JA plants.

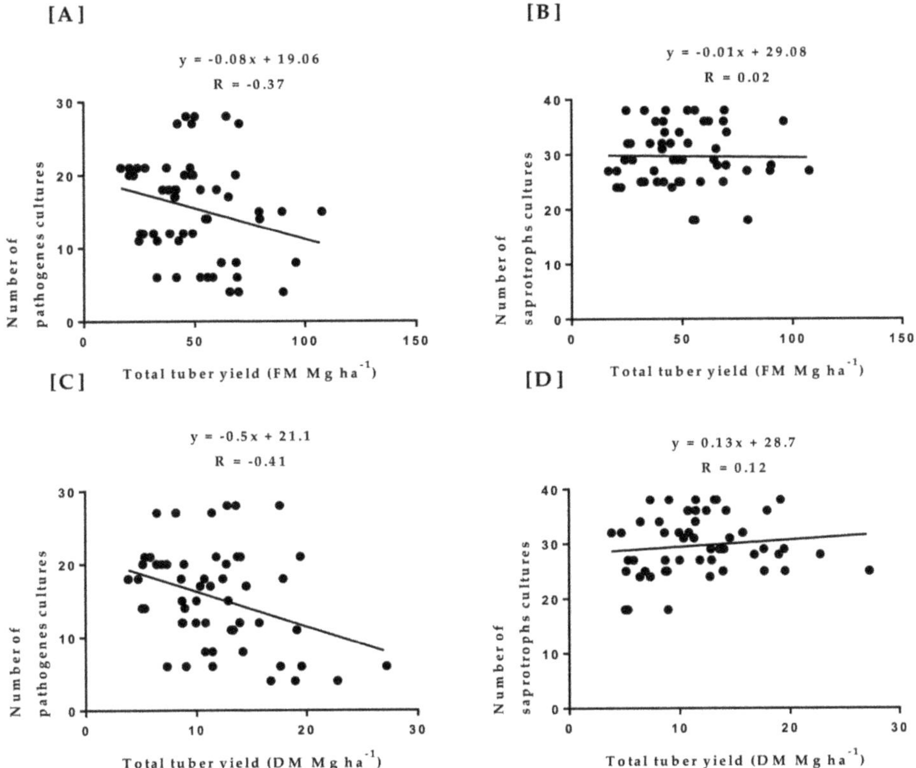

Figure 11. Analysis of linear regression between: total tuber yield (fresh matter yield) of Jerusalem artichokes and the number of cultures of pathogenic (**A**) and saprotrophic (**B**) fungal species; total tuber yield (dry matter yield) and the number of cultures of pathogenic (**C**) and saprotrophic (**D**) fungal species.

4. Conclusions

In the present study, JA cv. Topstar was characterized by the highest total tuber and above-ground biomass yields. An increase in the rate of mineral fertilizer applied to soil to 350 kg K_2O ha^{-1} did not affect the total tuber yields. Potassium fertilizer applied at the optimal rate of 150 kg K_2O ha^{-1} contributed to the greatest increase in the above-ground biomass yield of JA. Irrigation had a significant effect on total tuber and above-ground biomass yields, which increased by 59% and 42%, respectively, on average.

Phytopathological analyses of aerial plant parts revealed that fungi of the genera *Alternaria* and *Fusarium* were more prevalent in non-irrigated than in irrigated plots. In non-irrigated treatments, fungal pathogens were most frequently isolated from the leaves of cv. Violette de Rennes and were least prevalent in cv. Waldspindel. The severity of fungal infections increased only in cv. Violette Rennes fertilized with 250 kg K_2O ha^{-1}. Jerusalem artichoke tubers were most abundantly colonized by fungi of the genera *Penicillium*, *Fusarium*, *Alternaria*, *Botrytis cinerea*, and *Rhizopus*. The prevalence of fungal pathogens in tubers was similar in irrigated and non-irrigated plots. The severity of tuber infections was highest in cv. Waldspindel in non-irrigated treatments and in cv. Topstar in irrigated treatments. The applied potassium fertilizer rates exerted varied effects on tuber colonization by fungi. The analysis of the composition of pathogens and saprotrophs revealed that pathogenic species were predominant on Jerusalem artichoke leaves, which was confirmed by nearly all biodiversity indicators (*Relative frequency, Dominance, Species richness, Margalef index, Shannon–Wiener index* and, *Dominance index*). The number of cultures of pathogenic fungal species isolated from Jerusalem artichoke tubers exerted a negative influence on fresh and dry matter yield.

Further research involving long-term field experiments should be conducted in the future to examine the effects of fertilization and crop protection agents on the health, yield and quality of the produced crops.

Supplementary Materials: The following are available online at https://www.mdpi.com/2073-4395/11/2/234/s1, Table S1. The effect of irrigation and potassium fertilization on the total tuber yield (fresh matter basis) of the analyzed Jerusalem artichoke cultivars. Means with the same letters do not differ significantly at $p \leq 0.05$ in Tukey's test. Table S2. The effect of irrigation and potassium fertilization on the total tuber yield (dry matter basis) of the analyzed Jerusalem artichoke cultivars. Means with the same letters do not differ significantly at $p \leq 0.05$ in Tukey's test. Table S3. The effect of irrigation and potassium fertilization on the above-ground biomass yield (fresh matter basis) of the analyzed Jerusalem artichoke cultivars. Means with the same letters do not differ significantly at $p \leq 0.05$ in Tukey's test. Table S4. The effect of irrigation and potassium fertilization on the above-ground biomass yield (dry matter basis) of the analyzed Jerusalem artichoke cultivars. Means with the same letters do not differ significantly at $p \leq 0.05$ in Tukey's test.

Author Contributions: Conceptualization, methodology, B.B., methodology A.P., formal analysis, writing—original draft preparation, B.B., A.P., A.O. and K.J. All authors have read and agreed to the published version of the manuscript.

Funding: The project was financially co-supported by the Ministry of Science and Higher Education under the program entitled "Regional Initiative of Excellence" for the years 2019–2022, Project No. 010/RID/2018/19, amount of funding PLN 12000000.

Institutional Review Board Statement: Not applicable.

Informed Consent Statement: Not applicable.

Acknowledgments: The results presented in this paper were obtained as part of a comprehensive study financed by the University of Warmia and Mazury in Olsztyn (grant No. 20.610.020-110 and grant No. 20.610.004-110).

Conflicts of Interest: The authors declare no conflict of interest.

References

1. Praznik, W.; Cieślik, E.; Filipiak-Florkiewicza, A. Soluble dietary fibres in Jerusalem artichoke powders: Composition and application in bread. *Nahrung/Food* **2002**, *46*, 151–157. [CrossRef]
2. Panchev, I.; Delchev, N.; Kovacheva, D.; Slavov, A. Physicochemical characteristics of inulins obtained from Jerusalem artichoke (*Helianthus tuberosus* L.). *Eur. Food Res. Technol.* **2011**, *233*, 889–896. [CrossRef]
3. Yang, L.; He, Q.S.; Corscadden, K.; Udenigwe, C.C. The prospects of Jerusalem artichoke in functional food ingredients and bioenergy production. *Biotechnol. Rep.* **2015**, *5*, 77–88. [CrossRef]
4. Lakíc, Ž.; Balalíc, I.; Nožiníc, M. Genetic variability for yield and yield components in jerusalem artichoke (*Helianthus tuberosus* L.). *Genetika* **2018**, *50*, 45–57. [CrossRef]

5. Pan, L.; Sinden, M.R.; Kennedy, A.H.; Chai, H.; Watson, L.E.; Graham, T.L.; Kinghorn, A.D. Bioactive constituents of *Helianthus tuberosus* (Jerusalem artichoke). *Phytochem. Lett.* **2009**, *2*, 15–18. [CrossRef]

6. Yuan, X.; Gao, M.; Xiao, H.; Tan, C.; Du, Y. Free radical scavenging activities and bioactive substances of Jerusalem artichoke (*Helianthus tuberosus* L.) Leaves. *Food Chem.* **2012**, *133*, 10–14. [CrossRef]

7. Sawicka, B.; Skiba, D.; Pszczółkowski, P.; Aslan, I.; Sharifi-Rad, J.; Krochmal-Marczak, B. Jerusalem artichoke (*Helianthus tuberosus* L.) As a medicinal plant and its natural products. *Cell. Mol. Biol.* **2020**, *66*, 161–178. [CrossRef]

8. Soja, G.; Dersch, G.; Praznik, W.I. Harvest dates, fertilizer and varietal effects on yield, concentration and molecular distribution of fructan in Jerusalem artichoke (*Helianfhus tuberosus* L.). *J. Agron. Crop Sci.* **1990**, *165*, 181–189. [CrossRef]

9. Swanton, C.J.; Cavers, P.B.; Clementsl, D.R.; Mooret, M.J. The biolory of Canadian weeds. 101. *Helianthus tuberosusl. Can. J. Plant Sci.* **1992**, *72*, 1367–1382. [CrossRef]

10. Matias, J.; Gonzales, J.; Cabanillas, J.; Royano, L. Influence of NPK fertilisation and harvest date on agronomic performance of Jerusalem artichoke crop in the Guadiana Basin (Southwestern Spain). *Ind. Crop. Prod.* **2013**, *48*, 191–197. [CrossRef]

11. Ruttanaprasert, R.; Banterng, P.; Jogloy, S.; Vorasoot, N.; Kesmala, T.; Kanwar, R.S.; Holbrook, C.C.; Patanothai, A. Genotypic variability for tuber yield, biomass, and drought tolerance in Jerusalem artichoke germplasm. *Turk. J. Agric. For.* **2014**, *38*, 570–580. [CrossRef]

12. Stolarski, M.J.; Krzyżaniak, M.; Warmínski, K.; Tworkowski, J.; Szczukowski, S.; Olba-Ziety, E.; Gołaszewski, J. Energy efficiency of herbaceous crops production depending on the type of digestate and mineral fertilizers. *Energy* **2017**, *134*, 50–60. [CrossRef]

13. Paixão, S.M.; Alves, L.; Pacheco, R.; Silva, C.M. Evaluation of Jerusalem artichoke as a sustainable energy crop to bioethanol: Energy and CO_2eq emissions modeling for an industrial scenario. *Energy* **2018**, *150*, 468–481. [CrossRef]

14. Rossini, F.; Provenzano, M.E.; Kuzmanovíc, L.; Ruggeri, R. Jerusalem Artichoke (*Helianthus tuberosus* L.): A Versatile and Sustainable Crop for Renewable Energy Production in Europe. *Agronomy* **2019**, *9*, 528. [CrossRef]

15. Sawicka, B.; Skiba, D.; Bienia, B.; Kieltyka-Dadasiewicz, A.; Danilcenko, H. Jerusalem artichoke (*Helianthus tuberosus* L.) As energy raw material. In Proceedings of the 9th International Scientific Conference Rural Development, Research and Innovation for Bioeconomy, Kaunas, Lithuania, 26–28 September 2019; pp. 336–342. [CrossRef]

16. Denoroy, P. The crop physiology of *Helianthus tuberosus* L.: A model orientated view. *Biomass Bioenerg.* **1996**, *11*, 11–32. [CrossRef]

17. Paolini, R.; Del Puglia, S.; Abbate, V.; Copani, V.; Danuso, F.; De Mastro, G.; Losavio, N.; Marzi, V.; Molfetta, P.; Pignatelli, V.; et al. Produttività del topinambur (*Helianthus tuberosus* L.) in relazione a fattori agronomici diversi. *Agric. Ric.* **1996**, *163*, 126–144.

18. Danuso, F. Le colture per la produzione di inulina: Attualità eprospettive. *Riv. Agron.* **2001**, *35*, 176–187.

19. Izsáki, Z.; Kádi, G.N. Biomass acumulation and nutrient uptake of Jerusalem artichoke (*Helianthus tuberosus* L.). *Am. J. Plant Sci.* **2013**, *4*, 1629–1640. [CrossRef]

20. Gao, K.; Zhang, Z.; Zhu, T.; Coulter, J.A. The influence of flower removal on tuber yield and biomass characteristics of *Helianthus tuberosus* L. in a semi-arid area. *Ind. Crop. Prod.* **2020**, *150*, 112374. [CrossRef]

21. Soja, G.; Haunold, E. Leaf gas exchange and tuber yield in Jerusalem artichoke (*Helianthus tuberosus* L.) Cultivars. *Field Crop Res.* **1991**, *26*, 241–252. [CrossRef]

22. Westermann, D.T.; James, D.W.; Tindall, T.A.; Hurst, R.L. Nitrogen and potassium fertilization of potatoes: Sugars and starch. *Am. Potato J.* **1994**, *71*, 433–453. [CrossRef]

23. Monti, A.; Amaducci, M.T.; Venturi, G. Growth response, leaf gas exchange and fructans accumulation of Jerusalem artichoke (*Helianthus tuberosus* L.) As affected by different water regimes. *Eur. J. Agron.* **2005**, *23*, 136–145. [CrossRef]

24. Long, X.H.; Mehta, S.K.; Liu, Z.P. Efect of NO^-_3-N enrichment on seawater stress tolerance of jerusalem artichoke (*Helianthus tuberosus*). *Pedosphere* **2008**, *18*, 113–123. [CrossRef]

25. Kays, S.J.; Nottingham, S.F. *Biology and Chemistry of Jerusalem Artichoke: Helianthus tuberosus L.*; CRC Press; Taylor & Francis Group: London, UK, 2008; ISBN 9781420044959.

26. Koike, S.T. Southern blight of Jerusalem artichoke caused by *Sclerotium rolfsii.* in California. *Plant Dis.* **2004**, *88*, 769. [CrossRef]

27. Sennoi, R.; Jogloy, S.; Saksirirat, W.; Patanothai, A. Pathogenicity test of *Sclerotium rolfsii*, a causal agent of Jerusalem artichoke (*Helianthus tuberosus* L.) stem rot. *Asian J. Plant Sci.* **2010**, *95*, 281–284. [CrossRef]

28. Labergh, C.; Sackston, W.E. Adaptability and diseases of Jerusalem artichoke (*Helianthus tuberosus*) in Quebec. *Can. J. Plant Sci.* **1987**, *67*, 349–353. [CrossRef]

29. Dudhe, M.Y.; Bharsakle, S.S. Screening of germplasm line for resistance to *Alterneria* blight and necrosis disease in sunflower. *Crop Prot. Prod.* **2005**, *1*, 84–87.

30. Viriyasuthee, V.; Jogloy, S.; Saksirirat, W.; Saepaisan, S.; Gleason, M.L.; Chen, R.S. Biological Control of Alternaria Leaf Spot Caused by *Alternaria* spp. in Jerusalem Artichoke (*Helianthus tuberosus* L.) under Two Fertilization Regimes. *Plants* **2019**, *8*, 463. [CrossRef]

31. Zhao, Y.-Q.; Zhang, D.M.; Zhang, L.J.; Yu, X.Y.; Yu, H.R.; Shi, K.; Gao, K. First Report of Brown Spot on Jerusalem Artichoke (*Helianthus tuberosus*) Caused by Bipolaris zeae in China. *Plant Dis.* **2017**, *101*, 2146. [CrossRef]

32. Kosaric, N.; Cosentino, G.P.; Wieczorek, A.; Duvnjak, Z. The Jerusalem artichoke as an agricultural crop. *Biomass* **1984**, *5*, 1–36. [CrossRef]

33. AbdAl-Aziz, S.A.A.; El-Metwally, M.M.; Saber, I.A. Molecular identification of a novel inulinolytic fungus isolated from and grown on tubers of *Helianthus tuberosus* and statistical screening of medium components. *World J. Microb. Biot.* **2012**, *28*, 3245–3254. [CrossRef]

34. Stolzenburg, K. Anbau und Verwertung von Topinambur (*Helianthus tuberosus* L.). *Inf. furPflanzenproduktion Sonderh.* **2002**, *1*, 1.
35. PIORiN. Integrated Production Method for Potatoes. Main Inspectorate of Plant Health and Seed Inspection. 2014. Available online: http://www.piorin.gov.pl/30.01.2020 (accessed on 15 June 2020). (In Polish)
36. IUSS Working Group WRB. World Reference Base for Soil Resources. In *World Soil Resources Reports*, 2nd ed.; FAO: Roma, Italy, 2006.
37. Houba, V.J.G.; Van der Lee, J.J.; Novozamsky, I. Soil and Plant Analysis. Part 5B. In *Soil Analysis Procedure Other Procedure*; Wageningen Agriculture University: Wageningen, The Netherlands, 1995.
38. Ellis, M.B. *Dematiaceous Hyphomycetes*; Commonwealth Mycological Institute Kew: Surrey, UK, 1971.
39. Leslie, J.F.; Summerell, B.A. *The Fusarium Laboratory Manual*; Blackwell Publishing: Ames, IA, USA, 2006. [CrossRef]
40. Watanabe, T. Pictorial, Atlas of Soil and Seed Fungi. In *Morphologies of Cultured Fungi and Key to Species*, 2nd ed.; CRC Press: Boca Raton, FL, USA, 2002; p. 504. ISBN 9781439804193.
41. Du, W.; Yao, Z.; Li, J.; Sun, C.; Xia, J.; Wang, B.; Shi, D.; Ren, L. Diversity and antimicrobial activity of endophytic fungi isolated from *Securinega suffruticosa* in the Yellow River Delta. *PLoS ONE* **2020**, *10*, e0229589. [CrossRef]
42. Bac, S.; Koźmiński, C.; Rojek, M. *Agrometeorologia (Agrometeorology)*; Polish Scientific Publishers PWN: Warsaw, Polish, 1998. (In Polish)
43. Statsoft Inc. *Statistica (Data Analysis Software System)*, 10th ed.; Statsoft Inc.: Tulsa, OK, USA, 2011. Available online: http://www.statsoft.com (accessed on 1 April 2020).
44. Rodrigues, M.A.; Sousa, L.; Cabanas, J.E.; Arrobas, M. Tuber yield and leaf mineral composition of Jerusalem artichoke (*Helianthus tuberosus* L.) Grown under different cropping practices. *Span. J. Agric. Res.* **2007**, *5*, 545–553. [CrossRef]
45. Conde, J.R.; Lenorio, J.L.; Rodríguez-Maribona, B.; Ayerbet, L. Tuber yield of Jerusalem artichoke (*Helianthus tuberosus* L.) in relation to water stress. *Biomass Bioenerg.* **1991**, *1*, 137–142. [CrossRef]
46. Baldini, M.; Danuso, F.; Monti, A.; Amaducci, M.T.; Stevanato, P.; Mastro, G. Chichory and Jerusalem artichoke productivity in different areas of Italy, in relation to water availability and time of harvest. *Ital. J. Agron. Riv. Agron.* **2006**, *2*, 291–307. [CrossRef]
47. Kim, S.; Kim, C.H. Evaluation of whole Jerusalem artichoke (*Helianthus tuberosus* L.) For consolidated bioprocessing ethanol production. *Renew. Energ.* **2014**, *65*, 83–91. [CrossRef]
48. Pimsaen, W.; Jogloy, S.; Suriharn, B.; Kesmala, T.; Pensuk, V.; Patanothai, A. Genotype by Environment (GxE) Interactions for Yield Components of Jerusalem Artichoke (*Helianthus tuberosus* L.). *Asian J. Plant Sci.* **2010**, *9*, 11–19. [CrossRef]
49. Raso, E. Jerusalem Artichoke. Effect of Nitrogen-Potassium Fertilizing. *Terra e Sole* **1990**, *45*, 431–433.
50. Schittenhelm, S. Agronomic Performance of Root Chicory, Jerusalem Artichoke, and Sugarbeet in Stress and Nonstress Environments. *Crop Sci.* **1999**, *39*, 1815–1823. [CrossRef]
51. Gao, K.; Zhu, T.; Wang, Q. Nitrogen fertilization, irrigation, and harvest times affect biomass and energy value of *Helianthus tuberosus* L. *J. Plant Nutr.* **2016**, *39*, 1906–1914. [CrossRef]
52. Gunnarsson, I.B.; Svensson, S.E.; Johansson, E.; Karakashev, D.; Angelidaki, I. Potential of Jerusalem artichoke (*Helianthus tuberosus* L.) as a biorefinery crop. *Ind. Crop. Prod.* **2014**, *56*, 231–240. [CrossRef]
53. Niu, L.; Manxia, C.; Xiumei, G.; Xiaohua, L.; Hongbo, S.; Zhaopu, L.; Zed, R. Carbon sequestration and Jerusalem artichoke biomass under nitrogen applications in coastal saline zone in the northern region of Jiangsu, China. *Sci. Total Environ.* **2016**, *568*, 885–890. [CrossRef]
54. Viriyasuthee, W.; Saksirirat, W.; Saepaisan, S.; Gleason, M.L. Variability of Alternaria Leaf Spot Resistance in Jerusalem Artichoke (*Helianthus tuberosus* L.) Accessions Grown in a Humid Tropical Region. *Agronomy* **2019**, *9*, 268. [CrossRef]
55. Lagopodi, A.L.; Thanassoulopoulos, C.C. Effect of a leaf spot disease caused by *Alternaria alternata* on yield of sunflower in Greece. *Plant Dis.* **1998**, *82*, 41–44. [CrossRef]
56. Kgatle, M.G.; Flett, B.; Truter, M.; Ramusi, M.; Aveling, T. Distribution of Alternaria leaf blight of sunflowers caused by *Alternaria alternata* in South Africa. *J. Agric. Rural Dev. Tropics Subtropics* **2019**, *120*, 71–77. [CrossRef]
57. Kgatle, M.G.; Flett, B.; Truter, M.; Aveling, T.A.S. Control of Alternaria leaf blight caused by *Alternaria alternata* on sunflower using fungicides and *Bacillus amyloliquefaciens*. *Crop Prot.* **2020**, *132*, 105146. [CrossRef]
58. Cho, H.S.; Yu, S.H. Three *Alternaria* species pathogenic to sunflower. *Plant Pathol. J.* **2000**, *16*, 331–334.
59. Prasad, M.S.L.; Sujatha, M.; Rao, S.C. Analysis of cultural and genetic diversity in *Alternaria helianthi* and determination of pathogenic variability using wild *Helianthi* species. *J. Phytopathol.* **2009**, *157*, 609–617. [CrossRef]
60. Maldaner, I.C.; Heldwein, A.B.; Bortoluzzi, M.P.; Loose, L.H.; Lucas, D.D.P.; da Silva, J.R. Irrigation and fungicide application on disease occurrence and yield of early and late sown sunflower. *Rev. Bras. Eng. Agric. Amb.* **2015**, *19*, 630–635. [CrossRef]
61. Wright, P.R. Research into early senescence syndrome in cotton. *Better Crops Int.* **1998**, *12*, 14–16.
62. Blachinski, D.; Shtienberg, D.; Dinoor, A.; Kafkafi, U.; Sujkowski, L.S.; Zitter, T.A.; Fry, W.E. Influence of foliar application of nitrogen and potassium on Alternaria diseases in potato, tomato and cotton. *Phytoparasitica* **1996**, *24*, 281–292. [CrossRef]
63. Jansopa, C.; Jogloy, S.; Saksirirat, W. Inoculation with Sclerotium rolfsii, cause of stem rot in Jerusalem artichoke, under field conditions. *Eur. J. Plant Pathol.* **2016**, *146*, 47–58. [CrossRef]
64. Awad, A.A.M.; Ahmed, H.M.H. Impact of Organic Manure Combinations on Performance and Rot Infection of Stressed-Jerusalem Artichoke Plants. *Egypt. J. Soil Sci.* **2018**, *58*, 417–433. [CrossRef]
65. Snowden, A.L. *A Colour Allas of Post-Harvest Diseases and Disorders of Fruits and Vegetables: Vegetables*; Manson Publishing CRC Press: Boca Raton, FL, USA, 2010; Volume 2, p. 416.

66. Ezzat, A.E.S.; Ghoneem, K.M.; Saber, W.I.A.; Al-Askar, A.A. Control of wilt, stalk and tuber rots diseases using arbuscular mycorrhizal fungi, *Trichoderma* species and hydroquinone enhances yield quality and storability of Jerusalem artichoke (*Helianthus tuberosus* L.). *Egypt. J. Biol. Pest Control* **2015**, *25*, 11–22.
67. Ghoneem, K.M.; Saber, W.I.A.; El-Awady, A.A.; Rashad, Y.M.; Al-Askar, A.A. Alternative preservation method against Sclerotium tuber rot of Jerusalem artichoke using natural essential oils. *Phytoparasitica* **2016**, *44*, 341–352. [CrossRef]
68. Yang, S.P.; Du, G.L.; Tian, J.; Jiang, X.T.; Sun, X.M.; Li, Y.; Li, J.; Zhong, Q. First Report of Tuber Soft Rot of Jerusalem Artichoke (*Helianthus tuberosus*) Caused by *Rhizopus arrhizus* in Qinghai Province of China. *Plant Dis.* **2020**, *104*, 3265. [CrossRef]
69. Doehlemann, G.; Ökmen, B.; Zhu, W.; Sharon, A. Plant pathogenic fungi. *Microbiol. Spectr.* **2016**, *5*, FUNK-0023-2016. [CrossRef]
70. Sennoi, R.; Jogloy, S.; Saksirirat, W.; Kesmala, T.; Patanothai, A. Genotypic variation of resistance to southern stem rot of Jerusalem artichoke caused by *Sclerotium rolfsii*. *Euphytica* **2013**, *190*, 415–424. [CrossRef]

Article

Effect of Deficit Irrigation and Reduced N Fertilization on Plant Growth, Root Morphology and Water Use Efficiency of Tomato Grown in Soilless Culture

Ikram Ullah [1], Hanping Mao [1,*], Ghulam Rasool [2], Hongyan Gao [1], Qaiser Javed [3], Abid Sarwar [4] and Muhammad Imran Khan [5]

1 School of Agricultural Equipment Engineering, Jiangsu University, Zhenjiang 212013, China; ikramullah@ujs.edu.cn (I.U.); gaohy@ujs.edu.cn (H.G.)
2 College of Hydrology and Water Resources, Hohai University, Nanjing 210098, China; gr_uaf@yahoo.com
3 School of the Environment and Safety Engineering, Jiangsu University, Zhenjiang 212013, China; qjaved_uaf@yahoo.com
4 Department of Irrigation and Drainage, University of Agriculture, Faisalabad 38000, Pakistan; abid.sarwar@wsu.edu
5 Research Center of Fluid Machinery Engineering and Technology, Jiangsu University, Zhenjiang 212013, China; imrankhan7792@yahoo.com
* Correspondence: maohp@ujs.edu.cn; Tel.: +86-1351-169-5868

Citation: Ullah, I.; Mao, H.; Rasool, G.; Gao, H.; Javed, Q.; Sarwar, A.; Khan, M.I. Effect of Deficit Irrigation and Reduced N Fertilization on Plant Growth, Root Morphology and Water Use Efficiency of Tomato Grown in Soilless Culture. *Agronomy* 2021, 11, 228. https://doi.org/10.3390/agronomy11020228

Academic Editors: Ivan Francisco Garcia Tejero and Victor Hugo Durán-Zuazo
Received: 8 December 2020
Accepted: 22 January 2021
Published: 26 January 2021

Publisher's Note: MDPI stays neutral with regard to jurisdictional claims in published maps and institutional affiliations.

Abstract: This study was conducted to investigate the effects of various irrigation water (W) and nitrogen (N) levels on growth, root-shoot morphology, yield, and irrigation water use efficiency of greenhouse tomatoes in spring–summer and fall–winter. The experiment consisted of three irrigation water levels (W: 100% of crop evapotranspiration (ETc), 80%, and 60% of full irrigation) and three N application levels (N: 100%, 75%, and 50% of the standard nitrogen concentration in Hoagland's solution treatments equivalent to 15, 11.25, 7.5 mM). All the growth parameters of tomato significantly decreased ($p < 0.05$) with the decrease in the amount of irrigation and nitrogen application. Results depicted that a slight decrease in irrigation and an increase in N supply improved average root diameter, total root length, and root surface area, while the interaction was observed non-significant at average diameter of roots. Compared to the control, $W_{80} N_{100}$ was statistically non-significant in photosynthesis and stomatal conductance. The $W_{80} N_{100}$ resulted in a yield decrease of 2.90% and 8.75% but increased irrigation water use efficiency (IWUE) by 21.40% and 14.06%. Among interactions, the reduction in a single factor at $W_{80} N_{100}$ and $W_{100} N_{75}$ compensated the growth and yield. Hence, $W_{80} N_{100}$ was found to be optimal regarding yield and IWUE, with 80% of irrigation water and 15 mM of N fertilization for soilless tomato production in greenhouses.

Keywords: deficit irrigation; hydroponics; nitrogen; root growth; tomato; water saving

1. Introduction

Tomato is not only one of the world's most popular vegetable crops in taste but also a rich source of antioxidant lycopene, vitamins, organic acids, and essential minerals to keep us healthy [1]. As the leading tomato producer, recent total production of tomato in China has reached 6.05 million tons per year according to the report of the Food and Agriculture Organization of the United Nations (FAO) [2]. The tomato growth area in solar greenhouses is increasing rapidly in China to utilize the land and limited water resources more efficiently. Greenhouse technology added remarkable profits to farmers' income with the ease in managing the growth conditions compared to the field. To produce fresh tomatoes, tomatoes grown in greenhouses need far less water and N per kg of fresh tomato production than under conventional management but still, there is a large community of farmers using a high amount of fertilizers and water with the aim to increase yield [3]. However, the increased supply of water and fertilizers does not increase yield proportionally. As nitrogen (N) is a major constituent of fertilizers, the excessive nitrogen application may increase undesired plant leaf growth affecting the tomato yield instead. Therefore, it is imperative to optimize the amount of water and N supply for tomato production to maintain a balance among plant growth (root/shoot) and yield to enhance the water use efficiency (WUE).

Since water and N are intrinsically linked, suitable irrigation along with considerable inputs of nitrogen fertilizer is required to regulate plant physiology in results of plant water and nutrient uptake [4]. In an attempt to save water, deficit irrigation is a promising technique, and it increases WUE due to reduced transpiration caused by water stress. Some researchers have reported that deficit irrigation decreases tomato fruit yield [5,6] within an acceptable range but increased tomato quality [1,7,8].

Nitrogen is another important crop limiting factor for controlling physiological crop processes promoting plant growth and yield. Many studies reported that increased application of N improves plant water use efficiency, increasing leaf photosynthesis by promoting leaf and root development, and hence enhancing plant biomass production [9,10]. In contrast, excessive N application in tomato cultivation may result in high nitrate leaching, representing a high risk of nitrate pollution [3,7,11] and inhibiting crop transpiration [12,13]. Under limited nitrogen availability, root morphology plays an important role, which may increase root surface area or root elongation for nutrient acquisition. Such improvements may enhance the resource allocation to root development as a whole, leading to greater root–shoot ratios under nutrient-limited conditions [14]. This extensive root system promotes water uptake ability when the plant is susceptible to water stress [15]. However, further investigation on effects of reduced irrigation and nitrogen on root characteristics and yield of tomato grown in a soilless substrate is needed.

There are numerous studies about tomato yield, water use efficiency, and fruit quality responses to deficit irrigation [16,17] and nitrogen [18,19] individually. However, few studies were done on the effects of different amounts of irrigation water and nitrogen applications under soilless substrate on the shoot and root parameters. Therefore, it becomes imperative to employ appropriate strategies for the proper supply of nutrients with a decreased water amount to ensure proper root–shoot growth and enhance water use efficiency by saving water and nitrogen inputs. The aims of this research were to (i) explore the effects of various water and N levels on root–shoot morphology, (ii) determine yield and irrigation water use efficiency (IWUE), and (iii) study the relationship between photosynthetic parameters and water use efficiency under different W and N rates. The obtained results can recommend an optimal strategy for water and N management for greenhouse tomatoes.

2. Materials and Methods

2.1. Site Description

The pot experiments were conducted in two growing seasons from March to July (spring–summer, SS) and September to December (fall–winter, FW) in 2019 in a greenhouse located at Jiangsu University, Zhenjiang, PR China, (32.20° N, 119.45° E). Average indoor greenhouse relative humidity and temperature were 70.85%, 22.76 °C, and 77.89%, 20.20 °C, during growing seasons, respectively.

Tomato (*Solanum lycopersicum* L.) Hezuo 906 variety was used as crop material for this experiment. For both seasons (SS and FW), tomato seedlings were transplanted in pots on 1st March and 5th September, respectively. Plants were supported by a nylon cord vertically. Pruning was done to maintain the proper growth following the well-managed agronomic local practices. Plant density was maintained as 3 plants m^{-2}. Pots with height of 25 cm and diameter of 19 cm were filled with coarse perlite substrate (2–4 mm) to the height of 22.5 cm from the bottom. Some physical properties of perlite include bulk density (BD) 0.098 (g·cm^{-3}), effective pore space (EPS) 71.9% (*v/v*), and conventional container capacity (CCC) 49.3% (*v/v*).

2.2. Experiment Design and Method

The experiment was composed of three irrigation levels and three N levels. Three levels of irrigation water were applied based on crop evapotranspiration (ETc) in the control treatment, i.e., 100% ETc (W_{100}), 80% of W_{100} (W_{80}), and 60% of W_{100} (W_{60}), and three levels of N concentration were applied as a fertilization based on percent of the N available in full strength Hoagland's solution, i.e., 100% (N_{100}), 75% (N_{75}), and 50% (N_{50}). The plants were watered with Hoagland nutrient solution, which was modified in order to obtain three different N doses, by means of total or partial substitution of KNO_3 and $Ca(NO_3)_2$ by adding $CaCl_2$ and K_2SO_4 if necessary, where the tap water with electrical conductivity (EC) of 0.003 dSm^{-1} was used to prepare the nutrient solution. The measured EC of the full-strength Hoagland solution was 1.94 dSm^{-1}, whereas the EC for N_{75} and N_{50} were 1.69 dSm^{-1} and 1.44 dSm^{-1}, respectively. The pH of the nutrient solution was maintained between 5.5 and 6.5 by adding NaOH or HCl as needed. The treatments consisted of the application of different doses of N (15, 11.25, 7.5 mM). The experimental plan produced nine treatments with three replicates in each treatment ($3 \times 3 \times 3$), where each treatment consisted of 3 plants, as shown in Table 1. The randomized complete block design was used for the given set of pots in this experiment.

Table 1. The treatment detail of the experiment.

Treatments	Nitrogen (ppm)		
	N_{100}	N_{75}	N_{50}
W_{100}	$W_{100}\,N_{100}$	$W_{100}\,N_{75}$	$W_{100}\,N_{50}$
W_{80}	$W_{80}\,N_{100}$	$W_{80}\,N_{75}$	$W_{80}\,N_{50}$
W_{60}	$W_{60}\,N_{100}$	$W_{60}\,N_{75}$	$W_{60}\,N_{50}$

Uniform irrigation was applied for the first 10 days after transplanting the seedlings for proper establishment. The water consumed by plants was calculated based on the substrate water balance method in the control treatment. The amount of irrigation was applied and recorded to replace the consumed water in the control treatment ($W_{100}\,N_{100}$) in each irrigation event using the following equation.

$$ETc = I - L - \Delta S \tag{1}$$

where I is the total amount of irrigation water, L is the water leached after irrigation, and ΔS is the difference between the amount of substrate water stored before harvesting and treatment initiation. Moreover, there was no runoff in this experiment; however, a tiny amount of water leached out after a few irrigation events.

2.3. Measurements

2.3.1. Growth Measurements

Plant growth is defined by morphological measurement that includes the stem diameter, plant height, and plant leaf area. Manual calculation of plant height from each treatment was performed using a measuring tape (± 0.1 mm error). The height of each selected plant was taken from the base of the stem to the growing tip of the last leaf. Similarly, stem diameter of all plants was measured from each treatment using a Vernier caliper at the base of the stem. Leaf area was determined just before the final harvest. The area of pruned leaves during all growth stages was summed up and added to the measured leaf area at the time of harvest to get the total leaf area of the plant. The total leaf area of the tomato plants [7] was calculated based on study of Schwarz and Kläring [20].

$$A_L = aL_w^b \tag{2}$$

$$T.A_L = a \sum_{k=ti}^{tf} L_w^b \qquad (3)$$

where $T.A_L$ reflects the total leaf area of a plant, L_w is leaf width, $a = 0.2031$ and $b = 1.6738$ are the constant of integration and allometry coefficient, respectively, ti and tf are time limits from the initial time (day) of transplanting to final harvesting. At the end of the maturation stage, tomato plants from each treatment were selected, and fruit-free plants were used to assess fresh and dry plant biomass. Upon drying in an oven for 24 h at 105 °C, dry biomass including earlier collected leaves by pruning was measured to obtain the total dry biomass of plant from each treatment.

2.3.2. Root Morphology

Root morphological traits such as average diameter (AD, mm), total root length (TRL, cm), root surface area (RSA, cm^2), and root volume (RV, cm^3) were measured at the end of the harvesting. The substrate containing roots of tomato was taken out of the pot and immersed in water. The substrate clod was rinsed using pressurized tap water after an hour of immersion in water and live roots were removed after the substrate debris and dead roots flowed away with water. Live roots were spread evenly in a tray containing deionized water and scanned using a flatbed scanner (300 dpi). Root images were analyzed after scanning using WinRhizo image analysis software (Regent Instruments, Quebec, Canada) and the average diameter (AD) and total root length (TRL), root surface area (RSA) and root volume (RV) were calculated using this software.

2.3.3. Photosynthesis Parameters/Leaf Gas Exchange

In both seasons, photosynthesis (Pn) and stomatal conductance (gs) were determined at the fruiting stage, i.e., 57 days after transplantation (DAT) for SS, and 60 DAT in FW 2019, respectively. Pn and gs of the newly expanded leaves were measured from 9:00 to 11:00 in the morning using a Li-6400 portable photosynthesis system (Li-COR Biosciences Ltd., Cambridge, UK). The photosynthetic active radiation (PAR), temperature, and CO_2 concentration during the measurements were 800 $\mu mol\ m^{-2}\ s^{-1}$, 28 °C, and 500 $\mu mol\ mol^{-1}$, respectively.

2.3.4. Fruit Yield and Irrigation Water Use Efficiency

Tomato fruits were weighed separately from the first harvest until the last harvest and the total yield of each treatment was calculated at the end of the experiment. The total water applied as irrigation was determined as the total sum of applied water in each irrigation event during the whole experiment. Irrigation water use efficiency (IWUE) is the performance of tomato yield to applied irrigation water and it was calculated as [21]:

$$IWUE = \frac{Y}{W} \qquad (4)$$

where, Y indicates yield (ton ha^{-1}) and W represents water applied as irrigation (mm).

2.4. Data Analysis

Analysis of variance was attained using Statistix 8.1 statistical software. For means comparisons, least significant difference (LSD) was applied at $p < 0.05$ significance level.

3. Results

3.1. Plant Growth Parameter

Morphological traits such as plant height, stem diameter, and leaf area were used to represent plant growth. Both irrigation and N fertilization levels affected plant growth significantly, as shown in Table 2. When analyzed across the N fertilization, the plant height, stem diameter, and leaf area showed a significant decrease with the decrease in each irrigation level. Analysis across the irrigation levels showed that N_{100} and N_{75} plants had significantly higher stem diameter, plant height, and leaf area than did N_{50} plants. However, N_{100} and N_{75} had no significant difference with each other across irrigation levels in plant height.

Table 2. Effect of water (W) and N on stem diameter, plant height, and leaf area during spring–summer (SS) and fall–winter (FW) season.

Treatments	Stem Diameter (mm)		Plant Height (cm)		Leaf Area (cm^2)	
	SS	FW	SS	FW	SS	FW
Irrigation						
W_{100}	11.29 a	9.90 a	130.88 a	119.12 a	1496.5 a	1163.2 a
W_{80}	9.61 b	9.38 b	112.57 b	108.98 b	1275.7 b	1013.6 b
W_{60}	8.07 c	8.12 c	98.06 c	94.04 c	780.6 c	712.3 c
Fertilizer						
N_{100}	10.60 a	10.17 a	125.71 a	117.81 a	1382.4 a	1064.5 a
N_{75}	9.92 ab	10.11 a	111.11 ab	104.42 ab	1200.1 b	954.9 b
N_{50}	8.74 c	8.11 c	104.68 c	99.91 c	970.1 c	869.7 c
Interaction						
W_{100} N_{100}	12.13 a	11.30 a	153.50 a	130.43 a	1897.8 a	1299.2 a
W_{100} N_{75}	11.37 b	9.53 c	128.23 b	118.60 b	1561.9 b	1160 b
W_{100} N_{50}	10.37 c	8.86 de	110.90 c	108.33 c	1029.7 f	1030.4 c
W_{80} N_{100}	10.97 bc	10.27 b	129.37 b	120.70 b	1432.6 c	1141.6 b
W_{80} N_{75}	9.37 d	9.43 cd	107.10 d	105.17 cd	1272.6 d	1012.5 c
W_{80} N_{50}	8.50 e	8.40 e	101.23 ef	101.07 e	1121.8 e	886.8 d
W_{50} N_{100}	8.70 e	8.93 cde	94.27 g	102.30 de	816.9 g	752.8 e
W_{50} N_{75}	8.13 e	8.36 e	98 f	89.50 f	766 g	692.3 e
W_{50} N_{50}	7.36 f	7.07 f	101.90 e	90.33 f	758.9 g	691.7 e
ANOVA						
W	***	***	***	***	***	***
N	***	***	***	***	***	***
W × N	NS	NS	***	**	***	**

The data presented in the table are the mean of three replicates. According to the Least significant difference (LSD) test, values with different letters within the same columns are significantly different at $p < 0.05$. NS, **, *** are not significant, significant at $p \leq 0.01$, significant at $p \leq 0.001$, respectively.

Considering the interaction effect of irrigation and N fertilization, the treatment W_{100} N_{I00} was significantly higher in stem diameter, plant height, and leaf area than other interaction levels during both growing seasons, which was one of the treatments with the largest irrigation amount with maximum nitrogen input. Among the interaction of irrigation and N fertilization, stem diameter remained non-significant during both growing seasons, however, the plant height and leaf area were significant at $p < 0.05$ during SS, while these were significant at $p < 0.01$ during the FW season.

Total fresh and dry plant biomass were determined to evaluate the effects of different irrigation and N fertilization on plant growth, and results are shown in Table 3. Analysis across the N treatments, fresh plant biomass (PFM) and dry plant biomass (PDM) were decreased significantly, parallel with a decrease in irrigation level. Considering only N levels, N_{75} had no significant difference with N_{100} in dry plant biomass, while both were significantly higher than N_{50}. At the interaction effect of irrigation and N fertilization (W × N), the PFM and PDM showed significant differences during the SS and FW seasons. The obtained biomass results in this study are quite similar to results obtained by another study conducted on tomato growth by Rasool et al. [22].

Table 3. Effect of irrigation and fertilizer levels on plant fresh and dry biomass during spring–summer (SS) and fall–winter (FW) season.

Treatments	PFM (g·plant^{-1})		PDM (g·plant^{-1})	
	SS	FW	SS	FW
Irrigation				
W_{100}	364.89 a	345.78 a	115.26 a	109.11 a
W_{80}	315.11 b	305.33 b	109.39 b	105.33 b
W_{60}	254.89 c	247.67 c	83.78 c	85.96 c
Fertilizer				
N_{100}	357.44 a	315.44 a	106.14 a	102.08 a
N_{75}	317.78 b	294.44 b	106.69 a	100.62 a
N_{50}	259.67 c	288.89 b	95.59 b	97.70 b
Interaction				
W_{100} N_{100}	410.67 a	378.33 a	125.03 a	112.03 a
W_{100} N_{75}	382.00 b	329.00 b	118.23 b	109.73 a
W_{100} N_{50}	302.00 e	330.00 b	102.50 c	105.77 b
W_{80} N_{100}	362.67 c	336.00 b	106.33 c	105.57 b
W_{80} N_{75}	339.33 d	299.33 c	118.97 b	101.67 c
W_{80} N_{50}	243.33 f	280.67 cd	102.87 c	108.57 ab
W_{50} N_{100}	299.00 e	232.00 e	87.07 d	81.77 e
W_{50} N_{75}	232.00 f	238.33 e	82.87 de	88.17 d
W_{50} N_{50}	233.67 f	272.67 d	81.40 e	87.93 d
ANOVA				
W	***	***	***	***
N	***	***	***	***
W × N	***	***	***	***

PFM; fresh plant biomass, PDM; dry plant biomass. The data presented in the table are the mean of three replicates. According to the LSD test, values with different letters within the same columns are significantly different at $p < 0.05$. *** is significant at $p \leq 0.001$.

3.2. Root Morphology

Total root length, total root surface area, average diameter, and total root volume characteristics are described in Figures 1 and 2 under the results of irrigation and N fertilization experiments carried out during spring–summer and fall–winter.

Across the N-application rate, the effect of irrigation was significant on root morphological traits (TRL, RSA, AD, and RV). The TRL, RSA, and RV were significantly higher in W_{80} when compared with full irrigation (W_{100}) and severe deficit (W_{60}), while AD had no significant difference between W_{100} and W_{80}. Analysis across irrigation showed that N_{100} resulted in higher values of root traits among other N fertilization treatments. However, TRL, AD, and RV had no significant difference between N_{100} and N_{75}, except RSA, which had a significant difference among all N treatments.

The interaction of irrigation and N fertilization (W × N) was significant at TRL, RSA, and RV, except for AD which gave a non-significant difference during both seasons. The interaction of W_{80} N_{100} gave highest TRL and RSA, and had a significant difference when compared with other interactive pairs of irrigation and N during SS and FW. Similarly, RV was found almost the same in W_{80} N_{100} and W_{80} N_{75} with slightly higher values in W_{80} N_{75}, but there was no significant difference between both treatments.

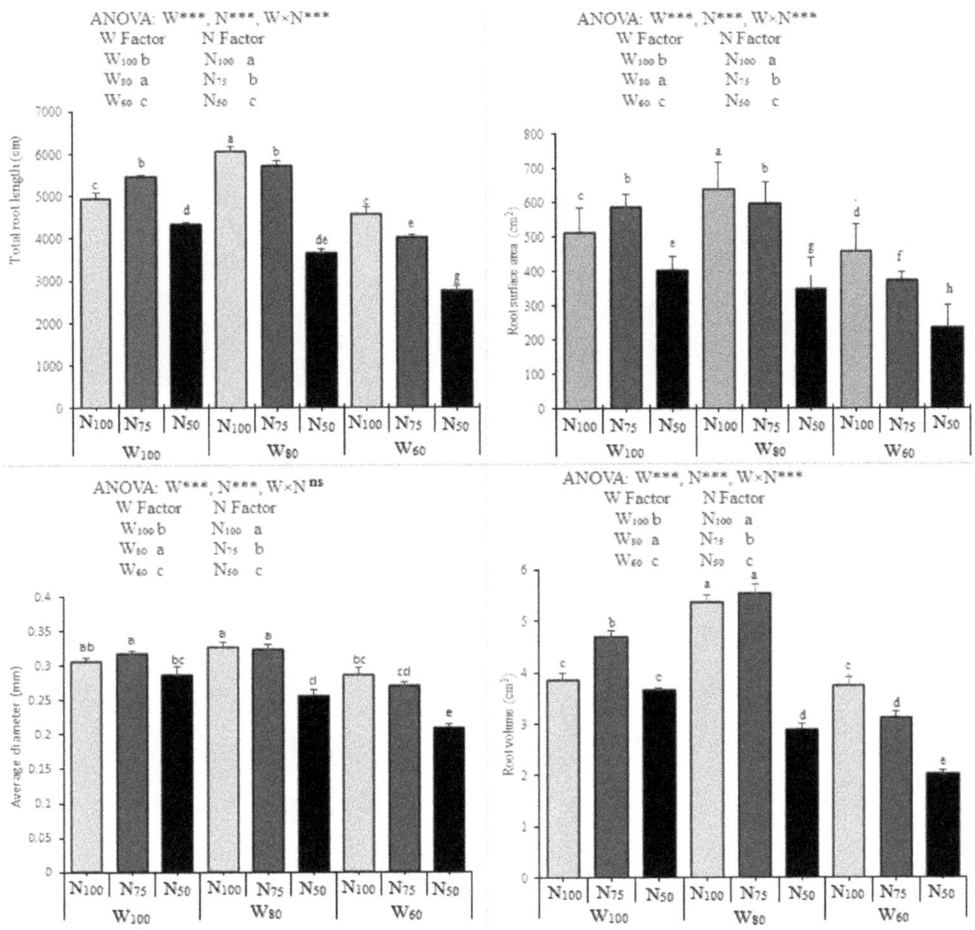

Figure 1. Effects of irrigation water (W) and nitrogen fertilization (N) on average diameter, total root length, root surface area, and root volume at the harvesting stage during spring–summer. Bars with different letters are significantly different at $p < 0.05$ according to LSD test. ns, *** are not significant at $p > 0.05$, significant at $p \leq 0.001$, respectively.

3.3. Photosynthesis (Pn) and Stomatal Conductance (gs)

Upon analysis across N fertilization, Pn and gs decreased significantly with each decreased level of irrigation. Pn and gs were found to be higher under full irrigation than those tomato plants under deficit water during both seasons, as shown in Table 4. Similarly, Pn and gs of tomato plants under high-N fertilization were higher than those of plants under low-N fertilization, irrespective of irrigation amount. Considering N fertilization overall, the pattern of Pn and gs were found to be parallel to irrigation except for gs at N_{100} and N_{75} during both seasons, where both treatments did not show a significant difference. Whatever the availability of N in the substrate, the increased availability of water in the rootzone has resulted in increased Pn. Compared to the control, the decrease in Pn was 7.72% and 36.77% in W_{80}, and 11.09% and 40.79% in W_{60} during SS and FW, respectively.

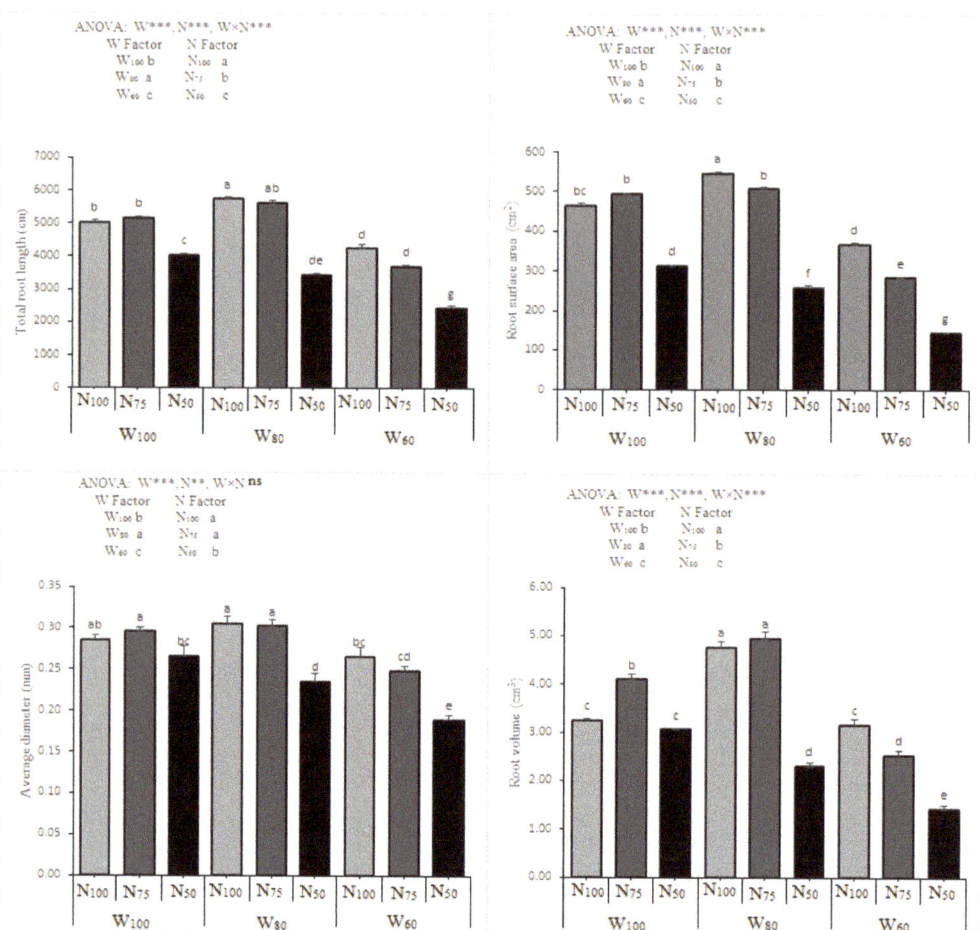

Figure 2. Effects of irrigation water (W) and nitrogen fertilization (N) on tomato total root length, root surface area, average diameter, and root volume at the harvesting stage during fall–winter. Bars with different letters are significantly dif-ferent at $p < 0.05$ according to LSD test. ***, ** indicate significance at $p < 0.001$ and $p < 0.01$, respectively, and ns indicates non-significance ($p > 0.05$).

Considering the interaction of irrigation and nitrogen (W × N), the response of Pn and gs were recorded statistically significant to irrigation and N fertilization. Conversely, W_{80} N_{100} did not induce any loss in Pn during SS and FW, and gs during the FW season, and remain statistically non-significant when compared to treatment W_{100} N_{100}, as shown in Table 4. The highest values of Pn and gs were found to be maximum under the interaction of the maximum level of water and nitrogen application.

Table 4. Photosynthetic rate (Pn) and stomatal conductance (gs) across irrigation levels with different nitrogen concentration applied levels during the spring–summer (SS) and fall–winter (FW) season.

Treatments	Pn $(\mu mol \cdot (CO_2) \cdot m^{-2} \cdot s^{-1})$		gs $mol \cdot (H_2O) \cdot m^{-2} \cdot s^{-1}$	
	SS	FW	SS	FW
Irrigation				
W_{100}	17.87 a	15.69 a	0.4533 a	0.2756 a
W_{80}	16.49 b	13.95 b	0.3262 b	0.2345 b
W_{60}	11.30 c	9.29 c	0.1943 c	0.1234 c
Fertilizer				
N_{100}	16.45 a	14.71 a	0.3881 a	0.2419 a
N_{75}	16.25 b	13.51 b	0.3697 ab	0.2346 ab
N_{50}	12.95 c	10.71 c	0.2492 c	0.1570 c
Interaction				
$W_{100} N_{100}$	19.53 a	17.82 a	0.5178 a	0.3294 a
$W_{100} N_{75}$	18.52 b	17.00 b	0.4419 c	0.2779 c
$W_{100} N_{50}$	11.29 g	13.05 c	0.2046 f	0.1184 g
$W_{80} N_{100}$	18.95 ab	17.76 ab	0.4839 b	0.3138 ab
$W_{80} N_{75}$	16.76 c	12.24 d	0.2698 e	0.2557 d
$W_{80} N_{50}$	13.05 f	12.05 d	0.2557 e	0.1344 f
$W_{50} N_{100}$	15.12 d	10.48 e	0.3582 d	0.1837 e
$W_{50} N_{75}$	14.18 e	9.54 f	0.2668 e	1699 e
$W_{50} N_{50}$	9.54 h	7.85 g	0.1224 g	0.1175 g
ANOVA				
W	***	***	***	***
N	***	***	***	***
W × N	***	***	***	***

The data presented in the table are the mean of three replicates. According to the LSD test, values with different letters within the same columns are significantly different at $p < 0.05$. *** is significant at $p \leq 0.001$.

3.4. Yield and Irrigation Water Use Efficiency Response to Water and Nitrogen Levels

As shown in Table 5, the tomato consumed a high water amount in the spring–summer season than in the fall–winter season. The highest tomato yield was observed in treatments with high water and N fertilization during the SS and FW seasons. The tomato yield decreased significantly ($p < 0.05$) with a decrease in irrigation amount, irrespective of N application. Similarly, when analyzed across the irrigation, the tomato yield increased with an increase in N-application and reached maximum values at N_{100} ($p < 0.05$).

The interaction effect of irrigation and N fertilization had a significant effect on yield ($p < 0.05$) in both seasons. Maximum yield was achieved at $W_{100} N_{100}$. The treatments of interaction, $W_{100} N_{75}$ and $W_{80} N_{100}$, did not induce significant loss of yield when compared to $W_{100} N_{100}$ during SS, while $W_{100} N_{100}$ remained significantly higher than other interaction treatments.

Considering IWUE, irrespective of N, the highest water use efficiency was recorded at an irrigation level of W_{80} and W_{60} in SS and FW, respectively. The highest IWUE was achieved in the highest N fertilization (N_{100}) when compared across the irrigation but there was no significant difference in the IWUE between N_{100} and N_{75} in FW.

During SS and FW, there was also a significant interaction between irrigation and N-fertilizer for IWUE. The interaction of $W_{80}N1_{00}$ gave maximum irrigation water use efficiency during both seasons. Davies et al. [23] reported watering strategies to decrease the transpiration rate in plants, leading to a decrease in leaf area and stomatal openings, eventually improving IWUE.

Table 5. Effect of W and N on water used, fruit yield, and irrigation water use efficiency.

Treatments	Irrigation Water (mm)		Fruit Yield (ton·ha^{-1})		Irrigation Water Use Efficiency	
	SS	FW	SS	FW	SS	FW
Irrigation						
W_{100}	245.98	222.51	92.23 a	81.03 a	37.50 c	36.41 c
W_{80}	196.79	178.00	85.21 b	72.48 b	43.30 a	40.72 b
W_{60}	147.59	133.50	59.38 c	55.89 c	40.24 b	41.86 a
Fertilizer						
N_{100}			85.95 a	76.17 a	43.60 a	42.99 a
N_{75}			83.04 b	72.49 b	40.49 b	41.15 ab
N_{50}			67.84 c	60.76 c	34.94 c	34.86 c
Interaction						
$W_{100}\ N_{100}$	245.98	222.51	100.82 a	89.71 a	40.98 c	40.32 d
$W_{100}\ N_{75}$			99.42 a	86.75 b	40.42 c	38.98 de
$W_{100}\ N_{50}$			76.47 c	66.62 e	31.09 e	29.94 g
$W_{80}\ N_{100}$	196.79	178.00	97.90 a	81.86 c	49.75 a	45.99 a
$W_{80}\ N_{75}$			84.90 b	71.52 d	43.14 b	40.18 d
$W_{80}\ N_{50}$			72.83 d	64.07 f	37.01 d	36.00 f
$W_{50}\ N_{100}$	147.59	133.50	59.13 f	56.96 g	40.07 c	42.66 c
$W_{50}\ N_{75}$			64.80 e	59.13 g	43.91 b	44.29 b
$W_{50}\ N_{50}$			54.22 g	51.57 h	36.73 d	38.63 e
ANOVA						
W	***	***	***	***	***	***
N	***	***	***	***	***	***
W × N	***	***	***	***	***	***

The data presented in the table are the mean of three replicates. According to the LSD test, values with different letters within the same columns are significantly different at $p < 0.05$. *** is significant at $p \leq 0.001$.

3.5. Relationship between Fruit Yield (FY), Photosynthesis (Pn), and Water Used (WU) under the Effect of Water and N Fertilization

The linear regression relationships of FY with water used, Pn, and gs during SS and FW are shown in Figure 3. Maximum linear regression (R^2) among FY and Pn was observed as compared to water in use. For FY and Pn, the largest linear regression coefficients (R^2) of 0.97 for SS and 0.98 for FW season were observed. The lowest values of R^2 were found to be at linear regression relationships of FY with WU.

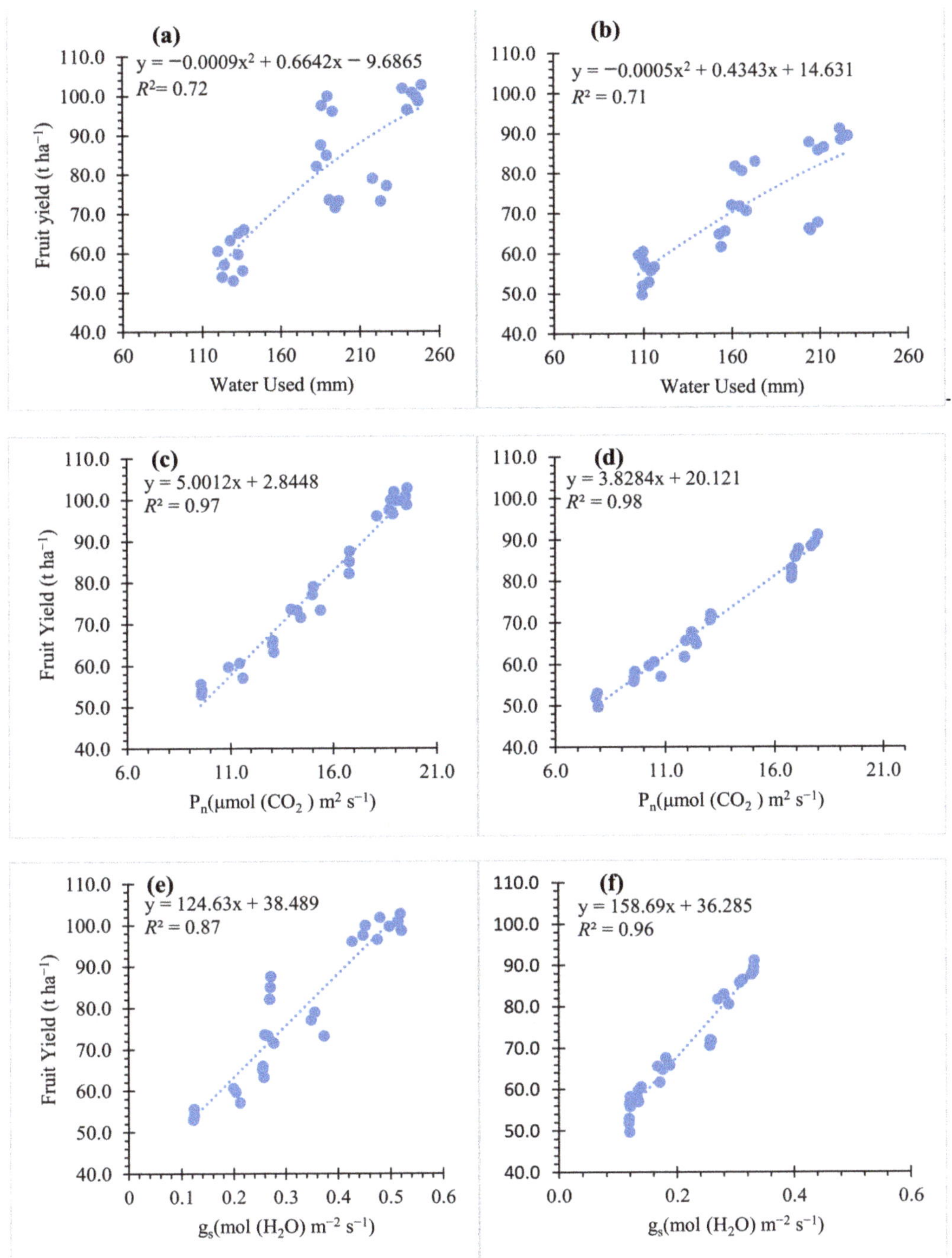

Figure 3. Regression models fitted for SS as (**a**) water used vs. fruit yield; (**c**) Pn vs. fruit yield; (**e**) gs vs. fruit yield, and for FW as (**b**) water used vs. fruit yield; (**d**) Pn vs. fruit yield; (**f**) gs vs. fruit yield.

4. Discussion

Increasing growth and yield rely on optimal water and N application [24]. Water is essential for plant growth and its deficit would impact plant growth and production, depending on severity and duration of water deficit. It is well recognized that shoot and root growth have a direct influence on crop yield under an optimal growth environment, whereas their response can vary under a deficit water condition as well as nitrogen amount. Water and N fertilization demonstrated significant effects on overall plant growth and photosynthetic capacity of tomato. Fresh and dry plant biomass, leaf area, and plant height showed maximum values at treatments without shortage of irrigation and N-application. Similar findings have been observed by Hou et al. [25] who found that stem diameter and plant height increased significantly with the increase in irrigation quantity at different growth stages of tomatoes. Rasool, Guo, Wang, Chen, and Ullah [22] stated comparable results and indicated that tomato biomass production was improved significantly due to the sufficient water supply during the crop maturity stage. Leaf area is an important growth parameter that serves for photosynthesis and gas exchange by allowing the PAR (photosynthetic active radiation) to fall on its surface area. The decreased leaf area could be due to stomatal closure and photosynthesis inhibition as a physiological response of plants to water and nitrogen deficit where flow of abscisic acid from roots to leaves tend to reduce stomatal aperture ultimately, causing reduction in leaf growth [26].

Being a primary organ, root characteristics are closely related to their water and nutrient absorption ability [27]. At the highest N-application, the W_{80} (a water deficit treatment) gave the maximum root average diameter, length, surface area, and volume, which is evidence that N-application enhances the root growth, and which can lead to an increased root biomass. This structural change in roots may have occurred due to reallocation of assimilates from shoot to root by plants under water stress, which resulted in decreased shoot growth and increased root system traits such as root length, root surface area, and root volume under reduced irrigation, as water is the main driver of resource allocation [28]. Moreover, N provided additive importance under water stress [29]. This could be because more photosynthetic products become conveyed to the root system under water stress, which improves root growth and nutrient and water absorption by enhancing root length, and finer root and deeper root spreading in the rootzone [30]. Changes in the root system to obtain more water and nutrients, the shoot system (reduced plant growth and leaf area), and physiology to trigger the underlying mechanisms are possible under deficit irrigation. However, the mechanisms through which the water deficit level of W_{80} might possibly regulate root growth were not fully revealed by the present study. The increase in IWUE may be possible due to the decrease in stomatal conductance caused by the increased signal transduction network of guard cells [31]. Our study was also supported by Wang et al. [32], who stated that the interaction of medium deficit irrigation and N-nutrients show a significantly improved root surface area and total root volume, and could make a plant drought resistant without significant loss of yield. In addition, a better nutrient and water environment in the root zone can produce abscisic acid in the xylem by improving root to shoot signals, which can regulate vegetative growth and stomata switch [33,34].

Pn and gs are affected by water and N-application [35]. In our study, Pn and gs showed a decline with the increase of water deficit and reducing N-nutrition. Comparable results were also stated by Chen et al. [36], who revealed that an increase in water stress increased abscisic acid in roots, resulting in low photosynthesis and stomatal conductivity with its transportation to canopy causing stomatal closure. Del Amor et al. [37] also found the reduction in gs was due to a decrease in water availability in tomato plants. In our findings, an increase in water and N supply resulted in an increase in photosynthesis and a bigger biomass, which ultimately resulted in increased tomato fruit yield. Moreover, Garcia et al. [38] observed that the reduction in gs due to water stress is also influenced by nitrogen application quantity. They stated that applying nitrogen at 60% did not show a significant reduction in gs, which confirmed the findings of our study that N_{75} did not decrease yield significantly compared to that measured at N_{100}. Water stress reduces photosynthesis in tomato leaves, decreasing dry material accumulation and reducing tomato growth, which in turn decreases yield [39].

Both irrigation and nitrogen fertilization are essential factors for tomato growth, and they both influence yield and IWUE, however, nitrogen provided an additive importance to sustain the yield under certain ranges of deficit irrigation. Yield decrease was observed more with a decrease in irrigation rather than nitrogen. Patane and Cosentino [40] also reported that a yield decrease under deficit irrigation was observed nearly twice as low compared to full irrigation. The non-significant difference in yield at W_{100} and W_{80} under N_{100} is evidence that N-application improves yield within a specific range of deficit irrigation [41], whereas further decreases in water or increases in nitrogen may lead to salt accumulation in the root zone, which could seriously affect the tomato yield [7].

Deficit watering has been described as a beneficial alternative strategy of water saving under an open field, as well as greenhouse cultivation in arid and semi-arid regions where water is the most limiting factor for crop cultivation. Traditionally, in situations of frequent water shortage and drought spells, it has been recommended to alleviate the adverse effects of water stress on crop yield and increase the efficiency of water efficiency [42]. Water use efficiency is a critical indicator at all scales, specifically during drought conditions when irrigation water is scarce. The highest irrigation water use efficiency was recorded at W_{80} N_{100} because of a non-significant difference in yield with the control, while water application was apparently reduced at W_{80} compared to W_{100}. Considering N-application, the highest water use efficiency was recorded at a higher nitrogen level. Among interactions, at the highest N, W_{80} resulted in the highest water use efficiency and N application reduced the impact of reduced water on yield and hence resulted in high IWUE. Therefore, it is very important to optimize nitrogen and irrigation water supplies to improve crop yield, and water and nitrogen use efficiency in substrate cultivation. Earlier research has also determined that water use efficiency decreases with a rise in irrigation water amount [43]. The best irrigation rate under three different nitrogen levels for optimum WUE and IWUE in cucumber is 0.8 Ep [44]. These results indicated that the reasonable application of nitrogen fertilizer should be emphasized. The highly efficient coordination of irrigation and nitrogen fertilization should also be stressed.

Along with deficit irrigation and reduced nitrogen application in greenhouses, WUE can be further improved by providing better climate control, CO_2 dosage, and its integration with other possible techniques like growth-promoting substances, grafting, and with the use of anti-transpirants. Hence, interactive effects of such potential applications with deficit irrigation and nitrogen can be studied in future in broader perspectives.

5. Conclusions

Plants grown under deficit irrigation must be well-fertilized to serve as a compensation factor so as not to suffer from double stress. The increased tomato root length, surface area, and root volume in W_{80} provided more absorption surface for N uptake, which could be a reason to improve the biomass, yield, and water use efficiency, and provided evidence for better utilization of high N under moderate water stress. Due to limited availability of water, relatively small deficit irrigations are found applicable with an acceptable risk of yield decrease. In the current study, highest tomato yield was found at the highest irrigation and N application. There was no significant difference in yield upon reducing one of both inputs, i.e., 20% irrigation reduction from full irrigation or 25% reduction in N applied in control, but a simultaneous decrease of irrigation and nitrogen could not maintain the yield and IWUE at its highest level. These findings illustrate the importance for evaluation of impact on tomato yield under various levels of the interaction between nitrogen and water for proper management practices for sustainable agro-production. As there was no significant difference in yield among W_{100} N_{100}, W_{100} N_{75}, and W_{80} N_{100}, hence, considering optimal tradeoffs between yield, water saving, and water use efficiency, the W_{80} N_{100} could be adopted as an optimal strategy where N_{100} can be used as a source to mitigate the effects of deficit irrigation and reduce the risk of yield loss where scarcity of water availability exists. Besides a water and nutrient saving strategy, future works can consider the integration of deficit irrigation and reduced nitrogen with growth promoting organisms or with the use of anti-transpirants to sustain tomato yield and improve fruit nutritional quality.

Author Contributions: Conceptualization, I.U. and H.M.; methodology, I.U. and H.G.; software, G.R.; validation, A.S. and M.I.K.; formal analysis, I.U. and G.R.; investigation, Q.J.; writing—original draft preparation, I.U.; writing—review and editing, I.U., H.M., and G.R.; visualization, H.G.; supervision, H.M.; funding acquisition, H.M. All authors have read and agreed to the published version of the manuscript.

Funding: This research was funded by "National Key R&D Projects, grant number 2018YFF0213600", "National Natural Science Foundation of China, grant number 61233006", "Natural Science Foundation of Jiangsu Province of China, grant number BK20180864", and "Jiangsu Synergy Innovation Center Program of Modern Agricultural Equipment and Technology, grant number 4091600028".

Institutional Review Board Statement: Not applicable.

Informed Consent Statement: Not applicable.

Data Availability Statement: Not applicable.

Conflicts of Interest: The authors declare no conflict of interest.

References

1. Wang, C.; Gu, F.; Chen, J.; Yang, H.; Jiang, J.; Du, T.; Zhang, J. Assessing the response of yield and comprehensive fruit quality of tomato grown in greenhouse to deficit irrigation and nitrogen application strategies. *Agric. Water Manag.* **2015**, *161*, 9–19. [CrossRef]
2. FAO. FAOSTAT (Food and Agriculture Organization of the United Nations). Available online: http://faostat3.fao.org/home/E (accessed on 6 February 2018).
3. Sun, Y.; Hu, K.; Fan, Z.; Wei, Y.; Lin, S.; Wang, J. Simulating the fate of nitrogen and optimizing water and nitrogen management of greenhouse tomato in North China using the EU-Rotate_N model. *Agric. Water Manag.* **2013**, *128*, 72–84. [CrossRef]
4. Zotarelli, L.; Scholberg, J.M.; Dukes, M.D.; Muñoz-Carpena, R.; Icerman, J. Tomato yield, biomass accumulation, root distribution and irrigation water use efficiency on a sandy soil, as affected by nitrogen rate and irrigation scheduling. *Agric. Water Manag.* **2009**, *96*, 23–34. [CrossRef]
5. Patanè, C.; Tringali, S.; Sortino, O. Effects of deficit irrigation on biomass, yield, water productivity and fruit quality of processing tomato under semi-arid Mediterranean climate conditions. *Sci. Hortic.* **2011**, *129*, 590–596. [CrossRef]
6. Shabbir, A.; Mao, H.; Ullah, I.; Buttar, N.A.; Ajmal, M.; Lakhiar, I.A. Effects of drip irrigation emitter density with various irrigation levels on physiological parameters, root, yield, and quality of cherry tomato. *Agronomy* **2020**, *10*, 1685. [CrossRef]
7. Ullah, I.; Hanping, M.; Chuan, Z.; Javed, Q.; Azeem, A. Optimization of irrigation and nutrient concentration based on economic returns, substrate salt accumulation and water use efficiency for tomato in greenhouse. *Arch. Agron. Soil Sci.* **2017**, *63*, 1748–1762. [CrossRef]
8. Ripoll, J.; Urban, L.; Staudt, M.; Lopez-Lauri, F.; Bidel, L.P.; Bertin, N. Water shortage and quality of fleshy fruits—Making the most of the unavoidable. *J. Exp. Bot.* **2014**, *65*, 4097–4117. [CrossRef]
9. Wang, L.; Palta, J.A.; Chen, W.; Chen, Y.; Deng, X. Nitrogen fertilization improved water-use efficiency of winter wheat through increasing water use during vegetative rather than grain filling. *Agric. Water Manag.* **2018**, *197*, 41–53. [CrossRef]
10. Hongyan, M.H.G.; Ni, J.; Gong, L.; Zhang, X. Effects of different fertilization on nutrient and growth of lettuce and models simulation. *J. Drain. Irrig. Mach. Eng.* **2020**, *38*, 1264–1269.
11. Rasool, G.; Guo, X.; Wang, Z.; Ali, M.U.; Chen, S.; Zhang, S.; Wu, Q.; Ullah, M.S. Coupling fertigation and buried straw layer improves fertilizer use efficiency, fruit yield, and quality of greenhouse tomato. *Agric. Water Manag.* **2020**, *239*, 106239. [CrossRef]
12. Matimati, I.; Verboom, G.A.; Cramer, M.D. Nitrogen regulation of transpiration controls mass-flow acquisition of nutrients. *J. Exp. Bot.* **2014**, *65*, 159–168. [CrossRef]
13. Hanping, M.; Ullah, I.; Jiheng, N.; Javed, Q.; Azeem, A. Estimating tomato water consumption by sap flow measurement in response to water stress under greenhouse conditions. *J. Plant Interact.* **2017**, *12*, 402–413. [CrossRef]
14. López-Bucio, J.; Cruz-Ramírez, A.; Herrera-Estrella, L. The role of nutrient availability in regulating root architecture. *Curr. Opin. Plant Biol.* **2003**, *6*, 280–287. [CrossRef]
15. Ding, L.; Lu, Z.; Gao, L.; Guo, S.; Shen, Q. Is Nitrogen a Key Determinant of Water Transport and Photosynthesis in Higher Plants Upon Drought Stress? *Front. Plant Sci.* **2018**, *9*, 1143. [CrossRef]
16. Hooshmand, M.; Albaji, M.; Ansari, N.A.Z. The effect of deficit irrigation on yield and yield components of greenhouse tomato (*Solanum lycopersicum*) in hydroponic culture in Ahvaz region, Iran. *Sci. Hortic.* **2019**, *254*, 84–90. [CrossRef]

17. Zhang, H.; Xiong, Y.; Huang, G.; Xu, X.; Huang, Q. Effects of water stress on processing tomatoes yield, quality and water use efficiency with plastic mulched drip irrigation in sandy soil of the Hetao Irrigation District. *Agric. Water Manag.* **2017**, *179*, 205–214. [CrossRef]
18. Truffault, V.; Ristorto, M.; Brajeul, E.; Vercambre, G.; Gautier, H. To stop nitrogen overdose in soilless tomato crop: A way to promote fruit quality without affecting fruit yield. *Agronomy* **2019**, *9*, 80. [CrossRef]
19. Rosa-Rodríguez, R.D.l.; Lara-Herrera, A.; Trejo-Téllez, L.I.; Padilla-Bernal, L.E.; Solis-Sánchez, L.O.; Ortiz-Rodríguez, J.M. Water and fertilizers use efficiency in two hydroponic systems for tomato production. *Hortic. Bras.* **2020**, *38*, 47–52. [CrossRef]
20. Schwarz, D.; Kläring, H.-P. Allometry to estimate leaf area of tomato. *J. Plant Nutr.* **2001**, *24*, 1291–1309. [CrossRef]
21. Yang, H.; Cao, H.X.; Hao, X.M.; Guo, L.J.; Li, H.Z.; Wu, X.Y. Evaluation of tomato fruit quality response to water and nitrogen management under alternate partial root-zone irrigation. *Int. J. Agric. Biol. Eng.* **2017**, *10*, 85–94. [CrossRef]
22. Rasool, G.; Guo, X.; Wang, Z.; Chen, S.; Ullah, I. The interactive responses of fertigation levels under buried straw layer on growth, physiological traits and fruit yield in tomato plant. *J. Plant Interact.* **2019**, *14*, 552–563. [CrossRef]
23. Davies, W.J.; Wilkinson, S.; Loveys, B. Stomatal control by chemical signalling and the exploitation of this mechanism to increase water use efficiency in agriculture. *New Phytol.* **2002**, *153*, 449–460. [CrossRef]
24. Liu, W.; Wang, J.; Wang, C.; Ma, G.; Wei, Q.; Lu, H.; Xie, Y.; Ma, D.; Kang, G. Root growth, water and nitrogen use efficiencies in winter wheat under different irrigation and nitrogen regimes in North China Plain. *Front. Plant Sci.* **2018**, *9*, 1798. [CrossRef]
25. Hou, M.; Zhu, L.; Jin, Q. Surface drainage and mulching drip-irrigated tomatoes reduces soil salinity and improves fruit yield. *PLoS ONE* **2016**, *11*, e0154799. [CrossRef]
26. Badr, M.; El-Tohamy, W.; Hussein, S.; Gruda, N. Tomato yield, physiological response, water and nitrogen use efficiency under deficit and partial root zone drying irrigation in an arid region. *J. Appl. Bot. Food Qual.* **2018**, *91*, 332–340.
27. Ristova, D.; Busch, W. Natural variation of root traits: From development to nutrient uptake. *Plant Physiol.* **2014**, *166*, 518–527. [CrossRef]
28. Khapte, P.; Kumar, P.; Burman, U.; Kumar, P. Deficit irrigation in tomato: Agronomical and physio-biochemical implications. *Sci. Hortic.* **2019**, *248*, 256–264. [CrossRef]
29. Basal, O.; Szabó, A. The Combined Effect of Drought Stress and Nitrogen Fertilization on Soybean. *Agronomy* **2020**, *10*, 384. [CrossRef]
30. Sharp, R.E.; Poroyko, V.; Hejlek, L.G.; Spollen, W.G.; Springer, G.K.; Bohnert, H.J.; Nguyen, H.T. Root growth maintenance during water deficits: Physiology to functional genomics. *J. Exp. Bot.* **2004**, *55*, 2343–2351. [CrossRef]
31. Schroeder, J.I.; Kwak, J.M.; Allen, G.J. Guard cell abscisic acid signalling and engineering drought hardiness in plants. *Nature* **2001**, *410*, 327–330. [CrossRef]
32. Wang, Y.; Zhang, X.; Chen, J.; Chen, A.; Wang, L.; Guo, X.; Niu, Y.; Liu, S.; Mi, G.; Gao, Q. Reducing basal nitrogen rate to improve maize seedling growth, water and nitrogen use efficiencies under drought stress by optimizing root morphology and distribution. *Agric. Water Manag.* **2019**, *212*, 328–337. [CrossRef]
33. Sarker, K.K.; Akanda, M.; Biswas, S.; Roy, D.; Khatun, A.; Goffar, M. Field performance of alternate wetting and drying furrow irrigation on tomato crop growth, yield, water use efficiency, quality and profitability. *J. Integr. Agric.* **2016**, *15*, 2380–2392. [CrossRef]
34. Chai, Q.; Gan, Y.; Zhao, C.; Xu, H.-L.; Waskom, R.M.; Niu, Y.; Siddique, K.H. Regulated deficit irrigation for crop production under drought stress. A review. *Agron. Sustain. Dev.* **2016**, *36*, 3. [CrossRef]
35. Mofokeng, M.; Steyn, J.; Du Plooy, C.; Prinsloo, G.; Araya, H. Growth of Pelargonium sidoides DC. in response to water and nitrogen level. *S. Afr. J. Bot.* **2015**, *100*, 183–189. [CrossRef]
36. Chen, J.; Kang, S.; Du, T.; Guo, P.; Qiu, R.; Chen, R.; Gu, F. Modeling relations of tomato yield and fruit quality with water deficit at different growth stages under greenhouse condition. *Agric. Water Manag.* **2014**, *146*, 131–148. [CrossRef]
37. Del Amor, F.M.; Ruiz-Sánchez, M.C.; Martínez, V.; Cerdá, A. Gas exchange, water relations, and Ion concentrations of salt-stressed tomato and melon plants. *J. Plant Nutr.* **2000**, *23*, 1315–1325. [CrossRef]
38. Garcia, A.; Marcelis, L.; García-Sánchez, F.; Nicolas, N.; Martínez, V. Moderate water stress affects tomato leaf water relations in dependence on the nitrogen supply. *Biol. Plant.* **2007**, *51*, 707–712. [CrossRef]
39. Li, Y.; Sun, Y.; Liao, S.; Zou, G.; Zhao, T.; Chen, Y.; Yang, J.; Zhang, L. Effects of two slow-release nitrogen fertilizers and irrigation on yield, quality, and water-fertilizer productivity of greenhouse tomato. *Agric. Water Manag.* **2017**, *186*, 139–146. [CrossRef]
40. Patanè, C.; Cosentino, S. Effects of soil water deficit on yield and quality of processing tomato under a Mediterranean climate. *Agric. Water Manag.* **2010**, *97*, 131–138. [CrossRef]
41. Gebremariam, M.; Tesfay, T. Optimizing Irrigation Water and N Levels for Higher Yield and Reduced Blossom End Rot Incidence on Tomato. *Int. J. Agron.* **2019**, *2019*, 1–10. [CrossRef]
42. Geerts, S.; Raes, D. Deficit irrigation as an on-farm strategy to maximize crop water productivity in dry areas. *Agric. Water Manag.* **2009**, *96*, 1275–1284. [CrossRef]
43. Chen, S.; Zhou, Z.-J.; Andersen, M.N.; Hu, T.-T. Tomato yield and water use efficiency—Coupling effects between growth stage specific soil water deficits. *Acta Agric. Scand. Sect. B Soil Plant Sci.* **2015**, *65*, 460–469. [CrossRef]
44. Zhang, H.-X.; Chi, D.-C.; Qun, W.; Jun, F.; Fang, X.-Y. Yield and quality response of cucumber to irrigation and nitrogen fertilization under subsurface drip irrigation in solar greenhouse. *Agric. Sci. China* **2011**, *10*, 921–930. [CrossRef]

Article

Effects of Drip Irrigation Emitter Density with Various Irrigation Levels on Physiological Parameters, Root, Yield, and Quality of Cherry Tomato

Abdul Shabbir [1,2], Hanping Mao [1,*], Ikram Ullah [1], Noman Ali Buttar [1], Muhammad Ajmal [1] and Imran Ali Lakhiar [1]

[1] Institute of Agricultural Engineering, Jiangsu University, Zhenjiang 212013, China; abdulshabbir@uaf.edu.pk (A.S.); ikram2155@gmail.com (I.U.); noman_buttar@yahoo.com (N.A.B.); engr.ajmal37@gmail.com (M.A.); 5103160321@stmail.ujs.edu.cn (I.A.L.)

[2] Department of Irrigation and Drainage, University of Agriculture, Faisalabad 3800, Pakistan

* Correspondence: maohp@ujs.edu.cn; Tel.: +86-1351-169-5868

Received: 5 October 2020; Accepted: 26 October 2020; Published: 30 October 2020

Abstract: Root morphology and its components' behavior could show a considerable response under multiple water application points per plant to help the ultimate effect of fruit yield and fruit quality. In this study, a comparison of a single emitter per plant was made with two, three, and four emitters per plant under drip irrigation and two irrigation levels (full irrigation 100% and deficit irrigation 75% of crop evapotranspiration) to investigate their effects on physiological parameters, root, yield, and their associated components for potted cherry tomato under greenhouse conditions in Jiangsu-China. The experimental results showed that the plants cultivated in the spring-summer planting season showed significantly higher results than the fall-winter planting season due to low temperatures in the fall-winter planting season. However, the response root length, root average diameter, root dry mass, leaf area index, photosynthetic rate, transpiration rate, fruit unit fresh weight, the number of fruits, and pH were increased by multiple emitters per plant over a single emitter per plant, but total soluble solids decreased. Besides, a decreasing trend was observed by deficit irrigation for both planting seasons, and vice versa for the case for tomato total soluble solids. Due to an increase in measured parameters for multiple emitters per plant over a single emitter per plant, the yield, water use efficiency, and water use efficiency biomass significantly increased by 18.1%, 17.6%, and 15.1%, respectively. The deficit irrigation caused a decrease in the yield of 5% and an increase in water use efficiency and water use efficiency biomass of 21.4% and 22.9%, respectively. Two, three, and four emitters per plant had no significant effects, and the obtained results were similar. Considering the root morphology, yield, water use efficiency, water use efficiency biomass, and fruit geometry and quality, two emitters per plant with deficit irrigation are recommended for potted cherry tomato under greenhouse conditions. The explanation for the increased biomass production of the plant, yield, and water use efficiency is that two emitters per plant (increased emitter density) reduced drought stress to the roots, causing increased root morphology and leaf area index and finally promoting the plant's photosynthetic activity.

Keywords: root components; yield components; fruit quality; deficit irrigation; leaf area index; harvest index; photosynthetic rate; transpiration rate; water use efficiency; greenhouse

1. Introduction

Tomato (*Solanum Lycopersicum* L.) is one of the most produced, popular, and nutritious crops worldwide [1,2]. It provides a nutrient needed for human health [3] and contains antioxidants like lycopene that play a crucial role in preventing cancer and cardiovascular diseases [4–6]. Due to tomato fruit's nutritional and health benefits, tomato producers are much more interested in enhancing its quality and production [7,8].

The fast increasing population has forced an increase in agricultural production that may be attained by consuming more water, thus making water resources run out [9]. Water is one of the main factors influencing crop yield under water-scarce conditions. The crop yield depends on water consumption during the reproductive stage [10,11]. Reducing the quantity of water used for irrigation can increase urban and industrial water use [12]. Thus, the primary research objective concerning sustainable agricultural development and agricultural-ecological balance is to enhance water use efficiency (WUE) [13,14].

In comparison to other gravity-driven and pressure-operated irrigation systems, the drip irrigation system has been accepted as the highest water-saving method for the cultivation of tomato [15] and many other horticultural crops. However, attempts to look for different and more efficient approaches are still ongoing. Water pillow irrigation had been compared with drip irrigation and resulted in improved yield, fruit quality, and WUE than drip irrigation [16]. Water deficit irrigation, mainly for horticultural crops, has been studied worldwide for saving water and, hence, improving WUE [17–22].

Soil water has been found to improve nutrient accessibility, and nutrients have been found to improve root growth and the productivity of crops [23]. Soil water and nutrient uptake are greatly affected by the root morphology and distribution [24]. Root diameter, root surface area, root length density, and root weight density are the primary root morphological features that directly affect the entire root system and, indirectly, the above-ground plant components [25,26].

The quantity of soil water consumed by plant roots depends primarily on soil water availability, the morphological and physiological characteristics of the roots, etc. [27,28]. The plants uptake water from a wetting zone produced by drip irrigation, but the drip irrigation activates only a side of the root zone. The shape and size of the wetting zone produced by an emitter is a function of soil hydraulic conductivity, soil texture, soil structure, and emitter flow rate [29,30]. Despite these factors, the number of irrigation spots (emitter density) on the soil surface may also impact soil moisture distribution and, in turn, root growth, root distribution, and root water uptake. The morphology of roots is affected by the soil water supply [31]. Due to reduced soil moisture stress, moisture is more readily available to roots under multiple irrigation points. An expanded root system supports crops by reducing moisture stress [28]. A study showed that emitters' location could significantly affect the distribution of moisture and salt in the substrate [32]. The use of multiple emitters per plant reduced the leaching fraction compared with that resulting from one emitter, resulting in enhancing the amount and uniformity of substrate moisture in gerbera cultivated within pots [33].

There have been few studies on root–yield relationships for tomato under variable water and nitrogen rates [34–37]. However, there have been scarce studies on the influences of root–yield relationships under variable drip irrigation emitter densities (number of emitters per plant) under water deficit conditions for cherry tomato crop. Therefore, the study was conducted to compare the effects of drip irrigation emitter density with deficit irrigation on root and root components, plant physiological parameters, yield and yield components, fruit quality, and water use efficiency for potted cherry tomato grown under greenhouse conditions.

2. Materials and Methods

2.1. Experimental Site and Materials

The current study was conducted at a Venlo-Type greenhouse located in the School of Agricultural Engineering, Jiangsu University, Zhenjiang, Jiangsu-China (31°56′ N, 119°10′ E) from 13 March to 14 July 2019 (124 days) as the spring-summer planting season (SS) and from 2 September to 31 December 2019 (121 days) as the fall-winter planting season (FW). The greenhouse is situated in a humid sub-tropical monsoon climatic zone with an average annual rainfall of 1058.8 mm, a relative humidity of 76%, and a mean annual air temperature of 15.5 °C [38]. Firm rooted seedlings, 30 days old, of cherry tomato ("fenxiaoke xt–12020") were transplanted on 13 March and 2 September 2019 for the SS and FW, respectively, with plant density equal to 3.84 per m². The experimental soil was clay loam with the physical properties shown in Table 1.

Table 1. Soil physical properties.

Soil Property	Particle Size Distribution (%)			Bulk Density (g cm⁻³)	Field Capacity (cm³ cm⁻³)	Organic Matter (g kg⁻¹)
	Sand	Silt	Clay			
Range	34%	23%	43%	1.31	0.36	31.23

2.2. Experimental Design

The experimental design involved 2 factors factorial under a randomized complete block design (RCBD) with 4 replications for both seasons. Each replication consisted of five plants, of which three representative plants were used for data collection and then their averaged values were used for data analysis. The test influencing parameters were a single emitter (N1) and multiple emitters per plant (N2, N3, and N4: 2, 3, and 4 emitters per plant, respectively) under the drip irrigation system with two irrigation levels (full irrigation, W1; 100% of crop evapotranspiration (ETc); and deficit irrigation, W2; 75% of ETc). The treatment details are given in Table 2. Moreover, a single emitter (as in the case of N1) was placed in the north-west quadrant of the pot, 2 emitters (as in the case of N2) were placed in opposite quadrants as north-west and south-east quadrants, 3 emitters (as in case of N3) were placed as, one emitter was placed in the north-west quadrant, and the other 2 emitters were placed at an angle of 120° from each other. In the case of N4, 4 emitters, they were placed in all 4 quadrants of the pot, i.e., north-east, north-west, south-east, and south-west. The placement of the emitters was arbitrary and not dependent on sun orientation. All the emitters were equally spaced from each other in the case of multiple emitters per plant (N2, N3, and N4) and were placed at a 9 cm radius from the plant, approximately midway between the plant and pot boundary. For the growing practice, the plastic pots (40 cm in diameter and 40 cm in height) were filled with approximately 53 kg of air-dried soil, sieved through a 5 mm mesh sieve, up to a height of 35 cm so that these may contain the roots for tomato plants; as [36] found, for the surface drip irrigation system, the surface soil layer with 0 to 15 cm depth contained a tomato root length density up to 70%–75% of the total. The pots were placed alternatively with a plant-to-plant spacing of 80 cm and row-to-row spacing of 25 cm. All the pots were placed on 3 to 5 cm gravels mixed with sand to avoid waterlogging in the bottom soil.

Table 2. Treatment details.

Treatments	Emitter Density (Number of Emitters Per Plant), N	Irrigation Level (% of Crop Evapotranspiration), W
N1W1 (control)	1	100
N2W1	2	100
N3W1	3	100
N4W1	4	100
N1W2	1	75
N2W2	2	75
N3W2	3	75
N4W2	4	75

The control treatment (NIW1) pots were used to find the quantity of water being applied by the weighing method. To avoid soil evaporation, initially, the pots were covered with a plastic sheet after saturation. When the drainage stopped, the pots were reweighed and considered to be at field capacity by weight. The mass increase was ignored because of the minimal increment between two successive waterings. The irrigation was applied when the accumulative pan (20 cm in diameter) evaporation (Ep) reached 20 mm [39]. The Ep was calculated by accumulating the values determined by counting the cup every morning at eight o'clock. Adjustable flow rate emitters were used to attain the same flow rate per plant (emitter flow rate decreased as the emitter density increased, keeping the time of irrigation the same). The emitters were tested for flow uniformity before the experiment. The water delivered to all plants under each treatment was controlled by a water flow meter mounted at the control unit.

The fertigation was done as per local practices. Nitrogen fertilizer was delivered as urea (46% nitrogen). Triple superphosphate (46% P_2O_5) and muriate of potash (60% K_2O) were used as phosphorus and potassium sources, respectively. In addition, 40% of nitrogen and all phosphorus and potassium were applied and mixed into the soil in powdered forms at the start of the experiment. The remaining nitrogen was supplied as 30% in the first week of fruit emergence and 30% in the first week of fruit maturing. The transplanted seedlings were immediately irrigated with the same water volume (1 L) for better establishment and to ensure seedling growth. Besides, the plants were pruned once a week.

2.3. Sampling and Measurements

For each replication of the control (N1W1), the crop evapotranspiration between two consecutive irrigations was calculated using the water balance equation (weight-based, using a ±1 g weighing indicator) reported by [40] using $ET_{c(CR)} = I + (W_n - W_{n+1})$, where $ET_{c(CR)}$ is the crop evapotranspiration of the control, I is the amount of applied irrigation water (L), W_n and W_{n+1} are the pot weights before the nth and n + 1th irrigation (kg), respectively, and $ETc_{(CR)}$ is considered the standard amount of water (W_s) to be applied in control treatment for the next irrigation. The low irrigation treatment was irrigated according to the assigned percentage of the ETc of the control ($ET_{c(CR)} \approx W_s$). A data logger (Hobo; Onset Computer, Pocasset, MA, USA) was used to measure the greenhouse air temperature and relative humidity at intervals of 30 s at 1.5 m above ground level, located in the center of the greenhouse.

The crop physiological parameters like the photosynthetic rate (Pn), transpiration rate (Tr), stomatal conductance (gs), and leaf intercellular CO_2 concentration (Ci) of fifteen leaves, chosen randomly per treatment, were measured with a portable photosynthesis system (LI-6400. Li-Cor, Lincoln, Nebraska, USA) on the 50th, 70th, and 100th days after transplanting at 9:00–11:00 h of local time on sunny days during the SS and FW. Finally, the instantaneous water use efficiency (WUE$_{leaf}$) was calculated as the ratio of photosynthetic rates and transpiration rates [41].

Fruit morphological parameters were determined upon each fruit harvest. A digital vernier caliper was used to measure the fruit diameter and fruit height, and fresh fruit weight was measured by a precision electronic scale (0.0001 g). To investigate the quality parameters, three fruits per plant with similar size, maturity, and no external defects were picked from each treatment. The fruits were peeled and mixed thoroughly. The blended paste was then squeezed with a muslin cloth to collect juice as a homogenized representative sample for each treatment. A refractometer (ABBE, WYA-2S, Lakeland, FL, USA) was used to measure total soluble solids (TSS, ° Brix), and pH was measured by a pH meter (WTW-InolabLevel 3 Terminal, Weilheim, Almanya, Germany) at 25 °C.

At the end of the planting seasons, three leaves per plant, 7th, 8th, and 9th from the top, were taken for leaf area determination. For root morphology, the roots with soil were dipped into water for 30 min to soften them, and the soil and roots were then poured onto a 100 mm mesh sieve. The roots were washed very gently with tap water and picked up by using tweezers. The roots and leaves were scanned (Figure 1) using an Epson Perfection V700 photo flatbed scanner (Seiko Epson Corp, Nagano, Japan). Later on, the scanned roots and leaves were then analyzed using WinRhizo software (Regent Instruments Inc., Ste-Foy, QC, Canada). The root length, root average diameter, root surface area, root volume, and leaf area were attained from the software.

Figure 1. Samples of scanned images for leaves and roots at the end of fall-winter planting season.

At the end of the experiment, the total water applied and fruit yield were recorded for all treatments. To attain dry mass, the samples for each plant component (the roots, stem, leaves, and fruits) were oven-dried separately at a temperature of 70 °C till a constant weight was obtained, and finally, a precision electronic scale (0.0001 g) was used to obtain the weight of the dry mass.

Based on the above measurements, specific root length (root length/root dry mass, m g^{-1}) [42], root fineness (root length/root volume, cm cm^{-3}), root tissue density (root dry mass/root volume, g cm^{-3}) [43], leaf area (0.348(leaf length x leaf width) + 33.85, cm^2, $R^2 = 0.99$, $P < 0.001$), leaf area index (leaf area/surface area occupied by a plant, -) [44], specific leaf area (leaf area/leaf dry mass, $cm^2\ g^{-1}$), leaf area ratio (leaf area/leaf and stem dry mass, $cm^2\ g^{-1}$), harvest index (yield/yield and biomass, -) [45], water use efficiency (yield/irrigation water, kg m^{-3}), and water use efficiency biomass (yield and biomass/irrigation water, kg m^{-3}) were calculated.

2.4. Statistical Analyses

The general linear model (GLM) in SPSS 16.0 software (SPSS Inc., Chicago, USA) was used for the analysis of variance (ANOVA) between seasons (S), emitter density (N), and irrigation level (W). All factors were considered as fixed effects. To compare means, the least significant difference (LSD) test was employed at a probability level of 0.05 when F-values were significant.

3. Results

3.1. Greenhouse Climate

The observed greenhouse climatic data during the spring-summer planting season (SS) and fall-winter planting season (FW) in 2019 are shown in Figure 2. The air temperature (T) and relative humidity (RH) during the SS ranged from 14.4 to 31.0 °C and 51.1 to 89.5%, respectively, with average values equal to 23.7 °C and 70.1%, respectively, while during the FW, T, and RH, they ranged from 6.5 to 28.8 °C and 53.3 to 91.1%, respectively, with average values equal to 18.7 °C and 73.5%, respectively. It is worth noting that T decreased gradually while the RH showed an upward trend in the FW and similar SS trends. A similar phenomenon was described in the same greenhouse in 2018 [38].

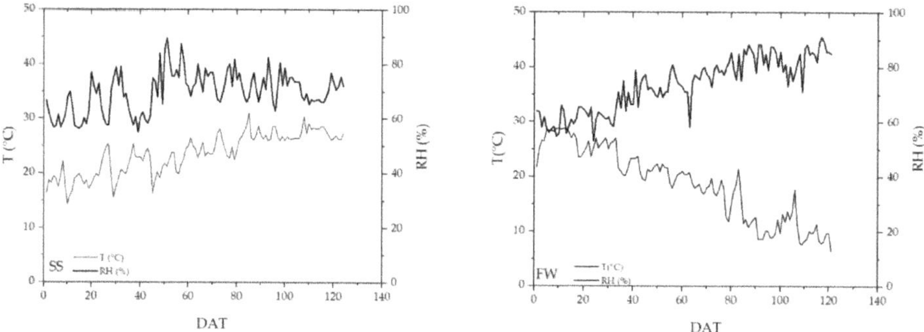

Figure 2. Variations in air temperature (T) and relative humidity (RH) observed for SS and FW. DAT, days after transplanting.

3.2. Root Morphology

Table 3 shows the root morphology and its associated components as affected by the emitter density, irrigation level, and planting season. The SS resulted in enhanced root length (21.7%), root average diameter (5.4%), root dry mass (15%), root surface area (28.2%), root volume (35.4%), and specific root length (6%), and reduced root fineness (6.3%) and root tissue density (15%) than the FW. All the factors had significant individual effects on all the roots measured and parameters calculated except for root average diameter (seasonal effect was significant only), root dry mass, and root fineness (seasonal effect was significant only). Overall, the effects of the emitter density and season were more significant than for irrigation level. All the root parameters increase with emitter density and decrease with deficit irrigation (W2) except for root fineness and root tissue density for which the effects were opposite.

3.3. Leaf Morphology and Yield

The effects of the emitter density, irrigation level, and planting season on leaf morphology, yield, and associated components are shown in Table 4. The SS performed better in terms of tomato plant leaf area (24.4%), leaf area index (24.4%), specific leaf area (13.2%), leaf area ratio (14%), yield (25.2%), harvest index (3.1%), and reduced (due to more irrigation water requirements in the SS than for FW) water use efficiency (WUE) (12.6%) and water use efficiency biomass (WUEB) (15.2%) than the FW. Harvest index was least affected by the treatments. All parameters increased with emitter density and decreased by W2 except WUE and WUEB, which increased with emitter density and W2 for both seasons. Compared to a single emitter per plant (N1), the increase in yield, harvest index, WUE (due to increased yield), and WUEB (due to increased total biomass) were 15.6%, 2.1%, 15%, and 12.7%, respectively, by two emitters per plant (N2). Multiple emitters per plant (N2, N3, and N4) resulted in similar values (with a nonsignificant difference) over N1. Compared to full irrigation (W1), the decrease in yield and harvest index were 5% and 2%, respectively, by W2. The increase in WUE and WUEB was 21.4% and 22.9%, respectively, for W2 over W1.

Table 3. Root morphology and its components as affected by the emitter density, irrigation level, and planting season.

Factors	Root Length	Root Average Diameter	Root Dry Mass	Root Surface Area	Root Volume	Specific Root Length	Root Fineness	Root Tissue Density
	(m)	(mm)	(g)	(m^2)	(cm^3)	(mg^{-1})	(cm cm^{-3})	(g cm^{-3})
Emitter density (N)								
N1	482.49 d	0.351 a	4.78 c	0.536 c	47.64 c	100.58 d	1048.6 a	0.105 a
N2	558.41 c	0.358 a	5.17 c	0.633 b	57.40 b	107.91 c	1008.0 a	0.094 b
N3	617.14 b	0.362 a	5.33 b	0.708 a	64.71 ab	115.54 b	991.3 a	0.086 bc
N4	667.83 a	0.362 a	5.39 a	0.766 a	70.31 a	123.47 a	987.7 a	0.081 c
Water Level (W)								
W1	617.13 a	0.361 a	5.27 a	0.707 a	64.58 a	116.51 a	992.9 a	0.086 b
W2	545.81 b	0.356 a	5.07 a	0.615 b	55.54 b	107.24 b	1024.9 a	0.096 a
Season (S)								
SS	638.41 a	0.368 a	5.53 a	0.743 a	69.05 a	115.11 a	955.6 b	0.086 b
FW	524.52 b	0.349 b	4.81 b	0.579 b	50.98 b	108.64 b	1062.2 a	0.097 a
ANOVA								
N	***	ns	***	***	***	***	ns	***
W	***	ns	*	***	**	***	ns	**
S	***	**	***	***	***	***	**	***
N * W	ns	ns	ns	ns	ns	ns	ns	ns
N * S	ns	ns	ns	ns	*	ns	ns	ns
W*S	ns	ns	ns	ns	ns	ns	ns	ns
N * W * S	ns	ns	ns	ns	ns	ns	ns	ns

N1, N2, N3, and N4; 1, 2, 3, and 4 emitters per plant, W1; full irrigation; W2; deficit irrigation, SS; spring-summer planting season, FW; fall-winter planting season. *; significant at $P < 0.05$, **; significant at $P < 0.01$, ***; significant at $P < 0.001$, ns; nonsignificant at $P > 0.05$. Values within the same columns that are accompanied by different letters vary significantly at $P < 0.05$.

3.4. Plant Physiological Parameters

The physiological parameters like photosynthetic rate (Pn), transpiration rate (Tr), stomatal conductance (gs), leaf intercellular CO_2 concentration (Ci), and instantaneous water use efficiency (WUE$_{leaf}$) under different treatments are given in Table 5. The SS had higher physiological parameter values (7%, 7.6%, 11.4%, 12.3%, and 25.4% higher Pn, Tr, gs, Ci, and WUE$_{leaf}$, respectively) than the FW due to higher temperatures in the SS. All parameters increased with emitter density and decreased with W2 except WUE$_{leaf}$ that increased with emitter density and also W2 for both seasons. The increase in Pn, Tr, gs, Ci, and WUE$_{leaf}$ for N2 over N1 was 6.2%, 6.4%, 10.8%, 12.9% and 17.6%, respectively. Multiple emitters per plant produced similar values with a nonsignificant difference over N1. The W2 caused a decrease in Pn, Tr, gs, and Ci by 7.2%, 10%, 13%, and 15.4%, respectively, and an increase in WUE$_{leaf}$ by 7.2%.

Table 4. Leaf morphology and tomato fruit yield as affected by the emitter density, irrigation level, and planting season.

Factors	Leaf Area	Leaf Area Index	Specific Leaf Area	Leaf Area Ratio	Yield	Harvest Index	WUE	WUEB
	(cm^2)	(-)	$(cm^2 \, g^{-1})$	$(cm^2 \, g^{-1})$	(kg plant^{-1})	(-)	(kg m^{-3})	(kg m^{-3})
Emitter density (N)								
N1	3672.7 b	2.92 b	151.52 c	79.01 c	1.86 b	0.661	41.04 b	62.23 b
N2	4541.9 a	3.62 a	176.15 b	92.16 b	2.15 a	0.675	47.20 a	70.11 a
N3	4844.0 a	3.86 a	184.20 ab	96.47 ab	2.20 a	0.676	48.46 a	71.87 a
N4	5104.7 a	4.06 a	192.94 a	100.75 a	2.24 a	0.676	49.17 a	72.89 a
Water Level (W)								
W1	4837.7 a	3.85 a	182.89 a	96.03 a	2.16 a	0.679	40.91 b	60.33 b
W2	4243.9 b	3.38 b	169.52 b	88.17 b	2.06 b	0.665	52.02 a	78.22 a
Season (S)								
SS	5034.7 a	4.01 a	187.08 a	98.10 a	2.35 a	0.682	43.33 b	63.56 b
FW	4047.0 b	3.22 b	165.33 b	86.10 b	1.88 b	0.662	49.60 a	74.99 a
ANOVA								
N	***	***	***	***	***	ns	***	***
W	**	**	**	**	**	ns	***	***
S	***	***	***	***	***	ns	***	***
N * W	ns	ns	ns	ns	ns	ns	ns	ns
N * S	ns	ns	ns	ns	ns	ns	ns	ns
W * S	ns	ns	ns	ns	ns	ns	ns	ns
N * W*S	ns	ns	ns	ns	ns	ns	ns	ns

N1, N2, N3, and N4; 1, 2, 3, and 4 emitters per plant, W1; full irrigation; W2; deficit irrigation, SS; spring-summer planting season, FW; fall-winter planting season, WUE; water use efficiency, WUEB; water use efficiency biomass. *; significant at $P < 0.05$, **; significant at $P < 0.01$, ***; significant at $P < 0.001$, ns; nonsignificant at $P > 0.05$. Values within the same columns that are accompanied by different letters vary significantly at $P < 0.05$.

Table 5. Physiological parameters as affected by the emitter density, irrigation level, and planting season.

Factors	Photosynthetic rate (Pn)	Transpiration Rate (Tr)	Stomatal Conductance (gs)	Leaf Intercellular CO$_2$ Concentration (Ci)	Instantaneous Water Use Efficiency (WUE$_{leaf}$ = Pn/Tr)
	(μmol (CO$_2$) m^2 s^{-1})	(mmol m^2 s^{-1})	(mol (H$_2$O) m^{-2} s^{-1})	(μmol mol^{-1})	(μmol mol^{-1})
Emitter density (N)					
N1	13.35 c	11.38 b	0.42 c	270.27 c	1.17 b
N2	15.07 b	12.11 a	0.49 b	299.34 b	1.25 a
N3	15.47 ab	12.34 a	0.52 a	310.49 ab	1.25 a
N4	15.81 a	12.44 a	0.53 a	320.63 a	1.27 a
Water Level (W)					
W1	15.44 a	12.93 a	0.51 a	318.44 a	1.19 b
W2	14.41 b	11.20 b	0.47 b	281.93 b	1.28 a
Season (S)					
SS	16.61 a	12.76 a	0.51 a	311.19 a	1.30 a
FW	13.24 b	11.37 b	0.47 b	289.17 b	1.17 b
ANOVA					
N	***	**	***	***	**
W	***	***	***	***	***
S	***	***	***	***	***
N * W	ns	ns	ns	ns	ns
N * S	ns	ns	ns	ns	ns
W * S	ns	ns	ns	ns	ns
N * W * S	ns	ns	ns	ns	ns

N1, N2, N3, and N4; 1, 2, 3, and 4 emitters per plant, W1; full irrigation; W2; deficit irrigation, SS; spring-summer planting season, FW; fall-winter planting season. The values are at 70 DAT (days after transplant). *; significant at $P < 0.05$, **; significant at $P < 0.01$, ***; significant at $P < 0.001$, ns; nonsignificant at $P > 0.05$. Values within the same columns that are accompanied by different letters vary significantly at $P < 0.05$.

3.5. Fruit Morphology and Fruit Quality

Table 6 shows the effects of the emitter density, irrigation level, and planting season on fruit morphology like unit fresh weight, number of fruits per plant, fruit diameter, fruit height, and fruit quality like total soluble solids (TSS) and pH for both the SS and FW. The SS produced increased fruit morphology (1%, 4.2%, 7.8%, and 16.1%, increase in unit fresh weight, number of fruits, fruit diameter, and fruit height, respectively) and decreased TSS (0.6%, nonsignificant) and pH (2.6% nonsignificant) over the FW. All the parameters increased with emitter density and decreased with W2, but the fruit TSS was vice versa. Compared to N1, N2 resulted in enhanced fruit morphology (unit fresh weight, number of fruits, fruit diameter, and fruit height by 1.5%, 3.5%, 4.3%, and 10.7%, respectively) and reduced TSS (3%) and increased pH (0.2%). Similar values were observed for multiple emitters per plant with no significant difference over N1. The W2 resulted in reduced fruit morphology (unit fresh weight, number of fruits, fruit diameter, and fruit height by 2.3%, 2.4%, 2.6%, and 3%, respectively). W2 increased TSS (11.7%) and reduced pH (1.3%).

Table 6. Fruit morphology and fruit quality as affected by the emitter density, irrigation level, and planting season.

Factors	Unit Fresh Weight	Number of Fruits Per Plant	Fruit Diameter	Fruit Height	Total Soluble Solids	pH
	(g)	(-)	(mm)	(mm)	(^{0}Brix)	(-)
Emitter density (N)						
N1	29.72 b	62.44 b	28.31 b	30.56	4.22 a	4.13
N2	30.99 ab	69.14 a	29.29 ab	31.00	4.10 ab	4.14
N3	31.62 a	69.50 a	29.96 a	31.72	4.05 ab	4.14
N4	31.92 a	69.88 a	30.31 a	31.91	4.02 b	4.15
Water Level (W)						
W1	31.45 a	68.51 a	29.90 a	31.66	3.84 b	4.16
W2	30.67 a	66.97 a	29.03 a	30.94	4.35 a	4.11
Season (S)						
SS	32.23 a	72.79 a	30.08 a	31.45	4.08 a	4.08
FW	29.90 b	62.69 b	28.86 b	31.15	4.11 a	4.19
ANOVA						
N	*	***	*	ns	ns	ns
W	ns	*	ns	ns	***	ns
S	***	***	**	ns	ns	ns
N * W	ns	ns	ns	ns	ns	ns
N * S	ns	ns	ns	ns	ns	ns
W * S	ns	ns	ns	ns	ns	ns
N * W * S	ns	ns	ns	ns	ns	ns

N1, N2, N3, and N4; 1, 2, 3, and 4 emitters per plant, W1; full irrigation; W2; deficit irrigation, SS; spring-summer planting season, FW; fall-winter planting season. *; significant at $P < 0.05$, **; significant at $P < 0.01$, ***; significant at $P < 0.001$, ns; nonsignificant at $P > 0.05$. Values within the same columns that are accompanied by different letters vary significantly at $P < 0.05$.

4. Discussion

The spring-summer planting season (SS) performed better than the fall-winter planting season (FW). Due to very low temperatures in the FW, 6.5 °C at the later stage in the month of December, it had a negative impact on the maximum fruit growth rate [46]. Several studies have shown that the tomato crop's optimum temperatures ranged between 20 and 24 °C to ensure further growth, flowering, and fruit maturation, and 12 and 36 °C were the temperature limits for growth [47]. Further, there were no significant effects on response parameters for two, three, and four emitters per plant, and the findings obtained were identical for both seasons.

Plant roots are the primary organ that absorbs water and nutrients, so their size and distribution in the soil are closely linked to their capacity for water and nutrient absorption [48]. Few earlier studies have found that water conditions in excess or deficit can change maize root zones' size and distribution, thus inhibiting root development [49]. The primary parameters that affect water and nutrient absorption are root diameter and length [50]. The present study indicated that root length, root average diameter, and root dry mass (Table 3) increased with emitter density and decreased with deficit irrigation (W2). The increase was attributed to the more uniform moisture distribution under increased emitter density that facilitated the root absorbing water easily (as the distance between water source and roots decreased) and reduced the drought stress to roots. This agrees with the previous study, which stated that root and shoot growth in weeping fig was significantly affected by both the irrigation water and several emitters per plant [51]. Furthermore, drought stress can prevent root growth and plays a vital role in the morphology of the maize crop root [52]. The increased root length with increased emitter density (Table 3) is attributed to the increased specific root length [42], increased root dry mass, and decreased root tissue density [43]. The effect of W2 is the opposite, which is in agreement with [53]. Besides, low-density tissues allow a fast relative growth rate and quick resource acquisition, as the plant can rapidly expand the leaf, stem, or root system with low investment in the dry matter [54–56]. In this study, in line with [57], the root fineness decreased due to the increased root dry mass with emitter density, and vice versa for W2.

The main physiological mechanism of plants is photosynthesis, which can provide 90% of plant biomass [58]. The primary explanation for most biomass and crop production differences is leaf photosynthesis [59]. In this study, the physiological parameters (Table 5) increased, due to an increase in leaf area index (Table 4), with emitter density and irrigation levels for both seasons (Table 5). A drop in intercellular carbon dioxide concentration can be attributed partly to metabolic elements due to the decline in photosynthesis [60] and partly due to the stomata's closure. The study results are in accordance with [22,61].

The leaf components, yield, and yield components (Table 4) increase with emitter density and decrease with W2. Previous studies have shown that, in line with the present findings, deficit irrigation reduces vegetative growth and fruit yield [62–65]. The explanation for this was that considerably decreased plant photosynthesis reduced the volume and energy of metabolites needed under water-stressed conditions for proper plant growth [62,66]. In a typical subtropical climate, the 80% crop evapotranspiration (ETc) irrigation regime induced a slight decrease in plant growth [67]. In the present study, in accordance with the above, the 75% ETc irrigation regime reduced yield by 4.6% only. The increase in yield (18.1% higher for multiple emitters per plant over a single emitter per plant (N1), Table 4) is due to an increase in mean unit fresh fruit weight and the number of fruits per plant (6% and 11.3%, respectively, higher for multiple emitters per plant over N1, Table 6). The increase in fruit unit fresh weight is due to increased fruit size (fruit diameter and fruit height, Table 6). As is known, plant vigor has been closely connected to the root system that supplies the shoot with water and nutrients. The reason is that the pattern of root growth and shoot development is strongly correlated [68], and both increase with emitter density in this study, which may be considered the key factor promoting increased yield (Table 4). The harvest index is increasing because the increase in fruit weight is higher as compared to the biomass weight with an increasing emitter density and vice versa for irrigation levels for both seasons. The harvest index decreases due to stress [69], which is consistent with these research results. The reported harvest index for cherry tomato is slightly more than [70], due to different growing conditions. The most critical indicator used to measure agricultural output is water use efficiency (WUE) [71,72]. The WUE and water use efficiency biomass (WUEB) (Table 4) increased with W2 because the numerator increased more than the denominator's reduction. According to [73], initially reducing irrigation water levels in tomatoes increases the water use efficiency, and reducing water level reduces water use efficiency.

Under deficit conditions, tomatoes' fruit quality can be increased [74]. The total soluble solids decreased with emitter density and irrigation level, and pH response was vice versa (Table 6). This is due to increased moisture availability to plant roots under increasing emitter density and irrigation level, ultimately decreasing water stress in soil. The fact that the accumulation of water in the fruit allows the dilution of fruit ingredients may explain this effect [75]. The results are in accordance with [65,76,77].

5. Conclusions

The spring-summer planting season performed better than the fall-winter planting season because of low temperatures in the fall-winter planting season. All the root and root components, leaf and leaf components, yield and yield components, plant physiology, and fruits morphology and pH showed an increasing trend with an increasing emitter density and decreasing trend with deficit irrigation except water use efficiency, water use efficiency biomass, instantaneous water use efficiency (increase with emitter density and deficit irrigation), root fineness, root tissue density, and total soluble solids (decrease with emitter density and increase with deficit irrigation). Overall, it was seen that two, three, and four emitters per plant had no significant effects, and the obtained results were similar under full and deficit irrigation. Bearing in mind the trade-off among plant fruit quality, fruit yield, and water use efficiency, the two emitters per plant with deficit irrigation was found as the best combination for the emitter density and irrigation level management approach for cherry tomato grown under greenhouse conditions within pots. It is recommended that the emitter placement around the plant (in this study at 9 cm radius) should be varied to study its effects on root development that will ultimately affect plant yield and water use efficiency.

Author Contributions: Investigation, A.S., H.M. and I.U.; funding acquisition, H.M.; writing—original draft, A.S.; writing—review and editing, A.S., N.A.B., M.A. and I.A.L. All authors have read and agreed to the published version of the manuscript.

Funding: This work was supported by Chinese Academy of Agricultural Sciences (CARS-23-C03).

References

1. Padayachee, A.; Day, L.; Howell, K.; Gidley, M. Complexity and health functionality of plant cell wall fibers from fruits and vegetables. *Crit. Rev. Food Sci. Nutr.* **2017**, *57*, 59–81. [CrossRef]
2. Sturm, R.; An, R. Obesity and economic environments. *CA A Cancer J. Clin.* **2014**, *64*, 337–350. [CrossRef] [PubMed]
3. Ilahy, R.; Piro, G.; Tlili, I.; Riahi, A.; Sihem, R.; Ouerghi, I.; Hdider, C.; Lenucci, M.S. Fractionate analysis of the phytochemical composition and antioxidant activities in advanced breeding lines of high-lycopene tomatoes. *Food Funct.* **2016**, *7*, 574–583. [CrossRef]
4. Heber, D. Colorful Cancer Prevention: A-Carotene, Lycopene, and Lung Cancer. *Am. J. Clin. Nutr.* **2000**, *72*, 901–902. [CrossRef]
5. Toor, R.K.; Savage, G.P. Antioxidant activity in different fractions of tomatoes. *Food Res. Int.* **2005**, *38*, 487–494. [CrossRef]
6. Zhang, Y.-J.; Gan, R.-Y.; Li, S.; Zhou, Y.; Li, A.-N.; Xu, D.-P.; Li, H.-B. Antioxidant phytochemicals for the prevention and treatment of chronic diseases. *Molecules* **2015**, *20*, 21138–21156. [CrossRef] [PubMed]
7. Gruda, N. Impact of environmental factors on product quality of greenhouse vegetables for fresh consumption. *Crit. Rev. Plant. Sci.* **2005**, *24*, 227–247. [CrossRef]
8. Flores, F.B.; Sanchez-Bel, P.; Estan, M.T.; Martinez-Rodriguez, M.M.; Moyano, E.; Morales, B.; Campos, J.F.; Garcia-Abellán, J.O.; Egea, M.I.; Fernández-Garcia, N. The effectiveness of grafting to improve tomato fruit quality. *Sci. Hortic.* **2010**, *125*, 211–217. [CrossRef]
9. Giuliani, M.M.; Nardella, E.; Gatta, G.; De Caro, A.; Quitadamo, M. Processing tomato cultivated under water deficit conditions: The effect of azoxystrobin. *III Int. Symp. Tomato Dis.* **2010**, *914*, 287–294. [CrossRef]
10. Merah, O. Potential importance of water status traits for durum wheat improvement under mediterranean conditions. *J. Agric. Sci.* **2001**, *137*, 139–145. [CrossRef]
11. Kato, Y.; Kamoshita, A.; Yamagishi, J. Preflowering abortion reduces spikelet number in upland rice (oryza sativa l.) under water stress. *Crop. Sci.* **2008**, *48*, 2389–2395. [CrossRef]
12. Naseri, A.; Rezania, A.R.; Albaji, M. Investigation of soil quality for different irrigation systems in lali plain, iran. *J. Food Agric. Environ.* **2009**, *7*, 955–960.

13. Beltrano, J.; Ronco, M.; Montaldi, E. Drought stress syndrome in wheat is provoked by ethylene evolution imbalance and reversed by rewatering, aminoethoxyvinylglycine, or sodium benzoate. *J. Plant. Growth Regul.* **1999**, *18*, 59–64. [CrossRef] [PubMed]
14. Shabbir, A.; Arshad, M.; Bakhsh, A.; Usman, M.; Shakoor, A.; Ahmad, I.; Ahmad, A. Apparent and real water productivity for cotton-wheat zone of punjab, pakistan. *Pak. J. Agri. Sci.* **2012**, *49*, 357–363.
15. Zhai, Y.; Shao, X.; Xing, W.; Wang, Y.; Hung, T.; Xu, H. Effects of drip irrigation regimes on tomato fruit yield and water use efficiency. *J. Food Agric. Environ.* **2010**, *8*, 709–713.
16. Gerçek, S.; Demirkaya, M.; Işik, D. Water pillow irrigation versus drip irrigation with regard to growth and yield of tomato grown under greenhouse conditions in a semi-arid region. *Agric. Water Manag.* **2017**, *180*, 172–177. [CrossRef]
17. Fereres, E.; Soriano, M.A. Deficit irrigation for reducing agricultural water use. *J. Exp. Bot.* **2006**, *58*, 147–159. [CrossRef]
18. Nagaz, K.; Masmoudi, M.; Ben Mechlia, N. Yield response of drip irrigated onion under full and deficit irrigation with saline water in arid regions of tunisia. *ISRN Agron.* **2012**, *212*, 562315. [CrossRef]
19. Sezen, S.M.; Yazar, A.; Daşgan, Y.; Yucel, S.; Akyıldız, A.; Tekin, S.; Akhoundnejad, Y. Evaluation of crop water stress index (cwsi) for red pepper with drip and furrow irrigation under varying irrigation regimes. *Agric. Water Manag.* **2014**, *143*, 59–70. [CrossRef]
20. Çolak, Y.B.; Yazar, A.; Sesveren, S.; Colak, I. Evaluation of yield and leaf water potantial (lwp) for eggplant under varying irrigation regimes using surface and subsurface drip systems. *Sci. Hortic.* **2017**, *219*, 10–21. [CrossRef]
21. Lu, J.; Shao, G.; Cui, J.; Wang, X.; Keabetswe, L. Yield, fruit quality and water use efficiency of tomato for processing under regulated deficit irrigation: A meta-analysis. *Agric. Water Manag.* **2019**, *222*, 301–312. [CrossRef]
22. Rasool, G.; Guo, X.; Wang, Z.; Chen, S.; Ullah, I. The interactive responses of fertigation levels under buried straw layer on growth, physiological traits and fruit yield in tomato plant. *J. Plant. Interact.* **2019**, *14*, 552–563. [CrossRef]
23. Besharat, S.; Nazemi, A.H.; Sadraddini, A.A. Parametric modeling of root length density and root water uptake in unsaturated soil. *Turk. J. Agric. For.* **2010**, *34*, 439–449.
24. Li, Q.; Dong, B.; Qiao, Y.; Liu, M.; Zhang, J. Root growth, available soil water, and water-use efficiency of winter wheat under different irrigation regimes applied at different growth stages in north china. *Agric. Water Manag.* **2010**, *97*, 1676–1682. [CrossRef]
25. Xia, J.; Liu, M.-Y.; Jia, S.-F. Water security problem in north china: Research and perspective. *Pedospher* **2005**, *15*, 563–575.
26. Fageria, N. Influence of dry matter and length of roots on growth of five field crops at varying soil zinc and copper levels. *J. Plant. Nutr.* **2005**, *27*, 1517–1523. [CrossRef]
27. Li, X.; Šimůnek, J.; Shi, H.; Yan, J.; Peng, Z.; Gong, X. Spatial distribution of soil water, soil temperature, and plant roots in a drip-irrigated intercropping field with plastic mulch. *Eur. J. Agron.* **2017**, *83*, 47–56. [CrossRef]
28. Wang, C.; Liu, W.; Li, Q.; Ma, D.; Lu, H.; Feng, W.; Xie, Y.; Zhu, Y.; Guo, T. Effects of different irrigation and nitrogen regimes on root growth and its correlation with above-ground plant parts in high-yielding wheat under field conditions. *Field Crop. Res.* **2014**, *165*, 138–149. [CrossRef]
29. Bar-Yosef, B. Advances in fertigation. In *Advances in Agronomy*; Elsevier: Amsterdam, The Netherlands, 1999; Volume 65, pp. 1–77.
30. Palomo, M.; Moreno, F.; Fernández, J.; Dıaz-Espejo, A.; Girón, I. Determining water consumption in olive orchards using the water balance approach. *Agric. Water Manag.* **2002**, *55*, 15–35. [CrossRef]
31. Chu, G.; Chen, T.; Wang, Z.; Yang, J.; Zhang, J. Reprint of "morphological and physiological traits of roots and their relationships with water productivity in water-saving and drought-resistant rice". *Field Crop. Res.* **2014**, *165*, 36–48. [CrossRef]
32. Ondrasek, G.; Romic, D.; Romic, M.; Tomic, F.; Mustac, I. Salt distribution in peat substrate grown with melon (cucumis melo l.). *Int. Symp. Grow. Media* **2005**, *779*, 307–312. [CrossRef]
33. Valdés, R.; Miralles, J.; Ochoa, J.; Bañón, S.; Sánchez-Blanco, M.J. The number of emitters alters salt distribution and root growth in potted gerbera. *HortScience* **2014**, *49*, 160–165. [CrossRef]

34. Qiu, R.; Du, T.; Kang, S. Root length density distribution and associated soil water dynamics for tomato plants under furrow irrigation in a solar greenhouse. *J. Arid Land* **2017**, *9*, 637–650. [CrossRef]

35. Wang, X.; Yun, J.; Shi, P.; Li, Z.; Li, P.; Xing, Y. Root growth, fruit yield and water use efficiency of greenhouse grown tomato under different irrigation regimes and nitrogen levels. *J. Plant. Growth Regul.* **2019**, *38*, 400–415. [CrossRef]

36. Zotarelli, L.; Scholberg, J.M.; Dukes, M.D.; Muñoz-Carpena, R.; Icerman, J. Tomato yield, biomass accumulation, root distribution and irrigation water use efficiency on a sandy soil, as affected by nitrogen rate and irrigation scheduling. *Agric. Water Manag.* **2009**, *96*, 23–34. [CrossRef]

37. Wang, X.; Xing, Y. Evaluation of the effects of irrigation and fertilization on tomato fruit yield and quality: A principal component analysis. *Sci. Rep.* **2017**, *7*, 350. [CrossRef]

38. Huang, S.; Yan, H.; Zhang, C.; Wang, G.; Acquah, S.J.; Yu, J.; Li, L.; Ma, J.; Darko, R.O. Modeling evapotranspiration for cucumber plants based on the shuttleworth-wallace model in a venlo-type greenhouse. *Agric. Water Manag.* **2020**, *228*, 105861. [CrossRef]

39. Hao, L.; Duan, A.-W.; Li, F.-S.; Sun, J.-S.; Wang, Y.-C.; Sun, C.-T. Drip irrigation scheduling for tomato grown in solar greenhouse based on pan evaporation in north china plain. *J. Integr. Agric.* **2013**, *12*, 520–531.

40. Kurunç, A.; Ünlükara, A. Growth, yield, and water use of okra (abelmoschus esculentus) and eggplant (solanum melongena) as influenced by rooting volume. *N. Z. J. Crop. Hortic. Sci.* **2009**, *37*, 201–210. [CrossRef]

41. Wang, W.-H.; Chen, J.; Liu, T.-W.; Chen, J.; Han, A.-D.; Simon, M.; Dong, X.-J.; He, J.-X.; Zheng, H.-L. Regulation of the calcium-sensing receptor in both stomatal movement and photosynthetic electron transport is crucial for water use efficiency and drought tolerance in arabidopsis. *J. Exp. Bot.* **2014**, *65*, 223–234. [CrossRef]

42. Oikeh, S.; Kling, J.; Horst, W.; Chude, V.; Carsky, R. Growth and distribution of maize roots under nitrogen fertilization in plinthite soil. *Field Crop. Res.* **1999**, *62*, 1–13. [CrossRef]

43. Ryser, P.; Lambers, H. Root and leaf attributes accounting for the performance of fast-and slow-growing grasses at different nutrient supply. *Plant and Soil* **1995**, *170*, 251–265. [CrossRef]

44. Donald, C.; Hamblin, J. The biological yield and harvest index of cereals as agronomic and plant breeding criteria. In *Advances in Agronomy*; Elsevier: Amsterdam, The Netherlands, 1976; Volume 28, pp. 361–405.

45. Gur, A.; Osorio, S.; Fridman, E.; Fernie, A.R.; Zamir, D. Hi2-1, a qtl which improves harvest index, earliness and alters metabolite accumulation of processing tomatoes. *Theor. Appl. Genet.* **2010**, *121*, 1587–1599. [CrossRef] [PubMed]

46. Wu, Y.; Yan, S.; Fan, J.; Zhang, F.; Xiang, Y.; Zheng, J.; Guo, J. Responses of growth, fruit yield, quality and water productivity of greenhouse tomato to deficit drip irrigation. *Sci. Hortic.* **2021**, *275*, 109710. [CrossRef]

47. Rosales, M.A. Producción y Calidad Nutricional en Frutos de Tomate Cherry Cultivados en dos Invernaderos Mediterráneos Experimentales: Respuestas Metabólicas y Fisiológicas. Ph.D. Thesis, Universidad de Granada, Granada, Spain, 2008.

48. Forde, B.; Lorenzo, H. The nutritional control of root development. *Plant. Soil* **2001**, *232*, 51–68. [CrossRef]

49. Liu, C.; Jin, S.; Zhou, L.; Jia, Y.; Li, F.; Xiong, Y.; Li, X. Effects of plastic film mulch and tillage on maize productivity and soil parameters. *Eur. J. Agron.* **2009**, *31*, 241–249. [CrossRef]

50. Ren, X.; Jia, Z.; Chen, X. Rainfall concentration for increasing corn production under semiarid climate. *Agric. Water Manag.* **2008**, *95*, 1293–1302. [CrossRef]

51. Valdés, R.; Ochoa, J.; Sánchez-Blanco, M.; Franco, J.; Bañón, S. Irrigation volume and the number of emitters per pot affect root growth and saline ion contents in weeping fig. *Agric. Agric. Sci. Procedia* **2015**, *4*, 356–364. [CrossRef]

52. Guo, X.; Kang, S.; Suo, L. Effects of regulated deficit irrigation on root growth in maize. *Irrig. Drain.* **2001**, *20*, 25–27.

53. Jha, S.K.; Gao, Y.; Liu, H.; Huang, Z.; Wang, G.; Liang, Y.; Duan, A. Root development and water uptake in winter wheat under different irrigation methods and scheduling for north china. *Agric. Water Manag.* **2017**, *182*, 139–150. [CrossRef]

54. Garnier, E.; Salager, J.-L.; Laurent, G.; Sonié, L. Relationships between photosynthesis, nitrogen and leaf structure in 14 grass species and their dependence on the basis of expression. *New Phytol.* **1999**, *143*, 119–129. [CrossRef]

55. Wahl, S.; Ryser, P. Root tissue structure is linked to ecological strategies of grasses. *New Phytol.* **2000**, *148*, 459–471. [CrossRef]

56. Hummel, I.; Vile, D.; Violle, C.; Devaux, J.; Ricci, B.; Blanchard, A.; Garnier, É.; Roumet, C. Relating root structure and anatomy to whole-plant functioning in 14 herbaceous mediterranean species. *New Phytol.* **2007**, *173*, 313–321. [CrossRef]

57. Sorgonà, A.; Abenavoli, M.R.; Cacco, G.; Gelsomino, A. Growth of tomato and zucchini seedlings in orange waste compost media: Ph and implication of dosage. *Compos. Sci. Util.* **2011**, *19*, 189–196. [CrossRef]

58. Wang, X.; Wang, W.; Huang, J.; Peng, S.; Xiong, D. Diffusional conductance to co2 is the key limitation to photosynthesis in salt-stressed leaves of rice (oryza sativa). *Physiol. Plant.* **2018**, *163*, 45–58. [CrossRef] [PubMed]

59. Takai, T.; Kondo, M.; Yano, M.; Yamamoto, T. A quantitative trait locus for chlorophyll content and its association with leaf photosynthesis in rice. *Rice* **2010**, *3*, 172–180. [CrossRef]

60. Lawlor, D.W.; Cornic, G. Photosynthetic carbon assimilation and associated metabolism in relation to water deficits in higher plants. *Plant Cell Environ.* **2002**, *25*, 275–294. [CrossRef] [PubMed]

61. Hanping, M.; Ullah, I.; Jiheng, N.; Javed, Q.; Azeem, A. Estimating tomato water consumption by sap flow measurement in response to water stress under greenhouse conditions. *J. Plant. Interact.* **2017**, *12*, 402–413. [CrossRef]

62. Agbna, G.H.; Dongli, S.; Zhipeng, L.; Elshaikh, N.A.; Guangcheng, S.; Timm, L.C. Effects of deficit irrigation and biochar addition on the growth, yield, and quality of tomato. *Sci. Hortic.* **2017**, *222*, 90–101. [CrossRef]

63. Kuşçu, H.; Turhan, A.; Demir, A.O. The response of processing tomato to deficit irrigation at various phenological stages in a sub-humid environment. *Agric. Water Manag.* **2014**, *133*, 92–103. [CrossRef]

64. Zhang, H.; Xiong, Y.; Huang, G.; Xu, X.; Huang, Q. Effects of water stress on processing tomatoes yield, quality and water use efficiency with plastic mulched drip irrigation in sandy soil of the hetao irrigation district. *Agric. Water Manag.* **2017**, *179*, 205–214. [CrossRef]

65. Hooshmand, M.; Albaji, M.; Ansari, N.A.Z. The effect of deficit irrigation on yield and yield components of greenhouse tomato (solanum lycopersicum) in hydroponic culture in ahvaz region, iran. *Sci. Hortic.* **2019**, *254*, 84–90. [CrossRef]

66. Kulkarni, M.; Phalke, S. Evaluating variability of root size system and its constitutive traits in hot pepper (capsicum annum l.) under water stress. *Sci. Hortic.* **2009**, *120*, 159–166. [CrossRef]

67. Mohawesh, O. Field evaluation of deficit irrigation effects on tomato growth performance, water-use efficiency and control of parasitic nematode infection. *South. Afr. J. Plant. Soil* **2016**, *33*, 125–132. [CrossRef]

68. Aguirre, L.; Johnson, D.A. Root morphological development in relation to shoot growth in seedlings of four range grasses. *Rangel. Ecol. Manag. J. Range Manag. Arch.* **1991**, *44*, 341–346. [CrossRef]

69. Flenet, F.; Bouniols, A.; Saraiva, C. Sunflower response to a range of soil water contents. *Eur. J. Agron.* **1996**, *5*, 161–167. [CrossRef]

70. Moccia, S.; Chiesa, A.; Oberti, A.; Tittonell, P. Yield and quality of sequentially grown cherry tomato and lettuce under long-term conventional, low-input and organic soil management systems. *Eur. J. Hortic. Sci.* **2006**, *71*, 183–191.

71. Chen, G.; Cao, H.; Liang, J.; Ma, W.; Guo, L.; Zhang, S.; Jiang, R.; Zhang, H.; Goulding, K.W.; Zhang, F. Factors affecting nitrogen use efficiency and grain yield of summer maize on smallholder farms in the north china plain. *Sustainability* **2018**, *10*, 363. [CrossRef]

72. Giuliani, M.; Nardella, E.; Gagliardi, A.; Gatta, G. Deficit irrigation and partial root-zone drying techniques in processing tomato cultivated under mediterranean climate conditions. *Sustainability* **2017**, *9*, 2197. [CrossRef]

73. Sharayeie, P.; Sobhani, A.; Rahimian, M. Effect of different levels of irrigation water and potassium fertilizer on water use efficiency and quality of tomato fruit of petvarli ch. *J. Agric. Eng. Res.* **2006**, *27*, 75–86.

74. Wang, C.; Gu, F.; Chen, J.; Yang, H.; Jiang, J.; Du, T.; Zhang, J. Assessing the response of yield and comprehensive fruit quality of tomato grown in greenhouse to deficit irrigation and nitrogen application strategies. *Agric. Water Manag.* **2015**, *161*, 9–19. [CrossRef]

75. Du, Y.-D.; Cao, H.-X.; Liu, S.-Q.; Gi, X.-B.; Cao, Y.-X. Response of yield, quality, water and nitrogen use efficiency of tomato to different levels of water and nitrogen under drip irrigation in northwestern china. *J. Integr. Agric.* **2017**, *16*, 1153–1161. [CrossRef]

76. Machado, R.M.; Maria do Rosàrio, G.O. Tomato root distribution, yield and fruit quality under different subsurface drip irrigation regimes and depths. *Irrig. Sci.* **2005**, *24*, 15–24. [CrossRef]
77. Zomorodi, S.; Emami, A. Study the effects of deficit irrigation on the yield, quality and storability of tomato. *J. Agric. Eng. Res.* **2006**, *27*, 19–30.

Publisher's Note: MDPI stays neutral with regard to jurisdictional claims in published maps and institutional affiliations.

Article

Improving Water Use Efficiency by Optimizing the Root Distribution Patterns under Varying Drip Emitter Density and Drought Stress for Cherry Tomato

Abdul Shabbir [1,2], Hanping Mao [1,*], Ikram Ullah [1], Noman Ali Buttar [1], Muhammad Ajmal [1] and Kashif Ali Solangi [1]

1   Institute of Agricultural Engineering, Jiangsu University, Zhenjiang 212013, China; abdulshabbir@uaf.edu.pk (A.S.); ikram2155@gmail.com (I.U.); noman_buttar@yahoo.com (N.A.B.); engr.ajmal37@gmail.com (M.A.); 5103180312@stmail.ujs.edu.cn (K.A.S.)
2   Department of Irrigation and Drainage, University of Agriculture, Faisalabad 38000, Pakistan
*   Correspondence: maohp@ujs.edu.cn; Tel.: +86-1351-169-5868

Abstract: The spatial distribution of root systems in the soil has major impacts on soil water and nutrient uptake and ultimately crop yield. This research aimed to optimize the root distribution patterns, growth, and yield of cherry tomato by using a number of emitters per plant. A randomized complete block design technique was adopted by selecting eight treatments with two irrigation regimes and four levels of emitters under greenhouse conditions. The experiment results showed that the root distribution extended over the entire pot horizontally and shifted vertically upwards with increased emitter density. The deficit irrigation resulted in reduced horizontal root extension and shifted the root concentrations deeper. Notably, tomato plants with two emitters per plant and deficit irrigation treatment showed an optimal root distribution compared to the other treatments, showing wider and deeper dispersion measurements and higher root length density and root weight density through the soil with the highest benefit–cost ratio (1.3 and 1.1 cm cm^{-3}, 89.8 and 77.7 µg cm^{-3}, and 4.20 and 4.24 during spring–summer and fall-winter cropping seasons, respectively). The increases in yield and water use efficiency (due to increased yield) were 19% and 18.8%, respectively, for spring–summer cropping season and 11.5% and 11.8%, respectively, for fall–winter cropping season, with two emitters per plant over a single emitter. The decrease in yield was 5.3% and 4%, and increase in water use efficiency (due to deficit irrigation) was 26.2% and 27.9% for spring-summer and fall-winter cropping seasons, respectively, by deficit irrigation over full irrigation. Moreover, it was observed that two, three, and four emitters per plant had no significant effects on yield and water use efficiency. Thus, it was concluded that two emitters per plant with deficit irrigation is optimum under greenhouse conditions for the cultivation of potted cherry tomatoes, considering the root morphology, root distribution, dry matter production, yield, water use efficiency, and economic analysis.

Keywords: root morphology; root length density; root weight density; root-shoot relationships; water use efficiency; benefit-cost ratio; greenhouse

Citation: Shabbir, A.; Mao, H.; Ullah, I.; Buttar, N.A.; Ajmal, M.; Solangi, K.A. Improving Water Use Efficiency by Optimizing the Root Distribution Patterns under Varying Drip Emitter Density and Drought Stress for Cherry Tomato. *Agronomy* **2021**, *11*, 3. https://dx.doi.org/10.3390/agronomy11010003

Received: 12 November 2020
Accepted: 21 December 2020
Published: 22 December 2020

Publisher's Note: MDPI stays neutral with regard to jurisdictional claims in published maps and institutional affiliations.

1. Introduction

Food demands are estimated to be doubled globally by 2050 due to the rapid increase in the population [1]. However, water resource availability has been further limited by changing climate, particularly in areas that favor good food production [2–4]. Therefore, water use efficiency (WUE) improvement has been a primary research topic related to sustainable agricultural production and agricultural-ecological balance [5,6]. Tomato (*Solanum lycopersicum* L.) is an important crop and can be found across the entire world for its nutritional value [7]. Studies have reported various biological and environmental conditions that can profoundly affect their production [8]. The soil water status and soil nutritional status are the most critical abiotic factors that can significantly affect the tomato vegetative and reproductive state. The effects of water and nutritional status have been studied in a series of experiments [9,10].

In addition, the leading irrigation technique with the most effective use of available water in irrigated agriculture is drip irrigation. To date, several research studies have been published on the superiority of drip irrigation over other techniques. However, attempts are still being made to search for new and more successful methods. Compared to drip irrigation, water pillow irrigation resulted in better yield, fruit quality, and WUE [11]. In other studies [12,13], water deficit irrigation was extensively studied to save water and thereby boost WUE, especially for the cultivation of several horticultural crops.

It has been found that soil water increases the accessibility of nutrients and that nutrients increase root growth and crop productivity [14]. The key root morphological features that directly influence the entire root system and indirectly influence the above-ground plant components are the root diameter, root surface area, root length density, and root weight density [15,16]. As we know, crop roots are an essential organ for the uptake of nutrients and water and have a crucial role in the ecosystem of plants and soils [17–19]. Besides, the interactions between soil moisture and crop rooting systems have thus been more intensively studied in recent times. Water scarcity reduces nutrient absorption and restricts root growth and distribution [20,21]. Previous research has shown that water conditions in surplus or deficit could alter the magnitude and distribution of maize root-zones, thus restricting root development [22]. As per spatial distribution of root, the root water uptake is spread over the root zone and regulated by climate demand and the spatial distribution of the availability of soil moisture and root density [23,24]. In relation to soil water's diverse distribution, root development and architecture show significant plasticity during the vegetative growth phase [25] and the final stage [26,27].

Just one side of the root zone is activated by drip irrigation. However, the soil hydraulic conductivity, soil texture, soil structure, and flow rate of the emitter affect the shape and extent of the wetting region formed by the drip emitter [28,29]. Despite these variables, the distribution of soil moisture and, in response, root morphology, root distribution, and root water uptake can also be influenced by the number of water emission points on the soil surface. Since moisture under many irrigation points is more readily accessible to roots, it may be attributed to reduced soil moisture stress. By reducing drought stress, a large root system supports the crops [30]. A study [31] has reported that moisture stress can suppress root morphology and plays a major role in root growth. Another study found that the emitter location can strongly affect the water and salt distribution in the substrate [32]. In addition, Valdés, R et al. revealed that compared to the use of one emitter, the use of multiple emitters per pot increased both the quantity and homogeneity of substrate water in gerbera grown in pots [33]. However, irrigation water quantity and multiple emitters per plant significantly affected the total volume of water applied, as well as root and shoot development in weeping fig [34]. The lower the emitter spacing, the higher the soil moisture distribution and uniformity and WUE and crop yield in drip irrigation systems [35]. Based on the review study, several studies concluded that much information is available on how the tomato root system responds to an irrigation system, irrigation scheduling, and fertilizer management to control tomato fruit yield and quality characteristics [10,36–38]. However, the effects of the emitter density (number of emitters per plant) and irrigation levels on the root distribution and root morphology concerning shoot morphology for cherry tomato crops are yet to be known.

Therefore, in this study, the emitter density under the drip system combined with full and deficit irrigation levels were considered to study the response of cherry tomato cultivated under greenhouse conditions. Moreover, the main objectives of the study were (i) to examine the effects of single and multiple emitters per plant drip irrigation under normal and water-deficit conditions on cherry tomato root morphology and distribution and (ii) to assess the relations between root and shoot morphology and yield for tomato plants. The overall research findings provide a baseline for managing water for cherry tomato production under greenhouse environments.

2. Materials and Methods

2.1. Site Description

The trials were performed as a spring-summer cropping season (SS) and fall-winter cropping season (FW) from 13 March to 14 July 2019 and from 2 September to 31 December 2019, respectively, at a Venlo-type greenhouse located in Zhenjiang City, Jiangsu Province, China ($31°56'$ N, $119°10'$ E). The study region was located in a humid subtropical monsoon climate zone with relative humidity, mean annual air temperature, and average annual rainfall of 76%, 15.5 °C, and 1058.8 mm, respectively [39]. The experimental area's soil was clay loam with a sand, silt, and clay particle distribution of 34%, 23%, and 43%, respectively. Furthermore, the soil had a bulk density, field capacity, and organic matter of 1.31 g cm^{-3}, 0.36 cm^3 cm^{-3}, and 31.23 g kg^{-1}, respectively. On 13 March and 2 September 2019, a 30-day-old seedling with firm roots of cherry tomato ('Fenxiaoke xt-12020') was transplanted to SS and FW, respectively, with a plant density of 3.84 m^{-2}. Moreover, the climatic data for both seasons are presented in Table 1. As can be seen in Table 1, the means of T and RH during SS (from 13 March to 14 July) equaled 23.66 °C and 70%, while during FW (from 2 September to 31 December), the averages were 18.66 °C and 73.5%, respectively.

Table 1. Air temperature and relative humidity observed for both cropping seasons.

Month	March	April	May	June	July	Sep	Oct	Nov	Dec
Mean temperature (°C)	18.25	21.04	23.80	27.22	27.43	26.58	21.77	16.62	10.12
Relative humidity (%)	60.58	66.90	75.95	72.13	70.72	60.73	70.68	78.48	83.57

2.2. Experimental Design

In this study, a total of eight treatments, including irrigation with 1, 2, 3, and 4 emitters per plant (N1, N2, N3, and N4) under full and deficit irrigation (W1; 100% and W2; 75% crop evapotranspiration) using drip irrigation, were selected as test-influencing factors. The emitters placement was as follows: one, two, three, and four emitters per plant were installed in the northwest, northwest, and southeast, one emitter in the northwest and two emitters at an angle of 120° from each other, and in the northeast, northwest, southeast, and southwest quadrants for N1, N2, N3, and N4, respectively. However, the location of the emitters was random and not concerned with the direction of the sun. All emitters were spaced evenly and were located at a radius of 9 cm from the plant, roughly halfway between the plant and pot borders. Moreover, the experimental design for both seasons was 2 factorial factors under a randomized complete block design (RCBD) with 4 replications. The plastic pots (diameter, 40 cm and height, 40 cm) were filled with 57 kg (43,960 cm^3) of air-dried soil per pot for cultivation practice. The used soil was first passed through a 5 mm mesh sieve, and the soil depth in the pots was maintained up to a height of 35 cm to accommodate the roots of tomato plants, following the recommendation of [10], which found that the 0 to 15 cm depth of the surface soil layer consists of tomato root length density (RLD) of up to 70–75% of the whole RLD for the surface drip irrigation system. The plant-to-plant and row-to-row distances were 80 and 25 cm, respectively. To prevent waterlogging in lower soil, all the pots were positioned on a gravel/sand mixture with a height of 3 to 5 cm. The fertigation was carried out as per local tradition. As nitrogen, phosphorus, and potassium sources, urea (46% nitrogen), triple superphosphate (46% P$_2$O$_5$), and muriate of potash (60% K$_2$O) were used, respectively. Additionally, at the beginning of the experiment, 40% of nitrogen and both phosphorus and potassium were added and blended into the soil in powdered forms. Two portions of the remaining 60% nitrogen were provided: 30% each in the first week of fruit emergence and in the first week of fruit ripening. The stages of crop growth were reported as DAT (days after transplantation). The plants were pruned on a weekly basis.

2.3. Sampling and Measurement

2.3.1. Irrigation and Crop Evapotranspiration

Irrigation was practiced at 8:00, when 20 mm of accumulative pan (20 cm diameter) evaporation (Ep) was achieved [40]. Adjustable flow rate emitters (with increased emitters per plant, the flow rate of emitter decreases to maintain the same irrigation period) were used to produce a uniform amount of water per plant. Before the experiment, the emitters were evaluated for flow rate uniformity. Under each treatment, the irrigation supplied to all pots was regulated using a water flow meter installed at the control unit.

The weight-based water balance equation [41,42] was used to calculate crop evapotranspiration ($ET_{c(CR)}$) by means of a $\pm$ 1 g weighing indicator for the control treatment (N1W1, single emitter per plants with 100% crop evapotranspiration). In the control treatment, $ET_{c(CR)}$ was assumed to be the normal water volume applied for the next irrigation, as no drainage was recorded after irrigation, so water applied for irrigation was taken equal to water loss in crop evapotranspiration. Because of the minimum rise between two consecutive waterings, the mass increase was overlooked. In accordance with the allocated proportion of the $ET_{c(CR)}$, the plants with deficit irrigation were irrigated.

2.3.2. Shoot Morphology, Yield, Water Use Efficiency, and Plant Dry Matter

Plant height (cm) (before the main tip) was measured using steel tape (1 mm). Plant stem diameter (mm) was measured using a digital vernier caliper (0.01 mm). The whole plant fruits were counted and weighed (g) using an electric weight balance (0.01 g). Moreover, for all treatments, the overall applied water (l) and tomato fruit yield (kg plant^{-1}) were reported at the end of the experiment. Fresh cherry tomato fruit matter was divided by the amount of water used to calculate water use efficiency (kg m^{-3}).

Plants were removed from all pots under each treatment at the end of the season, and the roots, stems, leaves, and fruits were collected separately. The samples were dried to achieve dry matter using an oven at 70 °C for each unit until a constant weight (g) was attained, and the precision indicator (0.0001 g) was finally used to obtain the dry matter.

2.3.3. Root Sampling and Morphological Characteristics

At the end of the cropping season, the entire soil column in the vertical direction (Z-axis) was divided into 5 layers, each with a thickness of 7 cm (Figure 1). Each 7 cm layer was cut into 7×7 cm small grids along X and Y axes (X and Y axes were taken along east–west and north-south directions, respectively). Each layer of 7 cm depth was divided into 25 grids of identical shape ($7 \times 7 \times 7$ cm = 343 cm^3), making a total of 125 grids (sampling units) per pot. The samples, symmetrically, were added up along the Y-axis (up to 35 cm) for horizontal root distribution and along the Z-axis (up to 35 cm) for vertical root distribution. The error caused by the reduced soil volume for the four boundary grids (lying partially out of the pot) and unconsidered soil volume at the 4 sides of the pot for each layer was neglected. The soil was cut horizontally and vertically with a sharp blade (Figure 2). To soften the root samples with soil, they were immersed in water for 30 min. The roots were separated from the soil very carefully using water. The roots with diameter < 2 mm only were scanned (because these roots were mainly responsible for soil water uptake [36]) using an Epson Perfection V700 photo flatbed scanner, and then WinRhizo software (Regent Instruments Inc., Quebec, QC, Canada) was used to get root length, root average diameter, root surface area, and root volume from the scanned root images. After that, root length density (RLD, root length/soil volume, cm cm^{-3}) and root weight density (RWD, root dry matter/soil volume, µg cm^{-3}) were calculated.

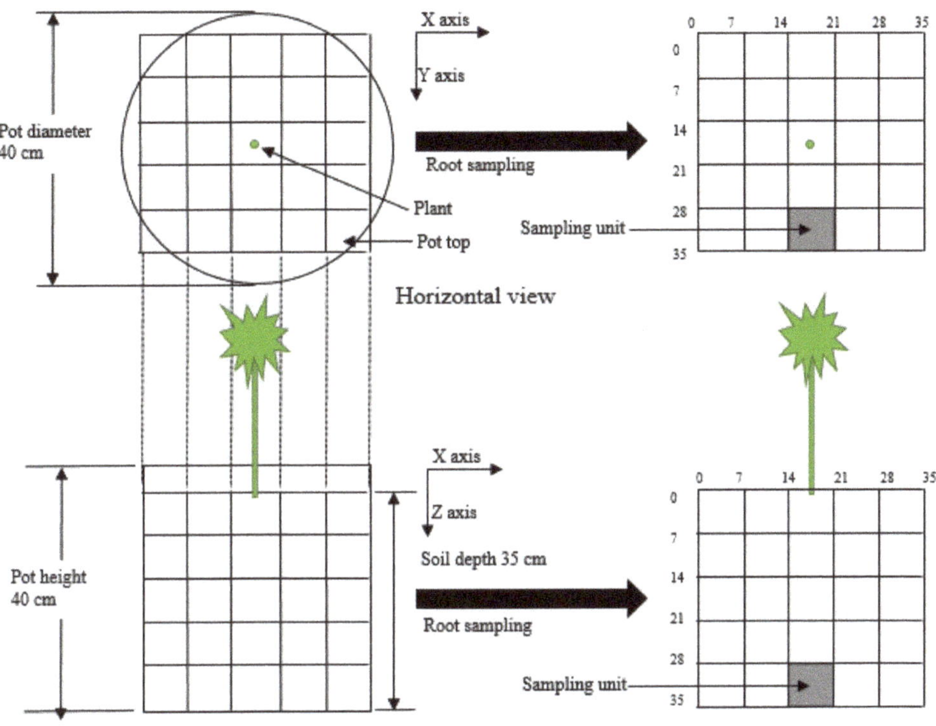

Figure 1. Diagram of a soil pot and root sampling method.

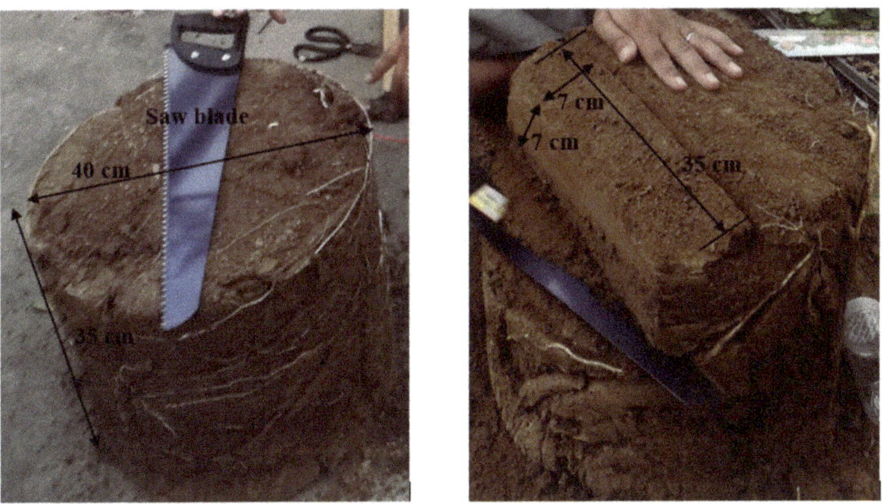

Figure 2. Soil ball and cutting soil ball into $7 \times 7 \times 7$ cm^3 sample grids.

2.3.4. Economic Optimization

To quantify the net benefits that each treatment produced, an economic analysis was carried out. The net benefits were measured as the difference between the overall costs of production and the total benefits per plant. Cherry tomato production costs ($ plant^{-1}) include system costs (emitters, the small pipe connecting the emitter and lateral joints, lateral joint, and arrow for emitter fixation, as these components vary with varying emitter density) and running costs (water cost, electricity cost, labor cost). Total benefits ($ plant^{-1}) were calculated by the product of the average market price ($ kg^{-1}) and yield (kg plant^{-1}). Finally, total benefits divided by total costs were calculated as the benefit-cost ratio.

2.4. Statistical Analyses

Taking season, emitter density, and irrigation level as fixed effects and including two-way (for an individual season) and three-way (to account for seasonal effect) interactions, the analysis of variance was carried out using the general linear model (GLM) in the SPSS 16.0 software (SPSS Inc., Chicago, IL, USA). The means were separated at $p < 0.05$ by the least significant difference test. Graphical representation was done using Origin Pro 2018 software (OriginLab Corporation, Northampton, MA, USA).

3. Results

3.1. Root Morphology

Figure 3 shows the effects of the emitter density, irrigation level, and cropping season on the root morphology. As can be seen from Figure 3, the seasonal effects were significant on all responses. However, there were significant individual effects of the emitter density and irrigation level on root length, root surface area, root volume, root length density, and root weight density, and non-significant effects on both SS's root average diameter and FW (except irrigation level effect that was significant for SS only). The SS caused an increase in root length, root surface area, root volume, root length density, and root weight density of 21.6%, 13.9%, 6.8%, 21.6%, and 15%, respectively, and a decrease of 3.3% in root average diameter compared to FW. All the root morphological responses were increased with increasing the emitter density and decreased with decreasing the irrigation levels except root average diameter, which was vice versa. Moreover, the treatments N4W1 and N1W2 produced the highest and lowest values in both seasons for all parameters except root average diameter, for which maximum and minimum was against N1W2 and N4W1, respectively.

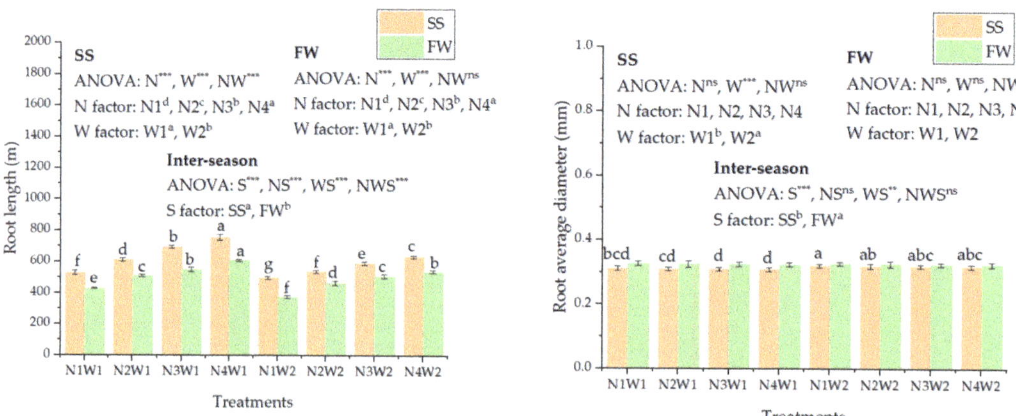

Figure 3. *Cont.*

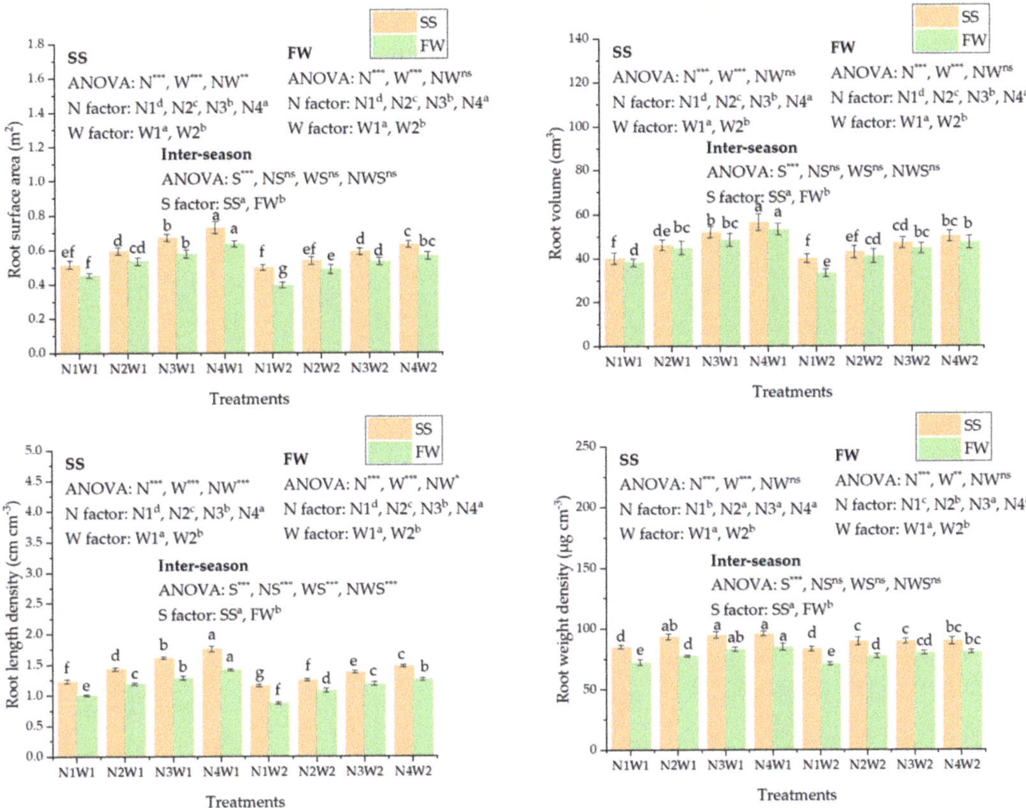

Figure 3. Effects of emitter density, irrigation level, and season on root morphology for both spring–summer (SS) and fall-winter (FW). N1, N2, N3, N4, W1, W2, SS, and FW: one emitter per plant, two emitters per plant, three emitters per plant, four emitters per plant, full irrigation, deficit irrigation, spring–summer cropping season, and fall-winter cropping season, respectively. *, ** and ***, ns; significant at $p < 0.05$, significant at $p < 0.01$, significant at $p < 0.001$, and non-significant at $p > 0.05$. Values that are followed by different letters within the same columns differ significantly at $p < 0.05$. Data are given in means ± standard deviations ($n = 4$) shown by vertical bars.

3.2. Plant Dry Matter Production

Figure 4 shows the effects of the emitter density, irrigation level, and season on cherry tomato plant dry matter production and root/shoot ratio. The experiment results in Figure 4 indicated that the SS resulted in more dry matter components (15%, 8.6%, 10%, 23.2%, and 18.6% more dry root, dry stem, dry leaves, dry fruit, and total dry matter, respectively) and root/shoot ratio (27.4%) than FW. The order of significance of responses for the treatment factors was S > N > W. The total and components of dry matter and root/shoot ratio increased with the increasing emitter density and decreased with deficit irrigation. However, N4W1 and N1W2 produced the highest (185.9 and 160.4 g total dry matter and 8.5 and 6.7% root/shoot ratio for SS and FW, respectively) and lowest (82.5 and 156.7 g total dry matter and 8.2 and 6.4% root/shoot ratio for SS and FW, respectively) values, respectively. Further, the root/shoot ratio was mainly affected by season.

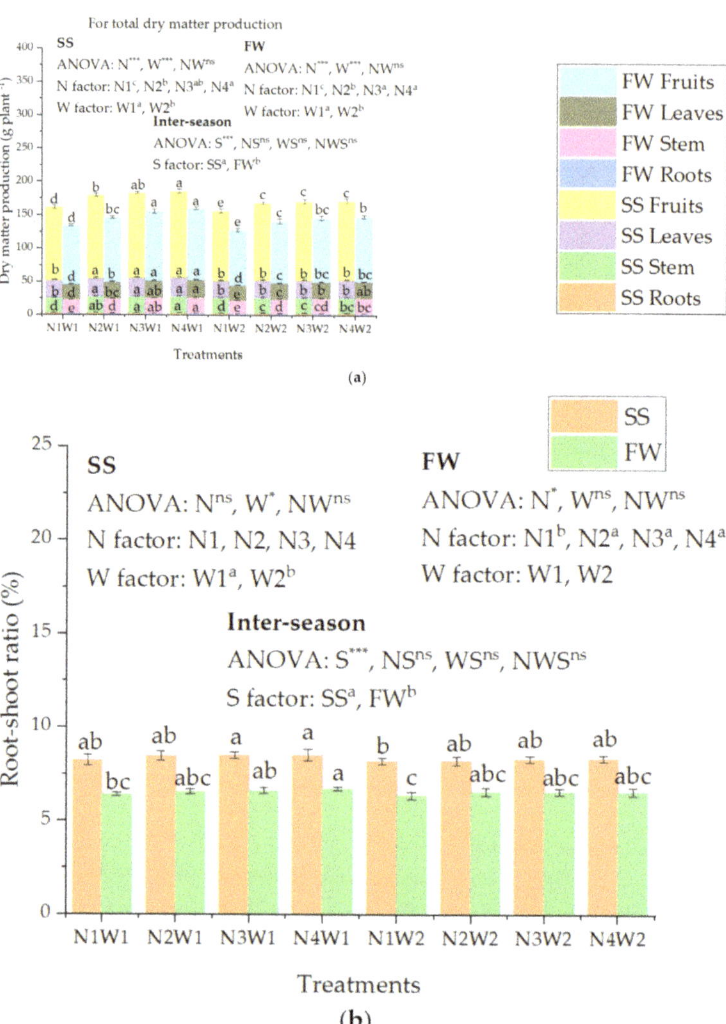

Figure 4. Effects of emitter density, irrigation level, and season on (**a**) dry matter production and (**b**) root/shoot ratio (root dry matter/(stem and leaves dry matter $\times$ 100, %) for both SS and FW. N1, N2, N3, N4, W1, W2, SS, and FW; one emitter per plant, two emitters per plant, three emitters per plant, four emitters per plant, full irrigation, deficit irrigation, spring-summer cropping season, and fall-winter cropping season, respectively. *, ***, ns: significant at $p < 0.05$, significant at $p < 0.001$, and non-significant at $p > 0.05$. Values that are followed by different letters within the same columns differ significantly at $p < 0.05$. Data were given in means $\pm$ standard deviations ($n = 4$) shown by vertical bars.

3.3. Root Distributions

Figures 5–8 show the isogram distributions of tomato root length density (RLD) distribution and root weight density (RWD) distribution both horizontally and vertically. Both emitter density and irrigation level had significant effects on the spatial distributions of cherry tomato RLD and RWD. The RLD and RWD distributions are shown for FW only because both seasons resulted in similar distribution patterns. The distributions expanded horizontally and focused greatly on the surface soil layer with increasing the emitter density due to improved root growth. For both the RWD and RLD, deficit irrigation led to smaller (horizontally) and deeper (vertically) concentration areas. Considering the emitter density (N), the average RLD and RWD were 1.1 cm cm^{-3} and 77.5 µg cm^{-3}, respectively, in W2 treatment, which were 10.6% and 3.8% lower compared to W1 treatment (1.2 cm cm^{-3} and 79.6 µg cm^{-3}), respectively. In comparison with the W factor, the N factor had more significant influences on the horizontal and vertical distribution of RLD and RWD, and root distribution responses varied with water levels (Figures 5–8). In the N1 treatments, both RLD and RWD concentration areas showed eccentricity horizontally (Figures 5 and 7) and dense and deeper extension vertically towards the emitter's region (Figures 6 and 8). This indicated that the tomato root system generated narrower and deeper distributions with steeper growth angles under the N1 conditions than multiple emitters per plant. Unlike horizontally, it was shown that the shape of RLD and RWD distributions showed negligible differences vertically among N2, N3, and N4 under both W conditions. Therefore, the N2W2 treatment presented an optimal root distribution throughout all N and W treatments, with an average of 1.1 cm cm^{-3} in RLD and 77.7 µg cm^{-3} in RWD, respectively. Its root dispersion range was wider horizontally across the entire pot (Figures 5 and 7) and deeper vertically throughout the soil profile (Figures 6 and 8).

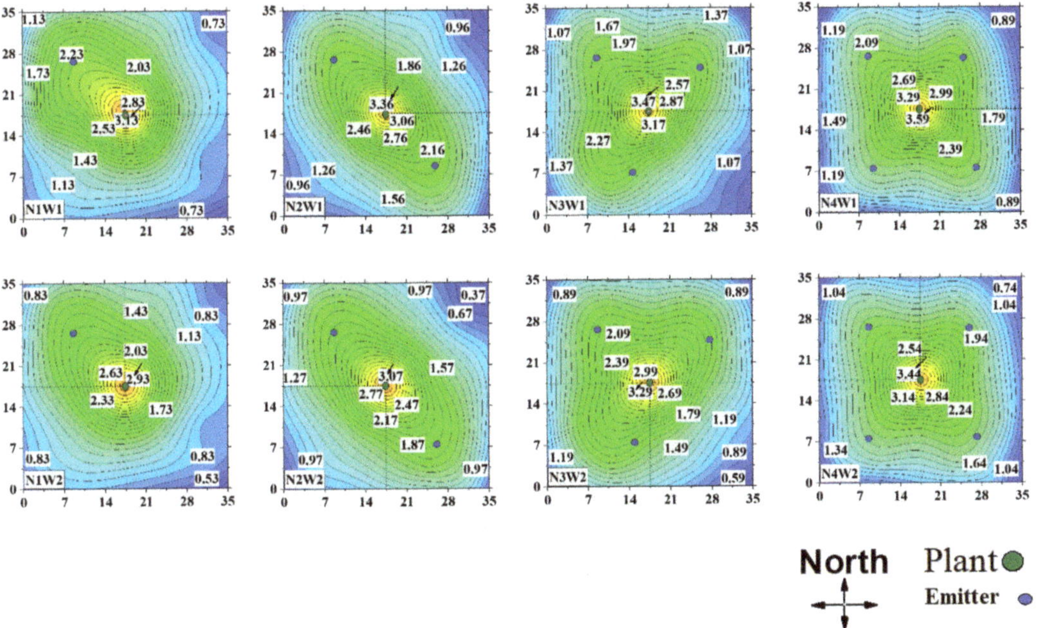

Figure 5. Effects of emitter density and irrigation levels on horizontal root length density (RLD, cm cm^{-3}) distribution for the fall-winter cropping season (at the end of the cropping season; the maximum value is marked by a solid arrow).

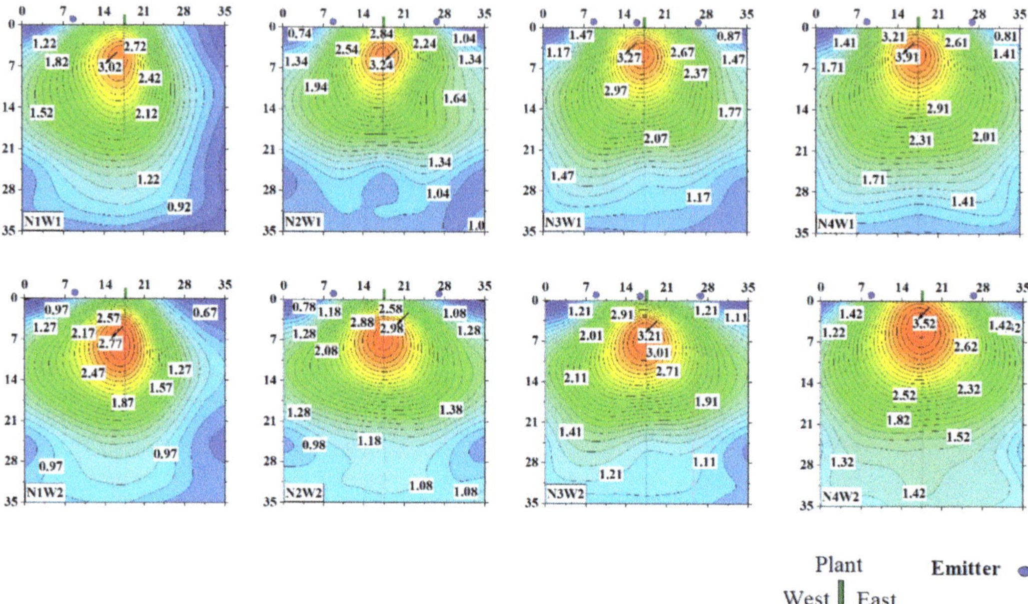

Figure 6. Effects of emitter density and irrigation levels on vertical root length density (RLD, cm cm^{-3}) distribution for the fall-winter cropping season (at the end of the cropping season; the maximum value is marked by a solid arrow).

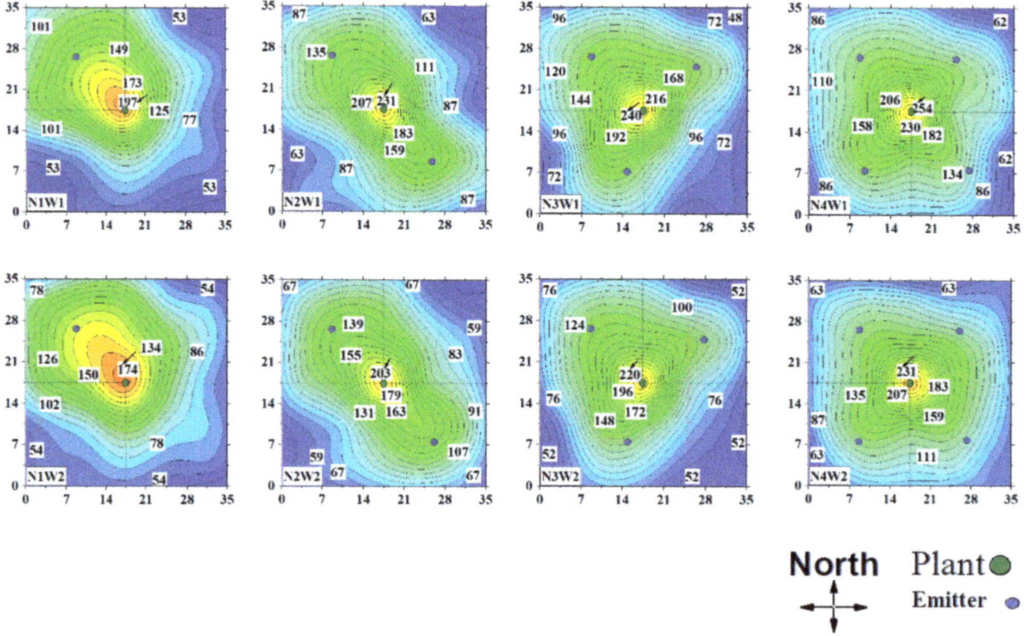

Figure 7. Effects of emitter density and irrigation levels on horizontal, vertical root weight density (RWD, µg cm^{-3}) distribution for the fall-winter cropping season (at the end of the cropping season; the maximum value is marked by a solid arrow).

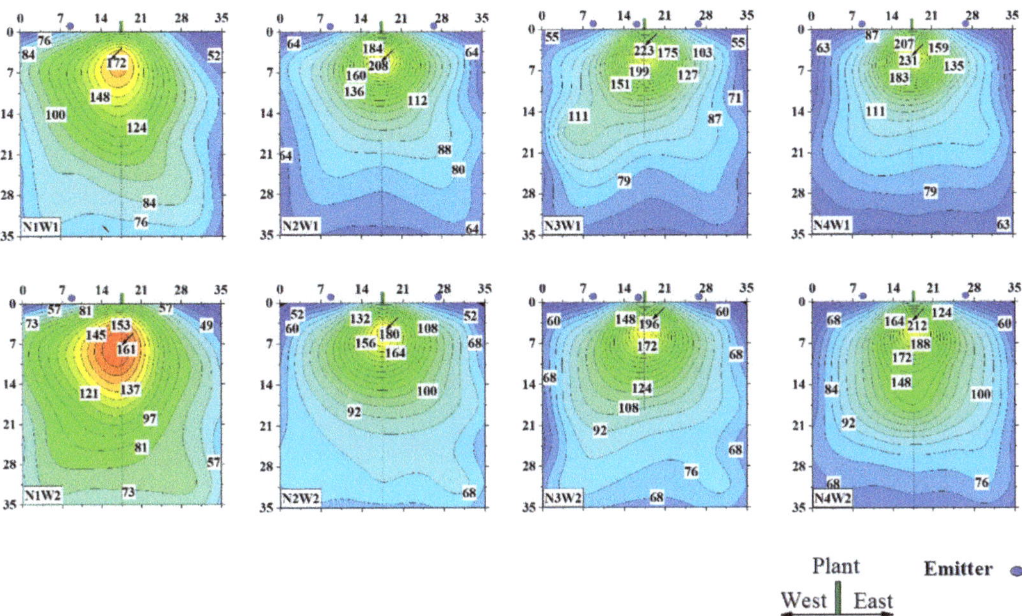

Figure 8. Effects of emitter density and irrigation levels on vertical root weight density (RWD, μg cm^{-3}) distribution for the fall-winter cropping season (at the end of the cropping season; the maximum value is marked by a solid arrow).

3.4. Shoot Morphology and Yield

The two seasons' analyzed results (Figure 9) showed that the SS performed better in terms of plant height (138.8–168.6 cm, 21.5%), stem diameter (7.5–9.1 mm, 21.2%), and yield (1.9–2.4 kg plant^{-1}, 25.2%) but reduced water use efficiency (49.6 to 43.3 kg m^{-3}) (WUE) (due to fewer water requirements in FW) (12.7%) than FW. The impact of N4W1 was found to be most significant, and the impact of N1W2 was least significant except for WUE, for which the maximum and minimum were against N4W2 and N1W1, respectively, for both seasons. The N4 and N1 effects were most and least significant on all parameters for both seasons. The increase in yield and WUE (due to increased yield) was 19% (2.0–2.4 kg plant^{-1}) and 18.8% (37.5–44.6 kg m^{-3}), respectively, for SS and 11.5% (1.7–1.9 kg plant^{-1}) and 11.8% (44.6–49.9 kg m^{-3}), respectively, for FW for N2 over N1. The increase in the corresponding values was similar for multiple emitters per plant over N1. Compared to W1, the decrease in yield was 5.3% (2.4–2.3 kg plant^{-1}) and 4% (1.9–1.8 kg plant^{-1}), and the increase in WUE (due to deficit irrigation) was 26.2% (38.3–48.3 kg m^{-3}) and 27.9% (43.5–55.7 kg m^{-3}) for SS and FW, respectively, by W2.

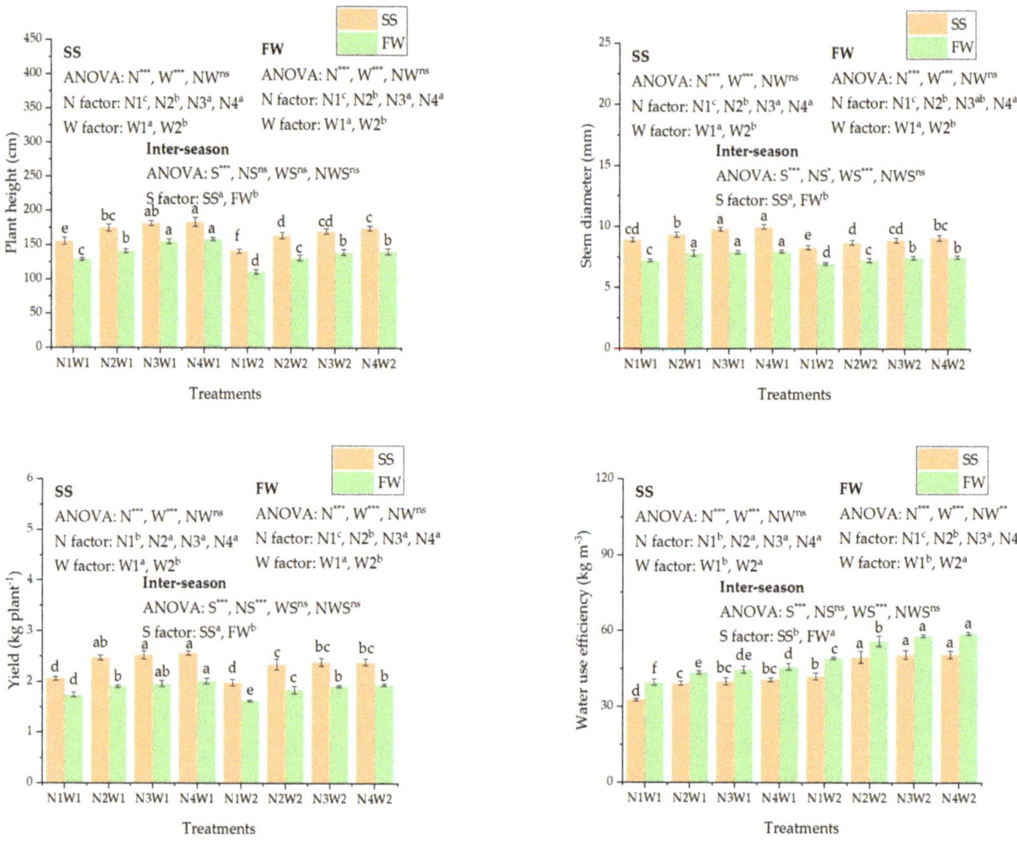

Figure 9. Effects of emitter density, irrigation level, and season on plant height, stem diameter, tomato yield, and water use efficiency for both SS and FW. N1, N2, N3, N4, W1, W2, SS, and FW: one emitter per plant, two emitters per plant, three emitters per plant, four emitters per plant, full irrigation, deficit irrigation, spring-summer cropping season, and fall-winter cropping season, respectively. *, **, ***, ns; significant at $p < 0.05$, significant at $p < 0.01$, significant at $p < 0.001$, and non-significant at $p > 0.05$. Values that are followed by different letters within the same columns differ significantly at $p < 0.05$. Data are given in means $\pm$ standard deviations ($n = 4$) shown by vertical bars.

3.5. Relationships between Root Growth, Shoot Growth, and Plant Yield

The relationships for leaf area index, plant height, stem diameter, plant yield, and total dry matter versus root morphological parameters like root length, average root diameter, root surface area, root volume, and root dry matter for both SS and FW are shown in Table 2. All the relationships were significant ($p < 0.001$). There were linear and quadratic relationships within each season, also for inter-season. The R^2 for root average diameter versus above ground parameters was lowest (0.487 (FW)–0.629 (SS)). For the most part, the relationships for root surface area and root volume versus above-ground parameters were quadratic. No relationship was found between the root morphology and WUE.

Table 2. Relationship between root morphology and plant above ground components for both SS and FW (n = 32).

Function	Model	R	R^2	Adj R^2	ANOVA F	Sig.
SS						
	The relation between leaf area index and					
Root length	$y = 0.00001x^2 + 0.021x - 5.14$	0.857	0.734	0.720	51.17	0.000
Average root diameter	$y = 129.79x^2 - 49.96x + 6.25$	0.912	0.832	0.823	91.54	0.000
Root surface area	$y = 6.79x - 0.30$	0.948	0.899	0.897	339.76	0.000
Root volume	$y = 0.064x\ 0.59$	0.974	0.949	0.948	708.62	0.000
Root dry matter	$y = 2.83x - 6.95$	0.975	0.950	0.948	716.77	0.000
	The relation between plant height and					
Root length	$y = 0.0004x^2 + 0.67x - 89.05$	0.931	0.866	0.859	119.51	0.000
Average root diameter	$y = 613.51x - 34.58$	0.810	0.657	0.648	72.65	0.000
Root surface area	$y = -140.29x^2 + 316.23x + 26.79$	0.964	0.929	0.925	242.38	0.000
Root volume	$y = -0.011x^2 + 2.47x + 71.37$	0.957	0.916	0.911	200.65	0.000
Root dry matter	$y = 52.22x - 33.75$	0.943	0.890	0.887	307.33	0.000
	The relation between stem diameter and					
Root length	$y = 0.008x + 4.35$	0.892	0.796	0.790	148.00	0.000
Average root diameter	$y = 30.11x - 0.83$	0.863	0.745	0.738	111.07	0.000
Root surface area	$y = 6.05x + 5.30$	0.962	0.925	0.923	471.58	0.000
Root volume	$y = 0.056x + 6.15$	0.969	0.940	0.938	590.58	0.000
Root dry matter	$y = 2.45x - 0.35$	0.961	0.924	0.921	458.78	0.000
	The relation between plant yield and					
Root length	$y = 0.000005x^2 + 0.009x - 1.24$	0.905	0.819	0.809	83.46	0.000
Root average diameter	$y = 8.66x - 0.519$	0.793	0.629	0.619	64.43	0.000
Root surface area	$y = -2.002x^2 + 4.48x + 0.34$	0.939	0.882	0.876	138.93	0.000
Root volume	$y = 0.0002x^2 + 0.035x + 0.97$	0.933	0.871	0.864	125.10	0.000
Root dry matter	$y = 0.754x - 0.574$	0.945	0.893	0.891	318.34	0.000
	The relation between total dry matter production and					
Root length	$y = 0.144x + 84.98$	0.872	0.760	.0754	120.25	0.000
Root average diameter	$y = 575.16x - 18.18$	0.885	0.783	0.777	136.80	0.000
Root surface area	$y = -80.16x^2 + 218.28 + 67.42$	0.959	0.919	0.915	210.66	0.000
Root volume	$y = -0.007x^2 + 1.90x + 9172.10$	0.974	0.949	0.946	342.41	0.000
Root dry matter	$y = -10.70x + 47.23$	0.994	0.987	0.987	2917.12	0.000
FW						
	The relation between leaf area index and					
Root length	$y = 0.008x - 1.01$	0.879	0.773	0.767	129.30	0.000
Average root diameter	$y = 126.91x^2 - 49.86x + 6.60$	0.889	0.790	0.778	69.45	0.000
Root surface area	$y = 7.18x - 0.33$	0.974	0.949	0.948	711.92	0.000
Root volume	$y = 0.072x + 0.40$	0.993	0.987	0.986	2793.26	0.000
Root dry matter	$y = 0.47x^2 - 0.572x + 0.14$	0.970	0.941	0.938	294.21	0.000
	The relation between plant height and					
Root length	$y = 0.214x + 31.96$	0.967	0.934	0.932	539.77	0.000
Average root diameter	$y = 542.17x - 31.45$	0.698	0.487	0.474	36.10	0.000
Root surface area	$y = -106.08x^2 + 271.90x + 31.36$	0.976	0.952	0.949	364.18	0.000
Root volume	$y = -0.014x^2 + 2.75x + 54.44$	0.946	0.894	0.889	156.48	0.000
Root dry matter	$y = 52.44x - 37.80$	0.926	0.858	0.855	230.06	0.000
	The relation between stem diameter and					
Root length	$y = 0.007x + 4.27$	0.822	0.676	0.667	79.19	0.000
Average root diameter	$y = 25.99x - 0.62$	0.928	0.860	0.857	234.01	0.000
Root surface area	$y = 5.75x + 4.70$	0.941	0.886	0.882	291.90	0.000
Root volume	$y = 0.0005x^2 + 0.098x + 4.49$	0.980	0.961	0.959	454.17	0.000
Root dry matter	$y = 1.92x + 1.08$	0.941	0.885	0.882	292.36	0.000

Table 2. *Cont.*

Function	Model	R	R^2	Adj R^2	ANOVA F	Sig.
	The relation between plant yield and					
Root length	$y = 0.002x + 0.87$	0.892	0.795	0.790	147.55	0.000
Root average diameter	$y = 6.36x - 0.13$	0.813	0.661	0.652	74.07	0.000
Root surface area	$y = -1.44x^2 + 3.085x + 1.17$	0.962	0.925	0.921	229.63	0.000
Root volume	$y = 0.0002x^2 + 0.032x\ 0.92$	0.964	0.930	0.926	244.28	0.000
Root dry matter	$y = 0.54x + 0.04$	0.952	0.907	0.904	369.77	0.000
	The relation between total dry matter production and					
Root length	$y = 0.171x\ 59.80$	0.939	0.882	0.879	284.73	0.000
Root average diameter	$y = 521.26x - 18.44$	0.815	0.665	0.656	75.38	0.000
Root surface area	$y = 138.38x + 76.82$	0.993	0.986	0.985	2600.12	0.000
Root volume	$y = -0.01x^2 + 2.20x + 75.79$	0.990	0.980	0.979	918.36	0.000
Root dry matter	$y = 45.90x - 9.33$	0.985	0.971	0.970	1266.11	0.000

3.6. Optimization of the Emitter Density and Irrigation Level

Table 3 shows that the total cost and total benefits increased with an increase in the emitter density and decrease with decreasing irrigation level. For N2, the net benefits and benefit–cost ratio were maximum, there was no significant effect of irrigation level on net income, but the benefit–cost ratio was significantly affected by the level of irrigation and was higher against W2. The optimization of the emitter density and irrigation level based on economic returns was calculated by calculating the benefit–cost ratio. Table 3 shows that the optimized emitter density and irrigation level management strategy is N2W2, giving the highest benefit–cost ratio, 4.20 and 4.24 for SS and FW, respectively.

Table 3. Effects of emitter density, irrigation level, and season on economic analysis for both SS and FW.

Factor	Total Costs ($ Plant⁻¹)		Total Benefits ($ Plant⁻¹)		Net Benefits ($ Plant⁻¹)		Benefit–Cost Ratio (-)	
Season	SS	FW	SS	SS	FW	FW	SS	FW
Emitter density (N)								
N1	1.12	0.85	$3.96 \pm 0.13b$	$3.30 \pm 0.14c$	$2.84 \pm 0.11c$	$2.45 \pm 0.07c$	$3.57 \pm 0.38b$	$3.91 \pm 0.30a$
N2	1.23	0.94	$4.71 \pm 0.20a$	$3.68 \pm 0.12b$	$3.48 \pm 0.15a$	$2.74 \pm 0.10a$	$3.88 \pm 0.37a$	$3.94 \pm 0.34a$
N3	1.44	1.16	$4.80 \pm 0.20a$	$3.79 \pm 0.09a$	$3.37 \pm 0.15ab$	$2.63 \pm 0.09b$	$3.36 \pm 0.24c$	$3.29 \pm 0.24b$
N4	1.54	1.26	$4.85 \pm 0.21a$	$3.87 \pm 0.11a$	$3.31 \pm 0.11b$	$2.61 \pm 0.08b$	$3.17 \pm 0.18d$	$3.08 \pm 0.19c$
Irrigation level (W)								
W1	1.46	1.14	$4.70 \pm 0.42a$	$3.73 \pm 0.22a$	$3.25 \pm 0.30a$	$2.59 \pm 0.12a$	$3.24 \pm 0.21b$	$3.31 \pm 0.34b$
W2	1.20	0.96	$4.45 \pm 0.37b$	$3.58 \pm 0.26b$	$3.25 \pm 0.27a$	$2.62 \pm 0.15a$	$3.75 \pm 0.37a$	$3.79 \pm 0.45a$
NW								
N1W1	1.25	0.94	$4.04 \pm 0.07d$	$3.41 \pm 0.10d$	$2.79 \pm 0.07c$	$2.47 \pm 0.10cd$	$3.23 \pm 0.06d$	$3.63 \pm 0.10b$
N2W1	1.36	1.03	$4.84 \pm 0.10ab$	$3.74 \pm 0.05b$	$3.48 \pm 0.10a$	$2.71 \pm 0.05ab$	$3.56 \pm 0.07c$	$3.63 \pm 0.05b$
N3W1	1.56	1.25	$4.93 \pm 0.17a$	$3.84 \pm 0.11ab$	$3.37 \pm 0.17ab$	$2.59 \pm 0.11bc$	$3.16 \pm 0.11de$	$3.07 \pm 0.09e$
N4W1	1.66	1.35	$5.01 \pm 0.09a$	$3.94 \pm 0.11a$	$3.35 \pm 0.09ab$	$2.59 \pm 0.11bc$	$3.02 \pm 0.05e$	$2.92 \pm 0.08f$
N1W2	0.99	0.76	$3.88 \pm 0.13d$	$3.18 \pm 0.03e$	$2.89 \pm 0.13c$	$2.42 \pm 0.03d$	$3.92 \pm 0.13b$	$4.18 \pm 0.04b$
N2W2	1.09	0.85	$4.58 \pm 0.21c$	$3.61 \pm 0.14c$	$3.49 \pm 0.21a$	$2.76 \pm 0.14a$	$4.20 \pm 0.19a$	$4.24 \pm 0.16a$
N3W2	1.31	1.07	$4.67 \pm 0.15bc$	$3.74 \pm 0.04b$	$3.36 \pm 0.15ab$	$2.67 \pm 0.04ab$	$3.57 \pm 0.12c$	$3.50 \pm 0.03c$
N4W2	1.41	1.17	$4.68 \pm 0.13bc$	$3.80 \pm 0.04b$	$3.27 \pm 0.13b$	$2.63 \pm 0.04ab$	$3.32 \pm 0.10d$	$3.25 \pm 0.03d$
Season (S)								
SS	1.33		$4.58 \pm 0.41a$		$3.25 \pm 0.28a$		$3.50 \pm 0.39b$	
FW	1.05		$3.66 \pm 0.25b$		$2.61 \pm 0.13b$		$3.55 \pm 0.46a$	
Analysis of variance								

Table 3. *Cont.*

Factor	Total Costs ($ Plant^{-1})		Total Benefits ($ Plant^{-1})		Net Benefits ($ Plant^{-1})		Benefit–Cost Ratio (-)	
Season	SS	FW	SS	SS	FW	FW	SS	FW
N			***	***	***	***	***	***
W			***	***	ns	ns	***	***
NW			ns	ns	ns	ns	**	**
S			***		***		*	
NS			***		***		***	
WS			ns		ns		ns	
NWS			ns		ns		ns	

N1, N2, N3, N4, W1, W2, SS, and FW; one emitter per plant, two emitters per plant, three emitters per plant, four emitters per plant, full irrigation, deficit irrigation, spring–summer cropping season, and fall–winter cropping season, respectively. *, ** , ***, ns: significant at $p < 0.05$, significant at $p < 0.01$, significant at $p < 0.001$, and non-significant at $p > 0.05$. Values that are followed by different letters within the same columns differ significantly at $p < 0.05$. Data are given in mean $\pm$ standard deviation (n = 4). The prices are $0.105 per emitter, $0.06 per arrow, $0.18 per meter small pipe, $0.195 per joint, $0.75 per m^{-3} water, and $0.075 per kWh, $3.6 per hour labor, and $1.95 per kg cherry tomato. The irrigation system is assumed to work for 2 seasons. Benefit–cost ratio (-): total benefits/total costs.

4. Discussion

Roots are involved in acquiring nutrients and soil water as a vital part of plant organs [43,44]. Root diameter and length are significant morphological parameters that affect soil water and nutrient absorption [45]. In this study, the root length (Figure 3) and root dry matter (Figure 4) were increased with increasing emitter density (24.2% and 31.5% (root length), and 9.5% and 12.4% (root dry matter) for spring–summer (SS) and fall–winter cropping season (FW), respectively) and irrigation level (12.8% and 10.2% (root length), and 4.5% and 2.6% (root dry matter) for SS and FW, respectively). The root average diameter was affected negligibly. This is because increasing the emitter density would reduce the drought stress to the root, and in turn, the root morphology is enhanced by the improved moisture distribution uniformity and wetting pattern around the plant roots [46]. The same phenomenon occurs as the irrigation level was increased from W2 to W1, which is in agreement with [31]. The main parameters for characterizing root systems are root length density (RLD) and root weight density (RWD) [47,48]. RLD and RWD increased with an increased emitter density and decrease with decreasing irrigation level (Figure 3), as was the case with root length (Figure 3) and root dry weight (Figure 4).

The RLD and RWD distributions expanded horizontally (Figures 5 and 7) and concentrated heavily into the topsoil layer (Figures 6 and 8) due to the increased emitter density and decreased emitter discharge (with increased emitters per plant, the flow rate of the emitter decreases to maintain the same irrigation period). This is in agreement with [35,49], which showed that the radial and vertical distributions of moisture were higher, respectively, at lower and higher flow rates. Further, [50] reported that the supply of soil water affects the morphology of the root. In well-watered conditions, the roots near the soil's surface are considered to be in the prime position of water and nutrient absorption [51,52]. Deficit irrigation resulted in smaller-concentration regions horizontally (Figures 5 and 7) and deeper vertically in the RLD and RWD (Figures 6 and 8) because the roots moved deeper in search of moisture. Several research studies reasoned the yield and water use efficiency enhancement to the enhanced deep soil layer water uptake and use [53,54] as restricted irrigation would encourage roots to expand into deeper soil layers [30,55]. Quinoa (Chenopodium quinoa Willd.) plasticity was investigated in dry and wet soil, and drier soil enhanced the tap root's growth more than wet soil to allow water to be absorbed in deeper soil [**56**]. Enlarged deep root biomass, lengthier seminal roots, and improved development of small root diameter at depths were included in root morphological parameters in drying soil [52]. The growth of the roots, especially the deep roots, had an important impact on the stable yield and, under drought conditions, absorbed greater quantities of water from the deep soil [57].

There may have been two explanations for the relationship between the root morphology and the emitter density and irrigation level. First, one reason for cherry tomato root propagation at different irrigation levels is the differences in soil bulk density [10,58]. Secondly, the finding that the root growth of increased emitter density and irrigation level is higher than that of single emitters per plant and deficit irrigation could be derived from the accumulation of dry matter (Figure 4), which has been attributed to plant roots offering a broader spatial range to absorb water and nutrients under increased emitter density and higher levels of irrigation [59,60]. Water stress can limit the development and distribution of roots in the soil [61] and lower the RLD [21]. Tomato plants with two emitters per plant with deficit irrigation treatment showed an optimal root distribution compared to the other emitter density and irrigation level treatment in the current research, describing wider and deeper dispersion measurements and higher RLD and RWD through the soil and thus showed less scarcity of plant growth reactions and increased yield and WUE with the highest benefit–cost ratio (4.20 and 4.24 (Table 3) for SS and FW, respectively). In view of root distribution and yield, optimization of emitter density and irrigation level based on economic benefits was achieved by measuring the net profits (3.49 and 2.76 $ plant^{-1} for SS and FW, respectively) and benefit–cost ratio (4.20 and 4.24 for SS and FW, respectively) (Table 3).

The total dry matter (Figure 4), tomato shoot morphology, and yield (Figure 9) increased with increasing emitters per plant (10.8% and 14.10% (root length), 17.9% and 20.3% (plant height), 8.1% and 7.9% (stem diameter), and 20.9% and 14.7% (yield) for SS and FW, respectively) and decreased with deficit irrigation (6.7% and 6.7% (root length), 6.6% and 10.5% (plant height), 7.9% and 5.4% (stem diameter), and 5.3% and 4.1% (yield) for SS and FW, respectively). This is due to the fact that the spatial distribution of root systems in a soil directly affects soil water and nutrient absorption and root morphology, thereby affecting crop growth and productivity **[62]**. Another reason is that plant nutrient absorption benefits from a good root structure, which will help achieve different leaf areas and dry weight [63,64]. Water-logged soil causes the leaf growth rate and dry matter accumulation by the shoots to decrease [65]. The reduced overall dry matter production and water consumption cause poorer cherry tomato yield under the drought stress [4,66,67]. Drought and waterlogging induce a drastic decrease in dry matter accumulation, causing low fruit yield [68–70]. Previous studies have shown that irrigation deficits have reduced vegetative growth and fruit yield, in line with current findings [71–74]. One reason for the reduced yield was that dramatically reduced photosynthesis of plants decrease the quantity and energy of metabolites necessary for appropriate plant growth under drought stress [71,75].

Plant performance has indeed been deeply related to the root system that provides water and nutrients to the shoot, as is well known. Highly significant results were found between root morphology and dry root matter versus plant abov- ground parameter (R^2 ranges from 0.487 to 0.993) (Table 2). The explanation for this is that root growth and shoot growth patterns are closely related [48,76–78], both growing in this study with an increased emitter density, which can be considered the main factor promoting increased yield. Greater root biomass, which led to greater yield and WUE, was significantly correlated with higher shoot biomass [37,79].

WUE is a crucial physiological parameter reflecting in a water-scarce area the capacity of crops to retain water as it integrates resistance to drought and great potential yield [55,80,81]. WUE increases with an increased emitter density (37.5–45.8 kg m^{-3} for SS and 44.6–52.5 kg m^{-3} for FW) and a decreased irrigation level (38.3–48.3 kg m^{-3} for SS and 43.5–55.7 kg m^{-3} for FW) (Figure 9).WUE is mainly a function of water input [37,82,83], which increases by deficit irrigation [4,84–88] and is the most significant measure of the agricultural production system [89]. It may be attributed to a decrease in the transpiration rate due to deficit irrigation, which eventually results in the stomata's partial closure [90,91].

The spring–summer cropping season resulted in the improved root (499–606.5 m (root length) and 3.4–3.9 g (root dry matter)) and shoot morphology (145.2–172.3 g (total dry matter), 6.6–8.4% (root/shoot ratio), 138.8–168.6 cm (plant height), and 7.5–9.1 mm (stem diameter)) and yield 1.9–2.4 kg plant^{-1} (yield), but reduced water use efficiency (49.6–43.3 kg m^{-3}) and benefit–cost–ratio (3.6–3.5) relative to fall–winter cropping season, but for both seasons, the root distribution patterns were the same. The explanation was that the constant low temperatures in the later winter period (6.5 °C in late December) negatively affected the maximum growth rate during the rapid growth period and consequently reduced the yield [92]. Numerous reports have said that the optimal temperature of the tomato crop varied around 20 and 24 °C to maintain greater fruit yield, and 12 and 36 °C have been the growth temperature limits [93]. Furthermore, for two, three, and four emitters per plant, there were no significant effects on response parameters, and the obtained results were identical for both seasons, except for root distribution patterns.

5. Conclusions

The spring–summer cropping season resulted in improved root and shoot morphology and dry matter production relative to the fall–winter cropping season, but for both seasons, the root distribution patterns were the same. The individual factors of the emitter density and irrigation level significantly affected the root distribution patterns along with root morphology, which affects shoot morphology and finally tomato yield and water use efficiency. All measured parameters were more sensitive to the emitter density except water use efficiency, which was more affected by irrigation level. In addition, the results indicated that plant height, stem diameter, fruit yield, and dry matter production were closely linked to root morphology and root dry matter. Compared with the other emitter density and irrigation level treatments in the present study, tomato plants with two emitters per plant with deficit irrigation treatment indicated an optimal root distribution, showing broader and deeper dispersion measures and greater root length density and root weight density through the soil, and hence presented lesser scarcity reactions of plants growth and gained improved yield and water use efficiency with the highest benefit–cost ratio (4.20 and 4.24 for spring–summer and fall–winter cropping seasons, respectively). Based on the results, it was concluded that among all the treatments, two emitters per plant with deficit irrigation is optimum under greenhouse conditions for the cultivation of potted cherry tomatoes, considering the root morphology, root distribution, dry matter production, yield, water use efficiency, and economic analysis. It is recommended to vary emitter location around the plant to investigate its impact on root distribution and growth, which will eventually affect plant yield.

Author Contributions: Investigation, A.S., H.M., and I.U.; funding acquisition, H.M.; writing—original draft, A.S.; writing—review and editing, A.S., N.A.B., M.A., and K.A.S. All authors have read and agreed to the published version of the manuscript.

Funding: This work was supported by the Chinese Academy of Agricultural Sciences (CARS-23-C03) and the Jiangsu Demonstration Project of Modern Agricultural Machinery Equipment and Technology (No. NJ2018-14).

Institutional Review Board Statement: "Not applicable" for studies not involving humans or animals.

Informed Consent Statement: "Not applicable" for studies not involving humans.

Data Availability Statement: Data is contained within the article. The data presented in this study are available in Tables 1–3 and Figures 3–9.

Conflicts of Interest: The authors declare no conflict of interest.

References

1. Tilman, D.; Balzer, C.; Hill, J.; Befort, B.L. Global food demand and the sustainable intensification of agriculture. *Proc. Natl. Acad. Sci. USA* **2011**, *108*, 20260–20264. [CrossRef]
2. Lakhiar, I.A.; Gao, J.; Syed, T.N.; Chandio, F.A.; Tunio, M.H.; Ahmad, F.; Solangi, K.A. Overview of the aeroponic agriculture—An emerging technology for global food security. *Int. J. Agric. Biol. Eng.* **2020**, *13*, 1–10. [CrossRef]
3. Lakhiar, I.A.; Jianmin, G.; Syed, T.N.; Chandio, F.A.; Buttar, N.A.; Qureshi, W.A. Monitoring and control systems in agriculture using intelligent sensor techniques: A review of the aeroponic system. *J. Sens.* **2018**, *2018*, 8672769. [CrossRef]

4. Patanè, C.; Tringali, S.; Sortino, O. Effects of deficit irrigation on biomass, yield, water productivity and fruit quality of processing tomato under semi-arid mediterranean climate conditions. *Sci. Hortic.* **2011**, *129*, 590–596. [CrossRef]

5. Beltrano, J.; Ronco, M.; Montaldi, E. Drought stress syndrome in wheat is provoked by ethylene evolution imbalance and reversed by rewatering, aminoethoxyvinylglycine, or sodium benzoate. *J. Plant Growth Regul.* **1999**, *18*, 59–64. [CrossRef] [PubMed]

6. Shabbir, A.; Arshad, M.; Bakhsh, A.; Usman, M.; Shakoor, A.; Ahmad, I.; Ahmad, A. Apparent and real water productivity for cotton-wheat zone of Punjab, Pakistan. *Pak. J. Agri. Sci* **2012**, *49*, 357–363.

7. Padayachee, A.; Day, L.; Howell, K.; Gidley, M. Complexity and health functionality of plant cell wall fibers from fruits and vegetables. *Crit. Rev. Food Sci. Nutr.* **2017**, *57*, 59–81. [CrossRef] [PubMed]

8. Hanping, M.; Ullah, I.; Jiheng, N.; Javed, Q.; Azeem, A. Estimating tomato water consumption by sap flow measurement in response to water stress under greenhouse conditions. *J. Plant Interact.* **2017**, *12*, 402–413. [CrossRef]

9. Wang, X.; Xing, Y. Evaluation of the effects of irrigation and fertilization on tomato fruit yield and quality: A principal component analysis. *Sci. Rep.* **2017**, *7*, 350. [CrossRef]

10. Zotarelli, L.; Scholberg, J.M.; Dukes, M.D.; Muñoz-Carpena, R.; Icerman, J. Tomato yield, biomass accumulation, root distribution and irrigation water use efficiency on a sandy soil, as affected by nitrogen rate and irrigation scheduling. *Agric. Water Manag.* **2009**, *96*, 23–34. [CrossRef]

11. Gerçek, S.; Demirkaya, M.; Işik, D. Water pillow irrigation versus drip irrigation with regard to growth and yield of tomato grown under greenhouse conditions in a semi-arid region. *Agric. Water Manag.* **2017**, *180*, 172–177. [CrossRef]

12. Nagaz, K.; Masmoudi, M.; Ben Mechlia, N. Yield response of drip irrigated onion under full and deficit irrigation with saline water in arid regions of tunisia. *ISRN Agron.* **2012**, *212*, 562315. [CrossRef]

13. Rasool, G.; Guo, X.; Wang, Z.; Chen, S.; Ullah, I. The interactive responses of fertigation levels under buried straw layer on growth, physiological traits and fruit yield in tomato plant. *J. Plant Interact.* **2019**, *14*, 552–563. [CrossRef]

14. Besharat, S.; Nazemi, A.H.; Sadraddini, A.A. Parametric modeling of root length density and root water uptake in unsaturated soil. *Turk. J. Agric. For.* **2010**, *34*, 439–449.

15. Xia, J.; Liu, M.-Y.; Jia, S.-F. Water security problem in north china: Research and perspective. *Pedosphere* **2005**, *15*, 563–575.

16. Fageria, N. Influence of dry matter and length of roots on growth of five field crops at varying soil zinc and copper levels. *J. Plant Nutr.* **2005**, *27*, 1517–1523. [CrossRef]

17. Spedding, T.; Hamel, C.; Mehuys, G.; Madramootoo, C. Soil microbial dynamics in maize-growing soil under different tillage and residue management systems. *Soil Biol. Biochem.* **2004**, *36*, 499–512. [CrossRef]

18. Zhang, S.; Sadras, V.; Chen, X.; Zhang, F. Water use efficiency of dryland maize in the loess plateau of china in response to crop management. *Field Crop. Res.* **2014**, *163*, 55–63. [CrossRef]

19. Forde, B.; Lorenzo, H. The nutritional control of root development. *Plant Soil* **2001**, *232*, 51–68. [CrossRef]

20. Araki, H.; Hamada, A.; Hossain, M.A.; Takahashi, T. Waterlogging at jointing and/or after anthesis in wheat induces early leaf senescence and impairs grain filling. *Field Crop. Res.* **2012**, *137*, 27–36. [CrossRef]

21. Li, Q.; Dong, B.; Qiao, Y.; Liu, M.; Zhang, J. Root growth, available soil water, and water-use efficiency of winter wheat under different irrigation regimes applied at different growth stages in north china. *Agric. Water Manag.* **2010**, *97*, 1676–1682. [CrossRef]

22. Liu, C.; Jin, S.; Zhou, L.; Jia, Y.; Li, F.; Xiong, Y.; Li, X. Effects of plastic film mulch and tillage on maize productivity and soil parameters. *Eur. J. Agron.* **2009**, *31*, 241–249. [CrossRef]

23. Albasha, R.; Mailhol, J.-C.; Cheviron, B. Compensatory uptake functions in empirical macroscopic root water uptake models–experimental and numerical analysis. *Agric. Water Manag.* **2015**, *155*, 22–39. [CrossRef]

24. Šimůnek, J.; Hopmans, J.W. Modeling compensated root water and nutrient uptake. *Ecol. Model.* **2009**, *220*, 505–521. [CrossRef]

25. Eapen, D.; Barroso, M.L.; Ponce, G.; Campos, M.E.; Cassab, G.I. Hydrotropism: Root growth responses to water. *Trends Plant Sci.* **2005**, *10*, 44–50. [CrossRef]

26. Hodge, A.; Berta, G.; Doussan, C.; Merchan, F.; Crespi, M. Plant root growth, architecture and function. *Plant Soil* **2009**, *321*, 153–187. [CrossRef]

27. Lynch, J. Root architecture and plant productivity. *Plant Physiol.* **1995**, *109*, 7. [CrossRef]

28. Bar-Yosef, B. Advances in fertigation. In *Advances in Agronomy*; Elsevier: Amsterdam, The Netherlands, 1999; Volume 65, pp. 1–77.

29. Palomo, M.; Moreno, F.; Fernández, J.; Díaz-Espejo, A.; Girón, I. Determining water consumption in olive orchards using the water balance approach. *Agric. Water Manag.* **2002**, *55*, 15–35. [CrossRef]

30. Wang, C.; Liu, W.; Li, Q.; Ma, D.; Lu, H.; Feng, W.; Xie, Y.; Zhu, Y.; Guo, T. Effects of different irrigation and nitrogen regimes on root growth and its correlation with above-ground plant parts in high-yielding wheat under field conditions. *Field Crop. Res.* **2014**, *165*, 138–149. [CrossRef]

31. Guo, X.; Kang, S.; Suo, L. Effects of regulated deficit irrigation on root growth in maize. *Irrig. Drain* **2001**, *20*, 25–27.

32. Ondrasek, G.; Romic, D.; Romic, M.; Tomic, F.; Mustac, I. *Salt Distribution in Peat Substrate Grown with Melon (Cucumis melo L.)*; International Symposium on Growing Media 779: Angers, France, 2005; pp. 307–312.

33. Valdés, R.; Miralles, J.; Ochoa, J.; Bañón, S.; Sánchez-Blanco, M.J. The number of emitters alters salt distribution and root growth in potted gerbera. *HortScience* **2014**, *49*, 160–165. [CrossRef]

34. Valdés, R.; Ochoa, J.; Sánchez-Blanco, M.; Franco, J.; Bañón, S. Irrigation volume and the number of emitters per pot affect root growth and saline ion contents in weeping fig. *Agric. Agric. Sci. Procedia* **2015**, *4*, 356–364. [CrossRef]

35. Bajpai, A.; Kaushal, A. Soil moisture distribution under trickle irrigation: A review. *Water Supply* **2020**, *20*, 761–772. [CrossRef]

36. Qiu, R.; Du, T.; Kang, S. Root length density distribution and associated soil water dynamics for tomato plants under furrow irrigation in a solar greenhouse. *J. Arid Land* **2017**, *9*, 637–650. [CrossRef]
37. Wang, X.; Yun, J.; Shi, P.; Li, Z.; Li, P.; Xing, Y. Root growth, fruit yield and water use efficiency of greenhouse grown tomato under different irrigation regimes and nitrogen levels. *J. Plant Growth Regul.* **2019**, *38*, 400–415. [CrossRef]
38. Machado, R.M.; Maria do Rosàrio, G.O. Tomato root distribution, yield and fruit quality under different subsurface drip irrigation regimes and depths. *Irrig. Sci.* **2005**, *24*, 15–24. [CrossRef]
39. Huang, S.; Yan, H.; Zhang, C.; Wang, G.; Acquah, S.J.; Yu, J.; Li, L.; Ma, J.; Darko, R.O. Modeling evapotranspiration for cucumber plants based on the shuttleworth-wallace model in a venlo-type greenhouse. *Agric. Water Manag.* **2020**, *228*, 105861. [CrossRef]
40. Hao, L.; Duan, A.-W.; Li, F.-S.; Sun, J.-s.; Wang, Y.-C.; Sun, C.-T. Drip irrigation scheduling for tomato grown in solar greenhouse based on pan evaporation in north china plain. *J. Integr. Agric.* **2013**, *12*, 520–531.
41. Hu, Y.; Buttar, N.A.; Tanny, J.; Snyder, R.L.; Savage, M.J.; Lakhiar, I.A. Surface renewal application for estimating evapotranspiration: A review. *Adv. Meteorol.* **2018**, *2018*, 1690714. [CrossRef]
42. Kurunç, A.; Ünlükara, A. Growth, yield, and water use of okra (abelmoschus esculentus) and eggplant (solanum melongena) as influenced by rooting volume. *N. Z. J. Crop Hortic. Sci.* **2009**, *37*, 201–210. [CrossRef]
43. Yang, C.; Yang, L.; Yang, Y.; Ouyang, Z. Rice root growth and nutrient uptake as influenced by organic manure in continuously and alternately flooded paddy soils. *Agric. Water Manag.* **2004**, *70*, 67–81. [CrossRef]
44. Wu, W.; Cheng, S. Root genetic research, an opportunity and challenge to rice improvement. *Field Crop. Res.* **2014**, *165*, 111–124. [CrossRef]
45. Ren, X.; Jia, Z.; Chen, X. Rainfall concentration for increasing corn production under semiarid climate. *Agric. Water Manag.* **2008**, *95*, 1293–1302. [CrossRef]
46. Shabbir, A.; Mao, H.; Ullah, I.; Buttar, N.A.; Ajmal, M.; Lakhiar, I.A. Effects of drip irrigation emitter density with various irrigation levels on physiological parameters, root, yield, and quality of cherry tomato. *Agronomy* **2020**, *10*, 1685. [CrossRef]
47. Mosaddeghi, M.; Mahboubi, A.; Safadoust, A. Short-term effects of tillage and manure on some soil physical properties and maize root growth in a sandy loam soil in western iran. *Soil Tillage Res.* **2009**, *104*, 173–179. [CrossRef]
48. Wang, Y.; Zhang, X.; Chen, J.; Chen, A.; Wang, L.; Guo, X.; Niu, Y.; Liu, S.; Mi, G.; Gao, Q. Reducing basal nitrogen rate to improve maize seedling growth, water and nitrogen use efficiencies under drought stress by optimizing root morphology and distribution. *Agric. Water Manag.* **2019**, *212*, 328–337. [CrossRef]
49. Shekhar, S.; Kumar, M.; Kumari, A.; Jain, S. Soil moisture profile analysis using tensiometer under different discharge rates of drip emitter. *Int. J. Curr. Microbiol. Appl. Sci.* **2017**, *6*, 908–917. [CrossRef]
50. Chu, G.; Chen, T.; Wang, Z.; Yang, J.; Zhang, J. Reprint of "morphological and physiological traits of roots and their relationships with water productivity in water-saving and drought-resistant rice". *Field Crop. Res.* **2014**, *165*, 36–48. [CrossRef]
51. Hopkins, W.G. *Introduction to Plant Physiology*; John Wiley and Sons: Hoboken, NJ, USA, 1999.
52. Becker, S.R.; Byrne, P.F.; Reid, S.D.; Bauerle, W.L.; McKay, J.K.; Haley, S.D. Root traits contributing to drought tolerance of synthetic hexaploid wheat in a greenhouse study. *Euphytica* **2016**, *207*, 213–224. [CrossRef]
53. Johnson, W.C. *Yield-Water Relationships of Summer-Fallowed Winter Wheat: A Precision Study in the Texas Panhandle*; Agricultural Research, Southern Region, Science and Education Administration, USDA: New Orleans, LA, USA, 1980; Volume 5.
54. Fei, G.; Zheng, L.-J.; Sun, X.-H.; Guo, X.-H.; Zhang, X.-L. Estimating distribution of water uptake with depth of winter wheat by hydrogen and oxygen stable isotopes under different irrigation depths. *J. Integr. Agric.* **2016**, *15*, 891–906.
55. Xu, C.; Tao, H.; Tian, B.; Gao, Y.; Ren, J.; Wang, P. Limited-irrigation improves water use efficiency and soil reservoir capacity through regulating root and canopy growth of winter wheat. *Field Crop. Res.* **2016**, *196*, 268–275. [CrossRef]
56. Alvarez-Flores, R.; Nguyen-Thi-Truc, A.; Peredo-Parada, S.; Joffre, R.; Winkel, T. Rooting plasticity in wild and cultivated andean chenopodium species under soil water deficit. *Plant Soil* **2018**, *425*, 479–492. [CrossRef]
57. White, R.G.; Kirkegaard, J.A. The distribution and abundance of wheat roots in a dense, structured subsoil–implications for water uptake. *Plantcell Environ.* **2010**, *33*, 133–148. [CrossRef] [PubMed]
58. Purushothaman, R.; Krishnamurthy, L.; Upadhyaya, H.D.; Vadez, V.; Varshney, R.K. Genotypic variation in soil water use and root distribution and their implications for drought tolerance in chickpea. *Funct. Plant Biol.* **2017**, *44*, 235–252. [CrossRef] [PubMed]
59. Alrajhi, A.; Beecham, S.; Hassanli, A. Effects of partial root-zone drying irrigation and water quality on soil physical and chemical properties. *Agric. Water Manag.* **2017**, *182*, 117–125. [CrossRef]
60. Li, Y.; Wang, L.; Xue, X.; Guo, W.; Xu, F.; Li, Y.; Sun, W.; Chen, F. Comparison of drip fertigation and negative pressure fertigation on soil water dynamics and water use efficiency of greenhouse tomato grown in the north china plain. *Agric. Water Manag.* **2017**, *184*, 1–8. [CrossRef]
61. Gajri, P.; Prihar, S.; Arora, V. Effects of nitrogen and early irrigation on root development and water use by wheat on two soils. *Field Crop. Res.* **1989**, *21*, 103–114. [CrossRef]
62. Amato, M.; Ritchie, J.T. Spatial distribution of roots and water uptake of maize (zea mays L.) as affected by soil structure. *Crop Sci.* **2002**, *42*, 773–780. [CrossRef]
63. Ayi, Q.; Zeng, B.; Liu, J.; Li, S.; van Bodegom, P.M.; Cornelissen, J.H.C. Oxygen absorption by adventitious roots promotes the survival of completely submerged terrestrial plants. *Ann. Bot.* **2016**, *118*, 675–683. [CrossRef]
64. Faucon, M.-P.; Houben, D.; Lambers, H. Plant functional traits: Soil and ecosystem services. *Trends Plant Sci.* **2017**, *22*, 385–394. [CrossRef]
65. Drew, M.; Sisworo, E. The development of waterlogging damage in young barley plants in relation to plant nutrient status and changes in soil properties. *New Phytol.* **1979**, *82*, 301–314. [CrossRef]

66. Badr, A.E.; Abuarab, M.E. Soil moisture distribution patterns under surface and subsurface drip irrigation systems in sandy soil using neutron scattering technique. *Irrig. Sci.* **2013**, *31*, 317–332. [CrossRef]
67. Costa, J.M.; Ortuño, M.F.; Chaves, M.M. Deficit irrigation as a strategy to save water: Physiology and potential application to horticulture. *J. Integr. Plant Biol.* **2007**, *49*, 1421–1434. [CrossRef]
68. Bisbis, M.B.; Gruda, N.; Blanke, M. Potential impacts of climate change on vegetable production and product quality—A review. *J. Clean. Prod.* **2018**, *170*, 1602–1620. [CrossRef]
69. Shao, G.; Cheng, X.; Liu, N.; Zhang, Z. Effect of drought pretreatment before anthesis and post-anthesis waterlogging on water relation, photosynthesis, and growth of tomatoes. *Arch. Agron. Soil Sci.* **2016**, *62*, 935–946. [CrossRef]
70. Sharma, H.S.; Fleming, C.; Selby, C.; Rao, J.; Martin, T. Plant biostimulants: A review on the processing of macroalgae and use of extracts for crop management to reduce abiotic and biotic stresses. *J. Appl. Phycol.* **2014**, *26*, 465–490. [CrossRef]
71. Agbna, G.H.; Dongli, S.; Zhipeng, L.; Elshaikh, N.A.; Guangcheng, S.; Timm, L.C. Effects of deficit irrigation and biochar addition on the growth, yield, and quality of tomato. *Sci. Hortic.* **2017**, *222*, 90–101. [CrossRef]
72. Kuşçu, H.; Turhan, A.; Demir, A.O. The response of processing tomato to deficit irrigation at various phenological stages in a sub-humid environment. *Agric. Water Manag.* **2014**, *133*, 92–103. [CrossRef]
73. Zhang, H.; Xiong, Y.; Huang, G.; Xu, X.; Huang, Q. Effects of water stress on processing tomatoes yield, quality and water use efficiency with plastic mulched drip irrigation in sandy soil of the hetao irrigation district. *Agric. Water Manag.* **2017**, *179*, 205–214. [CrossRef]
74. Hooshmand, M.; Albaji, M.; zadeh Ansari, N.A. The effect of deficit irrigation on yield and yield components of greenhouse tomato (solanum lycopersicum) in hydroponic culture in ahvaz region, iran. *Sci. Hortic.* **2019**, *254*, 84–90. [CrossRef]
75. Kulkarni, M.; Phalke, S. Evaluating variability of root size system and its constitutive traits in hot pepper (*Capsicum annum* L.) under water stress. *Sci. Hortic.* **2009**, *120*, 159–166. [CrossRef]
76. Aguirre, L.; Johnson, D.A. Root morphological development in relation to shoot growth in seedlings of four range grasses. *Rangel. Ecol. Manag. J. Range Manag. Arch.* **1991**, *44*, 341–346. [CrossRef]
77. Benjamin, J.; Nielsen, D.; Vigil, M.; Mikha, M.; Calderon, F. Water deficit stress effects on corn (*Zea mays*, L.) root: Shoot ratio. *Open J. Soil Sci.* **2014**, *2014*.
78. Mi, G.; Chen, F.; Wu, Q.; Lai, N.; Yuan, L.; Zhang, F. Ideotype root architecture for efficient nitrogen acquisition by maize in intensive cropping systems. *Sci. China Life Sci.* **2010**, *53*, 1369–1373. [CrossRef] [PubMed]
79. Zhang, X.; Chen, S.; Sun, H.; Wang, Y.; Shao, L. Root size, distribution and soil water depletion as affected by cultivars and environmental factors. *Field Crop. Res.* **2009**, *114*, 75–83. [CrossRef]
80. Fang, Q.; Ma, L.; Green, T.; Yu, Q.; Wang, T.; Ahuja, L. Water resources and water use efficiency in the north china plain: Current status and agronomic management options. *Agric. Water Manag.* **2010**, *97*, 1102–1116. [CrossRef]
81. Zhang, X.; Chen, S.; Sun, H.; Wang, Y.; Shao, L. Water use efficiency and associated traits in winter wheat cultivars in the north china plain. *Agric. Water Manag.* **2010**, *97*, 1117–1125. [CrossRef]
82. Coyago-Cruz, E.; Meléndez-Martínez, A.J.; Moriana, A.; Girón, I.F.; Martín-Palomo, M.J.; Galindo, A.; Pérez-López, D.; Torrecillas, A.; Beltrán-Sinchiguano, E.; Corell, M. Yield response to regulated deficit irrigation of greenhouse cherry tomatoes. *Agric. Water Manag.* **2019**, *213*, 212–221. [CrossRef]
83. Ullah, I.; Hanping, M.; Chuan, Z.; Javed, Q.; Azeem, A. Optimization of irrigation and nutrient concentration based on economic returns, substrate salt accumulation and water use efficiency for tomato in greenhouse. *Arch. Agron. Soil Sci.* **2017**, *63*, 1748–1762. [CrossRef]
84. Favati, F.; Lovelli, S.; Galgano, F.; Miccolis, V.; Di Tommaso, T.; Candido, V. Processing tomato quality as affected by irrigation scheduling. *Sci. Hortic.* **2009**, *122*, 562–571. [CrossRef]
85. Yang, L.; Qu, H.; Zhang, Y.; Li, F. Effects of partial root-zone irrigation on physiology, fruit yield and quality and water use efficiency of tomato under different calcium levels. *Agric. Water Manag.* **2012**, *104*, 89–94. [CrossRef]
86. Chen, J.; Kang, S.; Du, T.; Qiu, R.; Guo, P.; Chen, R. Quantitative response of greenhouse tomato yield and quality to water deficit at different growth stages. *Agric. Water Manag.* **2013**, *129*, 152–162. [CrossRef]
87. Akhtar, S.S.; Li, G.; Andersen, M.N.; Liu, F. Biochar enhances yield and quality of tomato under reduced irrigation. *Agric. Water Manag.* **2014**, *138*, 37–44. [CrossRef]
88. Hashem, M.S.; El-Abedin, T.Z.; Al-Ghobari, H.M. Rational water use by applying regulated deficit and partial root-zone drying irrigation techniques in tomato under arid conditions. *Chil. J. Agric. Res.* **2019**, *79*, 75–88. [CrossRef]
89. Giuliani, M.; Nardella, E.; Gagliardi, A.; Gatta, G. Deficit irrigation and partial root-zone drying techniques in processing tomato cultivated under mediterranean climate conditions. *Sustainability* **2017**, *9*, 2197. [CrossRef]
90. Bravdo, B.-A. *Physiological Mechanisms Involved in the Production of Non-Hydraulic Root Signals by Partial Rootzone Drying—A Review*; VII International Symposium on Grapevine Physiology and Biotechnology 689: Davis, CA, USA, 2004; pp. 267–276.
91. Sepaskhah, A.; Ahmadi, S. A review on partial root-zone drying irrigation. *Int. J. Plant Prod.* **2012**, *4*, 241–258.
92. Wu, Y.; Yan, S.; Fan, J.; Zhang, F.; Xiang, Y.; Zheng, J.; Guo, J. Responses of growth, fruit yield, quality and water productivity of greenhouse tomato to deficit drip irrigation. *Sci. Hortic.* **2021**, *275*, 109710. [CrossRef]
93. Rosales, M.A. Producción y Calidad Nutricional en Frutos de Tomate Cherry Cultivados en Dos Invernaderos Mediterráneos Experimentales: Respuestas Metabólicas y Fisiológicas. Ph.D. Thesis, Universidad de Granada, Granada, Spain, 2008.

Article

Evaluation of Soil Water Content Measurements with Capacitance Probes to Support Irrigation Scheduling in a "Red Beaut" Japanese Plum Orchard

Sandra Millán *, **Carlos Campillo**, **Antonio Vivas**, **María José Moñino** and **Maria Henar Prieto**

Centre for Scientific and Technological Research of Extremadura (CICYTEX), Department of Horticulture, Finca La Orden, Regional Government of Extremadura, Highway A-V, Km 372, 06187 Guadajira, Badajoz, Spain; carlos.campillo@juntaex.es (C.C.); antonio.vivas@juntaex.es (A.V.); mariajose.monino@juntaex.es (M.J.M.); henar.prieto@juntaex.es (M.H.P.)
* Correspondence: sandra.millan@juntaex.es; Tel.: +34-924-014-008

Received: 9 October 2020; Accepted: 8 November 2020; Published: 12 November 2020

Abstract: Advances in electromagnetic sensor technologies in recent years have made automated irrigation scheduling a reality through the use of state-of-the-art soil moisture sensing devices. However, correct sensor positioning and interpretation of the measurements are key to the successful implementation of these management systems. The aim of this study is to establish guidelines for soil moisture sensor placement to support irrigation scheduling, taking into account the physiological response of the plant. The experimental work was carried out in Vegas Bajas del Guadiana (Extremadura, Spain) on a drip-irrigated experimental orchard of the early-maturing Japanese plum cultivar "Red Beaut". Two irrigation treatments were established: control and drying. The control treatment was scheduled to cover crop water needs. In the drying treatment, the fruit trees were irrigated as in control, except in certain periods (preharvest and postharvest) in which irrigation was suspended (drying cycles). Over 3 years (2015–2017), a series of plant parameters were analyzed in relation to the measurements provided by a battery of frequency domain reflectometry probes installed in different positions with respect to tree and dripper: midday stem water potential (Ψstem), sap flow, leaf stomatal conductance, net leaf photosynthesis and daily fraction of intercepted photosynthetically active radiation. After making a comparison of these measurements as indicators of plant water status, Ψstem was found to be the physiological parameter that detected water stress earliest. The drying cycles were very useful to select the probe positions that provided the best information for irrigation management and to establish a threshold in the different phases of the crop below which detrimental effects could be caused to the crop. With respect to the probes located closest to the drippers, a drop in the relative soil water content (RSWC) below 0.2 would not be advisable for "non-stress" scheduling in the preharvest period. When no deficit irrigation strategies are applied in the postharvest period, the criteria are similar to those of preharvest. However, the probes located between the dripper at 0.15 and 0.30 m depth provide information on moderate water stress if the RSWC values falls below 0.2. The severe tree water stress was detected below 0.1 RSWC in probes located at 60 cm depth from this same position.

Keywords: midday stem water potential; sap flow; photosynthesis; stomatal conductance; FDR probes and daily fraction of intercepted photosynthetically active radiation

---

1. Introduction

The amount of water available for irrigation is limited [1], especially in the face of the increasing demand of a constantly growing world population which is predicted to rise to about 9.8 billion in

2050 [2]. Problems derived from a lack of water will likely increase if long-term global climate change predictions are correct. It has been reported that global mean land and ocean surface temperatures increased by 0.8 °C between 1888 and 2012 [3,4] and the worldwide average surface temperature has been predicted to increase by 1.4 to 5.8 °C by 2100 [3]. Furthermore, increases in evaporation and reductions in precipitation rates are expected [4], which will entail a reduction of the available water resources for the twenty-first century [5]. In this context, the adoption of efficient irrigation systems can help to decrease agricultural water consumption, improve farm profitability and reduce environmental impacts.

Despite constant advances in agriculture, it remains difficult to obtain accurate predictions of the crop water requirements of fruit trees in field conditions [6]. Nowadays, three methods can be used in this respect for proper irrigation scheduling: the water balance-based method, the method based on monitoring soil water content or potential and the method which uses plant water status as the reference for irrigation scheduling [1,7]. While each of these methods has its advantages and disadvantages, the water balance method is the most commonly applied. Soil water content monitoring does not always provide information about plant water status, as this depends on the complex relationship between soil, plant and atmosphere, and in addition, the uncertainty caused by the heterogeneous distribution of water in the soil needs to be taken into account. Despite this, the use of soil and plant measurements for irrigation scheduling assessment is a very attractive approach because they allow adaptation to specific plot and crop conditions. However, such an approach also entails greater complexity in terms of the collection, processing and interpretation of information. Today, through the use of state-of-the-art sensors and information and communications technologies, it is possible to integrate different scheduling methods and develop intelligent systems for irrigation automation or decision support for technicians and farmers [8–10].

Among the methods available for measuring soil water content (SWC), both gravimetry and neutron probe measurements are considered to be the most accurate. In both cases, measurements are time consuming and laborious. Moreover, neutron probes tend to be expensive, a radiation hazard, display insensitivity near the soil surface, give readings that vary due to changes in soil density [11], and require a trained operator due to the use of the radioactive source as well as extensive soil specific calibrations [12]. However, sensors are widely available that provide measurements with the desired frequency, have low maintenance needs and costs and are easily automatable. These include sensors based on frequency domain reflectometry (FDR), a technique to determine SWC which is based on the dielectric properties of the soil [13]. These probes require soil-specific calibration for accurate results [12], and are sensitive to air gaps, soil salinity, temperature, bulk density and clay content [14,15]. The proper positioning of the probes in the soil plays a key role in the quality of the information provided by them since each probe has a limited zone of influence. Due to the heterogeneity of environmental factors, SWC can also vary spatially [16,17]. In irrigated crops, SWC patterns in the root zone are dynamic and conditioned by numerous parameters including soil hydraulic properties, spatial heterogeneity, and the characteristics of the crop (e.g., rooting patterns) and the irrigation system that is employed (e.g., drip line spacing, emitter flow rate, irrigation dose) [18]. In drip irrigation systems, the spatial variability of the SWC formed under the emitters is higher due to the local application of irrigation water [19]. Consequently, the correct positioning and placement of the soil moisture probes are even more relevant in the case of drip irrigation systems [20].

When the SWC becomes limiting for the plant, it triggers a series of mechanisms that modify the plant's physiological processes in response to water stress. The physiological response of plants to water deficits depends on the severity as well as the duration of the stress. Only the most sensitive processes are altered by very mild stress, but as the water stress increases the changes intensify, and additional processes become affected in accordance with their relative sensitivity to the stress [21]. Water stress affects almost all plant functions, including photosynthesis and respiration, as well as having an impact on crop yield [22]. The extent of the effects depends on the interaction between SWC, the evaporative demand of the atmosphere and the sensitivity of each process to water stress.

Growth is one of the first physiological processes to be affected by water stress through a decline in tree canopy development. It is important to have reliable indicators of the water status of the crop for proper irrigation programming and irrigation strategy management. For various decades, the midday stem water potential (Ψstem) has been widely used to determine the water status of plants in many species [23], but especially in woody crops like the Japanese plum [24], as it is directly related to climatic and soil conditions. Stomatal conductance (Gs) is linked to the degree of the opening of the stomata. Usually, stomatal closure takes place during periods of drought to limit water loss by evapotranspiration, thus acting as an early physiological mechanism to reduce dehydration damage to water transport tissues [25,26]. Several works have also demonstrated the potential of Ψstem as a water status reference [27,28], mainly in species of anisohydric behavior as is the case of the Japanese plum cv. "Angeleno" [29]. Water stress can also affect the photosynthetic rate (Fn) of leaves, either through stomatic opening [30] or by intrinsically altering the photosynthesis process [31], and can also be considered a physiological water status indicator. Under mild to severe drought conditions, the basic plant organization structure could be damaged, giving rise to the inhibition of carbon assimilation and damaging photosynthetic apparatus [32]. The fraction of intercepted photosynthetically active radiation (FIPAR) determines the production of photosynthates, which influence plant growth, productivity and fruit quality [33]. One response of plants to water stress is to reduce the amount of intercepted radiation by reducing the rate of growth, modifying the leaf angle or accelerating the senescence of older leaves. However, using FIPAR as a reference for water status has its limitations as changes in the canopy can be slow or subtle, and so precise determinations are necessary and not always automatable. Casadesus et al. [34] observed that an automated irrigation system based on measurements of light interception by the canopy had slight deviations from actual irrigation requirements, but that these could be corrected through the use of additional measurements of, for example, air temperature or vapor pressure deficit. Sap flow (SF) is a continuous measurement related to daily transpiration, reaching its maximum value at solar midday and its minimum value during the night, coinciding with stomatal closure. Transpiration has an important role in physiology, the hydrological cycle and the global energy balance of crops in arid and semiarid regions [35–37]. Transpiration is controlled by the response of stomata to environmental factors such as solar radiation, vapor pressure deficit, air temperature, soil water availability and precipitation [36,38]. A progressive decrease in transpiration rates for a given crop and demand is indicative of a water stress situation [39]. Most of the aforementioned plant-based water stress indicators have different dimensions. For this reason, it is better to use the concept of signal intensity (SI), normalizing the absolute values of the indicator with respect to values in non-limiting soil water conditions [40]. The SI of the plant water indicator is a dimensionless variable, in which values above unity indicate that there is deficit irrigation and values equal to unity indicate that there is a lack of water stress [6]. In addition, the main characteristic that an indicator should have is sensitivity to water stress. Goldhamer and Fereres [41] defined the term sensitivity (S) as the ratio between the SI and the noise (coefficient of variation measurements for each indicator measured, CV).

Drought stress can cause serious damages in most crop plants including plum trees, but water excess can lead to root asphyxia, phytopathological problems, lower water use efficiency or unjustified increases in production costs. Due to the increasing water shortage worldwide, management of available soil water based on drought stress plant signals is becoming a crucial tool [42]. In the present study, we evaluate the relationship between the response of different plant physiological parameters to the dynamics of soil moisture in the face of increasing water stress and subsequent recovery in an early-maturing Japanese plum cultivar. The objective is to provide the necessary information to establish guidelines for the location and interpretation of soil moisture probes in automated or semi-automated irrigation scheduling systems, considering the possibility of using regulated deficit irrigation strategies. The aim of this work is to contribute to improving the usefulness of capacitance probes for the continuous measurement of SWC, as a reference or support for irrigation scheduling in fruit orchards. To do so, two key aspects are emphasized: (i) the selection of the most suitable points for the installation of the probes; and (ii) the interpretation of the measurements in relation to the physiological behavior of the tree in order to provide information for the decision-making process.

2. Materials and Methods

2.1. Experimental Plot Description and Climate

This work was carried out over three years (2015–2017) in a 1 ha orchard planted with early-maturing Japanese plum (*Prunus salicina Lindl.*) of the variety *"Red Beaut"*. The plot was located in an experimental farm of Badajoz, in southwestern Spain (latitude 38°51′19.06″ N, longitude 6°40′18.90″ W, datum WGS8), property of the Centre for Scientific and Technological Research of Extremadura (Regional Government of Extremadura). Plum trees were planted in the spring of 2005 with a 6 × 4 m spacing and in an east–west row orientation (5° toward the north). The soil of the plot is classified as an Anfisols according to the Soil Taxonomy [43], with slightly acidic pH values, low organic matter content (0.62%), high apparent density (Pa = 1.41 g cm^{-3}) and low cation exchange capacity (9.41 meq/100 g). Soil texture is loam, with an average 19.4% clay content, 40.2% silt content and 40.4% sand content. The soil was kept untilled and free of weeds through the application of herbicide treatments. Fertilization as well as control of pests or diseases were those commonly used in commercial orchard techniques.

The climate of the area is Mediterranean with a mild Atlantic influence, with a dry season from June to September (summer) and a wet season from October to May (winter) in which 80% of total precipitation falls. Average values in the area of reference evapotranspiration (ETo) and precipitation (P) were 1296 mm and 473 mm, respectively, for the 2007–2017 period. For that period, average maximum and minimum air temperatures were 23.38 °C and 9.42 °C, respectively. The hottest months are July and August. Maximum temperatures over 40 °C are recorded nearly every year, with peak values rarely over 45 °C. The coolest months are December and January. Temperatures below 0 °C are recorded every year, with minimum values rarely below −6 °C.

Sprouting was on 5 March 2015, 26 February 2016 and 2 March 2017. 31 October was considered the date of leaf fall in the three years.

2.2. Irrigation System, Irrigation Treatments and Experimental Design

Trees were irrigated daily using a drip system with a single lateral line per tree row located close to the base of the tree, with pressure-compensating drippers spaced at 1 m and with 4 l h^{-1} discharge rates (16 l/h/tree). The treatments were: control (CON) and drying (D). The experimental design of the plot was a completely randomized blocks with four replicates per treatment. Each experimental plot consisted of four adjacent rows, with each row containing four trees (16 trees/block). Data collection was carried out for the four central trees, with the other trees acting as guard trees including the four corner trees which were of a different cultivar (Figure 1a). In CON, trees were irrigated to cover crop water needs throughout the crop cycle. The irrigation dose applied was evapotranspiration (ETc) minus effective rainfall. The ETc was calculated following the procedure of Allen et al. [44], multiplying the reference evapotranspiration (ETo) by the crop coefficient (Kc). ETo was determined according to the Penman–Monteith method (ETo-PM) using the data obtained from a weather station located 200 m from the study plot (REDAREX), and the Kc was adjusted to the crop and local climatic conditions [45]. In the D treatments, fruit trees were irrigated as CON except in certain periods (preharvest and postharvest) in which irrigation was detained (drying) to induce a mild and severe water deficit in preharvest and postharvest, respectively (Table 1). The volume of applied water in each treatment was measured with a multi-jet meter with pulse output for remote reading (MTK, Zenner, Madrid, Spain) on a daily basis.

a)

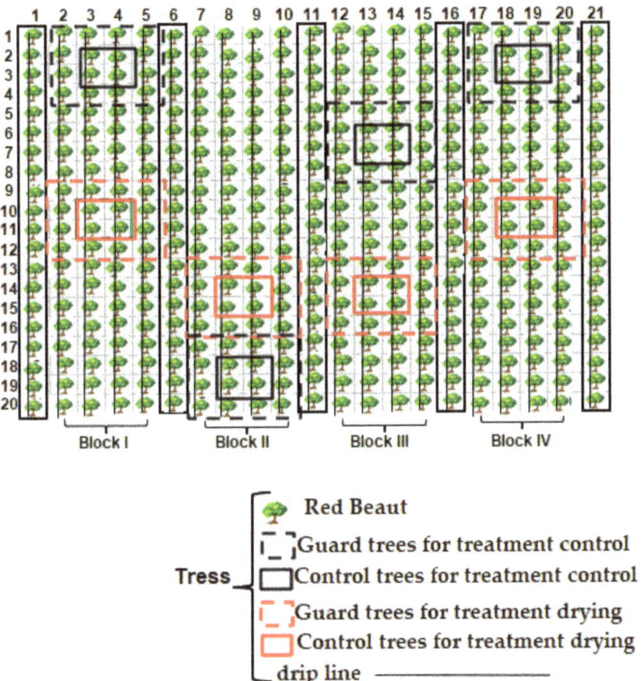

b)

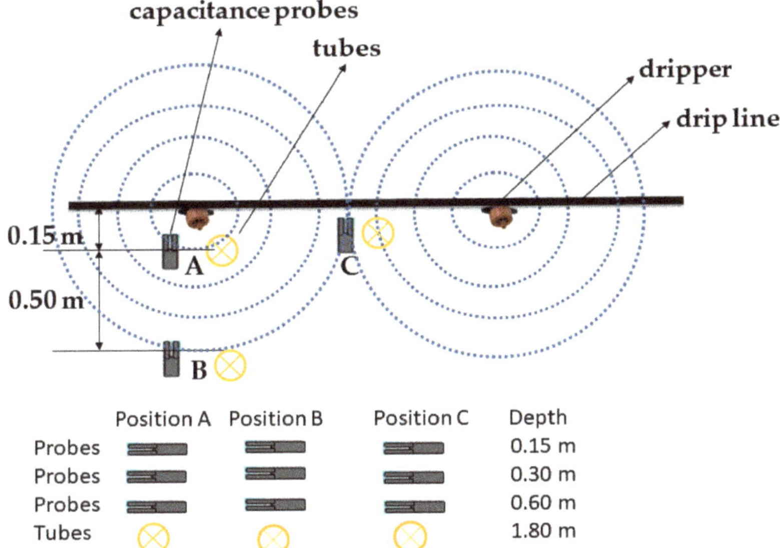

Figure 1. (**a**) Experimental design of the plot and (**b**) Localization of the neutron probe tubes and capacitance probes in the different positions and depths.

Table 1. Dates of the irrigation events.

	2015	2016	2017
ISD [1]	14 April 2015 (104)	19 April 2016 (109)	05 April 2017 (95)
IED [2]	15 October 2015 (288)	20 October 2016 (293)	08 November 2017 (312)
Drying-Pre		29 April 2016–14 June 2016 (119–165)	24 April 2017–31 May 2017 (114–151)
First Drying-Post	02 July 2015–27 July 2015 (183–208)	15 July 2016–27 July 2016 (196–208)	06 July 2017–01 August 2017 (187–213)
Second Drying-Post		12 August 2016–10 September 2016 (224–253)	

[1] Date when the irrigation season starts; [2] Date when the irrigation season ends; in brackets the day of the year.

2.3. Soil Water Content Probes

2.3.1. FDR Probes

In the D treatment, three trees in one block were selected to continuously monitor SWC. Nine 10HS capacitance probes (Decagon Devices Inc., Pullman, WA, USA) per tree were installed at different positions (position A, position B and position C) in relation to the dripper located under the tree canopy (Figure 1). Position A was located 0.15 m from the dripper to the alley, position B at 0.50 m from the dripper to the alley and position C between two drippers. In each position, the probes were installed at three depths: 0.15, 0.30 and 0.60 m. All probes used the general calibration for mineral soils proposed by the manufacturer and were connected to a datalogger (CR1000, Campbell Scientific Inc., Logan, UT, USA) which stored the data once every 5 min. The values obtained with each of the soil moisture probes were normalized every year calculating the RSWC. For this purpose, an interpolation was made to assign the value 1 to the highest value reached with each of them and the lowest value was assigned a 0.

2.3.2. Neutron Probes

Three other trees were selected in the D treatment to take the measurements with neutron probes (CPN 503DR Hydroprobe, CPN International Inc., Port Chicago Highway, CA, USA). Three access tubes (2.1 m long) were installed for each tree in the same position with respect to the dripper where the FDR probes were installed (Figure 1b). Soil water content (SWC) was monitored weekly throughout the irrigation season from a depth of 0.30 m to 1.8 m at intervals of 0.30 m. The neutron probe readings were calibrated according to the experimental equation:

$$\Theta = 0.00015 \times \text{N/SC} - 0.488 \ (R^2 = 0.97; p < 0.0001) \tag{1}$$

where Θ is the volumetric soil water content ($m^3 m^{-3}$), N is the neutron probe count reading and SC is the standard count reading.

2.4. Plant Measurements

2.4.1. Plant Water Status

The Ψstem was measured from two to three times a week from the beginning of the irrigation campaign to the end, between 13:00 and 15:00 h solar time, with a pressure chamber (Model 3005, Soil Moisture Equipment, Santa Barbara, CA, USA) during the drought periods, and once a week outside of these periods. Determinations were carried out on four trees per single plot: two mature and shaded leaves per tree were selected close to the base of the trunk and covered with aluminum foil at least 2 h before measurements started [23].

2.4.2. Photosynthesis and Leaf Stomatal Conductance

The Gs and Fn measurements were taken from just before the drought period until recovery from the drought period using an LI-6400XT device (Licor Inc., Lincoln, NE, USA) at midday (11:30 ± 13:30 h local solar time) every 2–3 days and on the same days as for Ψstem. Measurements were carried out on eight trees per treatment on four fully-developed young green leaves per tree on clear days, making a total of thirty-two replicated measurements per treatment. The device was calibrated before use on each occasion using factory calibration. In 2016, the LI-6400XT device broke down and Gs and Fn measurements could not be taken throughout the entire irrigation season.

2.4.3. Sap Flow

Sap flow was measured continuously using the compensation heat pulse method [46] combined with the calibrated average gradient technique [47] when sap velocities were low. Sap flow probes were installed on four trees of each treatment. The probes used were designed and produced at the Instituto de Agricultura Sostenible (CSIC, Córdoba, Spain). Each probe measures the heat pulse velocity at four depths in the xylem, spaced 10 mm apart, so that heat-pulse velocities were obtained at 5, 15, 25 and 35 mm below the cambium [47]. The probes were installed at 50 cm height from the soil, and measurements were taken continuously (every 30 min) from the date of installation in 2015 until the end of 2017. The system was controlled by a datalogger (CR1000, Campbell Scientific Inc., Logan, UT, USA). Due to variability in sap velocity and the thickness of conductive xylem around the perimeter of the tree [48], some uncertainty arises when the transpiration of each tree is calculated by azimuth integration of sap flow measurements obtained from a low number of probes. In order to solve this problem in treatment CON, the soil water balance method was applied in a dry period (8 days during spring) in which there was no rainfall to calibrate each sap flow probe.

$$ET = P + I - SR - DP \pm \Delta S - Es \tag{2}$$

$$\Delta S = \sum (\theta_i - \theta_{i-1}).\ Z \tag{3}$$

where ET is crop evapotranspiration, P is rainfall (mm), I is applied irrigation (mm), SR is surface runoff (mm), DP is deep percolation (mm), ΔS is the difference in soil-stored water at the beginning and end of a period, Es is soil evaporation, θ is soil water content (mm) and Z is the upper 1800 mm of the soil. The values of SR were considered negligible as the surface was flat and no runoff was observed. DP was considered null at the maximum observed depth (1.8 m). The values of θ were obtained from neutron probe measurements performed once a week throughout the irrigation period.

In order to estimate the soil evaporation coefficient (Ke) value, it was necessary to calculate a daily balance of the water present in the surface layer of the soil in order to determine the accumulated evaporation or depletion sheet.

$$De,I = De,i - 1 - Pi + SRi - Ii/fw + Esi/few + Tew,i + DPi \tag{4}$$

$$Esi = ke.ETo \tag{5}$$

where De,i is accumulated evaporation (exhaustion) after complete wetting at the end of day i [mm], De,i − 1 is accumulated evaporation (depletion), after complete wetting, originating from the exposed and wetted upper soil fraction at the end of day i − 1 (mm), fw is the fraction of the soil surface moistened through irrigation, Esi is evaporation on day i (mm), few is the exposed and wetted soil fraction, and Tew is transpiration that occurs in the exposed fraction and moistened from the surface layer of the soil on day i (mm). De,i and De,i − 1 were considered null because the topsoil was close to field capacity after irrigation. The value of fw was taken as 0.3 since a drip irrigation system was used. The value of few was taken as 0.3, and Tew was ignored as the crop in question was considered to be one with roots deeper than 0.6 m after a trial-pit had been dug and root depth visualized.

Assuming that transpiration (Ep) is practically equal to the ET, an estimate of total water transpired between day i and day f was obtained:

$$\alpha = Ep \times (i\text{-}f)/SF(i\text{-}f) \tag{6}$$

where SF is total sap flow accumulated between days i and f, Ep is transpiration and α is a constant that serves to calculate the calibrated transpiration on any day of the year through the following equation:

$$Ep = SF \times \alpha \tag{7}$$

The time evolution of sap flow allowed assessment of the transpiration coefficient (KT) calculated as the ratio of transpiration to ETo. The relative transpiration (RT) was determined as follows:

$$RT = K_{T,D}/K_{T,CON} \tag{8}$$

where K_T is the transpiration coefficient for the irrigation drying (D) and control (CON) treatments in the year of measurement.

The normalized values of RT provided us with a datum against which our observations of sap flow could be referenced. This datum also allowed us to interpret changes either in relation to a plant-induced behavioral modification, or a response due to changes in the environment of the soil.

2.4.4. The Canopy Photosynthetically Active Radiation Interception

The fraction of photosynthetically active radiation intercepted at solar noon by the canopy was measured weekly before the beginning of the drought period, during the drought periods and in the recovery from the drought period between 12:00 and 13:00 solar time with a linear ceptometer (probe length 80 cm; Accupar Linear PAR, Decagon Devices, Pullman, WA, USA). For 1 tree per treatment, the mean FIPAR (FIPARm) was taken as the average of 40 measurements taken at fixed positions by the ceptometer placed at soil level. The daily FIPAR (FIPARd) was calculated using the following linear relationship [49]:

$$FIPARd = 0.9427\ FIPARm + 0.0562\ (R^2 = 0.99;\ p < 0.0001) \tag{9}$$

2.5. Signal Intensity, Noise and Sensitivity

Signal intensity (SI) was calculated as the ratio between the values (V) of the D and CON treatments. $SI = V_D \times V_{CON}^{-1}$ in the case of Ψstem and FIPARd, and $SI = V_{CON} \times V_D^{-1}$ in the case of Fn, Gs and SF. To determine the noise, the coefficient of variation (CV) of the measurements for each indicator was used.

The sensitivity of the indicators was determined using two algorithms:

- Traditional method (S), as proposed by Goldhamer et al. [50]: S is always higher than 0, and the higher its value the greater the sensitivity.

$$S = SI/CV \tag{10}$$

- Corrected sensitivity (S*), as proposed by De la Rosa et al. [6]: The interpretation of the values obtained with this algorithm is as follows:

$$S^* = SI^{-1}/CV \tag{11}$$

(a). $S^* > 1$: indicates sensitivity to water deficit.
(b). $1 > S^* > 0$: The noise is greater than the increase in signal intensity. Therefore, there are no differences between treatments.
(c). $S^* = 0$: no differences between treatments, not sensitive to water deficit.

(d). $S^* < 0$: anomalous behavior.

2.6. Statistical Analysis

A means comparison with t-tests for independent samples was used for the statistical analysis of the data using the statistical package IBM SPSS version 24.0 for Windows (IBM Corp. Armonk, NY, USA).

3. Results

3.1. Climatic Conditions and Water Applied

In Table 2, it can be observed that the driest year from sprouting to leaf fall was 2017, which had the lowest rainfall and mean relative humidity and the highest annual ETo. The rainfall for this period was between 108 mm and 309 mm and during the pre-sprouting period (from leaf fall to sprouting) ranged from 158 to 438 mm. In 2016, pre-sprouting rainfall was high in relation to the average for the area. The highest mean temperature (Tmean) values were reached during the postharvest period coinciding with the summer months. Spring 2016 had a colder Tmean in the preharvest period and the Tmean was similar in the three years of study in the postharvest periods. The annual ETo-PM was similar in the three study years, with higher values in the postharvest period. The amount of water applied in the C treatment was 814 mm in 2017, which was notably higher than in the other years. The reduction of water applied in the D treatment compared to the CON treatment was 6%, 45% and 37% for 2015, 2016 and 2017, respectively.

Table 2. Mean temperature (Tmean), mean relative humidity (RHmean), accumulated rainfall, accumulated evapotranspiration (ETo) and irrigation applied in pre- and post-harvest periods and annually.

Year	Phases	Tmean	RHmean	ETo-PM	Irrigation (mm)		Rainfall (mm)	
		(°C)	(%)	(mm)	CON	D	LF-S [4]	S-LF [3]
	Pre [1]	19.6	56	276	168	172	158	272
2015	Post [2]	23.1	57	720	534	489		
	Annual	16.5	68	1310	702	661	370	
	Pre [1]	17.8	68	186	33	4	438	291
2016	Post [2]	23.7	54	730	570	327		
	Annual	16.4	71	1247	603	331	519.5	
	Pre [1]	19.7	56	346	245	66	214	108
2017	Post [2]	23.1	52	805	569	452		
	Annual	17.0	64	1383	814	519	284	

RH is relative humidity; ETo-PM is reference evapotranspiration calculated through the Penman-Monteith equation; CON is control treatment; D is drying treatment; [1] from the beginning of the irrigation season to harvest of the current year; [2] from harvest to the end of irrigation season of the current year. LF = leaf fall; S = sprouting; [3] period from the previous year's leaf fall to sprouting; [4] period from sprouting of current year to leaf fall.

3.2. Signal Intensity, Noise and Sensitivity

To compare the different responses to water deprivation in both continuously recorded plant-based measurement (SF) and discretely measured plant-based measurements (Ψstem, Fn, Gs and F$_{IPAR}$d) were calculated: signal intensity (SI), the signal noise evaluated as the coefficient of variation (CV), sensitivity (S) as SI/CV and corrected sensitivity (S*) as SI^{-1}/CV during the irrigation season (Tables 3–5).

Table 3. Mean values of each parameter in the CON and D treatments, signal intensity, coefficient of variation and sensitivity determined according to the traditional (S) and corrected (S*) methods in 2015.

Parameters 2015		A (182DOY)	SPost1 (208DOY)	RPost1 (288DOY)	Average
Ψstem	SI	1.38	1.53	1.19	1.37
	CV	0.18	0.19	0.15	0.17
	S	7.80	9.99	8.22	8.67
	S*	4.10	4.63	5.91	4.88
SF	SI	0.83	0.96	0.91	0.90
	CV	0.31	0.29	0.28	0.29
	S	2.67	3.34	3.24	3.08
	S*	3.86	3.61	3.89	3.79
Fn	SI	0.95	1.24	1.36	1.18
	CV	0.25	0.29	0.39	0.31
	S	3.77	4.82	3.4	4.00
	S*	15.67	17.99	6.64	13.43
Gs	SI	0.91	1.67	1.28	1.29
	CV	0.36	0.54	0.63	0.53
	S	2.50	3.31	2.72	2.84
	S*	1.10	0.63	0.62	0.78
FIPARd	SI	1.04	1.02	1.00	1.02
	CV	1.21	1.25	1.09	1.18
	S	0.86	0.81	0.94	0.87
	S*	0.96	0.98	0.99	0.98

A is the period previous to irrigation cut-off; SPost1 is the first period without irrigation in postharvest; RPost1 is the first recovery period in postharvest; Ψstem is midday stem water potential; SF is sap flow; Fn is net photosynthesis; Gs is stomatal conductance; FIPARd is the fraction of daily photosynthetically active radiation canopy interception; DOY is day of the year.

Table 4. Mean values of each parameter in the CON and D treatments, signal intensity, coefficient of variation and sensitivity determined according to the traditional (S) and corrected (S*) method in 2016.

Parameters 2016		A (118)	SPre (165)	RPre (195)	SPost1 (208)	RPost1 (210)	SPost2 (253)	RPost2 (293)	Average
Ψstem	SI	0.83	1.07	0.99	1.34	1.03	1.38	1.09	1.10
	CV	0.25	0.23	0.19	0.17	0.46	0.16	0.19	0.24
	S	3.37	5.72	5.04	7.96	2.91	14.71	9.03	6.96
	S*	4.89	4.65	5.25	4.47	2.35	8.08	7.6	5.33
SF	SI	0.82	0.74	0.75	0.64	0.87	0.86	0.8	0.78
	CV	0.50	0.44	0.6	0.62	0.55	0.47	0.5	0.53
	S	1.64	1.69	1.24	1.03	1.59	1.88	1.61	1.53
	S*	1.28	3.29	2.23	2.49	2.15	2.56	2.53	2.36
FIPARd	SI	0.98	1.06	1.04	1.03	1.07	1,00	0.94	1.02
	CV	0.88	0.87	0.85	0.87	0.89	0.96	1.25	0.94
	S	1.11	1.21	1.22	1.19	1.19	1.05	0.79	1.11
	S*	1.02	0.95	0.96	0.97	0.94	0.99	1.07	0.99

A is the period previous to irrigation cut-off; SPre is the period without irrigation in preharvest; RPre is the recovery period in preharvest; SPost1 is the first period previous to irrigation cut-off in postharvest; RPost1 is the first recovery period in postharvest; SPost2 is the second period previous to irrigation cut-off in postharvest; RPost2 is the second recovery in postharvest; day of the year in brackets. Ψstem is midday stem water potential; SF is sap flow; Fn is net photosynthesis; Gs is stomatal conductance; FIPARd is the fraction of daily photosynthetically active radiation canopy interception.

Table 5. Mean values of each parameter in the CON and D treatments, signal intensity, coefficient of variation and sensitivity determined according to the traditional (S) and corrected (S*) method in 2017.

Parameters 2017		A (113)	SPre (151)	RPre (186)	SPost1 (213)	RPost1 (312)	Average
Ψstem	SI	0.98	1.24	1.12	1.68	1.13	1.23
	CV	0.23	0.17	0.16	0.18	0.16	0.18
	S	4.61	7.69	7.68	9.23	7.44	7.33
	S*	4.69	5.18	6.24	3.78	5.62	5.10
SF	SI	0.65	0.79	0.76	1.07	0.87	0.828
	CV	0.47	0.35	0.37	0.39	0.36	0.388
	S	1.45	2.29	2.1	2.74	2.49	2.21
	S*	3.51	3.66	3.54	2.5	3.23	3.29
Fn	SI	0.96	1.16	0.95	1.22	1.18	1.09
	CV	0.41	0.32	0.26	0.23	0.32	0.31
	S	2.34	3.95	3.76	5.24	3.74	3.81
	S*	2.54	2.93	4.13	3.59	2.75	3.19
Gs	SI	0.94	1.5	1.01	1.7	1.53	1.34
	CV	0.46	0.46	0.76	0.63	0.57	0.58
	S	2.06	3.34	1.45	2.88	2.69	2.48
	S*	2.29	1.8	1.42	1.49	1.35	1.67
Fɪᴘᴀʀd	SI		0.99	0.97	0.97	0.91	0.96
	CV		1.07	0.9	0.8	0.83	0.90
	S		0.92	1.08	1.21	1.1	1.08
	S*		1.01	1.04	1.03	1.09	1.04

A is the period previous to irrigation cut-off; SPre is the period without irrigation in preharvest; RPre is the recovery period in preharvest; SPost1 is the first period previous to irrigation cut-off in postharvest; RPost1 is the first recovery period in postharvest; day of the year in brackets. Ψstem is midday stem water potential; SF is sap flow; Fn is net photosynthesis; Gs is stomatal conductance; Fɪᴘᴀʀd is the fraction of daily photosynthetically active radiation canopy interception.

In 2015, the SI of all parameters, except Fɪᴘᴀʀd, increased as water stress progressed and responded in the opposite direction when irrigation was restored. During the drying cycle, the highest SI values were 1.53 and 1.67, reached by Ψstem and Gs respectively. The lowest CV values were measured in Ψstem (between 0.15 and 0.19), while the rest of the plant parameters (Fn, SF, Gs and Fɪᴘᴀʀd) had values between 0.25 and 1.25. When sensitivity was calculated with the traditional method (S), Ψstem presented the highest values followed by Fn, SF, Gs and Fɪᴘᴀʀd. When corrected sensitivity (S*) was used, Fn had the highest values followed by Ψstem, SF, Fɪᴘᴀʀd and Gs.

In 2016, the SI of SF decreased in the three drying cycles, whereas Fɪᴘᴀʀd only decreased in the second and third drying cycles. In contrast, the SI of Ψstem increased in the three drying cycles, presenting the highest values in the third drying cycle. Once irrigation was restored again, the SI of Ψstem decreased. The lowest average CV value was obtained in Ψstem. The results obtained with S and S* were the same, with Ψstem presenting the highest values followed by SF and Fɪᴘᴀʀd.

In 2017, the SI of all the measurements increased in the first and second drying cycles, except for Fɪᴘᴀʀd. The highest values of SI were reached by Gs during the drying cycles. The lowest average CV value was measured in Ψstem (0.18), while the other plant water parameters (Fn, SF, Gs and Fɪᴘᴀʀd) presented the highest values (between 0.31 and 0.90). With respect to S, Ψstem presented the highest values followed by Fn, Gs, SF and Fɪᴘᴀʀd. When S* was used Ψstem also presented the highest values followed by SF, Fn, Gs and Fɪᴘᴀʀd.

3.3. Seasonal Dynamics

3.3.1. Stem Water Potential and Gas Exchange

In the three years of the experiment, the average Ψstem of the CON treatment was −0.80 MPa, with values higher than −0.90 MPa for most of the crop cycle. In 2015, the Ψstem values in both treatments were similar and lower during the preharvest period than in 2016 and 2017 and increased after the harvest. This was the year with the lowest pre-sprouting rainfall (LF-S). In 2016 and 2017, the initial (pre-irrigation) Ψstem of the trees was around −0.62 MPa with a declining trend as the crop season advanced, which was more pronounced in 2017 because of lower rainfall in the postharvest period.

In 2015 (Figure 2a), Ψstem for the CON treatment decreased from an initial value of −0.92 MPa at the start of the irrigation campaign to a final value of −0.81 MPa. The D treatment responded to the suppression of irrigation with a decrease in Ψstem to a minimum value of −1.32 MPa at 18 days after irrigation cut-off. Once irrigation was restored, Ψstem recovered to the CON values after 21 days.

In 2016 (Figure 2b), the Ψstem for the CON treatment fell from an initial value of −0.73 MPa to a final value of −0.94 MPa. With respect to the D treatment, Ψstem fell from initial values of −0.57 MPa to minimum values of −0.92 MPa 43 days after the first suppression of irrigation. In the second drought period, the minimum value reached was −1.19 MPa after 11 days without irrigation. In the third drought period, there followed a pronounced fall in Ψstem values, reaching a minimum value of −1.58 MPa after 21 days without irrigation. After irrigation was restored in the D treatment, the Ψstem recovered to the CON values at 8 days after the first and 14 days after the second and third drought period.

In 2017 (Figure 2c), the CON treatment had the highest Ψstem at the start of the season on DOY 95 (05 April 2017) and the lowest water status before the end of the season on DOY 225 (02 October 2017). In the preharvest drying cycle, the D treatment had the minimum Ψstem value just before rewatering (Figure 2c). The Ψstem fell from initial values of −0.61 MPa to minimum values of −1.06 MPa 36 days after the irrigation cut-off in preharvest. In the first postharvest drying cycle, Ψstem decreased sharply from −0.66 MPa to a minimum value of −1.60 MPa after 22 days without irrigation. These differences between the first and second drying-cycles were due to the spring rains that occurred in the second cycle and the higher evaporative demand. Once irrigation was restored again in the D treatment, Ψstem values recovered to the CON values at 13 days after the first drought period and 83 days after the second drought period.

For the 3 years of the evaluation, fluctuations in VPD consistently affected Ψstem values.

The values of Fn and Gs during the drying cycles are presented in Figure 3. In 2015, Fn and Gs values varied in the CON treatment, presenting a tendency to increase as the irrigation season progressed (Figure 3a,b). In this treatment, Fn and Gs values averaged 19.51 (μmolm^{-2}s^{-1}) and 0.25 (mmolm^{-2}s^{-1}), respectively. In the initial phase of the drying cycle, the suppression of irrigation in D caused a reduction in Fn and Gs values in relation to CON. The slight differences in Fn were not significant until DOY 196. In contrast, significant differences were found in Gs from DOY 191, reflecting an important degree of water use efficiency at leaf level. After restarting irrigation in D, Fn and Gs recovered to the CON values at 21 days on DOY 229.

Figure 2 shows the seasonal evolution of Ψstem and vapor pressure deficit (VPD) for each study year.

In 2017, the Fn and Gs values also varied over time. The two gas exchange parameters presented a similar seasonal pattern, with significant differences between treatments from DOY 128 in the preharvest drying cycle and from DOY 195 in the postharvest drying cycle. An increasing trend of Fn and Gs can be seen as the irrigation season advanced in the CON treatment, from an initial value of 10.90 (μmolm^{-2}s^{-1}) to a final value of 16.90 (μmolm^{-2}s^{-1}) for Fn and from an initial value of 0.11 (mmolm^{-2}s^{-1}) to a final value of 0.26 (mmolm^{-2}s^{-1}) for Gs. The D treatment responded to irrigation cut-off with a decrease in Fn and Gs to a minimum value of 9.58 (μmolm^{-2}s^{-1}) and

0.07 (mmolm^{-2}s^{-1}) at 14 days after the start of the preharvest drying cycle. In the postharvest drying cycle, the minimum value reached was 11.4 (μmolm^{-2}s^{-1}) on DOY 212 for Fn and 0.06 (mmolm^{-2}s^{-1}) for Gs after 25 days without irrigation.

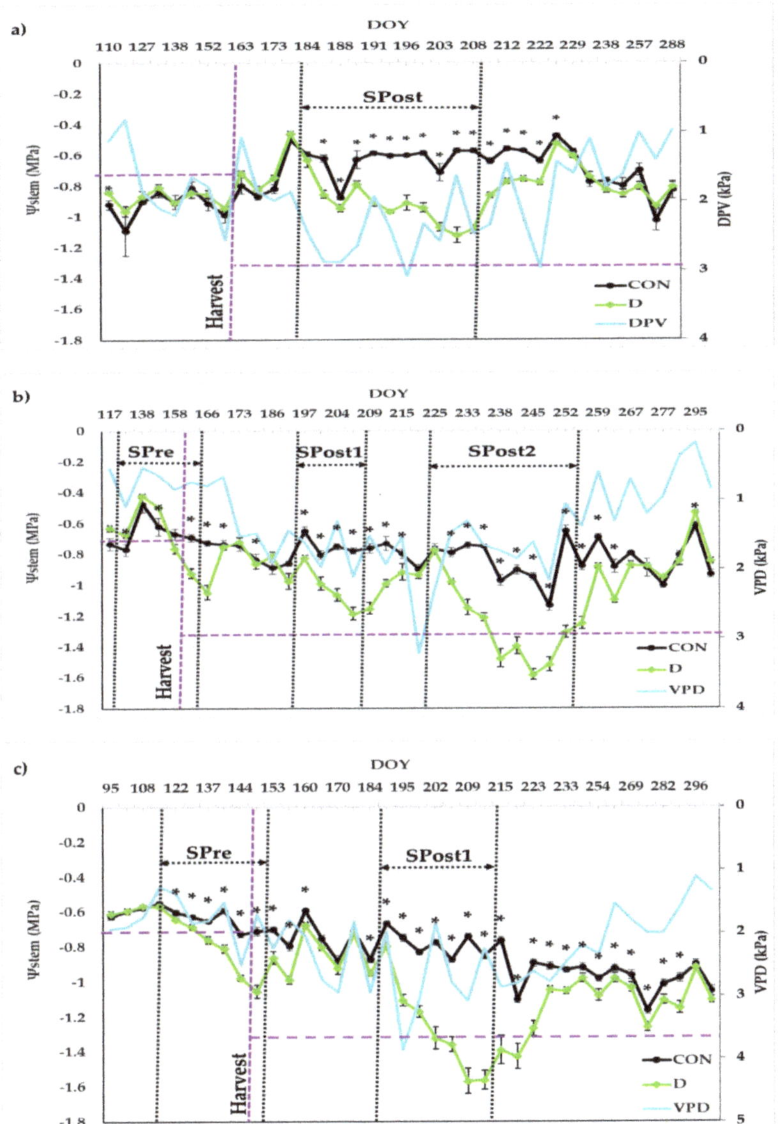

Figure 2. Seasonal patterns of vapor pressure deficit (VPD) and midday stem water potential (Ψstem) corresponding to the control (CON) and drying (D) treatments in "Red Beaut" plum trees during 2015 (**a**), 2016 (**b**), and 2017 (**c**). Each value is the mean of 32 measurements ± standard error. An asterisk (*) indicates statistically significant differences between treatments at $p < 0.05$. The vertical black dashed line indicates the start and end of the drying period. DOY is day of the year. The vertical violet dashed line indicates when harvest took place. The horizontal violet dashed line indicates the threshold established for each crop phase recommended by Samperio et al. [24]. Pre is the period from fruit set to harvest of the current year. Post is the period from harvest to the onset of leaf fall of the current year.

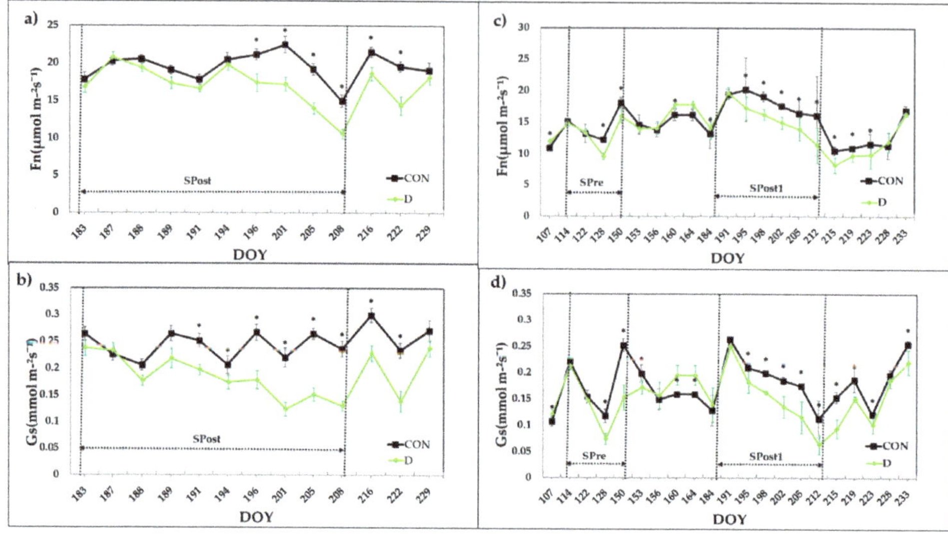

Figure 3. Seasonal pattern of: (**a**) net photosynthetic rate (Fn); (**b**) stomatal conductance (Gs) in 2015; (**c**) net photosynthetic rate (Fn) and (**d**) stomatal conductance (Gs) in 2017. Each value is the mean of 32 measurements ± standard error. An asterisk (*) indicates statistically significant differences between treatments at $p < 0.05$. The vertical black dashed line marks the start and end of the drying period. DOY is day of the year.

Table 6 shows the days that elapsed between irrigation cut-off and the existence of significant differences in Ψstem, Fn and Gs between the treatments in each drying cycle, and the days necessary for them to be equal again after restoring irrigation. In all drying cycles the first significant differences were always found in Ψstem after irrigation cut-off and ranged from 1 to 8 days. However, recovery after irrigation was restored was faster for Fn and Gs, ranging from 2 to 21 days. In 2017, it took 83 days of recovery for the Ψstem of D to equal that of CON.

Table 6. Period of time that elapses since each parameter detects water stress in each drying cycle and time that the D treatment takes to recover.

Parameters	Year	T_SPre (days)	T_RPre (days)	T_SPost1 (days)	T_RPost1 (days)	T_SPost2 (days)	T_RPost2 (days)
	2015			4	21		
Ψstem	2016	5	8	1	14	5	14
	2017	8	13	4	83		
Fn	2015			13	21		
	2017	14	2	8	15		
Gs	2015			8	21		
	2017	14	5	8	15		

T_SPre, T_SPost1 and T_SPost2 are the number of days that passed between the suppression of irrigation in the first, second and third drying cycles, respectively, and the detection of water stress; T_RPre, T_RPost1 and T_RPost2 are the number of days between the end of the first, second and third drying cycles, respectively, and the recovery of treatment D to the values of treatment CON.

3.3.2. Sap Flow

The relative transpiration (RT) for the 3 years of study is presented in Figure 4. Due to a failure in the sap flow probes, the RT was not calculated during a period of fifteen days in 2015, between DOY 153 and 167 (Figure 4a). Given the calibration procedure, the average values at the beginning of the experiment were close to 1. Once the drought period started, RT decreased to values below 1, reaching a minimum just before irrigation was restored. Soon after the restoration of irrigation, RT increased to values close to or even above 1 (Figure 4a).

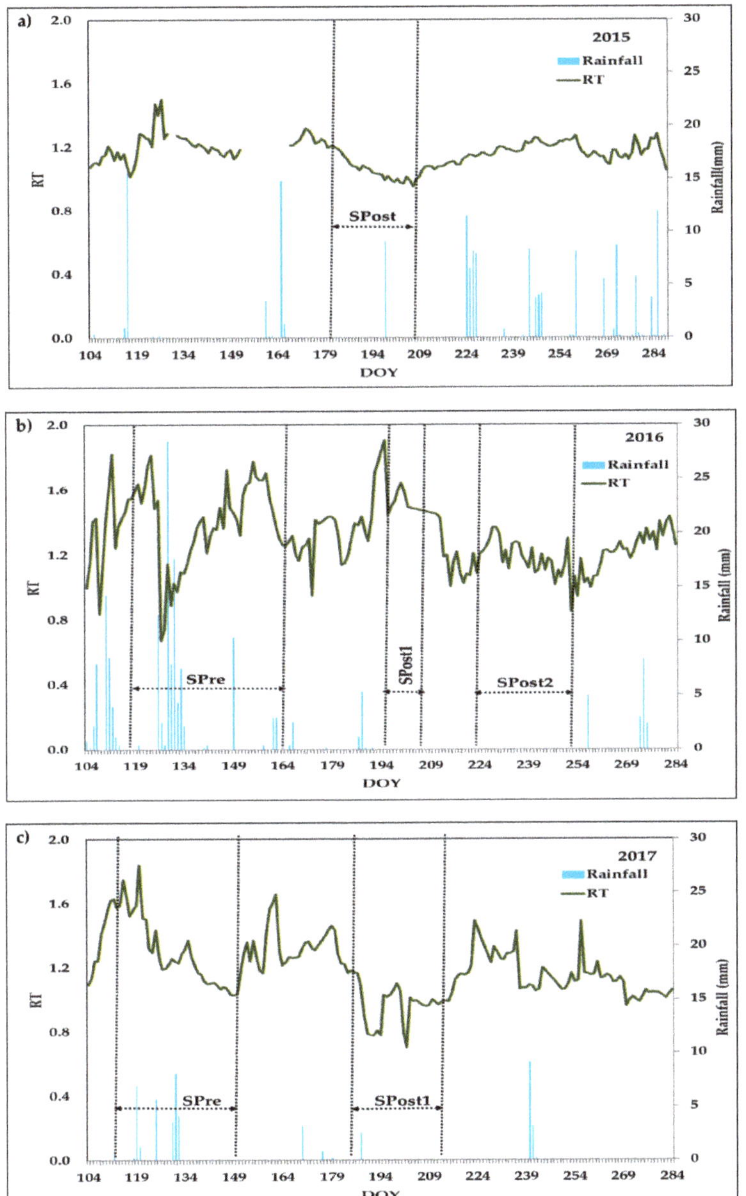

Figure 4. Evolution of relative transpiration (RT) for the 3 years of the study: (**a**) 2015, (**b**) 2016 and (**c**) 2017. Values correspond to the average of four trees. The vertical black dashed line marks the start and end of the drying period. DOY is day of the year.

In 2016 (Figure 4b), the daily values of RT fluctuated over time. Before the start of the different drying cycles, the average RT values were close to or even above 1, and after irrigation cut-off, they reached minimum values close to 0.6 on DOY 127, 1.44 on DOY 211 and 0.84 on DOY 252 for the first, second and third drying cycles, respectively. At the end of the irrigation season, RT was at a value above 1.

As in the previous years, in 2017 daily RT values fluctuated (Figure 4c). When the irrigation season began the average RT values were close to 1. Before the start of the first drying cycle, RT reached maximum values close to 1.8. Once the first drying cycle had started, RT decreased to values below 1, reaching a minimum just before the restoration of irrigation. In the first period of drought, RT was highly influenced by rainfall events. Just before the start of the second drying cycle, the RT values were close to 1. After the start of the second drying cycle, RT decreased, reaching a minimum value close to 0.7 on DOY 200. Once irrigation was restored, the RT values increased.

Despite the fact that RT responded to irrigation cut-off, constant fluctuations were observed throughout the vegetative cycle, indicating that other factors besides the availability of water in the soil were influencing the transpiration of the tree.

3.3.3. Soil Water Content (FDR Probes)

The seasonal evolution of RSWC, measured with FDR probes, is presented in Figure 5 for the D treatment in the different years of study. In Figure 5a–c, corresponding to 2015, only the period of the drying cycle is presented. Due to the calibration procedure that was carried out, the average values at the beginning of the drying cycle were close to 1. The probes located in position A (Figure 5a) were very sensitive to irrigation and had a very marked amplitude response between the maximum before the start of the drying cycle and the minimum at the end of the drying cycle. After irrigation cut-off, the RSWC fell very quickly in an initial stage between DOY 183 and DOY 191, and then slowed down to values close to 0 just before irrigation was restored. Soon after the restoration of irrigation, RSWC increased to values close to or even above 1 (Figure 5a). Probes located in position C (Figure 5c) at 0.15 and 0.30 m depth followed the same pattern as sensors located in position A, but these probes responded later to the suppression of irrigation. Probes located at 0.6 m depth (Figure 5c) showed a slower response to irrigation cut-off, and these probes presented a lower amplitude between the minimum just before irrigation cut-off and the maximum after its restoration. The probes located in position B (Figure 5b) were the last to respond to irrigation cut-off, having a progressive decrease in RSWC. After restarting irrigation, these probes detected almost no irrigation water.

In 2016 (Figure 5d–f), the year with the highest number of rain events during the vegetative cycle of the crop, RSWC fluctuated over time, representing the year with the highest number of rain events during the vegetative cycle of the crop. During the preharvest drying, the probes responded to the spring rains with an initial increase in RSWC after the suppression of irrigation (Figure 5d–f). In position A (Figure 5d), RSWC decreased to a minimum just before the restoration of irrigation. The RSWC values decreased slowly as depth increased. Just before the start of the first postharvest drying cycle, the RSWC values varied between 0.40 (probes located at 0.60 m depth) and 0.67 (probes located at 0.15 and 0.30 m depth). When the first postharvest drying cycle started, the decrease in RSWC varied between 0.17 (probes located at 0.6 m depth) and 0.44 (probes located at 0.15 and 0.30 m depth). In the last drying cycle, the three probes located in position A responded very fast to irrigation cut-off in the first 9 days before stabilizing. With respect to position C (Figure 5f), the progression of RSWC followed the same pattern as with the probes located in position A. However, only probes located at 0.30 m depth in position B (Figure 5e) followed the same pattern as the probes located in position A, but with less amplitude.

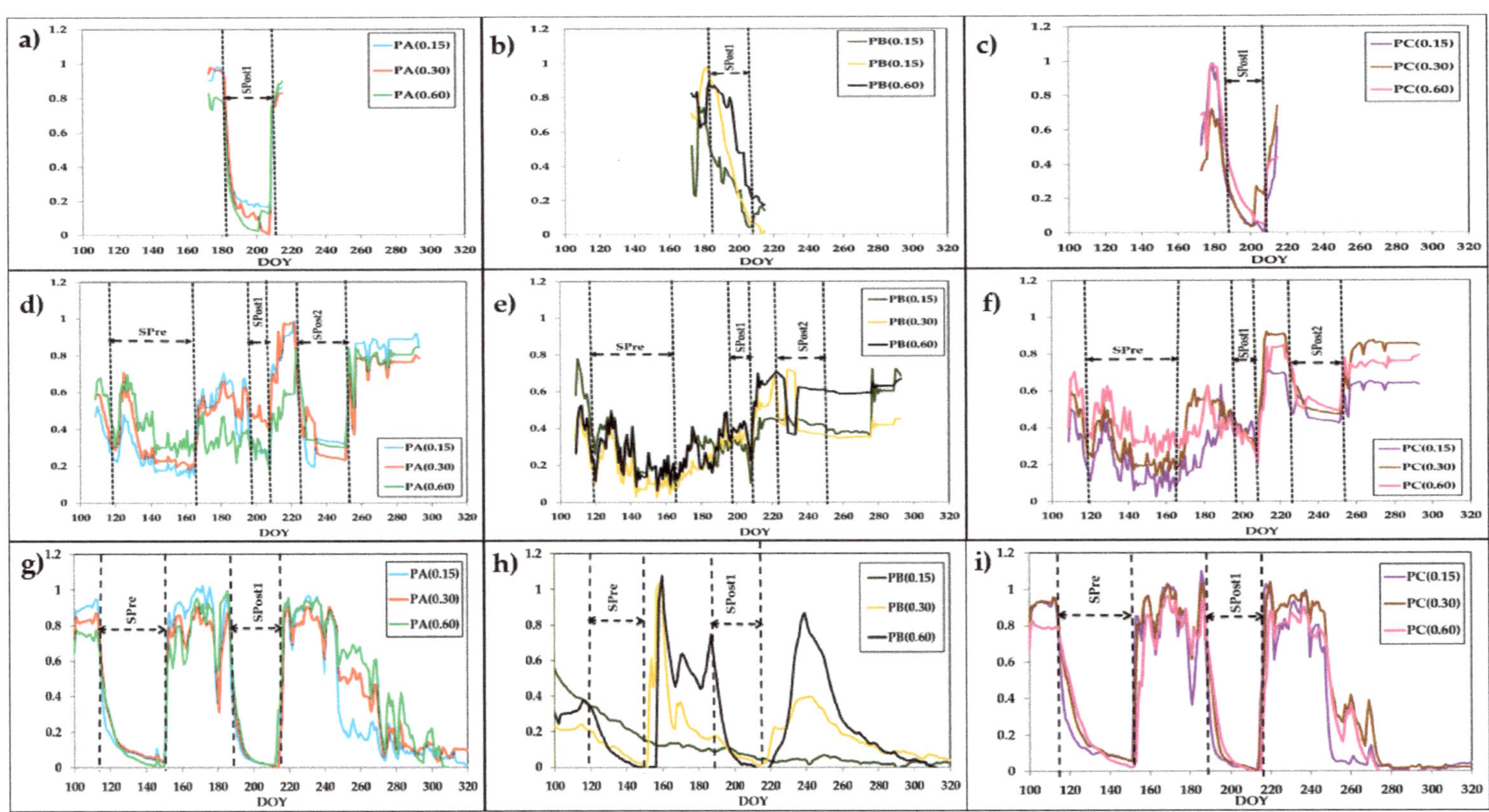

Figure 5. Seasonal evolution of relative soil water content (RSWC) measured by FDR probes located in different positions in the three years of study (**a–c**) 2015, (**d–f**) 2016 and (**g–i**) 2017. PA, PB and PC are positions A, B and C, respectively, where the capacitance probes were installed. The depth in meters to which the probe was installed is shown in brackets. Values are averages of three FDR sensors. The black vertical line corresponds to the beginning and end of each drying cycle. DOY, day of the year.

In 2017 (Figure 5g,i), after the beginning of the irrigation season, RSWC was higher than in 2016, the rainiest year. When the irrigation season began, the average RSWC values were close to 0.9 in both positions A and C. Before the preharvest drying cycle, RSWC reached maximum values close to 1. Once the first drying cycle started, RSWC decreased to values below 0, reaching a minimum just before irrigation was restored. Right before the start of the second drying cycle, the RSWC values were even above 1. From the start of the second drying cycle, RSWC decreased, reaching a minimum value close to 0 on DOY 213. Once irrigation was restored, the RSWC values increased until reaching a maximum value close to 1 on DOY 220. With respect to position B, the RSWC values of the probes located at 0.15 m depth decreased throughout the irrigation campaign from an initial value of 0.55 to a final value of 0. The probes located at 0.30 m and 0.60 m depth presented a similar pattern with RSWC decreasing progressively once the first and second drying cycles had started, but the probes located at 0.60 m responded later to the restoration of irrigation.

4. Discussion

4.1. Signal Intensity, Noise and Sensitivity

Many physiological processes in the plant can be affected by reducing soil water availability. Therefore, different plant measurements can be used as references for plant water status and to check the response to irrigation. For this purpose, they must be able to detect changes in water availability in short periods of time to optimize irrigation scheduling. In this work, we studied signal intensity (SI), the noise (CV) and the signal/noise ratio (S and S*) during the irrigation season in three different study years, both from indicators recorded continuously (SF) and from indicators recorded punctually (Ψstem, Fn, Gs and F$_{\text{IPAR}}$).

The similar SI of Ψstem and Gs to soil water depletion could be attributable to the fact that the main factor controlling Gs is xylem water potential, which determines the driving force for water transport between the bark and the xylem vessels [51]. In Japanese plum cv. "Angeleno", Blanco-Cipollone et al. [29] observed an anisohydric behavior and concluded that midday Ψstem can be a good reference for the water status of the plant in this species. De la Rosa et al. [6] studied maximum daily trunk shrinkage (MDS), trunk growth rate (TGR), Ψstem, Gs and Fn in extra-early nectarine trees and found that the plant indicator with the highest SI was Gs. However, in young almond trees, MSD was the indicator that presented the highest SI [52]. Other studies have also found that MSD had a higher SI than Ψstem in adult apple [53], young plum [54] and adult kaki [55]. In all these cases, MSD was more variable (higher CV) than Ψstem. In the present work, the CV values reached for Ψstem were also less variable than the other plant water indicators (Fn, SF, Gs and F$_{\text{IPAR}}$). F$_{\text{IPAR}}$ was the most variable indicator. Part of this variability may also be attributed to the fact that the F$_{\text{IPAR}}$ in fruit crops is influenced by planting density, the size and shape of the tree crown (which depends on tree age and the conduction system) and the leaf area index [56]. If to the high variability of FIPAR we add the time required to take the measurements, it turns out that this parameter is poorly adapted to support irrigation scheduling since it has to be adjusted by using some other measure due to deviations from the estimated water needs, as reported by Casadesus et al. [34]. The Gs and Fn leaf-to-leaf variability could be due to the effect on leaf transpiration of microenvironmental conditions [57], branch crop load [58], and leaf distance to fruit [59]. The SF also presented a high CV, which is probably caused by the variability that exists in sap velocity and the thickness of conductive xylem around the perimeter of the tree [48]. The Ψstem was the most sensitive indicator for detecting the initiation of stress in the three-year study and was the first one that showed significant differences between treatments, confirming that it is the most suitable indicator for early-maturing Japanese plum irrigation scheduling. Badal et al. [55] measured Ψstem, Gs, fruit growth rate and MDS in kaki trees and reported that Ψstem presented the highest S. Several authors have also observed that Ψstem shows the highest sensitivity values obtained by the traditional method, including in peach [60], pomegranate [61] and nectarine [6] trees. Tuccio et al. [62] measured the pre-dawn and midday leaf water potential, Ψstem, leaf temperature and Gs in one-year-old potted grapevines and identified Ψstem as the most sensitive indicator. However, the results obtained with S* indicate that Fn was the most sensitive indicator in 2015 and Ψstem in 2016 and 2017. The S is more influenced by the CV values, while the S* is influenced by the SI values as well as the CV values.

4.2. Plant Response to Soil Water Content Deprivation

The trees exhibited increasing water stress when irrigation was suppressed in the D treatment. According to Ψstem as the indicator of water status, the cut-offs of irrigation during the postharvest period led the plants to support more severe water stress than in preharvest: the plant had already used up rainwater reserves (outside the area of influence of the drippers), evaporative demand was higher, the period without irrigation was longer in some cases and the tree crown was fully formed. Samperio et al. [24] carried out previous studies on deficit irrigation strategies in an early-maturing Japanese plum crop (*Prunus salicina Lindl. cv. Red Beaut*) and recommended that Ψstem values should be above −0.7 MPa in the preharvest period and no less than −1.2 MPa in the postharvest period. However, Millán et al. [9] indicated that trees under regulated deficit irrigation could support more severe stress in the postharvest period, since the Ψstem values were above −1.3 MPa with no resulting loss in yield in "Red Beaut" Japanese plum cultivar. In our study, the Ψstem values were about −1.05 MPa in 2016 and 2017 during the drying cycles in the preharvest period in the D treatment. In the postharvest period, the Ψstem values in the D treatment were above −1.30 MPa in 2015, supporting more severe stress levels of around −1.60 MPa during the drying cycles in 2016 and 2017 (Figure 2). In this work, Ψstem during the drying cycles reached values lower than those recommended for this species and variety, both in pre- and post-harvest.

In order to efficiently use the measurements of soil water content provided by continuous measurement probes in the adjustment of irrigation scheduling, both to cover the needs of the crop and to establish deficit irrigation strategies, there are two key aspects: the location of the probes in relation to the dripper and the criteria for the interpretation of the measurements in the decision making process. Regarding the FDR probes, the driest year with the highest ETo was taken into account to establish a protocol for placing the probes in the optimum position. Figures 6 and 7 shows the temporal trends of soil moisture measured with FDR probes during the drying cycles of 2015 and 2017, which were the two years with the clearest response to irrigation cut-offs. The probes clearly react to the suppression of irrigation, with a sharp drop in RSWC immediately after each irrigation cut-off followed by a change in slope when irrigation is restored. This pattern was similar in the two drying cycles, although the second cycle presented a sharper fall of RSWC. The probes located furthest away from the dripper (position C at 0.3 and 0.6 m depth) and the probes located in position B at 0.15 m depth responded to the suppression of irrigation but with a greater delay in time than the rest of the probes. One approach to interpreting soil probe measurements is to analyze the physiological processes that are affected by variable water content in the soil (Tables 3–6). As can be seen in Figure 6, we identified a short period after irrigation cut-off in which the soil moisture content recorded by the probes closest to the dripper dropped rapidly, but no stress symptoms were detected in the tree. After that, in a second step, despite the existence of evidence of water stress (Ψstem), it did not affect the exchange of gases in the leaves (Gs and Fn). This period can be considered one of "low water stress level". When Fn and Gs are clearly affected it can be considered as "moderate stress" and below a Ψstem value of −1.2 MPa (Samperio et al., 2015) as "severe stress".

Now using 2017 as a reference, Figure 6 shows the evolution of RSWC in the different positions and the times when significant differences were detected between treatments in Ψstem, Fn and Gs. According to Samperio et al. [24], water stress should be avoided in this variety during the preharvest period (Ψstem = −0.7 MPa), which happened on DOY 128, and so it would not be advisable to allow a drop of RSWC below 0.2 in the probes located in position A in the preharvest period. In the postharvest period, the Ψstem detected significant differences between the two treatments 4 days before starting the second drying cycle on DOY 191. In the second drought period, significant differences between the two treatments were detected on DOY 195. For unstressed postharvest scheduling, the criteria would be similar to preharvest. Following the regulated deficit irrigation recommendations for this variety, Ψstem should not fall below −1.2 MPa in the postharvest period [24]. However, this happened on DOY 198 and RSWC fell to 0.1 according to the probes located in position C at 0.6 m.

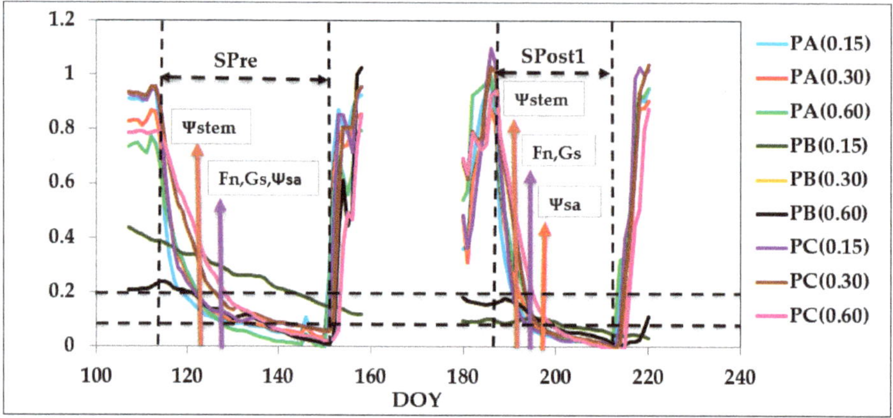

Figure 6. Temporal trends of soil moisture during the drying cycles in 2017 measured by FDR probes located in different positions. The black vertical line corresponds to the beginning and end of each drying cycle. The black horizontal lines correspond to the limits below which the RSWC value should not fall. DOY is day of the year; Ψstem is stem water potential; Fn is photosynthetic rate; Gs is stomatal conductance; Ψsa is the Ψstem recommended by Samperio et al. [24]. The arrows mark the soil moisture value when significant differences with the indicated physiological parameter are found or when the Ψsa is reached. PA, PB and PC are positions A, B and C, respectively, where the capacitance probes were installed. The depth in meters to which the probe was installed is shown in brackets. Values are averages of three FDR sensors.

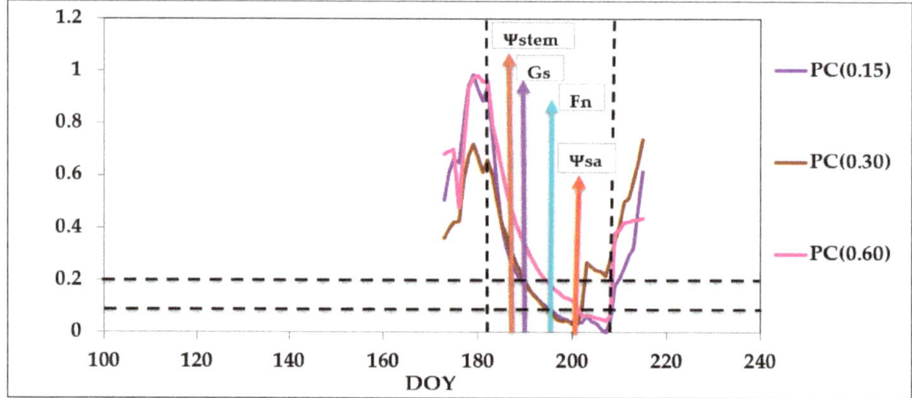

Figure 7. Temporal trends of soil moisture during the drying cycle in 2015 measured by FDR probes located in different positions. The black vertical lines correspond to the beginning and end of the drying cycle. The black horizontal lines correspond to the limits below which the RSWC value should not fall. DOY is day of the year; Ψstem is the stem water potential; Fn is the photosynthetic rate; Gs is stomatal conductance; Ψsa is the Ψstem recommended by Samperio et al. [24]. The arrows mark the soil moisture value when significant differences with the indicated physiological parameter are found or when the Ψsa is reached. PC is position C where the capacitance probes were installed. The depth in meters to which the probe was installed is shown in brackets. Values are averages of three FDR sensors.

In light of these results, the probes placed in the position between drippers (C) provide enough information to establish the irrigation levels: at 15 and 30 cm deep for non-stress situations (0.2 RSWC) and at 0.6 m deep for the lower limit recommended in postharvest (0.1 RSWC).

Figure 7 applies this criterion to the 2015 drying cycle: the limit of 0.2 applied to the two probes closest to the surface would mean a delay of 2 days in relation to the detection of stress by Ψstem; and the level of 0.2 for the deepest probe would anticipate by 1 day the minimum recommended level.

Many hypotheses have been proposed to explain the causes of variation in soil water extraction patterns by plants. For example, Atkinson [63] affirmed that the distribution of thin roots reflects the water extraction potential of a crop, and Nnyamah and Black [64] suggested that the water extraction pattern of a crop was similar to the distribution of thin roots when the soil water was not limiting. However, Clothier and Green [65] observed that root water uptake was more dependent on soil water availability than on thin roots distribution. Other studies suggest that soil water extraction by plants also depends on other factors such as the diameter of lateral roots [66], soil heterogeneity, root system structure, and the availability and partitioning of carbon in the roots [67], differences in xylem maturation and in the number and diameter of xylem vessels, as well as differences in the formation of endodermis and exodermis with the development of roots [68–71]. In this experiment, the fastest water extraction occurred in the areas closest to the dripper, repeating a similar extraction pattern, but displaced in time in more distant positions.

Several authors have used a mathematical model that simulates soil water dynamics for drip irrigation to identify the time stable representative positions (TSRPs) of the moisture sensors. It has been shown that in the case of uniform soil profiles, a single sensor can provide representative readings throughout the duration of the irrigation cycle. The optimum common TSRP is located 28 cm below the soil surface and 11 cm from the drip line [72]. Soulis and Elmaloglou [18] reported that, in soils with different layers, at least one sensor in each soil layer is required in order to provide representative readings. Furthermore, the optimum TSRPs are located 11, 15 and 19 cm below the soil surface and 10 and 16 cm from the dripline. Silva et al. [73] monitored time domain reflectometry probes inside the drainage lysimeter on different soil profiles and demonstrated that the optimum sensor position varied according to the development stage of the banana crop. However, this is the first time that an attempt has been made to establish guidelines for soil moisture sensor placement taking into account the physiological response of the plant. Although it is expected that the results obtained for these specific conditions may differ in other plantations, we propose the study of the dynamics recorded by soil moisture probes, and the relation to a sensitive measure of the tree water status as a criterion for the adjustment of irrigation schedules.

It should also be noted that while the neutron probe is considered the most effective method of measuring SWC, it was difficult in this study to establish a protocol to locate the tubes in an ideal position taking into account physiological plant measurements, since, with these probes, targeted measurements were obtained with a measurement interval of 7 days (data not presented).

5. Conclusions

This document addressed two fundamental issues: selecting a reliable indicator of the water status of the Japanese plum and, secondly, establishing a procedure for positioning and interpreting the soil moisture probe readings to support irrigation scheduling. The first of the points is a preliminary step for the second, since it is proposed that the water status of the tree be the evaluation criterion used for the installation and guidance of FDR probes.

The most sensitive parameter for detecting the initiation of stress was Ψstem with a lower CV, making it the most suitable indicator for early-maturing Japanese plum irrigation scheduling. However gaseous exchange recovered more quickly after irrigation was restored. Gs had a higher SI, close to that of Ψstem but with a higher CV. In this case, both parameters are appropriate when it is necessary to quantify the SI of the water stress supported. Fn, SF and FIPARD had lower sensitivity and greater difficulty in processing and interpreting these measures, been less recommendable as a support for irrigation scheduling.

The results of this study indicate the existence of high variability in soil water extraction by the plum crop in different locations of the soil profile. The drying cycles allowed the selection of the positions that best respond to the extraction of water by the tree. Additionally, they helped to establish a threshold in the different phases of the crop below which detrimental effects to the crop can be caused. We propose that this threshold be established through the physiological measures of Ψstem, Fn and Gs. In this experimental orchard, the positions closest to the drippers were the most suitable for monitoring "non-stress" schedules, while for "medium" and "severe" stress references are preferable positions further away from the drippers.

Author Contributions: S.M. (≈45%): Literature review; planning and execution of experimental work; processing of data; discussion of results and writing of the paper. C.C. (≈15%): discussion of results and final revision of the English. A.V. (≈10%) processing of data. M.J.M. (≈15%): discussion of results. M.H.P. (≈15%): project coordinator; planning of experimental work and revision of the paper. All authors have read and agreed to the published version of the manuscript.

Funding: This research was funded by INIA (RTA2013-00045-C04 project).

Acknowledgments: Junta de Extremadura (GR15112, Research Group AGA001 and CCESAGROS projects).

Conflicts of Interest: The authors declare no conflict of interest.

References

1. Romero, R.; Muriel, J.; García, I.; de la Peña, D.M. Research on Automatic Irrigation Control: State of the Art and Recent Results. *Agric. Water Manag.* **2012**, *114*, 59–66. [CrossRef]
2. Morales, F.; Ancín, M.; Fakhet, D.; González-Torralba, J.; Gámez, A.L.; Seminario, A.; Soba, D.; Ben Mariem, S.; Garriga, M.; Aranjuelo, I. Photosynthetic Metabolism Under Stressful Growth Conditions as a Bases for Crop Breeding and Yield Improvement. *Plants* **2020**, *9*, 88. [CrossRef] [PubMed]
3. Houghton, J.T. The Scientific Basis. In *Contribution of Working Group I to the Third Assessment Report of the Intergovernmental Panel on Climate Change*; Cambridge University Press: Cambridge, UK, 2001.
4. Edenhofer, O. *Climate Change 2014: Mitigation of Climate Change*; Cambridge University Press: Cambridge, UK, 2015.
5. Turral, H.; Burke, J.; Faurès, J. *Climate Change, Water and Food Security*; Food and Agriculture Organization of the United Nations (FAO): Rome, Italy, 2011.
6. De la Rosa, J.; Conesa, M.R.; Domingo, R.; Pérez-Pastor, A. A New Approach to Ascertain the Sensitivity to Water Stress of Different Plant Water Indicators in Extra-Early Nectarine Trees. *Sci. Hortic.* **2014**, *169*, 147–153. [CrossRef]
7. Osroosh, Y.; Peters, R.T.; Campbell, C.S.; Zhang, Q. Comparison of Irrigation Automation Algorithms for Drip-Irrigated Apple Trees. *Comput. Electron. Agric.* **2016**, *128*, 87–99. [CrossRef]
8. Casadesús, J.; Mata, M.; Marsal, J.; Girona, J. A General Algorithm for Automated Scheduling of Drip Irrigation in Tree Crops. *Comput. Electron. Agric.* **2012**, *83*, 11–20. [CrossRef]
9. Millán, S.; Casadesús, J.; Campillo, C.; Moñino, M.J.; Prieto, M.H. Using Soil Moisture Sensors for Automated Irrigation Scheduling in a Plum Crop. *Water* **2019**, *11*, 2061. [CrossRef]
10. Millán, S.; Campillo, C.; Casadesús, J.; Pérez-Rodríguez, J.M.; Prieto, M.H. Automatic Irrigation Scheduling on a Hedgerow Olive Orchard using an Algorithm of Water Balance Readjusted with Soil Moisture Sensors. *Sensors* **2020**, *20*, 2526. [CrossRef]
11. Zazueta, F.S.; Xin, J. Soil Moisture Sensors. *Soil Sci.* **1994**, *73*, 391–401.
12. Dobriyal, P.; Qureshi, A.; Badola, R.; Hussain, S.A. A Review of the Methods Available for Estimating Soil Moisture and its Implications for Water Resource Management. *J. Hydrol.* **2012**, *458*, 110–117. [CrossRef]
13. Robock, A.; Vinnikov, K.Y.; Srinivasan, G.; Entin, J.K.; Hollinger, S.E.; Speranskaya, N.A.; Liu, S.; Namkhai, A. The Global Soil Moisture Data Bank. *Bull. Am. Meteorol. Soc.* **2000**, *81*, 1281–1300. [CrossRef]
14. Evett, S.R. Soil Water Measurement by Time Domain Reflectometry. *Encycl. Water Sci.* **2003**, 894–898.
15. Erlingsson, S.; Baltzer, S.; Baena, J.; Bjarnason, G. Measurement techniques for water flow. In *Water in Road Structures*; Anonymous, Ed.; Springer: Dordrecht, The Netherlands, 2009; pp. 45–67.

16. Wang, Y.-N.; Fan, J.; Li, S.-Q.; Zeng, C.; Wang, Q.-J. Effects of Sensor's Laying Depth for Precision Irrigation on Growth Characteristics of Maturate Grapes. *Yingyong Shengtai Xuebao* **2012**, *23*, 2062–2068. [PubMed]

17. Dabach, S.; Shani, U.; Lazarovitch, N. The Influence of Water Uptake on Matric Head Variability in a Drip-Irrigated Root Zone. *Soil Tillage Res.* **2016**, *155*, 216–224. [CrossRef]

18. Soulis, K.X.; Elmaloglou, S. Optimum Soil Water Content Sensors Placement for Surface Drip Irrigation Scheduling in Layered Soils. *Comput. Electron. Agric.* **2018**, *152*, 1–8. [CrossRef]

19. Elmaloglou, S.; Soulis, K.X.; Dercas, N. Simulation of Soil Water Dynamics under Surface Drip Irrigation from Equidistant Line Sources. *Water Resour. Manag.* **2013**, *27*, 4131–4148. [CrossRef]

20. Nolz, R. A Review on the Quantification of Soil Water Balance Components as a Basis for Agricultural Water Management with a Focus on Weighing Lysimeters and Soil Water sensors/Ein Überblick Über Die Ermittlung Von Wasserhaushaltsgrößen Als Basis Für Die Landeskulturelle Wasserwirtschaft Mit Fokus Auf Lysimeter Und Bodenwassersensoren. *Die Bodenkult. J. Land Manag. Food Environ.* **2016**, *67*, 133–144.

21. Bradford, K.; Hsiao, T. Physiological responses to moderate water stress. In *Physiological Plant Ecology II*; Anonymous, Ed.; Springer: Berlin/Heidelberg, Germany, 1982; pp. 263–324.

22. Jiménez, S.; Dridi, J.; Gutiérrez, D.; Moret, D.; Irigoyen, J.J.; Moreno, M.A.; Gogorcena, Y. Physiological, Biochemical and Molecular Responses in Four Prunus Rootstocks Submitted to Drought Stress. *Tree Physiol.* **2013**, *33*, 1061–1075. [CrossRef]

23. Shackel, K.A.; Ahmadi, H.; Biasi, W.; Buchner, R.; Goldhamer, D.; Gurusinghe, S.; Hasey, J.; Kester, D.; Krueger, B.; Lampinen, B. Plant Water Status as an Index of Irrigation Need in Deciduous Fruit Trees. *HortTechnology* **1997**, *7*, 23–29. [CrossRef]

24. Samperio, A.; Prieto, M.H.; Blanco-Cipollone, F.; Vivas, A.; Moñino, M.J. Effects of Post-Harvest Deficit Irrigation in 'Red Beaut' Japanese Plum: Tree Water Status, Vegetative Growth, Fruit Yield, Quality and Economic Return. *Agric. Water Manag.* **2015**, *150*, 92–102. [CrossRef]

25. Sperry, J.S.; Wang, Y.; Wolfe, B.T.; Mackay, D.S.; Anderegg, W.R.; McDowell, N.G.; Pockman, W.T. Pragmatic Hydraulic Theory Predicts Stomatal Responses to Climatic Water Deficits. *New Phytol.* **2016**, *212*, 577–589. [CrossRef]

26. Martin-StPaul, N.; Delzon, S.; Cochard, H. Plant Resistance to Drought Depends on Timely Stomatal Closure. *Ecol. Lett.* **2017**, *20*, 1437–1447. [CrossRef] [PubMed]

27. Ruiz-Sanchez, M.C.; Pérez-Pastor, A.; Torrecillas, A.; Domingo, R. Regulated Deficit Irrigation in Apricot Trees. *Acta Hortic.* **2000**, *2*, 759–766. [CrossRef]

28. Bhusal, N.; Han, S.; Yoon, T. Impact of Drought Stress on Photosynthetic Response, Leaf Water Potential, and Stem Sap Flow in Two Cultivars of Bi-Leader Apple Trees (Malus× Domestica Borkh.). *Scientia Hortic.* **2019**, *246*, 535–543. [CrossRef]

29. Blanco-Cipollone, F.; Lourenço, S.; Silvestre, J.; Conceição, N.; Moñino, M.J.; Vivas, A.; Ferreira, M.I. Plant Water Status Indicators for Irrigation Scheduling Associated with Iso-and Anisohydric Behavior: Vine and Plum Trees. *Horticulturae* **2017**, *3*, 47. [CrossRef]

30. Medrano, H.; Escalona, J.M.; Bota, J.; Gulías, J.; Flexas, J. Regulation of Photosynthesis of C3 Plants in Response to Progressive Drought: Stomatal Conductance as a Reference Parameter. *Ann. Bot.* **2002**, *89*, 895–905. [CrossRef]

31. Wang, X.; Wang, W.; Huang, J.; Peng, S.; Xiong, D. Diffusional Conductance to CO_2 is the Key Limitation to Photosynthesis in Salt-stressed Leaves of Rice (Oryza Sativa). *Physiol. Plant.* **2018**, *163*, 45–58. [CrossRef]

32. Ashraf, M.; Harris, P.J. Photosynthesis under Stressful Environments: An Overview. *Photosynthetica* **2013**, *51*, 163–190. [CrossRef]

33. Ferree, D. Canopy Development and Yield Efficiency of' Golden Delicious' Apple Trees in Four Orchard Management Systems. *J. Am. Soc. Hort. Sci.* **1980**, *105*, 376–380.

34. Casadesus, J.; Mata, M.; Marsal, J.; Girona, J. Automated Irrigation of Apple Trees Based on Measurements of Light Interception by the Canopy. *Biosyst. Eng.* **2011**, *108*, 220–226. [CrossRef]

35. Deutscher, J.; Kupec, P.; Dundek, P.; Holík, L.; Machala, M.; Urban, J. Diurnal Dynamics of Streamflow in an Upland Forested Micro-Watershed during Short Precipitation-Free Periods is Altered by Tree Sap Flow. *Hydrol. Process.* **2016**, *30*, 2042–2049. [CrossRef]

36. Wang, X.; Liu, J.; Sun, Y.; Li, K.; Zhang, C. Sap Flow Characteristics of Three Afforestation Species during the Wet and Dry Seasons in a Dry-Hot Valley in Southwest China. *J. For. Res.* **2017**, *28*, 51–62. [CrossRef]

37. Kirschbaum, M.U.; McMillan, A.M. Warming and Elevated CO_2 have Opposing Influences on Transpiration. Which is More Important? *Curr. For. Rep.* **2018**, *4*, 51–71. [CrossRef]
38. Tie, Q.; Hu, H.; Tian, F.; Guan, H.; Lin, H. Environmental and Physiological Controls on Sap Flow in a Subhumid Mountainous Catchment in North China. *Agric. For. Meteorol.* **2017**, *240*, 46–57. [CrossRef]
39. Domingo, F.; Villagarcía, L.; Brenner, A.; Puigdefábregas, J. Evapotranspiration Model for Semi-Arid Shrub-Lands Tested Against Data from SE Spain. *Agric. For. Meteorol.* **1999**, *95*, 67–84. [CrossRef]
40. Goldhamer, D.; Fereres, E. Irrigation Scheduling of Almond Trees with Trunk Diameter Sensors. *Irrig. Sci.* **2004**, *23*, 11–19. [CrossRef]
41. Goldhamer, D.A.; Fereres, E. Irrigation Scheduling Protocols using Continuously Recorded Trunk Diameter Measurements. *Irrig. Sci.* **2001**, *20*, 115–125. [CrossRef]
42. Jones, H.G. Irrigation Scheduling: Advantages and Pitfalls of Plant-Based Methods. *J. Exp. Bot.* **2004**, *55*, 2427–2436. [CrossRef]
43. Soil Survey Division Staff. *Keys to Soil Taxonomy*, 8th ed.; U.S.D.A.-NRCS: Lincoln, NE, USA, 1998.
44. Allen, R.G.; Pereira, L.S.; Raes, D.; Smith, M. *Crop Evapotranspiration-Guidelines for Computing Crop Water Requirements-FAO Irrigation and Drainage Paper 56*; FAO: Rome, Italy, 1998; Volume 300, p. D05109.
45. Moñino, M.; Samperio, A.; Vivas, A.; Blanco-Cipollone, F.; Prieto, M. *Manual Práctico De Riego Ciruelo Japonés*; CICYTEX, Gobierno de Extremadura: Badajoz, Spain, 2014.
46. Swanson, R.; Whitfield, D. A Numerical Analysis of Heat Pulse Velocity Theory and Practice. *J. Exp. Bot.* **1981**, *32*, 221–239. [CrossRef]
47. Testi, L.; Villalobos, F.J. New Approach for Measuring Low Sap Velocities in Trees. *Agric. For. Meteorol.* **2009**, *149*, 730–734. [CrossRef]
48. López-Bernal, Á.; Alcántara, E.; Testi, L.; Villalobos, F.J. Spatial Sap Flow and Xylem Anatomical Characteristics in Olive Trees Under Different Irrigation Regimes. *Tree Physiol.* **2010**, *30*, 1536–1544. [CrossRef] [PubMed]
49. Samperio, A.; Moñino, M.J.; Marsal, J.; Prieto, M.H.; Stöckle, C. Use of CropSyst as a Tool to Predict Water use and Crop Coefficient in Japanese Plum Trees. Agric. *Water Manag.* **2014**, *146*, 57–68. [CrossRef]
50. Goldhamer, D.A.; Viveros, M. Effects of Preharvest Irrigation Cutoff Durations and Postharvest Water Deprivation on Almond Tree Performance. *Irrig. Sci.* **2000**, *19*, 125–131. [CrossRef]
51. Molz, F.J.; Klepper, B.; Browning, V.D. Radial Diffusion of Free Energy in Stem Phloem: An Experimental Study 1. *Agron. J.* **1973**, *65*, 219–222. [CrossRef]
52. Goldhamer, D.; Fereres, E.; Salinas, M. Can Almond Trees Directly Dictate their Irrigation Needs? *Calif. Agric.* **2003**, *57*, 138–144. [CrossRef]
53. Naor, A.; Cohen, S. Sensitivity and Variability of Maximum Trunk Shrinkage, Midday Stem Water Potential, and Transpiration Rate in Response to Withholding Irrigation from Field-Grown Apple Trees. *HortScience* **2003**, *38*, 547–551. [CrossRef]
54. Intrigliolo, D.S.; Castel, J.R. Continuous Measurement of Plant and Soil Water Status for Irrigation Scheduling in Plum. *Irrig. Sci.* **2004**, *23*, 93–102. [CrossRef]
55. Badal, E.; Buesa, I.; Guerra, D.; Bonet, L.; Ferrer, P.; Intrigliolo, D.S. Maximum Diurnal Trunk Shrinkage is a Sensitive Indicator of Plant Water, Stress in Diospyros Kaki (Persimmon) Trees. *Agric. Water Manag.* **2010**, *98*, 143–147. [CrossRef]
56. Robinson, T.L.; Lakso, A.N. Bases of Yield and Production Efficiency in Apple Orchard Systems. *J. Am. Soc. Hort. Sci.* **1991**, *116*, 188–194. [CrossRef]
57. Hsiao, T.C. Plant Responses to Water Stress. *Annu. Rev. Plant Physiol.* **1973**, *24*, 519–570. [CrossRef]
58. Marsal, J.; Girona, J.; Basile, B.; Dejong, T. Heterogeneity in Fruit Distribution and Stem Water Potential Variations in Peach Trees under Different Irrigation Conditions. *J. Hortic. Sci. Biotechnol.* **2005**, *80*, 82–86. [CrossRef]
59. Palmer, J. Effects of Varying Crop Load on Photosynthesis, Dry Matter Production and Partitioning of Crispin/M. 27 Apple Trees. *Tree Physiol.* **1992**, *11*, 19–33. [CrossRef] [PubMed]
60. Goldhamer, D.A.; Fereres, E.; Mata, M.; Girona, J.; Cohen, M. Sensitivity of Continuous and Discrete Plant and Soil Water Status Monitoring in Peach Trees Subjected to Deficit Irrigation. *J. Am. Soc. Hort. Sci.* **1999**, *124*, 437–444. [CrossRef]
61. Intrigliolo, D.S.; Puerto, H.; Bonet, L.; Alarcón, J.; Nicolas, E.; Bartual, J. Usefulness of Trunk Diameter Variations as Continuous Water Stress Indicators of Pomegranate (Punica Granatum) Trees. *Agric. Water Manag.* **2011**, *98*, 1462–1468. [CrossRef]

62. Tuccio, L.; Piccolo, E.L.; Battelli, R.; Matteoli, S.; Massai, R.; Scalabrelli, G.; Remorini, D. Physiological Indicators to Assess Water Status in Potted Grapevine (Vitis Vinifera L.). *Sci. Hortic.* **2019**, *255*, 8–13. [CrossRef]

63. Atkinson, D. The Distribution and Effectiveness of the Roots of Tree Crops. *Hortic. Rev.* **1981**.

64. Nnyamah, J.U.; Black, T. Rates and Patterns of Water Uptake in a Douglas-Fir Forest. *Soil Sci. Soc. Am. J.* **1977**, *41*, 972–979. [CrossRef]

65. Clothier, B.E.; Green, S.R. Rootzone Processes and the Efficient Use of Irrigation Water. *Agric. Water Manag.* **1994**, *25*, 1–12. [CrossRef]

66. Thaler, P.; Pagès, L. Periodicity in the Development of the Root System of Young Rubber Trees (Hevea Brasiliensis Müell. Arg.): Relationship with Shoot Development. *Plant Cell Environ.* **1996**, *19*, 56–64. [CrossRef]

67. Lecompte, F.; Pagès, L.; Ozier-Lafontaine, H. Patterns of Variability in the Diameter of Lateral Roots in the Banana Root System. *New Phytol.* **2005**, *167*, 841–850. [CrossRef]

68. Steudle, E.; Frensch, J. Water Transport in Plants: Role of the Apoplast. *Plant Soil* **1996**, *187*, 67–79. [CrossRef]

69. Barrowclough, D.E.; Peterson, C.A.; Steudle, E. Radial Hydraulic Conductivity along Developing Onion Roots. *J. Exp. Bot.* **2000**, *51*, 547–557. [PubMed]

70. Watt, M.; Magee, L.J.; McCully, M.E. Types, Structure and Potential for Axial Water Flow in the Deepest Roots of Field-Grown Cereals. *New Phytol.* **2008**, *178*, 135–146. [PubMed]

71. Draye, X.; Kim, Y.; Lobet, G.; Javaux, M. Model-Assisted Integration of Physiological and Environmental Constraints Affecting the Dynamic and Spatial Patterns of Root Water Uptake from Soils. *J. Exp. Bot.* **2010**, *61*, 2145–2155. [CrossRef]

72. Soulis, K.X.; Elmaloglou, S. Optimum Soil Water Content Sensors Placement in Drip Irrigation Scheduling Systems: Concept of Time Stable Representative Positions. *J. Irrig. Drain. Eng.* **2016**, *142*, 04016054. [CrossRef]

73. Da Silva, A.J.P.; Coelho, E.F.; Filho, M.A.C.; De Souza, J.L. Water Extraction and Implications on Soil Moisture Sensor Placement in the Root Zone of Banana. *Sci. Agric.* **2018**, *75*, 95–101. [CrossRef]

Publisher's Note: MDPI stays neutral with regard to jurisdictional claims in published maps and institutional affiliations.

 *agronomy*

Article

Physiological Responses of Apple and Cherry In Vitro Culture under Different Levels of Drought Stress

Zuzana Kovalikova [1,†], **Petra Jiroutova** [2,*,†] , **Jakub Toman** [1], **Dominika Dobrovolna** [1] and **Lenka Drbohlavova** [2]

[1]   Department of Biology, Faculty of Science, University of Hradec Kralove, Rokitanskeho 62, 50003 Hradec Kralove, Czech Republic; zuzana.kovalikova@uhk.cz (Z.K.); jakub.toman@uhk.cz (J.T.); dominika.dobrovolna@uhk.cz (D.D.)

[2]   Research and Breeding Institute of Pomology Ltd., Holovousy 129, 50801 Hořice, Czech Republic; lenka.drbohlavova@vsuo.cz

*   Correspondence: petra.jiroutova@vsuo.cz

†   These authors contributed equally to this work.

Received: 5 October 2020; Accepted: 28 October 2020; Published: 30 October 2020

Abstract: Drought stress is a serious threat. Therefore, improvements in crop productivity under conditions of limited water availability are vital to keep global food security. Apples and cherries belong to the most produced fruit worldwide. Thus, searching for their tolerant or resistant cultivars is beneficial for crop breeders to produce more resistant plants. We studied five apple ("Malinové holovouské", "Fragrance", "Rubinstep", "Idared", "Car Alexander") and five cherry ("Regina", "Napoleonova", "Kaštánka", "Sunburst", "P-HL-C") cultivars for their adaptation in response to progressive drought stress. The reaction of an in vitro culture to osmotic stress simulated by increasing polyethylene glycol (PEG) concentration in medium was evaluated through the morphological (fresh and dry weight, water content, leaf area), physiological (chlorophyll and carotenoids content), and biochemical (reactive oxygen species and malondialdehyde content) parameters. Drought-like stress negatively affected the water content, leaf areas, and chlorophyll content in both fruit species. Oxidative status and membrane damage of plants under water deficiency conditions occurred to be important indicators of stress tolerance mechanism. Cherries exhibited higher hydrogen peroxide levels compared to apples, whereas their malondialdehyde values were generally lower. The overall results indicated wide tolerance range to water deficit among apple and cherry in vitro culture as well as among cultivars within single plant species.

Keywords: drought stress; in vitro culture; apple; cherries; oxidative stress

1. Introduction

Drought stress is a major environmental stress negatively affecting growth, development, and the agricultural production of many plants worldwide. Low water availability can be chronic in dry climate regions or unpredictable due to the changes in weather conditions during the period of plant development. It is expected that the areas suffering from water deficiency will be increasing due to global warming. Thus, searching for drought tolerant plant species and cultivars could be beneficial for crop breeders to produce more resistant plants, which could help to maintain the food security under the conditions of the warming world [1]. Selection or breeding of drought resistant fruit species and cultivars in combination with new approaches in effective use of water are considered to improve the crop production and quality under the drought stress conditions [2].

The responses of plants to water deficit depend upon the intensity and duration of the stress conditions as well as plant species/cultivar and its stage of development. In plants, drought negatively affects several physiological processes including photosynthesis, respiration, nutrient uptake, and metabolism and causes a complex of physio-biochemical responses [3].

Under water deficit conditions, water uptake by the roots is impaired. This causes the reduction in turgor pressure, resulting in suppression of cell elongation and expression growth [4]. Decreased leaf size is one of the first and most obvious plant responses to water stress. It is caused by both decreased cell development and reduced photosynthesis rate. Reduction in the leaf area associated with decrease in the transpiration surface presents a very important water saving mechanism. Recovery leaf growth is dependent on phases of leaf development and on intensity of stress conditions. Stress occurring at the early phase can be recovered. However, this phenomenon does not occur in later stages of development. Severe drought stress stops both leaf development and the production of new leaves, and accelerates the leaf senescence [5].

Reduction in fresh and dry biomass is another typical physiological response of plants to drought stress that represents unfavorable impact of water stress, especially on crop plants. On the contrary, maintaining fresh and dry weight under water deficit conditions is a desirable character trait of plants, especially of agriculturally important crops because of its impact on yield [6].

Drought stress also affects photosynthetic pigment composition in terms of changing the ratio of chlorophyll *a* and *b*, and carotenoids [7]. A significant reduction in chlorophyll content under water stress conditions has been reported for many plant species [8–10]. Aside from their essential role in photosynthesis, carotenoids are important non-enzymatic antioxidants playing an essential role in response to various stress conditions [11]. Hence, their level and ratio can increase as a reaction to drought stress [4].

A decrease in leaf size, biomass, photosynthetic pigments, and water availability leads to reduction in the photosynthesis rate, resulting in absorption of more light energy than could be consumed by photosynthetic carbon fixation. This excess energy has the potential to cause an oxidative stress and increase the production of reactive oxygen species (ROS) such as hydrogen peroxide (H_2O_2) or superoxide radicals (O^{2-}). These species react with proteins and lipids, causing damage of cellular structures and the photosynthesis related metabolism [12]. For example, ROS destroy the cellular membrane through lipid peroxidation. During this process, malondialdehyde (MDA) as a product of acid peroxidation is accumulated in the plant. Thus, the determination of the MDA content can be used as a measure of membrane damage-induced oxidative stress during water stress [13].

Polyethylene glycol (PEG) is described as water-soluble, high molecular weight polymers, widely used to induce water stress in higher plants through lowering the water potential of the nutrient solution [14]. PEG is widely used for identification of drought tolerant genotypes under stimulated osmotic stress [14,15]. The major advantage of using in vitro based techniques with the PEG enriched medium is the rapid screening of diverse plant cultures in laboratory conditions. On the other hand, the main limitation of this method is simulation of drought stress conditions by application of osmotic stress, contrary to withhold irrigation methods based on whole plants experiments [16]. Although, the results of Kautz et al. [17] indicate that physiological responses of plants subjected to PEG-induced osmotic stress are similar to physical water deficit, PEG cannot be considered as an unconditional equivalent for natural drought. The objective of this in vitro study was evaluation of selected morphological, physiological, and biochemical characteristics in five different apple ("Malinové holovouské", "Fragrance", "Rubinstep", "Idared", "Car Alexander") and five cherry ("Regina", "Napoleonova", "Kaštánka", "Sunburst", "P-HL-C") cultivars under drought-like stress conditions induced by polyethylene glycol (PEG). We focused on expanding the available information on selected cultivars (Table 1) with respect to their drought tolerance.

Table 1. Main features of examined apple and cherry cultivars in relation to abiotic stress.

Cultivar (Apple)	Main Feature in Relation to Abiotic Stress	Cultivar (Cherry)	Main Feature in Relation to Abiotic Stress
Malinové holovouské	Low drought resistance [18]. Sufficient frost resistance [19].	Regina	High resistance to rain-induced splitting [20].
Fragrance	Resistant to winter and spring frosts [21].	Napoleonova	Low resistance to rain-induced splitting [20]. Tolerant to drought [22].
Rubinstep	Resistant to winter and spring frosts [23].	Kaštánka	High resistance to rain-induced splitting [20].
Idared	Sensitive to winter frosts [24].	Sunburst	Very resistant to frosts. High resistance to rain-induced splitting [25].
Car Alexander	Very low drought resistance. High frost resistance [19].	P-HL-C	Resistant to winter frosts [26].

2. Materials and Methods

2.1. Plant Material and Experimental Conditions

Selected genotypes of apple (*Malus* × *domestica*) and cherry (*Prunus avium*) were produced in vitro from donor shoots collected in the Research and Breeding Institute of Pomology Holovousy Ltd., Czech Republic. In vitro explants were cultured on 25 mL of modified solid (7.0 g L^{-1} agar) Murashige and Skoog (MS) medium [27] with 6-aminobenzylpurine (BAP) as the shoot growth stimulant at a concentration of 1.5 mg L^{-1} in 100 mL Erlenmeyer flasks capped with aluminum foil. pH was adjusted to 5.7 before autoclaving at 121 °C for 15 min. Growth room conditions were: temperature 22 ± 1 °C, photoperiod 16 h light/8 h dark. Viable cultures were transferred to fresh MS medium at 30-day intervals. Osmotic stress was induced by adding polyethylene glycol (PEG-6000) at a concentration of 0, 5, 10, 25, and 50 g L^{-1} to the basal medium (which is equivalent to 0%, 0.5%, 1%, 2.5%, and 5% (*w/v*) PEG-6000), labelled as PEG0, PEG5, PEG10, PEG25, and PEG50, respectively. All parameters were measured in plants cultivated at a rate 4 pieces per flask on media (with or without PEG-6000) for 30 days. At least three independent measurements were carried out for each cultivar and each concentration.

2.2. Determination of Fresh and Dry Weight, Water Content, Leaf Area

Each explant cultivated under the same conditions was weighed on analytical balance (AS 220.R2, Radwag, Poland) for determination of fresh weight. For determination of both dry weight and moisture content, explants grown under the same conditions were pooled to reach fresh weight 0.7–1.3 g and then analyzed by a moisture analyzer (MB27, Ohaus, NJ, USA). The leaves were photographed (PowerShot G16, Canon, Japan) for leaf area determination immediately after collection one by one from the explant and the entire leaf area of the single explant was measured by using ImageJ software [28].

2.3. Determination of the Chlorophyll and Carotenoids Content

The contents of chlorophyll *a*, chlorophyll *b*, and total carotenoids were determined in methanolic extracts analyzed at 666, 653, and 480 nm, respectively. Contents were calculated according to the equation published in ref. [29].

2.4. Determination of Reactive Oxygen Species and Malondialdehyde

Homogenates in potassium phosphate buffer (50 mM, pH 7.0) were used for determination of reactive oxygen species. The content of hydrogen peroxide was determined using the $TiCl_4$ method (410 nm) with H_2O_2 as a standard. Superoxide radical was measured as a nitrite produced by the reaction in a mixture of homogenate, 10 mM hydroxylamine, 17 mM sulfanilamid, 7 mM α-naphtylamine, and diethyl ether (530 nm). Sodium nitrite was used for calibration [30].

The extent of membrane lipid peroxidation was expressed as the amount of MDA. Trichloroacetic acid (TCA) homogenates were mixed with TCA-thiobarbituric acid (TBA) solution and heated for 30 min at 90 °C. The samples were analyzed after rapid cooling at a wavelength of 532 nm. The MDA contents were calculated using the extinction coefficient of MDA-TBA complex (155 mM^{-1} cm^{-1}) and expressed as μmol MDA g^{-1} FW [30].

2.5. Statistical Analysis

Statistical differences among the treatments and cultivars were evaluated using analysis of variance (ANOVA) followed by Tukey test ($p < 0.05$) in Minitab v. 19 software (Minitab LLC, Coventry, UK). Number of replications (n) in tables/figures denotes individual samples measured for each parameter.

3. Results

3.1. Plant Growth, Water Content, and Leaf Area

The overall appearance of the in vitro explants in both fruit species was affected by the increasing concentration of PEG. Figure 1 shows in the most pronounced manifestations that leaf drooping, wilting, browning, and reduction in area were induced by water stress in both plant species. These symptoms occurred earlier in apples, whereas the cherries displayed better growth and more green leaves. The growth of in vitro plant explants was determined based on fresh weight (FW), dry weight (DW), water content, and leaf area. Our results are summarized in Figure 2 (detailed data are given in Supplementary Materials—TablesSupplementary Materials—Tables S1 and S2).

Figure 1. Effect of different concentration of polyethylene glycol (PEG-6000) on the growth and appearance of the in vitro culture of apple cultivar "Idared" (**A**) and cherry cultivar "Sunburst" (**B**); Numbers indicate PEG concentration in g L^{-1}.

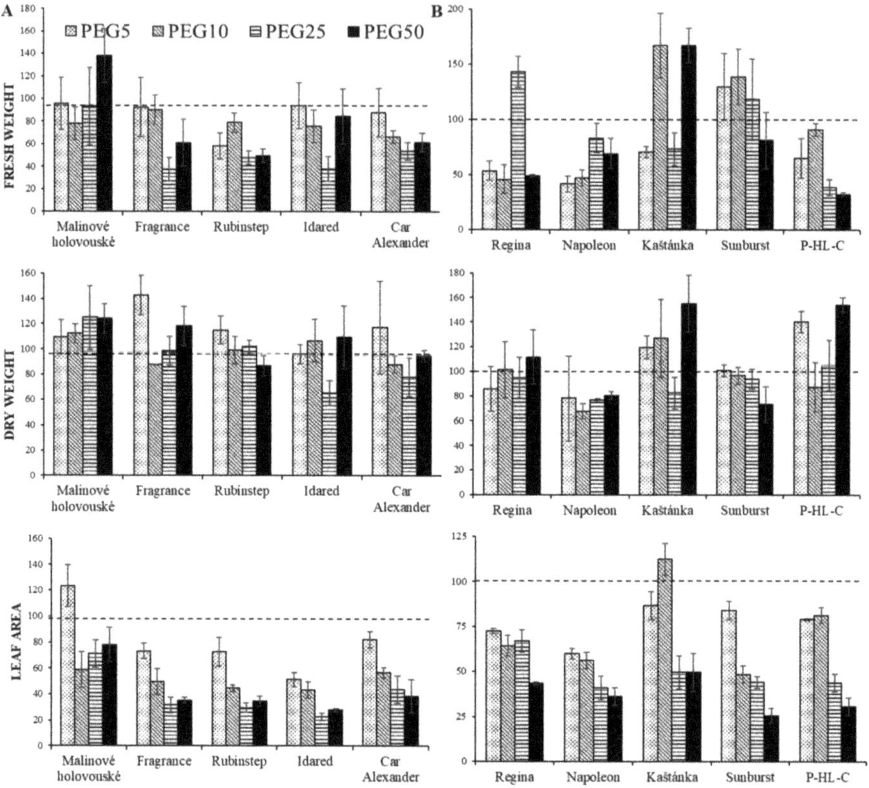

Figure 2. Effect of different concentration of PEG-6000 on content of fresh weight (g), dry weight (g), and leaf area (mm^2) in in vitro culture of apple (**A**) and cherry (**B**) cultivars. All bar values were recalculated relative to the compound content in untreated samples taken as 100% (dashed line). Error bars represent means ± SDs (n = 3).

A decrease in FW with increasing concentration of PEG in comparison with plants grown in medium containing no PEG (PEG0) was observed for apple cultivars most significantly for "Fragrance" in PEG25 and PEG50, for "Rubinstep" in PEG25 and PEG50, for "Idared" in PEG25, and for "Car Alexander" in PEG25. Within the cultivars themselves, the lowest FW in PEG0 plants was recorded for cultivar "Malinové holovouské" for which we also observed the lowest decrease. In contrast, the FW of PEG0 plants was the highest for the "Rubinstep" and "Car Alexander" cultivars and, at the same time, the decrease was more pronounced with the increasing concentration of PEG. A considerable variability in FW content was monitored for cherry plants, meaning it was not possible to unambiguously determine the interdependence between increasing PEG concentration and FW. A significant decrease in FW compared to PEG0 plants was observed in cultivars "Regina" (PEG 5, PEG 10, and PEG25), "Napoleonova" (PEG5 and PEG10), and "P-HL-C" (PEG 5, PEG25, and PEG50).

The decrease in FW was directly related to the decrease in the leaf area. Increasing PEG concentration led to a significant decrease in the leaf area of all tested cultivars, notably at PEG25 and PEG50. Figure 1 illustrated the most significant decrease (compared to control plants) observed for the apple cultivar "Idared" which was 4.2 times. The decrease (compared to control plants) in cherry cultivar "Sunburst" was 4 times. Overall, cherries exhibited larger leaf areas compared to apples, but the area decrease was more pronounced. The smallest leaf area was observed in "Malinové holovouské" and at the same time, the lowest decrease in leaf area was recorded here. The water content in the tissues was also significantly negatively affected with an increase in PEG concentration.

3.2. Chlorophyll and Carotenoids Content

Drought stress induced by PEG negatively influenced the content of chlorophylls in all cultivars of both tested species. Figure 3 and Table S3 show that increased concertation of PEG in apple cultivars resulted in simultaneous depression of chlorophyll *a* and *b*. The lowest level of chlorophyll *a* in PEG0 plants was observed in the cultivar "Fragrance", whereas the "Car Alexander" featured the highest 2 times higher concentration. The most significant decrease of 53% was observed in "Rubinstep". A similar trend was also observed for chlorophyll *b*. For the cherry cultivars, the concentrations of chlorophyll *a* and *b* were generally higher compared to apple cultivars (Table S4). At the same time, the differences in chlorophyll levels between individual cultivars were not significant. The highest value of chlorophyll *a* in PEG0 plants was observed in "Kaštánka" while the 1.2 times lower one in the "P-HL-C" cultivar. An approximately 1.4 times decrease in the content due to increased PEG concentration was similar within all cultivars. The decrease in contents of chlorophyll *b* was always the most pronounced in PEG5 plants, with the exception of the cultivar "Kaštánka", where it was observed at the concentration PEG10.

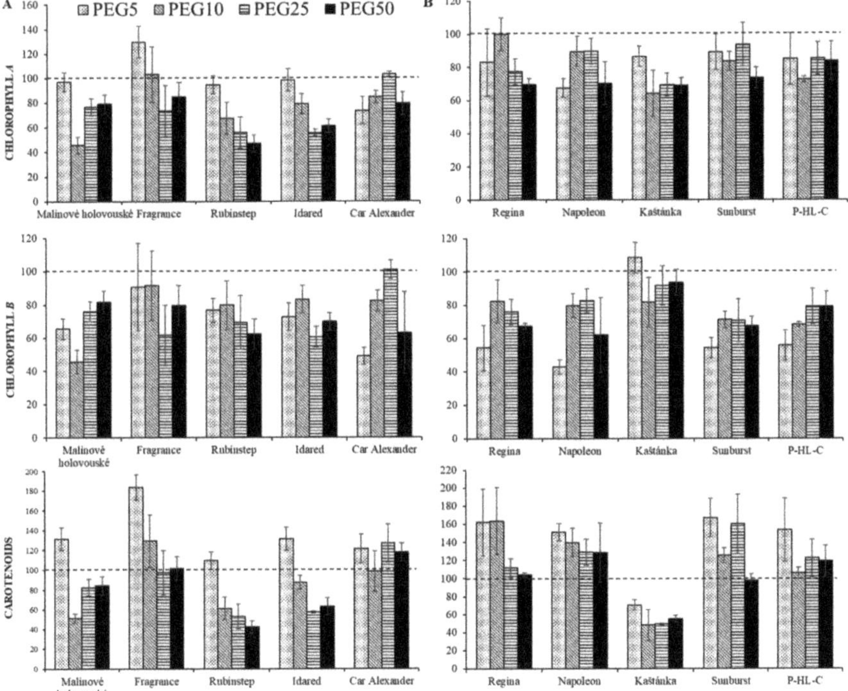

Figure 3. Effect of different concentration of PEG-6000 on content of chlorophyll *a* (mg g^{-1} FW), chlorophyll *b* (mg g^{-1} FW) and total carotenoids (mg g^{-1} FW) in in vitro culture of apple (**A**) and cherry (**B**) cultivars. All bar values were recalculated relative to the compound content in untreated samples taken as 100% (dashed line). Error bars represent means ± SDs (n = 3).

Simulated water deficit affected the content of total carotenoids differently as demonstrated in Figure 3 (Tables S3 and S4). Compared to PEG0, all apple cultivars displayed an increase in total carotenoids when the lowest concentration of PEG (PEG5) was used with most significant effect only in "Fragrance". A similar accumulation trend was also recorded for cherries. A significant increase was observed only in "Sunburst". In contrast, the highest values were reached in "Kaštánka" PEG0 plants. Higher PEG concentrations led in apples to a decrease in carotenoids content below the level of control plants (PEG0).

3.3. Oxidative Status and Membrane Damage

Figure 4 showed that PEG-induced osmotic stress slightly affected the overall oxidation status of the plants. H_2O_2 levels were generally higher in both PEG0 and stressed cherry plants compared to apples. The apple cultivars differed considerably from each other. Relatively low values measured for control and stressed plants were recorded for "Fragrance", "Car Alexander", and "Idared". However, a 2.7 times H_2O_2 increase due to PEG was pronounced. On the contrary, relatively high values with their increase comparable to cherries, were found in the cultivar "Malinové holovouské". Due to PEG treatment, a simultaneous increase in amounts of superoxide radicals was typical of all cultivars. The only exception was the cultivar "Napoleonova" which featured a decrease below the values monitored in PEG0 plants. Low levels of superoxide radical in PEG0 plants were recorded in apple cultivars "Fragrance" and "Idared", but at the same time, the most noticeable increase was recorded here, 2.5 times (PEG25) and 3.1 times (PEG50), respectively. Within cherries, the most significant increase in superoxide radical was recorded in the cultivar 'P-HL-C', amounting for 1.8 times compared to PEG0 plants.

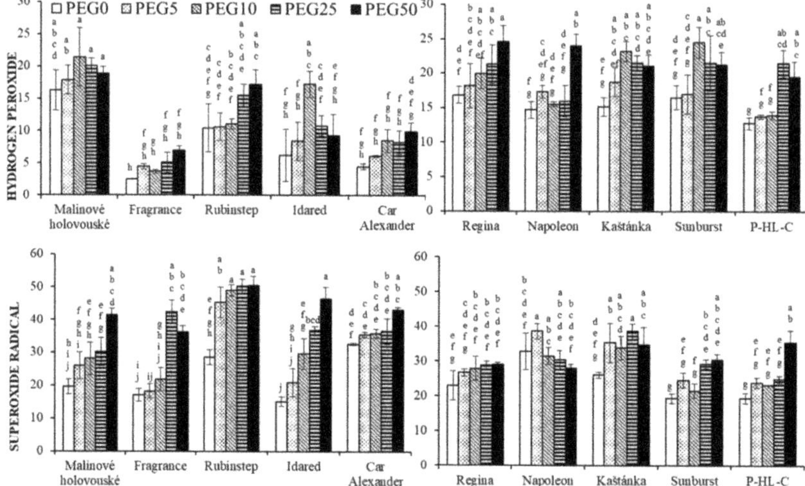

Figure 4. Effect of different concentration of PEG-6000 on content of hydrogen peroxide (μmol g^{-1} FW) and superoxide radicals (μg g^{-1} FW) in in vitro culture of apple and cherry cultivars. Error bars represent means $\pm$ SDs (n = 3). Values within column, followed by the same letter(s), are not significantly different according to Tukey's test ($p < 0.05$).

The accumulation of reactive oxygen species led to damage of membrane lipids expressed as a change in the content of MDA (Figure 5). The values in the PEG0 plants did not differ significantly within the apple cultivars. The highest increase of 1.7 times for PEG50 was observed in "Malinové holovouské". The MDA levels in cherries were generally lower for both non-treated and PEG treated cultivars. However, even the highest values observed in stressed plants did not reach the PEG0 values found for apple cultivars. The increase due to the increasing concentration of PEG was 2.3 and 2.7 times more pronounced for "Kaštánka" and "Napoleonova", respectively.

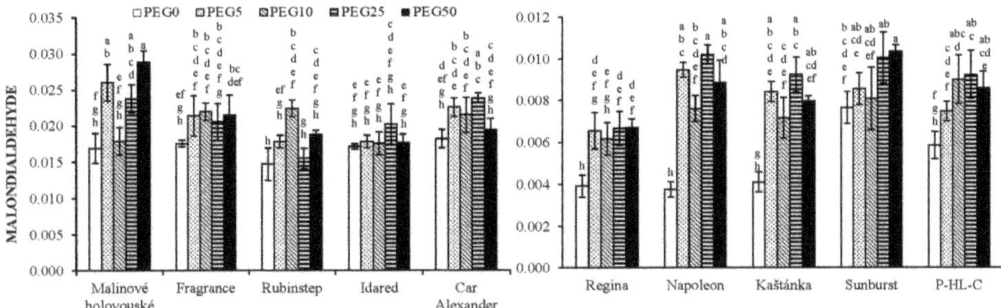

Figure 5. Effect of different PEG-6000 concentration on content of malondialdehyde (μmol g^{-1} FW) in in vitro culture of apple and cherry cultivars. Error bars represent means ± SDs (n = 3). Values within column, followed by the same letter(s), are not significantly different according to Tukey's test ($p < 0.05$).

4. Discussion

Apple and cherry are important fruit trees grown predominantly in temperate regions of the world. Limited water availability is a serious threat of production. Therefore, improvements in crop productivity under drought conditions are vital to keep global food security. The reactions of plants to the limited water regime are usually monitored at several levels, based on selected morphological, physiological, and biochemical parameters that have proven to be good indicators of drought sensitivity/tolerance. We tested, via induced osmotic stress, drought tolerance of five apple and five cherry cultivars by examining their physiological and biochemical parameters under progressive drought-like stress conditions.

Plant wilting, leaf twisting, and reduction in water content, are among the primary manifestations of drought stress. Shortage in water content and the associated decrease in nutrient availability lead to a reduction in cell division and cell proliferation resulted in overall reduction of leaf area [31]. A significant reduction in leaf area and total fresh and dry weight with increasing level of water stress was observed [32,33]. Moreover, several studies confirmed with sorghum [34], guava [35], grass pea [36], apples [37], and almonds [38] that drought tolerant genotypes displayed less damage than the susceptible counterparts. The results of our study demonstrate a negative relationship between water stress and growth parameters such as FW, DW, and leaf area although the response of tested cultivars differed in comparison to previously published studies. Cultivars "Malinové holovouské" and "Car Alexander" are thought to be varieties of low drought resistance [18]. We confirmed this only for "Car Alexander" in which we observed a significant decrease in FW and leaf area. However, these differences were minimal for "Malinové holovouské". "Napoleonova" is reported as a tolerant cherry to drought [22]. Here, we recorded a significant decrease in FW and leaf area but the water content remained constant.

Structural damages to chloroplasts due to ROS formation and/or photodegradation of the pigments probably led to loss of chlorophylls in dehydrated plants. The rootstock chloroplasts became deform in the water stressed apple, stacking of grana was less frequent, and thylakoids were loosened and distorted [37]. Decrease in photosynthetic pigments resulting in leaf yellowing or in an extreme case in leaf necrosis are common visual symptoms under severe water deficit. We noticed in our experiments a reduction in chlorophyll that was accompanied by yellowing and browning of the leaves at the highest concentrations of PEG as shown in Figure 1. Similar responses were also reported in cherry [39] and *Prunus* [40]. However, none of these studies focused in detail on individual types of chlorophylls. A detailed analysis of pigments is given in Šircelj et al. [41] who noticed a significant decrease in chlorophyll *a* and *b* after severe water deficit for apple cultivar "Jonagold Wilmuta". Similar to our study, mild stress led to a significant increase in carotenoids, specifically β-carotene [41] that is a major defense to generation of singlet oxygen in photosynthetic tissue through direct quenching of triplet chlorophyll. Severe stress reduced β-carotene level [6]. Of the other pigments, there was an increase in content of zeaxanthin, a member of the xanthophyll cycle which being a stimulant of non-photochemical quenching and lipid-protective antioxidant plays a direct photoprotective role [41]. Simultaneously with a decrease in chlorophyll, fluorescence of the chlorophyll also changed, namely the maximum photochemical efficiency and effective PSII quantum yield [12,42]. The drop in content of photosynthetic pigments accompanied with inhibition of chlorophyll fluorescence, electron transport rate, and photochemical quenching is a reliable indicator of limited capacity of the photosynthetic system [43]. A decrease in the photosynthetic rate, transpiration rate, and stomatal conductance has been observed in several species [40–42].

One of the many manifestations of negative environmental stimuli is the formation of ROS, including hydrogen peroxide, superoxide, and hydroxyl radicals. Their overproduction can lead to protein degradation, lipid peroxidation, and enzyme inhibition [44]. Besides their toxic properties, they often appear as signaling molecules in mediating adequate defense responses [45]. We found that water limitation caused significant growth in H_2O_2 and superoxide radical concentrations in all cultivars of both species. Enhanced production of ROS was also detected by others in apple leaves [12,37] or cherry rootstock [39].

The degree of oxidative damage can usually be expressed as a change in MDA content, a product of ROS-mediated oxidation of polyunsaturated membrane lipids. Oxidative damage to membranes disrupts the complexity of cellular structures [37]. Electron flux occurs [2], which in turn disrupts the homeostasis of the internal environment and the complexity of individual metabolic pathways. Here, a gradual increment of MDA content as the effects of water shortage indicates that this parameter was directly related to drought similar to previous reports [39,46]. It is well known that different plants and even cultivars of a single plant respond to stress differently. This can result in a different accumulation of monitored ROS and MDA as in the present study. Zhong et al. [46] studied five kiwifruit species that differed significantly in MDA accumulation. Drought-sensitive *Malus hupehensis* apples had a higher H_2O_2 rate and superoxide radical generation and subsequent lipid peroxidation in comparison with drought-tolerant *M. prunifolia* [37]. In our case, the increase in H_2O_2 and MDA content was more pronounced for cherries. This means that cherries reacted more sensitively compared to apples. On the other hand, the increment in the superoxide radical was more pronounced in apple cultivars. These differences can be related to the different activity of antioxidant enzymes and other antioxidants. For a better understanding, it is necessary to determine the activity of enzymes such as catalase, ascorbate peroxidase, superoxide dismutase, as well as levels of other antioxidants such as ascorbic acid, phenolic substances, and total plant antioxidant activity.

In vitro tissue culture represents a biotechnological technique that has been used to genetically improve cultivars and rootstocks. Another option enabling an increase in the stress resistance in many plant species is the so-called polyploidization. For example, Zhang et al. [47] showed that autotetraploid apple cultivar "Hanfu" and "Gala" had, under drought stress, higher relative water content and chlorophyll fluorescence parameters, and lower MDA and proline levels compared to diploid apple.

5. Conclusions

Our work proposes the use of in vitro cultivation of fruit plants with subsequent evaluation of selected physiological parameters as a useful tool in the search for drought resistant or tolerant fruit cultivars. In this study, we examined the reaction of in vitro apple and cherry culture to drought-like stress simulated by enhancing polyethylene glycol concentration in the growing medium. The water content, leaf areas, and chlorophyll content in both fruit species was negatively affected by water deficit conditions. Interestingly, under drought stress, cherries and apples showed different changes in

ROS production that could be related to different antioxidant strategies. Based on the data from our study, we can only roughly estimate the tolerance order of individual cultivars as follows: for apple "Malinové holovouské" > "Fragrance" > "Car Alexander" ≥ "Rubinstep" > "Idared"; and for cherry "Napoleonova" > "Regina" > "Kaštánka" ≥ "P-HL-C" ≥ "Sunburst". However, it should be noted the conclusion is very preliminary and future research focused on determination of the antioxidant enzyme activity and levels of antioxidants will be helpful for better understanding of the mechanism of drought stress in the apple and cherry trees. As well as future research focused on implementation of the acquired knowledge in vitro to ex vitro environment would be beneficial.

Supplementary Materials: The following are available online at http://www.mdpi.com/2073-4395/10/11/1689/s1, Table S1: Effect of different concentration of PEG-6000 on fresh weight, dry weight, water content, and leaf area of in vitro culture of apple cultivars. Data are means ± SD (n = 3). Values within column, followed by the same letter(s), are not significantly different according to Tukey's test ($p < 0.05$)., Table S2: Effect of different concentration of PEG-6000 on fresh weight, dry weight, water content, and leaf area of in vitro culture of cherry cultivars. Data are means ± SD (n = 3). Results of statistic are the same as in Table S1, Table S3: Effect of PEG-6000 concentration on chlorophylls and total carotenoids contents in in vitro culture of apple cultivars. Data are means ± SDs (n = 3). Results of statistic are the same as in Table S1, Table S4: Effect of PEG-6000 concentration of on chlorophylls and total carotenoids contents in in vitro culture of cherry cultivars. Data are means ± SDs (n = 3). Results of statistic are the same as in Table S1.

Author Contributions: Conceptualization, Z.K. and P.J.; methodology, Z.K. and P.J.; investigation, Z.K., P.J., J.T. and D.D.; resources, P.J. and L.D.; data curation, Z.K. and P.J.; writing—original draft preparation, Z.K., P.J. and L.D.; writing—review and editing, Z.K. and P.J.; supervision Z.K. and P.J.; All authors have read and agreed to the published version of the manuscript.

Funding: This research was funded by Technology Agency of the Czech Republic (programme ZETA 2), grant number TJ02000066 (Research of laboratory method for prediction of tolerance of fruit crops to drought).

Acknowledgments: The authors would like to thank Jiřina Rachotová for her technical assistance on this research.

Conflicts of Interest: The authors declare no conflict of interest and the funders had no role in the design of the study; in the collection, analyses, or interpretation of data; in the writing of the manuscript, or in the decision to publish the results.

References

1. Alizadeh, V.; Shokri, V.; Soltani, A.; Yousefi, M.A. Effects of Climate Change and Drought-Stress on Plant Physiology. *Int. J. Adv. Biol. Biom. Res* **2015**, *3*, 38–42.

2. Bolat, I.; Dikilitas, M.; Ikinci, A.; Ercisli, S.; Tonkaz, T. Morphological, physiological, biochemical characteristics and bud success responses of myrobolan 29 c plum rootstock subjected to water stress. *Can. J. Plant Sci.* **2016**, *96*, 485–493. [CrossRef]

3. Tuna, A.L.; Kaya, C.; Ashraf, M. Potassium sulfate improves water deficit tolerance in melon plants grown under glasshouse conditions. *J. Plant Nutr.* **2010**, *33*, 1276–1286. [CrossRef]

4. Jaleel, C.A.; Manivannan, P.; Wahid, A.; Farooq, M.; Al-Juburi, H.J.; Somasundaram, R.; Panneerselvam, R. Drought stress in plants: A review on morphological characteristics and pigments composition. *Int. J. Agric. Biol.* **2009**, *11*, 100–105.

5. Anjum, S.A.; Ashraf, U.; Zohaib, A.; Tanveer, M.; Naeem, M.; Ali, I.; Tabassum, T.; Nazir, U. Growth and developmental responses of crop plants under drought stress: A review. *Zemdirb. Agric.* **2017**, *104*, 267–276. [CrossRef]

6. Farooq, M.; Wahid, A.; Kobayashi, N.; Fujita, D.; Basra, S.M.A. Plant drought stress: Effects, mechanisms and management. *Agron. Sustain. Dev.* **2009**, *29*, 185–212. [CrossRef]

7. Mibei, E.K.; Ambuko, J.; Giovannoni, J.J.; Onyango, A.N.; Owino, W.O. Carotenoid profiling of the leaves of selected African eggplant accessions subjected to drought stress. *Food Sci. Nutr.* **2016**, *5*, 113–122. [CrossRef]

8. Khayatnezhad, M.; Gholamin, R. The effect of drought stress on leaf chlorophyll content and stress resistance in maize cultivars (Zea mays). *Afr. J. Microbiol. Res.* **2012**, *6*, 2844–2848. [CrossRef]

9. Li, R.; Guo, P.; Michael, B.; Stefania, G.; Salvatore, C. Evaluation of Chlorophyll Content and Fluorescence Parameters as Indicators of Drought Tolerance in Barley. *Agric. Sci. China* **2006**, *5*, 751–757. [CrossRef]

10. Meher; Shivakrishna, P.; Ashok Reddy, K.; Manohar Rao, D. Effect of PEG-6000 imposed drought stress on RNA content, relative water content (RWC), and chlorophyll content in peanut leaves and roots. *Saudi J. Biol. Sci.* **2018**, *25*, 285–289. [CrossRef]

11. Young, A.; Lowe, G. Carotenoids—Antioxidant Properties. *Antioxidants* **2018**, *7*, 28. [CrossRef] [PubMed]

12. Wang, Z.; Li, G.; Sun, H.; Ma, L.; Guo, Y.; Zhao, Z.; Gao, H.; Mei, L. Effects of drought stress on photosynthesis and photosynthetic electron transport chain in young apple tree leaves. *Biol. Open* **2018**, *7*, bio035279. [CrossRef] [PubMed]

13. Farooq, M.; Wahid, A.; Lee, D.-J.; Cheema, S.A.; Aziz, T. DROUGHT STRESS: Comparative Time Course Action of the Foliar Applied Glycinebetaine, Salicylic Acid, Nitrous Oxide, Brassinosteroids and Spermine in Improving Drought Resistance of Rice: Improving rice drought tolerance. *J. Agron. Crop Sci.* **2010**, *196*, 336–345. [CrossRef]

14. Bhadra, S.; Roy, B.; Ghimiray, T.S. Polyethyleneglycol mediated rapid in vitro screening of rice (*Oryza sativa* L.) genotypes for drought tolerance. *Ind. Jrnl. Gen. Plnt. Bree.* **2017**, *78*, 142. [CrossRef]

15. Sakthivelu, G.; Akitha Devi, M.K.; Giridhar, P.; Rajasekaran, T.; Nedev, T.; Kosturkova, G. Drought-induced alterations in growth, osmotic and in vitro regeneration of soybean cultivars. *Genet. Appl. Plant Physiol.* **2008**, *34*, 103–112.

16. Osmolovskaya, N.; Shumilina, J.; Kim, A.; Didio, A.; Grishina, T.; Bilova, T.; Keltsieva, O.A.; Zhukov, V.; Tikhonovich, I.; Tarakhovskaya, E.; et al. Methodology of Drought Stress Research: Experimental Setup and Physiological Characterization. *IJMS* **2018**, *19*, 4089. [CrossRef]

17. Kautz, B.; Noga, G.; Hunsche, M. PEG and drought cause distinct changes in biochemical, physiological and morphological parameters of apple seedlings. *Acta Physiol. Plant.* **2015**, *37*, 162. [CrossRef]

18. Říha, J. *České Ovoce–díl III. Jablka*; České nakladatelství: Praha, Czech Republic, 1919.

19. Boček, S. *Ovocné Dřeviny v Krajině: Pilotní vZdělávací Program, Hostětín 2007/8: Sborník Přednášek a Seminárních Prací*; ZO ČSOP Veronica: Brno, Czech Republic, 2008; ISBN 978-80-904109-2-3.

20. Richter, M. *Malý Obrazový Atlas Odrůd Ovoce. 3*; TG tisk: Lanškroun, Czech Republic, 2004; ISBN 978-80-903487-2-1.

21. *Nove Odrudy Ovoce = New Cultivars of Fruit*; Vyzkumny a slechtitelsky ustav ovocnarsky Holovousy: Holovousy, Czech Republic, 2007; ISBN 978-80-87030-04-2.

22. Kutina, J. *Pomologický Atlas 1*; Brázda: Praha, Czech Republic, 1991; ISBN 80-209-0089-6.

23. Blažek, J. Odrůda jabloně Rubinstep. *Vědecké Práce Ovocnářské* **2001**, *17*, 163–165.

24. Kutina, J.; Holeček, S. *Pomologický Atlas 2*; Brázda: Praha, Czech Republic, 1992; ISBN 978-80-209-0192-7.

25. Lane, W.D.; Schmid, H. Lapins and Sunburst sweet cherry. *Can. J. Plant Sci.* **1984**, *64*, 211–214. [CrossRef]

26. Blažková, J. *Pěstování Třešní na Slabě Rostoucích Podnožích*; Výzkumný a šlechtitelský Ústav Ovocnářský Holovousy: Holovousy, Czech Republic, 2005; ISBN 978-80-902636-8-0.

27. Murashige, T.; Skoog, F. A Revised Medium for Rapid Growth and Bio Assays with Tobacco Tissue Cultures. *Physiol. Plant.* **1962**, *15*, 473–497. [CrossRef]

28. ImageJ Home Page. Available online: http://rsbweb.nih.gov/ij (accessed on 30 October 2020).

29. Wellburn, A.R. The spectral determination of chlorophyll *a* and *b*, as well as total Carotenoids, using various solvents with spectrophotometers of different resolutions. *J. Plant. Physiol.* **1994**, *144*, 307–313. [CrossRef]

30. Ducaiova, Z.; Sajko, M.; Mihalicova, S.; Repcak, M. Dynamics of accumulation of coumarin-related compounds in leaves of *MatriCaria chamomilla* after methyl jasmonate elicitation. *Plant Growth Regul.* **2016**, *79*, 81–94. [CrossRef]

31. Tardieu, F. Plant response to environmental conditions: Assessing potential production, water demand, and negative effects of water deficit. *Front. Plant Physiol.* **2013**, *4*, 1–11. [CrossRef]

32. Karimi, S.; Hojati, S.; Eshghi, S.; Moghaddam, R.N.; Jandoust, S. Magnetic exposure improves tolerance of fig 'Sabz' explants to drought stress induced in vitro. *Sci. Hortic.* **2012**, *137*, 95–97. [CrossRef]

33. Turhan, H.; Baser, I. In vitro and in vivo water stress in sunflower. *Helia* **2004**, *27*, 227–236. [CrossRef]

34. Tsago, Y.; Andargie, M.; Takele, A. In vitro selection of sorghum (*Sorghum bicolor* (L) Moench) for polyethylene glycol (PEG) induced drought stress. *Plant Sci. Today* **2014**, *1*, 62–68. [CrossRef]

35. Abouzaid, E.; El-Sayed, E.S.N.; Mohamed, E.S.A.; Youseff, M. Molecular Analysis of Drought Tolerance in Guava Based on In Vitro PEG Evaluation. *Trop. Plant Biol.* **2016**, *9*, 73–81. [CrossRef]

36. Piwowarczyk, B.; Kamińska, I.; Rybiński, W. Influence of PEG Generated Osmotic Stress on Shoot Regeneration and Some Biochemical Parameters in Lathyrus Culture. *Czech J. Genet. Plant Breed.* **2014**, *50*, 77–83. [CrossRef]

37. Wang, W.; Liang, D.; Li, C.; Hao, Y.; Ma, F.; Shu, F. Influence of drought stress on the cellular ultrastructure and antioxidant system in leaves of drought-tolerant and drought-sensitive apple rootstocks. *Plant Physiol. Bioch.* **2012**, *51*, 81–89. [CrossRef]

38. Gikloo, S.T.; Elhami, B. Physiological and morphological responsesof two almond cultivars to drought stress and cycocel. *Int. Res. J. Appl. Bas. Sci.* **2012**, *3*, 1000–1004.

39. Sivritepe, N.; Erturk, U.; Yerlikaya, C.; Turkan, I.; Bor, M.; Ozdemir, F. Response of the cherry rootstock to water stress induced in vitro. *Biol. Plant.* **2008**, *52*, 573–576. [CrossRef]

40. Jiménez, S.; Dridi, J.; Gutiérrez, D.; Moret, D.; Irigoyen, J.J.; Moreno, M.A.; Gogorcena, Y. Physiological, biochemical and molecular responses in four *Prunus* rootstocks submitted to drought stress. *Tree Physiol.* **2013**, *33*, 1061–1075. [CrossRef]

41. Šircelj, H.; Tausz, M.; Grill, D.; Batič, F. Detecting different levels of drought stress in apple trees (*Malus domestica* Borkh.) with selected biochemical and physiological parameters. *Sci. Hortic.* **2007**, *113*, 362–369. [CrossRef]

42. Liu, B.; Liang, J.; Tang, G.; Wang, X.; Liu, F.; Zhao, D. Drought stress affects on growth, water use efficiency, gas exchange and chlorophyll fluorescnce of *Juglans* rootstocks. *Sci. Hortic.* **2019**, *250*, 230–235. [CrossRef]

43. Baker, N.R.; Rosenqvist, E. Applications of chlorophyll fluorescence can improve crop production strategies: An examination of future possibilities. *J. Exp. Bot.* **2004**, *55*, 1607–1621. [CrossRef]

44. Appel, K.; Hirt, H. Reactive oxygen species: Metabolism, oxidative stress, and signal transduction. *Annu. Rev. Plant Biol.* **2004**, *55*, 373–399. [CrossRef] [PubMed]

45. Pérez-Pérez, M.E.; Lemaire, S.D.; Crespo, J.L. Reactive oxygen species and autophagy in plants and algae. *Plant Physiol.* **2012**, *160*, 156–164. [CrossRef] [PubMed]

46. Zhong, Y.-P.; Li, Z.; Bai, D.-F.; Qi, X.-J.; Chen, J.-Y.; Wei, C.-G.; Lin, M.-M.; Fang, J.-B. In Vitro Variation of Drought Tolerance in Five *Actinidia* Species. *J. Am. Soc. Hortic. Sci.* **2018**, *143*, 226–234. [CrossRef]

47. Zhang, F.; Xue, H.; Lu, X.; Zhang, B.; Wang, F.; Ma, Y.; Zhang, Z. Autotetraploidization enhances drought stress tolerance in two apple cultivars. *Trees* **2015**, *29*, 1773–1780. [CrossRef]

Publisher's Note: MDPI stays neutral with regard to jurisdictional claims in published maps and institutional affiliations.

 *agronomy*

Article

Morphological, Physiological, and Biochemical Impacts of Different Levels of Long-Term Water Deficit Stress on *Linum album* Ky. ex Boiss. Accessions

Reza Kiani [1], Vahideh Nazeri [1,*], Majid Shokrpour [1] and Christophe Hano [2]

[1] Department of Horticultural Sciences, College of Agriculture & Natural Resources, University of Tehran, P.O. Box 4111, Karaj 3158777871, Iran; kianireza37@ut.ac.ir (R.K.); shokrpour@ut.ac.ir (M.S.)

[2] Laboratoire de Biologie des Ligneux et des Grandes Cultures, INRAE USC1328, University of Orleans, CEDEX 2, 45067 Orléans, France; hano@univ-orleans.fr

* Correspondence: nazeri@ut.ac.ir

Received: 16 November 2020; Accepted: 13 December 2020; Published: 14 December 2020

Abstract: *Linum album* (Ky. ex Boiss.) is an important medicinal plant that produces compounds such as the well-known anticancer lignan podophyllotoxin and fatty acids. Despite its high medicinal value, it has not yet been studied in detail under agricultural conditions. This study was conducted to evaluate the morphological, phenological, and physiological responses of six *L. album* accessions under different levels of water deficit treatments (100%, 75%, 50%, and 25% available water) in pot conditions. Based on the results, some of the morphological characteristics of the response to water deficit were established. Accessions UTLA7, UTLA9, and UTLA10 showed a higher seed yield and dry weight of the vegetative part. There was a substantial difference in the occurrence of phenological stages in the accessions. The maturation process was accelerated in plants under stress conditions, and accession UTLA9 completed its complete growth cycle faster than the other accessions. The physiological responses of the different accessions did not show the same pattern on the basis of the characteristics studied, and significant differences were observed depending on the trait and accession. Among the most important results of this study was the diversity of responses in different accessions. Based on these results, it is recommended that morphological features (such as seed yield per plant, plant height, number of inflorescences per plant, shoot and root dry weight) be used to select tolerant accessions for the desired product.

Keywords: abiotic stress; *Linum album* Ky. ex Boiss.; morphological properties; phenology; pigments; diversity

1. Introduction

Linum album Ky. ex Boiss. is a perennial medicinal plant belonging to the Linaceae family. This species is an endemic plant in Iran, where it grows in the northwest, west, and central regions. The flowering and maturing stage of this plant lasts from May to July [1,2]. *L. album* contains important lignan compounds such as podophyllotoxin and 6-methoxypodophyllotoxin, which have antiviral and antitumoral properties [3]. Podophyllotoxin, a well-known lignan, serves as the unique starting compound for the semisynthesis of the leading anticancer drug Etoposide (VP16) and its derivatives (class of topoisomerase II inhibitors) used in a dozen anticancer chemotherapy treatments [4,5]. These drugs are on the list of essential medicines of the World Health Organization (WHO) [5]. However, the availability of podophyllotoxin is restricted because it is still exclusively extracted from the rhizomes of *Podophyllum* plants growing in wild forests in Asia. The supply of

Podophyllum plants is rather limited since the occurrence of these plant species is scarce and they require a long growth period (five to seven years) before harvest [5,6]. Nowadays, this species is endangered by overcollection, which exceeds its regeneration capacity, and a lack of cultivation. Consequently, *Podophyllum* is listed in Appendix II of the Convention on International Trading of Endangered Species (CITES), and Etoposide has been identified by the French Drug Agency (ANSM, Agence National de Securité du Médicament) on the list of products "out of stock" several times since late 2016 [5]. Chemical synthesis of podophyllotoxin is difficult due to the presence of four contiguous chiral centers and the presence of a base-sensitive *trans*-lactone moiety. As neither chemical production nor extraction from in vitro plant cultures is economically competitive with the extraction of podophyllotoxin from *Podophyllum* roots, alternatives are sought such as other natural sources [5]. In recent decades, several putative alternative sources of podophyllotoxin have been identified, including *Cupressaceae, Lamiaceae, Linaceae, Podophyllaceae*, and *Polygalaceae* [5]. Some *Linum* species are considered to be a promising alternative source of podophyllotoxin given their lignan accumulation capacities [5–10]. The *Linum* genus comprises more than 230 species, largely distributed among temperate and subtropical climates [5], and species from the *Sylinum* section, including *L. album*, have been reported to accumulate high amounts of podophyllotoxin and its derivatives in their aerial parts, roots, and seeds [5,11,12]. *L. album* seeds constitute some of the richest alternative sources of podophyllotoxin and its derivatives [11,12]. *L. album* seeds also accumulate fatty acid compounds such as palmitic, stearic, oleic, linoleic, and linolenic acid [13]. However, contrary to its congener, the common flax *L. usitatissimum*, little is known about the agronomical performance of *L. album*, in particular its response to drought stress.

Plants face a large number of biotic (fungi, viruses, and insects) and abiotic (drought and salinity) stress in their environment. Environmental stresses confine crop yield and create many changes in their molecular processes such as variation in metabolite profile [14]. Growth is complete through cell division, enlargement, and differentiation and depends on genetic, physiological, ecological, and morphological events and their complex interactions. The quality and quantity of plant growth depend on these events, which are affected by water deficit. Water deficit stress is a condition of insufficient water availability, caused by intermittent to continuous periods without irrigation [15]. In Iran, more than 75% of the arid and semiarid regions have been classified as water-deficient regions [16]. The outcome of water deficit is limitations in the distribution and survival of plants in arid and semiarid regions [17–19]. These areas include most areas where *L. album* grows [1]. Water deficit condition, the most important stress in plants, conducts an unusual increase in reactive oxygen species (ROS) production [20]. ROS can damage the cell membrane and increase the production of malondialdehyde (MDA) content [21]. Another important that effect inhibits the growth and photosynthetic abilities of plants is the loss of balance between the production of ROS and the antioxidant defense [22,23], which causes the accumulation of ROS, leading to oxidative damage to proteins, membrane lipids, and other cellular components [24]. Plants can use both enzymatic and nonenzymatic systems to control ROS production. The enzymatic antioxidant processes involved the activity of enzymes such as catalase (CAT), superoxide dismutase (SOD), ascorbate peroxidase (APX), and polyphenol oxidase (PPO) [20,25,26]. The antioxidant defense system consists of nonenzymatic components, such as ascorbate, glutathione, proline, glycine betaine, amino acids, and/or phenolic compounds [20]. Water deficit stress also affects the rate of plant growth and development [27,28]. Under water stress conditions, plants complete their life cycle faster than under normal conditions; consequently, crop growth stages have a short duration, with fewer days to accumulate assimilates during the life cycle, and the production of biomass is reduced [29–31]. Crops have a definite temperature requirement before they attain certain phenological stages. Accumulative heat units and systems were adopted for determining the dates of the flowering and maturity of different field crops [28,32]. However, susceptibility to drought varies as a function of the phenological stage, depending on plant species and genotypes, and considerable inter- and intraspecific variations can be observed [33,34]. Abscisic acid (ABA) is a key phytohormone involved in the control of many physiological processes such as seed dormancy and germination and in the response to many abiotic and biotic signals [35,36]. ABA is known as a key stress-signaling hormone, acting in the regulation of stomatal closure, the synthesis of compatible osmolytes, and in the upregulation of genes, leading to adaptive responses [35]. Proline is an active protective osmolyte for plants formed as a result of oxidative stress by free-radical stimulation [37,38] in response to different environmental stresses [39,40]. Glycine betaine is another essential compatible solute found in plants, animals, and bacteria in response to water deficit stress [29,41–43]. Many studies have shown that glycine betaine plays an important role in improving plant tolerance under many abiotic stresses, including drought stress [44]. In addition to the direct protective functions of glycine betaine, either by beneficial effects on enzyme and membrane integrity or as a compatible solute, it can also indirectly protect cells from environmental stress by participating in signal transduction pathways [45].

To date, most of the studies focusing on *L. album* have been designed to enhance lignan compounds in in vitro cultures. Various techniques, such as optimizing the culture medium [46], the use of elicitors [47], or inducing polyploidy [48], have been successful in increasing the accumulation of lignans in various *L. album* in vitro culture. However, insufficient research has been conducted under greenhouse or field conditions. The main objective of the present study is to investigate and understand the morphological, phenological, and physiological responses of six different *L. album* accessions to different levels of water deficit stress (100%, 75%, 50%, and 25% available water) in pot conditions.

2. Materials and Methods

2.1. Plant and Soil Materials

This study was a continuation of a project on morphological and physiological variations in different populations of *L. album* in the west of Iran [13,49]. To study the effect of water deficit stress on some valuable characteristics of this species, six superior accessions were selected and subjected to the treatments (Table 1, seeds of accession UTLA12 were obtained from the seed gene bank at the Forest and Rangeland Research Institute in Tehran, Iran).

Table 1. *L. album* seed source locations.

Accession Code	Latitude (N)	Longitude (E)	Altitude (m)	Voucher Number [a]
UTLA1	34°13′56″	48°57′25″	1904	6426
UTLA6	34°55′50″	48°11′34″	2176	6425
UTLA7	34°41′12″	48°38′02″	2124	6430
UTLA9	34°46′11″	48°43′17″	1955	6427
UTLA10	34°22′45″	48°40′02″	1721	6428
UTLA12	32°54′11″	50°4′39″	2630	-

[a] Department of Horticultural Sciences Herbarium, University of Tehran.

Seeds were treated with 1000 ppm gibberellic acid for 24 h to overcome seed dormancy and germinated in a plastic germination tray containing coco-peat in April 2018 [49]. The growing media consisted of a mixture of field soil, sand, and leaf mold (formed from decaying leaves to improve soil structure and water retention) in an equal ratio (Table 2).

Table 2. Edaphic parameters of soil.

pH	EC (Ds/m)	Silt (%)	Clay (%)	Sand (%)	Soil CLASS	FC (%)	PWP (%)	OC (%)	Total N (%)	Usable K (mg/kg)	Usable P (mg/kg)
8.1	2.8	21	11	68	Sandy loam	24.98	12.52	3.18	0.25	397	47.6

The leaf mold was provided by the botanic garden of the University of Tehran. Sixty days after germination, uniformly sized seedlings were randomly selected and transplanted into the pots (one seedling per pot). After a one-year growth period, in January 2019, the dried shoots of plants were uniformly cut at 1 cm above soil level. All steps were performed in open-field conditions at the College of Agriculture and Natural Resources, University of Tehran (daily temperature conditions are provided in Table S1).

2.2. Water Stress Treatment

Irrigation of plants during the first year of cultivation was performed continuously to the extent of field capacity. Water deficit stress (different levels of irrigation consisting of 100 (control), 75, 50, and 25% of plant available water (AW = FC − PWP)) was applied in the second year of cultivation. The weight method was used, and the pots were weighed every 48 h. Due to nonuniformity in the growth of plants, flower bud emergence was considered the criterion for the onset of water deficit stress. The experiment was performed as a factorial experiment in a randomized complete block design (RCBD) with three replications and three observations (three pots) in each replication.

2.3. Determination of Phenological Stages

Phenological stages of accessions were recorded individually from the beginning of the growth of the first plant in the second year of growth (2 March 2019). The occurrence of phenological stages was reported based on growing degree days (GDD). The following formula [49] was used to calculate the GDD [50].

$$\text{GDD} = \sum [(\text{Tmax} - \text{Tmin})/2] - \text{Tbase with Tbase} = 2.67$$

2.4. Relative Water Content

Relative water content (RWC) was determined by the procedure outlined by Turner [51]. From each sample in the water stress treatments, 15 fully extended leaves were removed from the plant stem, weighed (FW), and floated on double distilled water for 24 h at 4 °C. Turgid leaves were quickly weighed, the weight of the samples was considered the turgidity weight (TW). The samples were oven-dried at 70 °C for 48 h and reweighed to obtain the dried leaf weight (DW). The relative water content was calculated by the following equation.

$$\text{RWC (\%)} = [(\text{FW} - \text{DW})/(\text{TW} - \text{DW})] \times 100$$

2.5. Chlorophyll and Carotenoid Contents

For all physiological experiments, samples were taken from fully mature leaves during the maturing period. Samples from each replicate and treatment were separately mixed, ground, and stored at −80 °C.

To measure leaf chlorophyll content, 500 mg of the frozen powdered sample was mixed with 10 mL of 95% ethanol. The homogenized sample mixture was centrifuged at 8000 rpm for 15 min. Supernatants were read by a microplate spectrophotometer (BioTek Eon, Winooski, VT, USA) at 664 nm for chlorophyll-a (Ch a), 649 for chlorophyll-b (Ch b), and 470 nm for carotenoids [52]. The amounts of chlorophyll and carotenoid were calculated by the following formulas.

$$\text{Ch a } (\mu g/mL) = 13.36_{A664} - 5.19_{A649}$$

$$\text{Ch b } (\mu g/mL) = 27.43_{A649} - 8.12_{A664}$$

$$\text{Carotenoids } (\mu g/mL) = (1000A470 - 2.13\text{Ch a} - 97.63\text{Ch b})/209$$

2.6. Determination of Proline and Glycine Betaine

Proline content was determined using the method established by Bates et al. [53]. A 500 mg amount of leaves was commixed in 10 mL of 3% sulfosalicylic acid, and the mixture was centrifuged at 10,000 rpm for 10 min. Then, 2 mL of the supernatant was added to 2 mL of an acid-ninhydrin solution and 2 mL of glacial acetic acid in a tube. The tubes were incubated in a bain-marie at 100 °C for 1 h. The reaction was stopped in ice. The reaction mixture was extracted with 4 mL of toluene and vortexed for 15–20 s. The tubes were allowed to stand for at least 20 min in darkness at room temperature for the separation of toluene from the aqueous phase. The toluene phase was then collected in tubes, and the absorbance at 520 nm was measured with a microplate spectrophotometer (BioTek Eon, Winooski, VT, USA). The proline concentration was determined according to the standard curve of proline.

The amount of glycine betaine was evaluated according to Grieve and Grattan [54]. A 250 mg amount of leaf powder (dried leaves) was shaken with 10 mL of deionized water for 48 h at 25 °C. The extracts were filtered using filter paper, diluted (1:1) with H_2SO_4 (2N), and cooled in ice water for 60 min. A 0.2 mL volume of cold KI-I2 was then added to the samples and softly mixed. The tubes were kept at 4 °C for 16 h and centrifuged at 10,000 rpm for 15 min at 0 °C. The supernatant was attentively discarded, and periodide crystals were dissolved in 9 mL of 1,2-dichloroethane. After 2 h, the value of absorbance at 365 nm was evaluated using a spectrophotometer. The amount of glycine betaine was calculated according to the standard curve of glycine betaine.

2.7. Measurement of Electrolyte Leakage and Malondialdehyde Contents

Malondialdehyde determination began by homogenizing 500 mg of fresh leaves in 5 mL of 10% trichloroacetic acid (TCA). In the next step, samples were centrifuged at 12,000 rpm for 10 min at 4 °C. A 2 mL volume of supernatant was added to 4 mL of 0.6% thiobarbituric acid (TBA, in 10% TCA) and incubated at 100 °C in a bain-marie for 15 min. The samples were cooled at room temperature, and the absorbance of the supernatant was measured at 450, 532, and 600 nm using a microplate spectrophotometer (BioTek Eon, Winooski, VT, USA). The MDA contents were calculated by the following formula [55]:

$$\text{MDA } (\mu\text{mol g}^{-1} \text{ FW}) = 6.45 \, (A_{532} - A_{600}) - 0.56 \, A_{450}$$

Electrolyte leakage (EL) was estimated using a conductivity meter. Fresh leaf samples were cut into 10 pieces of equal sizes, each with a 4.5 mm diameter. The samples were then dipped in 15 mL of distilled water and shaken at 100 rpm for 24 h at room temperature. The initial electrical conductivity (EC1) of the solution was recorded. The tubes were then autoclaved at 120 °C for 20 min. After cooling the tubes to room temperature, the final EC (EC2) was recorded [56]. Finally, the percentage of ion leakage was calculated by the following equation:

$$\text{EL } (\%) = [\text{EC1/EC2}] \times 100$$

2.8. Enzymatic Antioxidant Activity

For enzyme assays, frozen leaf samples were ground to a fine powder with liquid nitrogen and extracted with 50 mM phosphate buffer (pH = 7.0). The extracts were centrifuged at 4 °C for 15 min at 13,000 rpm. The supernatant was then collected and used for the protein content assay and enzyme activities. Protein extraction was performed according to Bradford [57] using bovine serum albumin as a standard. Catalase activity was determined using the spectrophotometric method (BioTek Eon, Winooski, VT, USA) according to Hadwan [58]. Total guaiacol peroxidase activity was determined according to Plewa et al. [59]. The reaction mixture included 3000 µL of 50 mM phosphate buffer (pH = 7), 10 µL of 30% hydrogen peroxide, 3 µL of 200 µM guaiacol solution, and 100 µL of enzymatic extract. The addition of enzyme extract started the reaction, and the increase in absorbance was recorded at 470 nm for 4 min (Perkin Elmer, Waltham, MA, USA; UV-VIS Spectrophotometer, LAMBDA EZ201). The activity of ascorbate peroxidase was measured according to Ranieri et al. [60]. The reaction mixture contained 600 µL of 0.1 mM EDTA, 1500 µL of 50 mM phosphate buffer (pH = 7), 400 µL of 0.5 mM ascorbic acid, 400 µL of 30% hydrogen peroxide, and 100 µL enzyme extract. Enzyme activity assays were recorded at 470 nm for 4 min (Perkin Elmer, Waltham, MA, USA; UV-VIS Spectrophotometer, LAMBDA EZ201, USA).

2.9. Abscisic Acid (ABA) Extraction and Quantification

A 250 mg amount of dried leaf samples (sampling was performed at the end of flowering stages) was extracted in 2 mL of aqEtOH (80%) solvent. The extracts were sonicated by an ultrasonic bath for 60 min (USC1200TH, Prolabo, Fontenay-sous-Bois, France) with a maximal heating power of 400 W (i.e., acoustic power of 1 W/cm^2). The extract supernatant was filtered through 0.45 µm nylon syringe membranes. ABA was quantified using a Phytodetek ABA ELISA kit (Agdia) and (±) *cis-trans* ABA (Sigma-Aldrich, Saint-Quentin Fallavier, France) as a standard.

2.10. Statistical Analysis

All data were subjected to analysis of variance (ANOVA) using SAS V.9.2 software followed by an LSD test with $p < 0.01$ as the significant differences between means. The results are presented as means ± SE (standard errors). Pearson's correlation was investigated with IBM SPSS Statistics 23.0. Cluster analysis was performed by Ward's method. Factor analysis was investigated by the principal components extraction method (factor rotation was performed by Varimax with the Kaiser normalization method) using the IBM SPSS Statistics 23.0 software (Chicago, IL, USA). The principal component analysis triplot was drawn based on the factor score of the first 3 components.

3. Results

3.1. Morphological Properties

The response of the six *L. album* accessions to different irrigation levels showed a significant difference in most traits, including the seed width, weight of 1000 seeds, inflorescence length, main branch length, plant height, leaf length, flower diameter, root length, and fresh and dry weight of shoots and roots (Table S2).

As shown in Table 3, under normal irrigation conditions (100% AW), the highest seed yield per plant was observed in accession UTLA1 (1.86 g) and the lowest was observed in accessions UTLA10 and UTLA12 (1.19 and 1.24, respectively). However, the weight of 1000 grains in accession UTLA1 was the lowest (3.12) and higher in accessions UTLA10 and UTLA12 (3.67 and 3.81, respectively) than other accessions. The number of flowers per inflorescence in accession UTLA9 (12.81) was significantly higher than other accessions. Plant height in accessions UTLA9 and UTLA12 (36.5 and 36.17) was higher than the other accessions. The highest shoot and root dry weight was observed in accessions UTLA9 and UTLA10, respectively, whereas accession UTLA1 showed the lowest shoot and root dry weight with a significant difference.

Under severe water stress treatment (25% AW), the highest seed yield per plant was seen in accession UTLA9 (1.9 g), and UTLA7 showed the lowest yield (0.5). The weight of 1000 seeds was the highest in accession UTLA10 (3.86) and the lowest in UTLA9 (2.98). Interestingly, the sample that showed the highest seed yield per plant under normal conditions and severe stress had the lowest 1000-seed weight, which indicates that the seeds were smaller in size but larger in number.

As observed in plant height, the number of flowers per inflorescence in the 25% AW treatment decreased in all accessions compared to the control. Accession UTLA9 showed the highest plant height (30.5 cm). The dry weight of the shoots and roots in accessions UTLA10 and UTLA12 was higher than in the other accessions. Accession UTLA1 had the lowest dry weight of shoots and roots under control conditions and the lowest weight under severe stress levels.

Table 3. Effect of water deficit stresses on the morphological traits of *L. album* accessions.

Accession/ Irrigation	Number of Seeds per Capsule	Seed Yield per Plant (g)	Seed Width (mm)	Weight of 1000 Seeds (g)	Number of Flowers in Inflorescence	Number of Mature Capsules	Inflorescence Length (cm)	Length of the Main Branch (cm)	Plant Height (cm)
UTLA1									
100% AW	8.7 ± 0.15	1.86 ± 0.03	2.35 ± 0.03	3.12 ± 0.1	7.33 ± 0.51	6.76 ± 0.57	13.24 ± 0.38	14.83 ± 0.44	27.5 ± 0.76
75% AW	8.31 ± 0.18	0.61 ± 0.04	2.29 ± 0.06	3.03 ± 0.14	11.67 ± 0.67	11 ± 1	13.33 ± 1.2	13.33 ± 0.67	23.83 ± 0.6
50% AW	8.5 ± 0.69	0.6 ± 0.06	2.25 ± 0.03	3.12 ± 0.16	10.08 ± 1.08	9.42 ± 1.39	15.11 ± 0.49	11 ± 0.58	24.67 ± 2.19
25% AW	7.56 ± 0.34	0.71 ± 0.06	2.31 ± 0.01	3.36 ± 0.16	8.08 ± 0.65	6.25 ± 0.8	17.75 ± 0.43	13.5 ± 1.04	25 ± 0.5
UTLA6									
100% AW	7.55 ± 0.62	1.34 ± 0.16	2.43 ± 0.06	3.42 ± 0.16	11.39 ± 1.52	10.56 ± 1.44	15.37 ± 1.81	15.5 ± 1.32	29.33 ± 1.45
75% AW	7.92 ± 0.52	0.96 ± 0.09	2.36 ± 0.04	3.51 ± 0.12	9.33 ± 0.95	8.83 ± 0.98	17.49 ± 1.5	17 ± 1.04	34.67 ± 0.73
50% AW	7.04 ± 0.61	1.01 ± 0.1	2.48 ± 0.06	3.87 ± 0.13	12.18 ± 0.84	11.16 ± 0.63	15.48 ± 1.58	14.67 ± 0.73	30.17 ± 1.64
25% AW	9.06 ± 0.48	1.33 ± 0.09	2.55 ± 0.01	3.7 ± 0.12	9.94 ± 1.06	8.44 ± 0.29	12.59 ± 0.6	15.83 ± 0.6	27.83 ± 1.3
UTLA7									
100% AW	8.49 ± 0.18	1.47 ± 0.15	2.46 ± 0.02	3.57 ± 0.15	8.94 ± 0.78	8.49 ± 0.7	16.45 ± 0.78	17.33 ± 0.6	31.5 ± 1.53
75% AW	9.23 ± 0.31	2.79 ± 0.08	2.4 ± 0.07	3.43 ± 0.16	8.78 ± 0.91	7.89 ± 1.02	18.78 ± 0.94	21 ± 0.58	36.17 ± 0.73
50% AW	8.59 ± 0.33	1.24 ± 0.13	2.48 ± 0.06	3.66 ± 0.08	8.87 ± 0.38	7.88 ± 0.06	15 ± 0.86	19.33 ± 1.09	31.83 ± 1.17
25% AW	7.44 ± 0.06	0.55 ± 0.07	2.49 ± 0.07	3.39 ± 0.14	7.21 ± 0.74	6.07 ± 0.97	12.88 ± 0.74	17.67 ± 0.17	28.33 ± 0.33
UTLA9									
100% AW	8.96 ± 0.13	1.43 ± 0.13	2.31 ± 0.05	3.38 ± 0.12	12.81 ± 0.78	11.72 ± 0.64	23.25 ± 0.59	17.33 ± 0.88	36.5 ± 1.32
75% AW	8.74 ± 0.41	2.27 ± 0.18	2.54 ± 0.07	3.84 ± 0.11	10.74 ± 0.9	10.3 ± 0.71	21.66 ± 1.28	18.5 ± 0.29	36.17 ± 1.88
50% AW	6.92 ± 0.36	1.16 ± 0.14	2.42 ± 0.09	3.71 ± 0.04	10.92 ± 0.84	9.19 ± 0.91	19.87 ± 0.13	16.5 ± 1.04	34.67 ± 0.67
25% AW	8.96 ± 0.14	1.91 ± 0.02	2.18 ± 0.05	2.98 ± 0.06	8.56 ± 0.29	6.73 ± 0.91	15.67 ± 1.18	14 ± 0.5	30.5 ± 1.76
UTLA10									
100% AW	8 ± 0.19	1.19 ± 0.08	2.37 ± 0.06	3.67 ± 0.04	9 ± 0.96	8.89 ± 0.87	17.2 ± 0.31	18.17 ± 0.88	31.5 ± 1
75% AW	8.83 ± 0.33	1.32 ± 0.05	2.46 ± 0.08	3.67 ± 0.08	10.97 ± 0.51	9.72 ± 0.36	17.63 ± 0.59	18.83 ± 0.93	33.17 ± 0.93
50% AW	8.44 ± 0.48	1.31 ± 0.13	2.39 ± 0.04	3.49 ± 0.09	8.78 ± 1.61	7.69 ± 0.54	14 ± 0.51	20.43 ± 0.74	30.17 ± 1.48
25% AW	9.05 ± 0.22	1.48 ± 0.06	2.42 ± 0.01	3.86 ± 0.16	7.78 ± 0.22	6.41 ± 0.13	14.72 ± 1.07	17.83 ± 0.44	29.67 ± 1.3
UTLA12									
100% AW	8.39 ± 0.29	1.24 ± 0.1	2.39 ± 0.04	3.81 ± 0.15	9.58 ± 1.05	9.4 ± 0.97	18.97 ± 0.8	19.67 ± 0.88	36.17 ± 0.73
75% AW	8.72 ± 0.24	1.38 ± 0.29	2.38 ± 0.07	3.93 ± 0.13	10.22 ± 0.67	9.44 ± 0.73	14.17 ± 0.73	13.33 ± 0.6	22.17 ± 1.01
50% AW	8.86 ± 0.56	0.45 ± 0.06	2.46 ± 0.08	3.91 ± 0.08	7.42 ± 0.3	6.83 ± 0.17	16.92 ± 0.92	15.33 ± 0.88	28 ± 1.53
25% AW	8.34 ± 0.6	0.59 ± 0.07	2.41 ± 0.07	3.68 ± 0.07	7.02 ± 0.39	6.11 ± 0.36	13.88 ± 0.51	15.83 ± 0.6	25.83 ± 1.42
LSD 1%	1.5	0.44	0.2	0.47	3.05	2.95	2.68	2.96	4.67
UTLA1									
100% AW	16.07 ± 1.27	5.33 ± 0.33	1.44 ± 0.29	34 ± 0.58	26.6 ± 0.81	21.85 ± 1.47	5.92 ± 0.42	24.11 ± 3.2	3.44 ± 0.32
75% AW	19.08 ± 0.58	3 ± 0.58	1.56 ± 0.06	32.33 ± 0.67	23.5 ± 0.87	27.4 ± 0.35	7.68 ± 0.19	46.05 ± 2.42	5.19 ± 0.67
50% AW	20.07 ± 0.23	2.33 ± 0.33	2.67 ± 0.17	34.33 ± 0.6	27.5 ± 0.87	17.38 ± 0.36	4.46 ± 0.86	21.17 ± 4.2	2.32 ± 0.62
25% AW	19.98 ± 0.59	2 ± 0.58	4.22 ± 0.4	30.17 ± 0.44	32 ± 0.58	10 ± 1.15	3.7 ± 0.17	20.61 ± 0.8	3.05 ± 0.03

Table 3. *Cont.*

Accession/ Irrigation	Number of Seeds per Capsule	Seed Yield per Plant (g)	Seed Width (mm)	Weight of 1000 Seeds (g)	Number of Flowers in Inflorescence	Number of Mature Capsules	Inflorescence Length (cm)	Length of the Main Branch (cm)	Plant Height (cm)
UTLA6									
100% AW	18.45 ± 1.12	5.83 ± 0.44	1.39 ± 0.46	32.78 ± 0.62	25 ± 0.58	28.19 ± 0.73	10.41 ± 0.43	56.9 ± 5.12	6.33 ± 0.88
75% AW	21.97 ± 0.38	4.67 ± 0.88	1.22 ± 0.22	34.56 ± 0.68	27.05 ± 1.18	39.95 ± 1.7	14.05 ± 1.18	48.35 ± 5.98	8.1 ± 0.64
50% AW	24.33 ± 0.45	4 ± 0.58	3 ± 0.58	31 ± 0.76	23.55 ± 0.89	26.6 ± 0.58	9.62 ± 0.07	48.06 ± 6.35	8.37 ± 0.21
25% AW	21.07 ± 0.52	3.67 ± 0.88	5.28 ± 0.31	33.45 ± 0.48	26.57 ± 1.26	18.7 ± 0.81	7.78 ± 0.05	38 ± 4.09	5.93 ± 0.16
UTLA7									
100% AW	21.19 ± 0.74	6.33 ± 0.67	1.11 ± 0.4	36.87 ± 1.04	30.05 ± 1.18	56.8 ± 1.27	19.44 ± 1	62.47 ± 10.91	10.56 ± 1.17
75% AW	23.78 ± 1.01	6.5 ± 0.5	1.78 ± 0.22	34.11 ± 0.67	29 ± 2	32.33 ± 1.59	13.63 ± 0.89	87.05 ± 26.03	9.36 ± 2.6
50% AW	22.08 ± 1.29	4.67 ± 0.67	3.56 ± 0.29	33.17 ± 1.09	24.88 ± 1.07	32.8 ± 2.19	10.45 ± 0.2	56.66 ± 0.81	7.91 ± 1.12
25% AW	21.36 ± 0.5	3.83 ± 0.73	4.67 ± 0.38	32.72 ± 1.07	35.15 ± 0.61	23.2 ± 1.33	7.85 ± 0.78	28.8 ± 0.8	5.4 ± 0.29
UTLA9									
100% AW	22.32 ± 0.22	4.5 ± 0.29	1.11 ± 0.11	41 ± 0.76	33.5 ± 1.44	42.03 ± 1.6	15.47 ± 1.63	65.69 ± 5.31	9.03 ± 0.56
75% AW	20.9 ± 0.57	7.67 ± 0.33	1.89 ± 0.11	36.08 ± 1.23	28.05 ± 0.66	44.15 ± 0.66	15.91 ± 0.17	68.84 ± 5.45	11.26 ± 0.86
50% AW	20.67 ± 1.15	4.67 ± 0.33	2.66 ± 0.33	38.07 ± 0.95	26.55 ± 1.82	19.15 ± 2.17	6.51 ± 1.03	27.23 ± 0.77	3.46 ± 0.31
25% AW	20.57 ± 1.16	4.67 ± 0.33	3.78 ± 0.4	32.61 ± 0.81	30.88 ± 1.77	19.6 ± 2.66	6.45 ± 0.72	27.57 ± 7.34	3.79 ± 0.79
UTLA10									
100% AW	22.37 ± 0.9	3.67 ± 0.67	1.22 ± 0.4	32.72 ± 0.64	32.5 ± 0.87	43.76 ± 1.25	15.02 ± 1.58	91.24 ± 19.24	11.02 ± 1.63
75% AW	23.48 ± 0.67	6.33 ± 0.67	1.39 ± 0.06	30.43 ± 0.74	31 ± 1.15	37.28 ± 0.85	10.86 ± 0.42	56.65 ± 0.4	8.02 ± 0.09
50% AW	20 ± 0.75	5 ± 0.58	2.67 ± 0.44	35.28 ± 0.55	27.25 ± 0.43	14.55 ± 0.32	4.24 ± 0.14	28.45 ± 1.47	3.78 ± 0.15
25% AW	24.77 ± 1.13	5.67 ± 0.67	4.17 ± 0.25	34.83 ± 0.88	25.75 ± 1.24	28.62 ± 2.08	9.9 ± 0.44	58.89 ± 6.21	8.92 ± 0.72
UTLA12									
100% AW	20.98 ± 0.28	4.67 ± 0.67	3.11 ± 0.49	34.5 ± 1.04	27.1 ± 0.52	25.78 ± 0.85	9.9 ± 0.76	38.02 ± 2.21	4.15 ± 0.24
75% AW	21.68 ± 0.41	4.33 ± 0.67	3.22 ± 0.28	33.39 ± 0.56	26 ± 0.58	13.2 ± 0.64	4.66 ± 0.32	35.93 ± 3.89	4.53 ± 0.48
50% AW	21.83 ± 0.93	1.67 ± 0.33	2.78 ± 0.29	31.5 ± 0.76	31.2 ± 0.69	13.65 ± 0.95	5.13 ± 0.38	37.85 ± 3.38	4.25 ± 0.37
25% AW	22.97 ± 0.88	2.67 ± 0.33	4 ± 0.51	33.55 ± 0.78	28.55 ± 0.32	30.72 ± 1.36	11.31 ± 0.25	51.37 ± 1.11	6.62 ± 0.73
LSD (1%)	3.12	2.12	1.29	3.07	4.08	4.51	2.81	30.58	3.18

Values are given as mean ± SE.

In response to reduced irrigation levels, the number of seeds per capsule was either decreased (accessions UTLA1 and UTLA7), increased (accession UTLA10), or not significantly modified (accessions UTLA6, UTLA9, and UTLA12). Seed yield per plant in accessions UTLA1, UTLA7, and UTLA12 were significantly decreased in response to water deficit stress. On the contrary, seed width was not affected by water stress (Table 3). The maximum and minimum seed widths were observed in accessions UTLA7 and UTLA1, respectively. The weight of 1000 seeds and the number of flowers per inflorescence showed a significant decrease only in accession UTLA9. Water deficit led to a significant decrease in the mature capsules number in accessions UTLA9 and UTLA12. The length of the inflorescence decreased in all accessions except for UTLA1, which, interestingly, showed a substantial increase in the length of the inflorescence with an increase in stress. The length of the main branch in UTLA7 and UTLA9 and the height of the plant in UTLA6, UTLA7, and UTLA9 showed a significant decrease under reduced irrigation levels. Although other accessions were not particularly affected by water deficit stress, leaf length in UTLA1 showed a significant increase. By increasing the water deficit stress level, the number of inflorescences per plant in accessions UTLA1, UTLA6, UTLA7, and UTLA12 showed a significant decrease. As expected, the number of chlorotic leaves in all accessions increased significantly with increasing water stress levels. The highest flower diameter was observed in accession UTLA9, although a significant decrease in flower diameter occurred with decreasing irrigation. As the stress level increased, the root length decreased in accession UTLA10 (-26%) and increased significantly in UTLA1 (+20%). By reducing the irrigation level, shoot fresh weight in all accessions except UTLA12 decreased. The shoot dry weight showed a significant decrease in UTLA6, UTLA7, UTLA9, and UTLA10. Root fresh and dry weight in UTLA7 and UTLA9 showed a significant decrease by increasing stress levels (Table 3).

The results indicated that the capsule diameter and length of the seeds and the leaves under water deficit stress did not show significant changes (Table S2). However, a significant difference was observed between accessions. The largest capsule diameter was observed in UTLA12 (5.86 mm) and the lowest in UTLA1 (5.41) (Figure 1). Minimum and maximum seed and leaf lengths were observed in accessions UTLA1 (6.55) and UTLA10 (8.06), respectively (Figure 1).

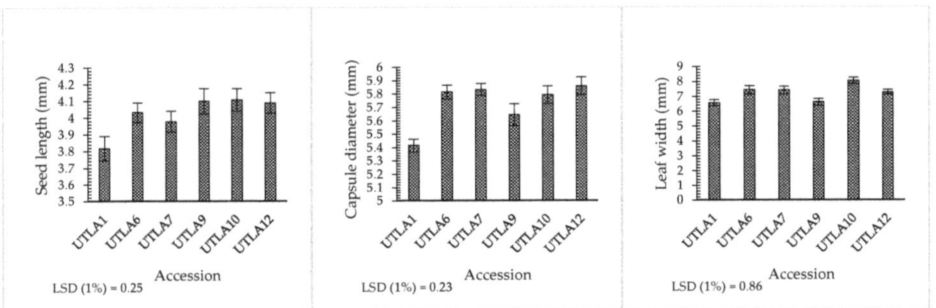

Figure 1. Mean of morphological traits in *L. album* accessions. Values are given as mean ± SE.

The capsule-to-flower ratio was affected by water deficit stress, whereas the response of the accessions did not show a significant difference (Table S2). Figure 2 shows the changes in the capsule-to-flower ratio under four irrigation conditions. The lowest percentage of the flower-to-capsule ratio (0.82) with a significant difference compared with other levels was observed at the 25% AW condition.

3.2. Phenological Stages

The occurrence of different phenological stages in *L. album* accessions under water deficit stress showed a significant difference (Table S3). The growth of UTLA10 plants began earlier than other accessions (after receiving 47.52 degree days). Plants of accession UTLA12 required more heat units

(64.28) to start growing. Plants of accession UTLA6 entered the bud and flowering stages earlier than other accessions (after receiving 179.90 and 278.45), whereas accession UTLA12 required more heat units to enter these stages (209.54 and 302.71, respectively). Maturation took 648.22 degree days in UTLA10 plants (the longest maturing period), whereas UTLA9 plants (the shortest maturing period) completed their life cycle by receiving 618.82 degree days (Figures 3 and 4).

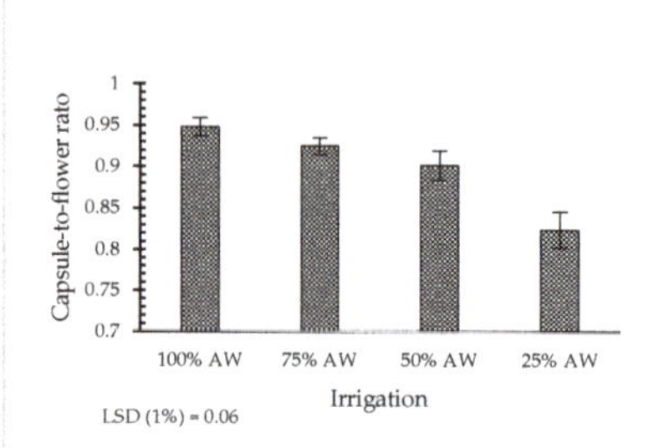

Figure 2. Effect of water deficit stresses on the capsule-to-flower ratio of *L. album*. Values are given as mean ± SE.

Figure 3. Phenological stages in *L. album*. Starting leaf stage (**a**), budding (**b**), flowering (**c**), and maturity (**d**).

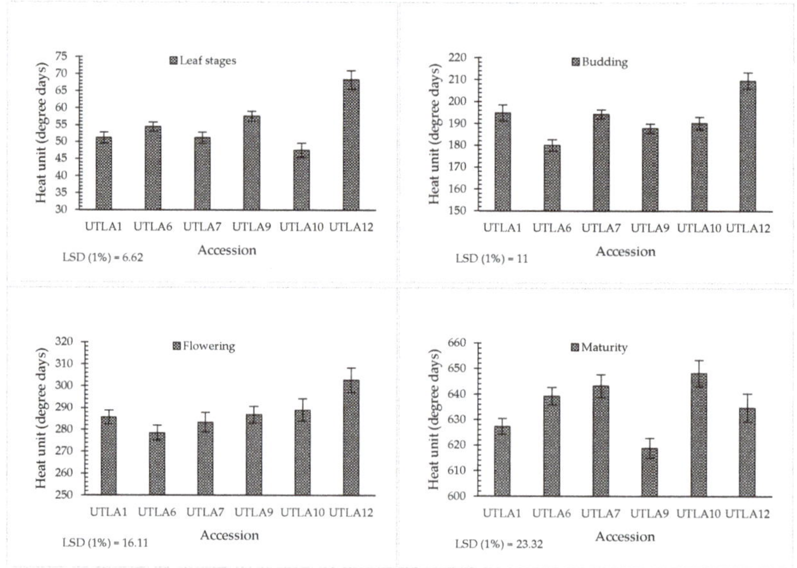

Figure 4. Phenological stages of *L. album* accessions. Values are given as mean ± SE.

Among the reported phenological stages, only maturity was affected by water deficit stress (Table S3). Plants under the 25% AW condition matured significantly faster than other levels of irrigation (Figure 5).

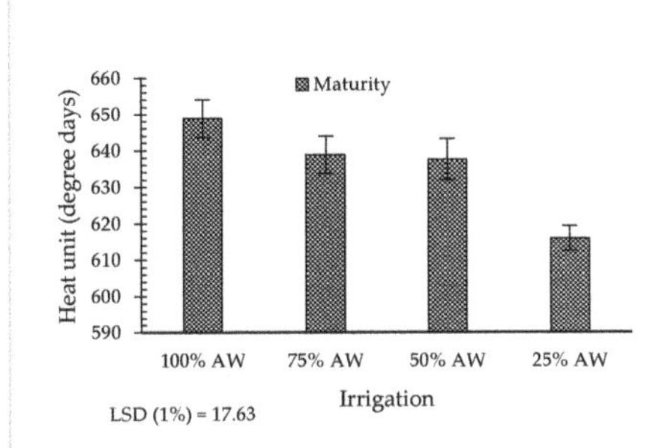

Figure 5. Effect of water deficit stresses on the maturity of *L. album*. Values are given as mean ± SE.

3.3. RWC, Chlorophyll, and Carotenoid Contents

Based on the results of ANOVA the different accessions showed different physiological responses to irrigation levels (Table S4). In response to the increasing water stress level, the RWC decreased in accessions UTLA6, UTLA7, and UTLA10, increased in UTLA1 and UTLA12, and was observed without any significant change in UTLA9 compared to the control (Table 4). Chlorophyll-a and chlorophyll-b as well as carotenoids showed significant decreases in response to increasing stress levels in all accessions except accessions UTLA6 and UTLA12. Under normal irrigation conditions (control), the highest amount of chlorophyll-a and chlorophyll-b was observed in accession UTLA7, whereas under severe stress, accession UTLA6 showed the highest amounts. Under severe stress (25% AW), accession UTLA1 showed the lowest amount of pigments, whereas accession UTLA7 showed the highest amounts (Table 4).

3.4. Proline and Glycine Betaine

A significant increase in proline levels was observed in all accessions in the 25% AW treatment compared to the control. Under the 25% AW treatment, accessions UTLA6 and UTLA7 showed the lowest (4.19 µmol/g FW) and highest (6.01) amount of proline, respectively. With increasing stress levels, the amount of glycine betaine in accessions UTLA1 and UTLA9 showed a significant decrease, whereas an increase was observed for accessions UTLA10 and UTLA12. The highest amount of glycine betaine (249.64 µmol/g DW) was observed in accession UTLA10 under the 25% AW treatment (Table 4 and Table S3).

3.5. Electrolyte Leakage and Malondialdehyde

The results revealed that the water deficit stress in accessions UTLA1, UTLA6, UTLA9, and UTLA12 resulted in a significant increase in ion leakage, whereas the ion leakage decreased in accession UTLA10 and there was no significant change in accession UTLA7. In accessions UTLA1, UTLA9, and UTLA10, exacerbation of dehydration caused a significant increase in MDA. Accessions UTLA6 and UTLA12 did not show significant changes (Table 4).

3.6. Abscisic Acid

Total ABA content in leaves is reported in Table 4. Increased water stress levels increased the amount of ABA in all accessions with the exception of accession UTLA7. In accession UTLA1, the amount of ABA in the 25% AW treatment was 2.25-fold higher than the control, whereas accessions UTLA6 and UTLA7 did not show significant differences. In accessions UTLA9, UTLA10, and UTLA12, ABA content increased significantly in response to increasing water stress levels.

Table 4. Effect of water deficit stresses on physiological traits of *L. album* accessions.

Accession/ Irrigation	RWC (%)	Chlorophyll-a (µg/mL)	Chlorophyll-b (µg/mL)	Carotenoids (µg/mL)	Proline (µmol/g FW)	Glycine Betaine (µmol/g DW)	Electrolyte Leakage (%)	MDA (µmol/g FW)	ABA (pmol/g DW)
UTLA1									
100% AW	77.05 ± 1.98	19.3 ± 0.6	7.22 ± 0.25	5.45 ± 0.15	2.64 ± 0.04	228.75 ± 7.27	55.55 ± 3.71	0.102 ± 0.023	150.89 ± 3.45
75% AW	80.74 ± 0.91	16.47 ± 1.48	5.51 ± 0.53	4.99 ± 0.35	5.62 ± 0.3	166.48 ± 5.82	65.7 = 1.08	0.113 ± 0.0143	177.75 ± 11.72
50% AW	80.67 ± 2.14	17.5 ± 0.19	5.55 ± 0.46	5.19 ± 0.12	3.67 ± 0.23	189.09 ± 10.66	78.21 ± 2.56	0.136 ± 0.0053	134.8 ± 13.83
25% AW	85.11 ± 1.11	13.2 ± 1.82	4.72 ± 0.51	3.87 ± 0.41	5.78 ± 0.15	130.19 ± 5.25	65.59 ± 2.56	0.149 ± 0.005	340.01 ± 26.85
UTLA6									
100% AW	84.61 ± 0.54	20.38 ± 0.27	7.43 ± 0.1	5.47 ± 0.09	4.99 ± 0.22	169.59 ± 4.04	62.54 ± 0.83	0.175 ± 0.0021	101.59 ± 5.86
75% AW	84.07 ± 0.73	19.3 ± 0.19	7.01 ± 0.11	5.37 ± 0.05	5.07 ± 0.21	122.69 ± 6.7	75.82 ± 0.64	0.151 ± 0.0088	89.62 ± 5.55
50% AW	80.43 ± 0.62	19.18 ± 0.33	6.93 ± 0.15	5.37 ± 0.08	4.06 ± 0.32	141.62 ± 1.76	67.9 ± 1.06	0.155 ± 0.0109	135.71 ± 13.3
25% AW	74.37 ± 1.16	20.44 ± 0.63	7.59 ± 0.11	5.72 ± 0.19	4.19 ± 0.05	168.04 ± 0.14	72.66 ± 1.16	0.154 ± 0.0108	112.73 ± 1.54
UTLA7									
100% AW	84.95 ± 0.92	22.68 ± 0.74	9.11 ± 0.05	5.97 ± 0.19	5.28 ± 0.2	199.09 ± 6.55	64.52 ± 1.89	0.184 ± 0.0085	136.87 ± 15.44
75% AW	88.36 ± 0.94	20.88 ± 0.46	7.76 ± 0.19	6.12 ± 0.14	5.22 ± 0.12	164.51 ± 5.83	74.67 ± 1.28	0.12 ± 0.0101	137.79 ± 15.53
50% AW	86.01 ± 1.1	20.15 ± 0.77	7 ± 0.32	6.47 ± 0.21	4.44 ± 0.18	152.06 ± 13.22	63.27 ± 1.55	0.174 ± 0.0102	109.44 ± 5.3
25% AW	77.38 ± 1.23	18.88 ± 0.3	6.86 ± 0.36	5.76 ± 0.04	6.01 ± 0.13	209.33 ± 5.82	60.9 ± 2.78	0.128 ± 0.0099	113.19 ± 3.21
UTLA9									
100% AW	76.81 ± 0.42	21.19 ± 0.1	7.75 ± 0.29	6.13 ± 0.1	5.15 ± 0.08	227.11 ± 11.27	57.45 ± 1.65	0.157 ± 0.0099	103.23 ± 6.11
75% AW	83.33 ± 1.44	21.62 ± 1	7.92 ± 0.19	5.96 ± 0.09	5.04 ± 0.14	193.48 ± 2.17	75.89 ± 3.42	0.139 ± 0.0079	90.38 ± 5.29
50% AW	82.63 ± 1.25	18.99 ± 0.3	6.79 ± 0.2	5.28 ± 0.07	4.38 ± 0.21	171.31 ± 8.41	71.27 ± 1.53	0.208 ± 0.008	115.82 ± 4.89
25% AW	75.82 ± 1.2	17.53 ± 0.33	5.81 ± 0.59	5.01 ± 0.17	5.4 ± 0.15	180.33 ± 4.68	67.71 ± 1.27	0.203 ± 0.014	139.39 ± 6.16
UTLA10									
100% AW	83.43 ± 0.52	21.44 ± 0.56	8.03 ± 0.19	5.5 ± 0.15	4.49 ± 0.14	182.79 ± 7.82	67.52 ± 1.11	0.112 ± 0.0006	101.87 ± 1.88
75% AW	84.92 ± 0.18	20.42 ± 0.61	7.43 ± 0.43	5.64 ± 0.29	5.23 ± 0.17	185.98 ± 2.01	73.89 ± 1.4	0.085 ± 0.0073	122.38 ± 4.71
50% AW	84.18 ± 1.06	17.58 ± 0.47	5.73 ± 0.28	5.1 ± 0.05	5.44 ± 0.16	190.73 ± 6.47	41.19 ± 3.54	0.18 ± 0.0039	117.18 ± 6.21
25% AW	78.57 ± 0.42	16.82 ± 0.13	6.25 ± 0.35	5.44 ± 0.16	5.93 ± 0.15	249.64 ± 4.68	58.36 ± 0.9	0.182 ± 0.0066	155.81 ± 11.55
UTLA12									
100% AW	83.59 ± 2.92	17.58 ± 0.79	6.1 ± 0.32	5.25 ± 0.23	4.5 ± 0.1	168.78 ± 4	60 ± 1.85	0.131 ± 0.0079	117.36 ± 9.29
75% AW	85.46 ± 0.92	17.14 ± 0.67	6.17 ± 0.09	4.64 ± 0.17	3.21 ± 0.05	215.52 ± 7.89	67.95 ± 1.57	0.122 ± 0.009	150.82 ± 11.34
50% AW	87.5 ± 1.24	15.28 ± 0.65	5.58 ± 0.21	4.69 ± 0.19	5.72 ± 0.16	215.93 ± 8.4	80.17 ± 0.35	0.146 ± 0.0067	204.98 ± 10.91
25% AW	82.18 ± 1.32	18.8 ± 0.48	6.72 ± 0.1	5.87 ± 0.14	5.46 ± 0.08	195.9 ± 7.14	85.14 ± 3.61	0.14 ± 0.0068	174.2 ± 7.06
LSD (1%)	4.536	2.011	1.156	0.562	0.658	25.7	7.768	0.0375	37.79

Values are given as mean ± SE.

3.7. Enzymatic Antioxidant Activity

Figure 6 presents the enzymatic antioxidant activity of *L. album* accessions. In response to a decreased irrigation level, the catalase enzyme activity increased (+76.9%) in accession UTLA1, whereas it showed a significant decrease in accessions UTLA6 (−107%), UTLA7 (−133%), and UTLA10 (−142%) (Figure 6). The guaiacol peroxidase enzyme activity of *L. album* accessions showed substantial variation in response to the different irrigation levels (Table S4). Accession UTLA1 showed the highest guaiacol peroxidase enzyme activity, whereas UTLA7 showed the lowest activity. The activity of this enzyme also showed a significant increase with increasing stress levels in accessions UTLA6 and UTLA10. The response of different accessions based on the ascorbate peroxidase enzyme activity showed a significant difference. Decreases were observed in accessions UTLA7 and UTLA10, whereas significant increases were observed in accessions UTLA9 and UTLA12 (Figure 6).

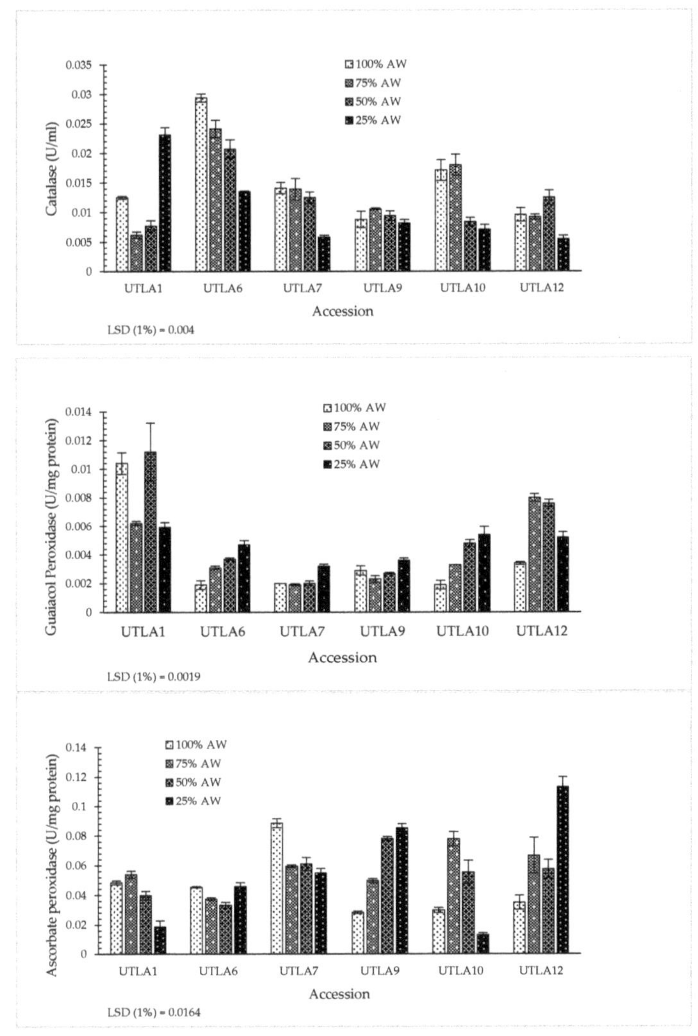

Figure 6. Effect of water deficit stresses on the enzymatic antioxidant activity of *L. album* accessions. Values are given as mean ± SE.

3.8. Correlations between Traits

Significant correlations were found between the studied traits. A positive correlation coefficient of the number of inflorescences with seed yield ($r = 0.81$), the number of mature capsules ($r = 0.96$), seed length with leaf length ($r = 0.65$), weight of 1000 seeds ($r = 0.72$), correlations of shoot dry weight with root dry weight ($r = 0.90$), chlorophyll-a and chlorophyll-b with the number of inflorescences ($r = 0.65$ and 0.64), shoot dry weight ($r = 0.79$ and 0.83), and root dry weight ($r = 0.72$ and 0.77) were observed. Guaiacol peroxidase activity with plant height ($r = -0.72$) and shoot dry weight ($r = -0.70$), ABA content with chlorophyll-a, chlorophyll-b, and carotenoids ($r = -0.75$, -0.60, and -0.69) were negatively correlated (Table S5).

3.9. Cluster Analysis and Factor Analysis

Accessions were divided into three different clusters based on hierarchical clustering analysis (Figure 7). Accessions UTLA6, UTLA10, and UTLA12 were placed in a separate group. The accessions of this cluster showed relative response similarity in traits such as inflorescence length, plant height, flower and capsule diameter, seed length, guaiacol peroxidase activity, and maturity. Accessions UTLA7 and UTLA9 showing a similar response (in seed yield per plant, plant height, number of inflorescences per plant, root length, shoot dry weight, chlorophyll-a and chlorophyll-b, carotenoids, seeds per capsule, weight of 1000 seeds, and ascorbate peroxidase activity) were grouped in the same cluster (Figure 7). Accession UTLA1 was placed on its own in a separate cluster, this accession showed the lowest value in most of the measured traits, except for ABA content and guaiacol peroxidase activity.

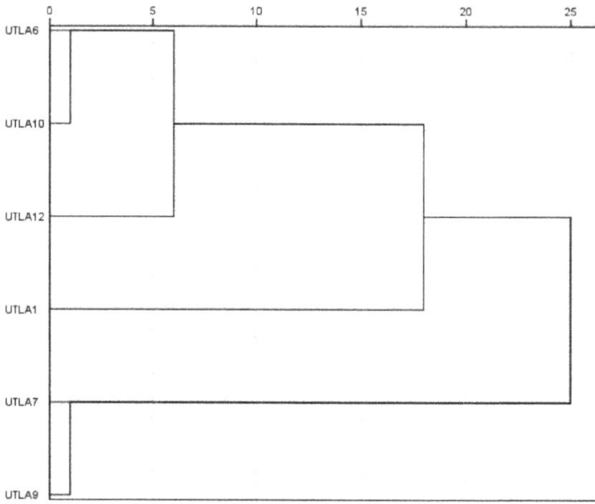

Figure 7. Cluster analysis dendrogram of *L. album* accessions.

The factor analysis results of *L. album* accessions showed that five components explain 100% of the total variance (Table S6). The components accounted for 39.07, 18.35, 16.48, 14.19, and 11.92% of the total variance, respectively. The first component was the most important factor in justifying the total variance with traits such as seed yield per plant, length of the main branch, plant height, number of inflorescences per plant, shoot fresh and dry weight, root fresh and dry weight, chlorophyll-a and chlorophyll-b, carotenoids, proline, guaiacol peroxidase, and ABA, with loading factors of 0.84, 0.81, 0.85, 0.91, 0.98, 0.99, 0.87, 0.93, 0.97, 0.98, 0.99, 0.75, -0.96, and -0.83, respectively (Table S6).

As observed with the hierarchical clustering analysis, accession UTLA1 was placed in a different category in this principal component analysis (Figure 8). The lower performance of this accession in the effective traits of the first component can be considered the main reason for this difference (Table S6). Interestingly, accession UTLA6, with the effective traits in the second component, including catalase (highest), glycine betaine (lowest), flowering time (earliest), root length (lowest), and the number of seeds per capsule (lowest), was significantly different from other accessions. Accession UTLA7 generally had the highest value in traits related to dry matter production in vegetative organs, such as main branch length, fresh and dry weight of shoots and roots, as well as chlorophyll and proline content. As a consequence, it was isolated from the other accessions in the resulting figure plot (Table S6 and Figure 8).

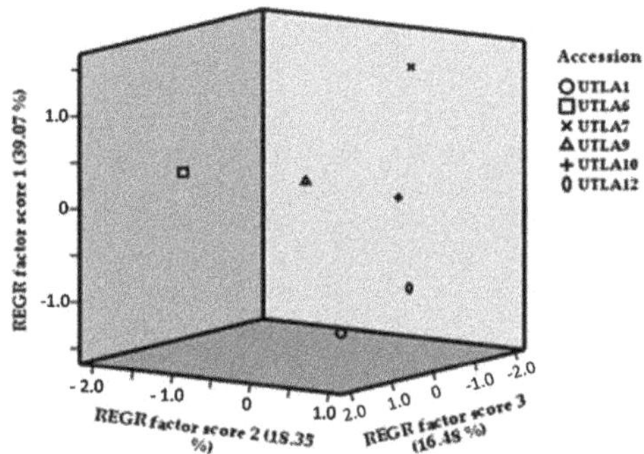

Figure 8. Principal component analysis triplot of *L. album* accessions.

4. Discussion

In this study, the effect of different levels of water deficit stress on six *L. album* accessions was examined based on various morphological, physiological, and biochemical traits. The results revealed that there was a significant difference between accessions in all the studied traits. However, water deficit, an important factor in plant growth and development, led to different responses in the different accessions. On the basis of the results of the analysis of morphological traits, and given that the experiment was conducted under fully uniform conditions, the high genetic diversity of accessions can be considered the main cause of significant differences, as in previous studies, where the existence of diversity among *L. album* accessions has been stated [13,61,62]. In addition to genetic factors in the occurrence of traits, other explanations for the existence of different responses include the existence of different mechanisms to avoid dehydration stress, including the use of large molecules (mucilage) to adjust the osmotic potential not yet known to wild plant species, such as *L. album*.

Cell growth is among the most drought-sensitive physiological processes due to the reduction in turgor pressure [15]. As revealed by the results of morphological traits, the effect of water deficit stress on accessions was different, and differences were also observed within each accession in their responses under different irrigation levels. For water stress response, severity, duration, and timing of stress, as well as responses of plants after stress removal and the interaction between stress and other factors, are extremely important [63]. Plant growth is typically severely affected by the water deficit. At the morphological level, shoots and roots are the most affected and both are the main components of plant adaptation to drought, as correlated in this study. Plants usually restrict the number and area of leaves in response to drought stress only to reduce water allocation at the expense of loss of yield [64]. As roots are the source of soil water, root growth, density, growth, and size are the main responses of plants to drought stress [65].

Based on the results of this study, the capsule-to-flower ratio in the 25% AW treatment showed a substantial decrease, which may have a serious impact on crop yields. Decreased grain growth in wheat due to reduced sucrose synthase activity [66] and the increased frequency of kernel abortion due to water deficit during pollination in corn [67,68] have been reported. The acceleration of the final stage in seed abortion for plants subjected to this stress tends to be a survival mechanism. Furthermore, the filling period was reduced because of earlier physiological maturity. This shorter filling period reduced seed growth [69]. As a consequence, having different dry matter partitioning patterns in response to different environmental conditions might vary plant behavior. For example, accession UTLA1 showed the highest seed yield per plant under normal irrigation conditions but the lowest vegetative yield.

A significant effect of the genetic background in *L. album* accessions was observed in the occurrence of different phenological stages, whereas water deficit stress only affected the time of maturity. The effects of drought range from morphological to molecular levels and are evident at any phenological stages. Plants that experienced stress during flower and pod development had a shorter period of organ appearance. This seemed to be due to an increase in the progression rates of reproductive organs under stress. Water stress applied at preanthesis reduced time to anthesis, whereas at postanthesis, it shortened the grain-filling period in triticale genotypes [70]. In summary, plants can escape water deficit stress by shortening their growth duration and avoid stress with the maintenance of high tissue water potential either by reducing water loss from plants or improving water uptake or both. Some plants may reduce their surface area by leaf shedding or producing smaller leaves [24].

Drought stress reduces the relative water content of leaves and is used as a reliable method for measuring the osmotic stress status [71]. RWC is commonly used for the measurement of plant water status in terms of the physiological and biochemical consequences of water deficit in plant cells [72]. It has been reported that plants with higher RWC are more resistant to water deficit stress [73]. In this study, accessions UTLA1 and UTLA12 had the highest RWC under the lowest irrigation level conditions. However, considering the performance traits, we found that these two accessions did not provide acceptable performance. Therefore, RWC cannot be a suitable trait for select tolerant *L. album* accessions (at least for the considered accessions).

Decreased levels of chlorophyll-a and chlorophyll-b have been reported in flaxseed (*Linum usitatissimum*) under dehydration stress [74], in agreement with the decrease in chlorophyll-a, chlorophyll-b, and carotenoid content observed in the *L. album* accessions in the present study. In most plant species, water deficit stress decreases the level of chlorophyll-a, chlorophyll-b, and total chlorophyll [18,75,76]. Drought stress induced changes in photosynthetic pigments and components [77], damaged photosynthetic apparatus [22], and decreased activities of the Calvin cycle enzymes, which are important causes of reduced crop yield [78]. In this study, a positive relationship between chlorophyll-a and chlorophyll-b content and yield-related traits (i.e., number of inflorescences, shoot dry weight, and root dry weight) was observed. For example, accession UTLA7 showed a higher chlorophyll content under stress conditions, and subsequently showed the highest vegetative yield (shoot and root dry weight).

Proline, an important compatible solute, is a general response of higher plants, algae, animals, and bacteria to low water potential [79,80]. In plants, its synthesis in leaves at low water potential is caused by a combination of increased biosynthesis and slow oxidation in mitochondria. Despite some controversy, many physiological roles have been assigned to free proline. Proline is considered a stabilizer of macromolecules, a sink for excess reductant, and a store of carbon and nitrogen for use after relief of water deficit [68]. As observed in our study, many experiments have reported an increase in proline levels in plants under water stress conditions in different plants [67,81–84]. Under severe water stress conditions, the amount of proline in accession UTLA7 showed a significant increase. Given the higher vegetative yields of this accession than others, increasing the amount of proline can be seen as a factor to boost plant growth in this accession.

Glycine betaine (GB) was reported to accumulate in response to stress in many plants, including sugar beet (*Beta vulgaris*), spinach (*Spinacia oleracea*), barley (*Hordeum vulgare*), and wheat (*Triticum aestivum*) [85–87]. In these species, tolerant genotypes accumulate more GB than sensitive genotypes in response to water stress. This, however, is not a general association, and it is most likely that the relationship between GB accumulation and stress tolerance is species- or even genotype-specific. As in *L. album*, the response of each accession was different. Here, with increasing stress levels, the amount of GB increased in accession UTLA10 but decreased in accessions UTLA1 and UTLA9. Therefore, this trait cannot be considered decisive for *L. album* drought tolerance.

The determination of MDA concentration, a membrane lipid peroxidation product, is used for quantifying the level of membrane peroxidation that leads to ion leakage [88]. Under stress conditions, the unsaturated fatty acids of the cell membranes are impressed by free radicals and form a chain reaction of lipid peroxidation [89]. In this study, ion leakage and MDA content were relatively low in plants under normal irrigation treatment but increased as water stress intensified in accessions UTLA1 and UTLA9, indicating the loss of cell stability and viability [90,91]. Zarrinabadi et al. [83] reported increased MDA content in pot marigold (*Calendula officinalis*) genotypes under water deficit conditions. Stress = sensitive species exhibit a sharper increase in lipid peroxidation than regular species under water deficit stress [92,93]. Here, ion leakage decreased in accession UTLA10 in response to the increasing water stress level, whereas no significant decrease in MDA was observed in other accessions. Since MDA is the final product of lipid peroxidase, various studies have reported that lower MDA content within genotypes indicates the greater antioxidant activities alongside resistances to arid conditions [94,95].

When plants are faced with water deficit conditions, in most cases, ABA levels increase as a result of increased synthesis [96]. A variation in the response of *L. album* accessions to water deficit stress based on their ABA accumulation was observed in the present study. Accessions UTLA2 and UTLA3 did not show any significant change in their ABA content under different irrigation levels, whereas in accessions UTLA1, UTLA4, UTLA,5 and UTLA6, ABA content was significantly increased. This increase was even more pronounced in accession UTLA1, and as a result, its root length was significantly higher after the 25% AW treatment. ABA affects the relative growth rates of different parts of the plant, such as raising the root-to-shoot dry weight ratio, inhibiting the development of the leaf area and growing deeper roots [97]. Here, we reported that the amount of ABA showed a negative correlation with pigments (chlorophyll-a, chlorophyll-b, and carotenoids) content. Under water stress conditions, apoplastic pH increases resulting in increased retention of ABA, serving as a signal for stomatal closure, accompanied by a reduction of transpiration in leaves, which is a significant water conservation response [52,98].

In the present study, the activity of antioxidant enzymes varied as a function of both the genetic background (i.e., accessions) and stress levels (i.e., %AW). A variation in the response mechanism of the antioxidant enzyme activity was clearly observed. ROS production is affected by the severity and duration of stress, species, genotype, and the developmental stage of the plant [99,100]. It is also affected by the ability of the plant to adapt to stress conditions [101]. When the defense response is unable to neutralize high levels of ROS, oxidative stress occurs. In this study, the response of accession UTLA1 to increasing water stress level was an increase in CAT activity, whereas APX activity decreased. However, in accessions UTLA6 and UTLA10, CAT activity decreased and POX activity increased. Interestingly, accession UTLA12 showed a decrease in CAT activity and an increase in APX activity. Catalase is the main antioxidant enzyme that scavenges the oxidant H_2O_2 by decomposition to oxygen and water [102]. In drought-tolerant genotypes, catalase activity increased under water stress conditions, which could be an adaptive mechanism to ROS [103]. Ascorbate peroxidase is a key antioxidant enzyme in plants [104], which catalyzes the conversion of H_2O_2 into H_2O. It has been reported that ascorbate peroxidase activity increases alongside other enzymes under drought stress [105] Askari and Ehsanzadeh [81] studied different drought treatments on fennel genotypes, and reported a higher increase of antioxidant enzymes, in particular catalase, in drought-tolerant genotypes. But in our research, and at least in accession UTLA1, this does not seem to be the case. The same effect of water deficit on the antioxidant content has been illustrated in medicinal plants such as peppermint (*Mentha piperita*), Oregano (*Origanum vulgare*), and marigold genotypes [83,106,107].

Based on the results of factor analysis, it is possible to determine the most effective variables to identify more tolerant accessions in future studies. Due to the placement of most of the influential traits for the vegetative and reproductive yield in *L. album*, the first PCA component can be considered a yield-influencing factor.

5. Conclusions

Evaluating the response of plant accessions under different environmental conditions can help to understand the growth characteristics of plants. Overall, significant diversity was observed in the different *L. album* accessions, which is very important for the advancement of breeding programs. A decrease in the response to increased water deficit stress was observed in most of the morphological traits studied here. Under normal irrigation conditions, accession UTLA1 is the best for seed production, and accessions UTLA7 and UTLA10 are better for the production of vegetative sections. Under severe stress conditions, however, UTLA9 appears to be more suitable for seed production, whereas UTLA10 and UTLA12 may be used for the production of vegetative parts. In general, accessions UTLA7, UTLA9, and UTLA10 had higher vegetative and seed yields and therefore are recommended for use in future breeding research. Water deficit stress accelerates the maturation of the plant, and accession UTLA9 completes its growth cycle faster; therefore, if the duration of the growth period is an important factor for production, this accession can be considered an ideal option for seed production. On the contrary, if the vegetative yield performance during the flowering stage is examined, accession UTLA6 may be recommended due to early entry into the flowering stage. Genetic variation between accessions was also confirmed by physiological responses. The examination of physiological characteristics confirms the enhanced efficiency of accessions UTLA7 and UTLA9. Overall, accession UTLA1 showed unique characteristics and responses among the accessions, whereas accessions UTLA7 and UTLA9 were almost identical in their response. The use of leaf length traits and chlorophyll content to estimate seed yield and shoot yield may be an appropriate marker. Traits such as seed yield per plant, plant height, number of inflorescences per plant, shoot and root dry weight, chlorophyll-a and chlorophyll-b, carotenoids, proline, guaiacol peroxidase, and ABA were obtained as important and discriminant traits for identifying tolerant *L. album* accessions. According to this analysis, it is recommended to use morphological traits to distinguish higher accessions.

To summarize, different accessions were studied and proposed for the advancement of breeding programs. Here, the productivity of *L. album* plants tends to be influenced by genotype rather than water deficit stress. However, we note that the water requirements of the *L. album* accessions also tend to be essential to the optimum achievement of some agronomical traits.

Supplementary Materials: The following are available online at http://www.mdpi.com/2073-4395/10/12/1966/s1. Table S1—Daily temperature conditions of field; Table S2—Analysis of variance of morphological traits; Table S3—Analysis of variance of phenological traits; Table S4—Analysis of variance of physiological traits; Table S5—Correlation between traits by Pearson method; Table S6—Factor analysis matrix of *L. album* accessions.

Author Contributions: Conceptualization: V.N., M.S.; data curation: R.K.; formal analysis: M.S. and R.K.; funding acquisition: V.N., M.S. and C.H.; investigation: R.K.; methodology: V.N., M.S., C.H. and R.K.; project administration: V.N. and M.S.; resources: V.N., M.S., C.H. and R.K.; software: R.K.; supervision; V.N. and M.S.; validation: V.N., M.S., C.H. and R.K.; visualization: R.K.; writing—original draft: R.K.; writing—review and editing: V.N., M.S., C.H. and R.K. All authors have read and agreed to the published version of the manuscript.

Funding: This research was supported by grants from University of Tehran and Région Centre-Val de Loire (Acti-LIN).

Acknowledgments: R.K. acknowledges research fellowship from Ministry of Science, Research and Technology of Iran.

Conflicts of Interest: The authors declare no conflict of interest.

References

1. Rechinger, K.H. Linaceae. In *Flora Iranica*; Akademische Druck-u. Verlagsanstalt: Graz, Austria, 1974; Volume 106, pp. 1–20.
2. Sharifnia, F.; Asadi, M. A phenetic study of the genus *Linum* L. (Linaceae) in Iran. *Iran. J. Bot.* **2002**, *9*, 135–139.

3. Vardapetyan, H.R.; Kirakosyan, A.B.; Oganesyan, A.A.; Penesyan, A.R.; Alfermann, W.A. Effect of various elicitors on lignan biosynthesis in callus cultures of *Linum austriacum. Russ. J. Plant Physiol.* **2003**, *50*, 297–300. [CrossRef]

4. Koulman, A.; Quax, W.J.; Pras, N. Podophyllotoxin and related lignans produced by plants. In *Biotechnology of Medicinal Plants: Vitalizer and Therapeutic*; Ramawat, K.G., Ed.; Science Publishers: New York, NY, USA, 2004; pp. 225–266.

5. Mikac, S.; Markulin, L.; Drouet, S.; Corbin, C.; Tungmunnithum, D.; Kiani, R.; Kabra, A.; Abassi, B.H.; Renouard, S.; Fuss, E.; et al. Bioproduction of Anticancer Podophyllotoxin and Related Aryltretralin-Lignans in Hairy Root Cultures of *Linum Flavum* L. In *Plant Cell and Tissue Differentiation and Secondary Metabolites*; Ramawat, K., Ekiert, H., Goyal, S., Eds.; Springer: Cham, Switzerland, 2020; pp. 1–38.

6. Bedir, E.; Khan, I.; Moraes, R.M. Bioprospecting for Podophyllotoxin. In *Trends New Crop and New Uses*; ASHS Press: Alexandria, VA, USA, 2002; pp. 545–549.

7. Huynh Cong, L.; Dauwe, R.; Lequart, M.; Vinchon, S.; Renouard, S.; Fliniaux, O.; Colas, C.; Corbin, C.; Doussot, J.; Hano, C.; et al. Kinetics of glucosylated and non-glucosylated aryltetralin lignans in *Linum* hairy root cultures. *Phytochemistry* **2015**, *115*, 70–78. [CrossRef] [PubMed]

8. Doussot, J.; Mathieu, V.; Colas, C.; Molinie, R.; Corbin, C.; Montguillon, J.; Moreno, Y.; Banuls, L.; Renouard, S.; Lamblin, F.; et al. Investigation of the lignan content in extracts from *Linum*, *Callitris* and *Juniperus* species in relation to their in vitro antiproliferative activities. *Planta Med.* **2017**, *83*, 574–581. [CrossRef]

9. Renouard, S.; Corbin, C.; Drouet, S.; Medvedec, B.; Doussot, J.; Colas, C.; Maunit, B.; Bhambra, A.S.; Gontier, E.; Jullian, N.; et al. Investigation of *Linum flavum* (L.) hairy root cultures for the production of anticancer aryltetralin lignans. *Int. J. Mol. Sci.* **2018**, *19*, 990. [CrossRef]

10. Malik, S.; Bilba, O.; Gruz, J.; Arroo, R.R.J.; Strnad, M. Biotechnological approaches for producing aryltetralin lignans from Linum species. *Phytochem. Rev.* **2014**, *13*, 893–913. [CrossRef]

11. Schmidt, T.J.; Klaes, M.; Sendker, J. Lignans in seeds of *Linum* species. *Phytochemistry* **2012**, *82*, 89–99. [CrossRef]

12. Farkya, S.; Bisaria, V.S.; Srivastava, A.K. Biotechnological aspects of the production of the anticancer drug podophyllotoxin. *Appl. Microbiol. Biotechnol.* **2004**, *65*, 504–519. [CrossRef]

13. Nazeri, V.; Kiani, R.; Rezaei, K.; Kalvandi, R. Diversity study of some ecological, morphological and fatty acid profile of *Linum album* Ky. ex Boiss. *Iran. J. Med. Aromat. Plants* **2017**, *33*, 168–183.

14. Akula, R.; Ravishankar, G.A. Influence of abiotic stress signals on secondary metabolites in plants. *Plant Signal. Behav.* **2011**, *6*, 1720–1731. [CrossRef]

15. Taiz, L.; Zeiger, E.; Møller, I.M.; Murphy, A. *Fisiologia e Desenvolvimento Vegetal*; Artmed Editora: Porto Alegre, Brazil, 2017; pp. 731–759.

16. Tabari, H.; Talaee, P.H. Temporal variability of precipitation over Iran: 1966–2005. *J. Hydrol.* **2011**, *396*, 313–320. [CrossRef]

17. Lu, C.; Zhang, J. Effects of water stress on photosystem II photochemistry and its thermostability in wheat plants. *J. Exp. Bot.* **1999**, *50*, 1199–1206. [CrossRef]

18. Liu, C.; Liu, Y.; Guo, K.; Fan, D.; Li, G.; Zheng, Y.; Yu, L.; Yang, R. Effect of drought on pigments, osmotic adjustment and antioxidant enzymes in six woody plant species in karst habitats of southwestern China. *Environ. Exp. Bot.* **2011**, *71*, 174–183. [CrossRef]

19. Tabari, H.; Abghari, H.; Hosseinzadeh Talaee, P. Temporal trends and spatial characteristics of drought and rainfall in arid and semiarid regions of Iran. *Hydrol. Process.* **2012**, *26*, 3351–3361. [CrossRef]

20. Gill, S.S.; Tuteja, N. Reactive oxygen species and antioxidant machinery in abiotic stress tolerance in crop plants. *Plant Physiol. Biochem.* **2010**, *48*, 909–930. [CrossRef] [PubMed]

21. Choudhury, F.K.; Rivero, R.M.; Blumwald, E.; Mittler, R. Reactive oxygen species, abiotic stress and stress combination. *Plant J.* **2017**, *90*, 856–867. [CrossRef] [PubMed]

22. Fu, J.; Huang, B. Involvement of antioxidants and lipid peroxidation in the adaptation of two cool-season grasses to localized drought stress. *Environ. Exp. Bot.* **2001**, *45*, 105–114. [CrossRef]

23. Reddy, A.R.; Chaitanya, K.V.; Vivekanandan, M. Drought-induced responses of photosynthesis and antioxidant metabolism in higher plants. *J. Plant Physiol.* **2004**, *161*, 1189–1202. [CrossRef]

24. Farooq, M.; Wahid, A.; Kobayashi, N.; Fujita, D.B.S.M.A.; Basra, S.M.A. Plant drought stress: Effects, mechanisms and management. In *Sustainable Agriculture*; Springer: Dordrecht, The Netherlands, 2009; pp. 153–188.

25. Talbi, S.; Romero-Puertas, M.C.; Hernández, A.; Terrón, L.; Ferchichi, A.; Sandalio, L.M. Drought tolerance in a Saharian plant Oudneya africana: Role of antioxidant defences. *Environ. Exp. Bot.* **2015**, *111*, 114–126. [CrossRef]

26. Dietz, K.J.; Mittler, R.; Noctor, G. Recent progress in understanding the role of reactive oxygen species in plant cell signaling. *Plant Physiol.* **2016**, *171*, 1535–1539. [CrossRef]
27. Hall, A.E. *Crop Responses to Environment*; CRC Press: Boca Raton, FL, USA, 2000; pp. 137–150.
28. Sikder, S. Accumulated heat unit and phenology of wheat cultivars as influenced by late sowing heat stress condition. *J. Agric. Rural Dev.* **2009**, *7*, 59–64. [CrossRef]
29. Wahid, A.; Gelani, S.; Ashraf, M.; Foolad, M.R. Heat tolerance in plants: An overview. *Environ. Exp. Bot.* **2007**, *61*, 199–223. [CrossRef]
30. Fischer, R.A. Crop Temperature Modification and Yield Potential in a Dwarf Spring Wheat 1. *Crop Sci.* **1976**, *16*, 855–859. [CrossRef]
31. Nemeskéri, E.; Neményi, A.; Bőcs, A.; Pék, Z.; Helyes, L. Physiological factors and their relationship with the productivity of processing tomato under different water supplies. *Water* **2019**, *11*, 586. [CrossRef]
32. Bishnoi, O.P.; Singh, S.; Niwas, R. Effect of temperature on phenological development of wheat (*Triticum aestivum*) crop in different orientation. *Indian J. Agric. Sci.* **1995**, *65*, 211–214.
33. Powell, W. *Abiotic Stresses: Plant Resistance through Breeding and Molecular Approaches*; Ashraf, M., Harris, P., Eds.; The Haworth Press: Binghamton, NH, USA, 2005; pp. 367–373.
34. Wollenweber, B.; Porter, J.R.; Schellberg, J. Lack of interaction between extreme high-temperature events at vegetative and reproductive growth stages in wheat. *J. Agron. Crop Sci.* **2003**, *189*, 142–150. [CrossRef]
35. Finkelstein, R. Abscisic acid synthesis and response. *Arab. Book Am. Soc. Plant Biol.* **2013**, *11*, e0166. [CrossRef]
36. Bartels, D.; Sunkar, R. Drought and salt tolerance in plants. *Crit. Rev. Plant Sci.* **2005**, *24*, 23–58. [CrossRef]
37. Moustakas, M.; Sperdouli, I.; Kouna, T.; Antonopoulou, C.I.; Therios, I. Exogenous proline induces soluble sugar accumulation and alleviates drought stress effects on photosystem II functioning of *Arabidopsis thaliana* leaves. *Plant Growth Regul.* **2011**, *65*, 315. [CrossRef]
38. Zaher-Ara, T.; Boroomand, N.; Sadat-Hosseini, M. Physiological and morphological response to drought stress in seedlings of ten citrus. *Trees* **2016**, *30*, 985–993. [CrossRef]
39. Hong, Z.; Lakkineni, K.; Zhang, Z.; Verma, D.P.S. Removal of feedback inhibition of Δ1-pyrroline-5-carboxylate synthetase results in increased proline accumulation and protection of plants from osmotic stress. *Plant Physiol.* **2000**, *122*, 1129–1136. [CrossRef] [PubMed]
40. Deuschle, K.; Funck, D.; Hellmann, H.; Däschner, K.; Binder, S.; Frommer, W.B. A nuclear gene encoding mitochondrial Δ1-pyrroline-5-carboxylate dehydrogenase and its potential role in protection from proline toxicity. *Plant J.* **2001**, *27*, 345–356. [CrossRef] [PubMed]
41. Ahmadian Chashmi, N.; Sharifi, M.; Behmanesh, M. Lignan enhancement in hairy root cultures of *Linum album* using coniferaldehyde and methylenedioxycinnamic acid. *Prep. Biochem. Biotechnol.* **2016**, *46*, 454–460. [CrossRef] [PubMed]
42. Venkatesan, A.; Chellappan, K.P. Accumulation of proline and glycine betaine in Ipomoea pes-caprae induced by NaCl. *Biol. Plant.* **1998**, *41*, 271–276. [CrossRef]
43. Mansour, M.M.F. Nitrogen containing compounds and adaptation of plants to salinity stress. *Biol. Plant.* **2000**, *43*, 491–500. [CrossRef]
44. Yang, W.J.; Rich, P.J.; Axtell, J.D.; Wood, K.V.; Bonham, C.C.; Ejeta, G.; Rhodes, D. Genotypic variation for glycinebetaine in sorghum. *Crop Sci.* **2003**, *43*, 162–169. [CrossRef]
45. Quan, R.; Shang, M.; Zhang, H.; Zhao, Y.; Zhang, J. Improved chilling tolerance by transformation with betA gene for the enhancement of glycinebetaine synthesis in maize. *Plant Sci.* **2004**, *166*, 141–149. [CrossRef]
46. Subbarao, G.V.; Wheeler, R.M.; Levine, L.H.; Stutte, G.W. Glycine betaine accumulation, ionic and water relations of red-beet at contrasting levels of sodium supply. *J. Plant Physiol.* **2001**, *158*, 767–776. [CrossRef]
47. Javadian, N.; Karimzadeh, G.; Sharifi, M.; Moieni, A.; Behmanesh, M. In vitro polyploidy induction: Changes in morphology, podophyllotoxin biosynthesis, and expression of the related genes in *Linum album* (Linaceae). *Planta* **2017**, *245*, 1165–1178. [CrossRef]
48. Tashackori, H.; Sharifi, M.; Chashmi, N.A.; Safaie, N.; Behmanesh, M. Induced-differential changes on lignan and phenolic acid compounds in *Linum album* hairy roots by fungal extract of Piriformospora indica. *Plant Cell Tissue Organ Cult.* **2016**, *127*, 187–194. [CrossRef]
49. Kiani, R. Evaluation of Morphological and Phytochemical Diversity of *Linum album* Ky. ex Bioss., an Endemic Medicinal Plant in Iran. Master's Thesis, Department of Horticultural Sciences College of Agriculture and Natural Resources, University of Tehran, Tehran, Iran, 2016.

50. Miller, P.; Lanier, W.; Brandt, S. *Using Growing Degree Days to Predict Plant Stages*; Ag/Extension Communications Coordinator, Communications Services, Montana State University-Bozeman: Bozeman, MO, USA, 2001; Volume 59717, pp. 994–2721.
51. Turner, N.C. Crop water deficits: A decade of progress. In *Advances in Agronomy*; Academic Press: Cambridge, MA, USA, 1986; Volume 39, pp. 1–51.
52. Sumanta, N.; Haque, C.I.; Nishika, J.; Suprakash, R. Spectrophotometric analysis of chlorophylls and carotenoids from commonly grown fern species by using various extracting solvents. *Res. J. Chem. Sci.* **2014**, *2231*, 606X.
53. Bates, L.S.; Waldren, R.P.; Teare, I.D. Rapid determination of free proline for water-stress studies. *Plant Soil* **1973**, *39*, 205–207. [CrossRef]
54. Grieve, C.M.; Grattan, S.R. Rapid assay for determination of water soluble quaternary ammonium compounds. *Plant Soil* **1983**, *70*, 303–307. [CrossRef]
55. Hodges, D.M.; DeLong, J.M.; Forney, C.F.; Prange, R.K. Improving the thiobarbituric acid-reactive-substances assay for estimating lipid peroxidation in plant tissues containing anthocyanin and other interfering compounds. *Planta* **1999**, *207*, 604–611. [CrossRef]
56. Nayyar, H.; Walia, D.P. Water stress induced proline accumulation in contrasting wheat genotypes as affected by calcium and abscisic acid. *Biol. Plant.* **2003**, *46*, 275–279. [CrossRef]
57. Bradford, M.M. A rapid and sensitive method for the quantitation of microgram quantities of protein utilizing the principle of protein-dye binding. *Anal. Biochem.* **1976**, *72*, 248–254. [CrossRef]
58. Hadwan, M.H. Simple spectrophotometric assay for measuring catalase activity in biological tissues. *BMC Biochem.* **2018**, *19*, 7. [CrossRef]
59. Plewa, M.J.; Smith, S.R.; Wagner, E.D. Diethyldithiocarbamate suppresses the plant activation of aromatic amines into mutagens by inhibiting tobacco cell peroxidase. *Mutat. Res. Fundam. Mol. Mech. Mutagenesis* **1991**, *247*, 57–64. [CrossRef]
60. Ranieri, A.; D'urso, G.; Nali, C.; Lorenzini, G.; Soldatini, G.F. Ozone stimulates apoplastic antioxidant systems in pumpkin leaves. *Physiol. Plant.* **1996**, *97*, 381–387. [CrossRef]
61. Kiani, R.; Nazeri, V.; Kalvandi, R. Investigation of genetic diversity in some populations of flax (*Linum album* Ky. ex Bioss.) using morphological traits. In Proceedings of the 4th National Congress on Medicinal Plants, Tehran, Iran, 12–13 May 2015.
62. Kiani, R.; Nazeri, V.; Rezaei, K.; Kalvandi, R. Evaluation of morphological diversity in *Linum album* Ky. ex Boiss. populations in Karaj conditions. In Proceedings of the 11th Iranian Horticultural Science, Urmia, Iran, 26 August 2019.
63. Plaut, Z. Plant exposure to water stress during specific growth stages. In *Encyclopedia of Water Science*; Stewart, B.A., Howell, T., Eds.; CRC Press: Boca Raton, FL, USA, 2003; pp. 673–685.
64. Schuppler, U.; He, P.H.; John, P.C.; Munns, R. Effect of water stress on cell division and Cdc2-like cell cycle kinase activity in wheat leaves. *Plant Physiol.* **1998**, *117*, 667–678. [CrossRef]
65. Kavar, T.; Maras, M.; Kidrič, M.; Šuštar-Vozlič, J.; Meglič, V. Identification of genes involved in the response of leaves of *Phaseolus vulgaris* to drought stress. *Mol. Breed.* **2008**, *21*, 159–172. [CrossRef]
66. Ahmadi, A.; Baker, D.A. The effect of water stress on the activities of key regulatory enzymes of the sucrose to starch pathway in wheat. *Plant Growth Regul.* **2001**, *35*, 81–91. [CrossRef]
67. Morgan, P.W. Effects of abiotic stresses on plant hormone systems. In *Stress Responses in Plants: Adaptation and Acclimation Mechanisms*; Alscher, R.G., Cumming, J.R., Eds.; Wiley-Liss, Inc.: New York, NY, USA, 1991; pp. 113–146.
68. Ober, E.S.; Setter, T.L.; Madison, J.T.; Thompson, J.F.; Shapiro, P.S. Influence of water deficit on maize endosperm development: Enzyme activities and RNA transcripts of starch and zein synthesis, abscisic acid, and cell division. *Plant Physiol.* **1991**, *97*, 154–164. [CrossRef] [PubMed]
69. Sionit, N.; Kramer, P.J. Effect of water stress during different stages of growth of soybean. *Agron. J.* **1977**, *69*, 274–278. [CrossRef]
70. Estrada-Campuzano, G.; Miralles, D.J.; Slafer, G.A. Genotypic variability and response to water stress of pre-and post-anthesis phases in triticale. *Eur. J. Agron.* **2008**, *28*, 171–177. [CrossRef]
71. Uzilday, B.; Turkan, I.; Sekmen, A.H.; Ozgur, R.E.N.G.İ.N.; Karakaya, H.C. Comparison of ROS formation and antioxidant enzymes in Cleome gynandra (C4) and Cleome spinosa (C3) under drought stress. *Plant Sci.* **2012**, *182*, 59–70. [CrossRef]

72. Barrs, H.D.; Weatherley, P.E. A re-examination of the relative turgidity technique for estimating water deficits in leaves. *Aust. J. Biol. Sci.* **1962**, *15*, 413–428. [CrossRef]

73. Arjenaki, F.G.; Jabbari, R.; Morshedi, A. Evaluation of drought stress on relative water content, chlorophyll content and mineral elements of wheat (*Triticum aestivum* L.) varieties. *Int. J. Agric. Crop Sci.* **2012**, *4*, 726–729.

74. Guo, R.; Hao, W.; Gong, D. Effects of water stress on germination and growth of linseed seedlings (*Linum usitatissimum* L), photosynthetic efficiency and accumulation of metabolites. *J. Agric. Sci.* **2012**, *4*, 253. [CrossRef]

75. Talebi, R.; Ensafi, M.H.; Baghebani, N.; Karami, E.; Mohammadi, K. Physiological responses of chickpea (*Cicer arietinum*) genotypes to drought stress. *Environ. Exp. Biol.* **2013**, *11*, 9–15.

76. Pirzad, A.; Shakiba, M.R.; Zehtab-Salmasi, S.; Mohammadi, S.A.; Darvishzadeh, R.; Samadi, A. Effect of water stress on leaf relative water content, chlorophyll, proline and soluble carbohydrates in *Matricaria chamomilla* L. *J. Med. Plants Res.* **2011**, *5*, 2483–2488.

77. Anjum, F.; Yaseen, M.; Rasul, E.; Wahid, A.; Anjum, S. Water stress in barley (*Hordeum vulgare* L.). II. Effect on chemical composition and chlorophyll contents. *Pak. J. Agric. Sci.* **2003**, *40*, 45–49.

78. Monakhova, O.F.; Chernyad'ev, I.I. Protective role of kartolin-4 in wheat plants exposed to soil draught. *Appl. Biochem. Microbiol.* **2002**, *38*, 373–380. [CrossRef]

79. Zhu, J.K. Salt and drought stress signal transduction in plants. *Annu. Rev. Plant Biol.* **2002**, *53*, 247–273. [CrossRef]

80. Wahid, A.; Close, T.J. Expression of dehydrins under heat stress and their relationship with water relations of sugarcane leaves. *Biol. Plant.* **2007**, *51*, 104–109. [CrossRef]

81. Askari, E.; Ehsanzadeh, P. Osmoregulation-mediated differential responses of field-grown fennel genotypes to drought. *Ind. Crop. Prod.* **2015**, *76*, 494–508. [CrossRef]

82. Dhar, P.; Ojha, D.; Kar, C.S.; Mitra, J. Differential response of tossa jute (*Corchorus olitorius*) submitted to water deficit stress. *Ind. Crop. Prod.* **2018**, *112*, 141–150. [CrossRef]

83. Zarrinabadi, I.G.; Razmjoo, J.; Mashhadi, A.A.; Boroomand, A. Physiological response and productivity of pot marigold (*Calendula officinalis*) genotypes under water deficit. *Ind. Crop. Prod.* **2019**, *139*, 111488. [CrossRef]

84. Kaur, G.; Asthir, B.J.B.P. Proline: A key player in plant abiotic stress tolerance. *Biol. Plant.* **2015**, *59*, 609–619. [CrossRef]

85. Fallon, K.M.; Phillips, R. Responses to water stress in adapted and unadapted carrot cell suspension cultures. *J. Exp. Bot.* **1989**, *40*, 681–687. [CrossRef]

86. McCue, K.F.; Hanson, A.D. Drought and salt tolerance: Towards understanding and application. *Trends Biotechnol.* **1990**, *8*, 358–362. [CrossRef]

87. Rhodes, D.; Hanson, A.D. Quaternary ammonium and tertiary sulfonium compounds in higher plants. *Annu. Rev. Plant Biol.* **1993**, *44*, 357–384. [CrossRef]

88. Heath, R.L.; Packer, L. Photoperoxidation in isolated chloroplasts: I. Kinetics and stoichiometry of fatty acid peroxidation. *Arch. Biochem. Biophys.* **1968**, *125*, 189–198. [CrossRef]

89. Gaweł, S.; Wardas, M.; Niedworok, E.; Wardas, P. Malondialdehyde (MDA) as a lipid peroxidation marker. *Wiad. Lek.* **2004**, *57*, 453–455. [PubMed]

90. Masia, A. Physiological effects of oxidative stress in relation to ethylene in postharvest produce. In *Postharvest Oxidative Stress in Horticultural Crops*; Food Products Press: New York, NY, USA, 2003; pp. 165–197.

91. Xu, Z.Z.; Zhou, G.S.; Wang, Y.L.; Han, G.X.; Li, Y.J. Changes in chlorophyll fluorescence in maize plants with imposed rapid dehydration at different leaf ages. *J. Plant Growth Regul.* **2008**, *27*, 83–92. [CrossRef]

92. Khoshro, H.H.; Taleei, A.; Bihamta, M.R.; Shahbazi, M.; Abbasi, A. Expression analysis of the genes involved in osmotic adjustment in bread wheat (*Triticum aestivum* L.) cultivars under terminal drought stress conditions. *J. Crop Sci. Biotechnol.* **2013**, *16*, 173–181. [CrossRef]

93. Shanjani, P.S.; Izadpanah, M.; Mohamadpour, M.R. Effects of water stress on germination of yarrow populations (*Achillea* spp.) from different bioclimatic zones in Iran. *Plant Breed. Seed Sci.* **2014**, *68*, 39. [CrossRef]

94. Dhanda, S.S.; Sethi, G.S.; Behl, R.K. Indices of drought tolerance in wheat genotypes at early stages of plant growth. *J. Agron. Crop Sci.* **2004**, *190*, 6–12. [CrossRef]

95. Yousfi, N.; Slama, I.; Ghnaya, T.; Savoure, A.; Abdelly, C. Effects of water deficit stress on growth, water relations and osmolyte accumulation in *Medicago truncatula* and *M. laciniata* populations. *Comptes Rendus Biol.* **2010**, *333*, 205–213. [CrossRef]

96. Davies, W.J.; Jones, H.G. Physiology and Biochemistry of Abscisic Acid. *J. Exp. Bot.* **1991**, *42*, 7–17. [CrossRef]

97. Sharp, R.E.; Wu, Y.; Voetberg, G.S.; Saab, I.N.; LeNoble, M.E. Confirmation that abscisic acid accumulation is required for maize primary root elongation at low water potentials. *J. Exp. Bot.* **1994**, *45*, 1743–1751. [CrossRef]
98. Turner, N.C.; Wright, G.C.; Siddique, K.H.M. Adaptation of grain legumes (pulses) to water-limited environments. *Adv. Agron.* **2001**, *71*, 194–233.
99. Abedi, T.; Pakniyat, H. Antioxidant enzymes changes in response to drought stress in ten cultivars of oilseed rape (*Brassica napus* L.). *Czech J. Genet. Plant Breed.* **2010**, *46*, 27–34. [CrossRef]
100. Caverzan, A.; Casassola, A.; Brammer, S.P. Antioxidant responses of wheat plants under stress. *Genet. Mol. Biol.* **2016**, *39*. [CrossRef] [PubMed]
101. Miller, G.A.D.; Suzuki, N.; Ciftci-Yilmaz, S.U.L.T.A.N.; Mittler, R.O.N. Reactive oxygen species homeostasis and signalling during drought and salinity stresses. *Plant Cell Environ.* **2010**, *33*, 453–467. [CrossRef] [PubMed]
102. Mittler, R. Oxidative stress, antioxidants and stress tolerance. *Trends Plant Sci.* **2002**, *7*, 405–410. [CrossRef]
103. Tunc-Ozdemir, M.; Miller, G.; Song, L.; Kim, J.; Sodek, A.; Koussevitzky, S.; Shintani, D. Thiamin confers enhanced tolerance to oxidative stress in Arabidopsis. *Plant Physiol.* **2009**, *151*, 421–432. [CrossRef]
104. Orvar, B.L.; Ellis, B.E. Transgenic tobacco plants expressing antisense RNA for cytosolic ascorbate peroxidase show increased susceptibility to ozone injury. *Plant J.* **1997**, *11*, 1297–1305. [CrossRef]
105. Shigeoka, S.; Ishikawa, T.; Tamoi, M.; Miyagawa, Y.; Takeda, T.; Yabuta, Y.; Yoshimura, K. Regulation and function of ascorbate peroxidase isoenzymes. *J. Exp. Bot.* **2002**, *53*, 1305–1319. [CrossRef]
106. Rahimi, Y.; Taleei, A.; Ranjbar, M. Long-term water deficit modulates antioxidant capacity of peppermint (*Mentha piperita* L.). *Sci. Hortic.* **2018**, *237*, 36–43. [CrossRef]
107. Morshedloo, M.R.; Salami, S.A.; Nazeri, V.; Craker, L.E. Prolonged Water Stress on Growth and Constituency of Iranian of Oregano (*Origanum vulgare* L.). *J. Med. Act. Plants* **2017**, *5*, 7–19.

Publisher's Note: MDPI stays neutral with regard to jurisdictional claims in published maps and institutional affiliations.

Article

Cultivar Dependent Impact on Yield and Its Components of Young Almond Trees under Sustained-Deficit Irrigation in Semi-Arid Environments

Saray Gutiérrez-Gordillo [1,*], Víctor Hugo Durán Zuazo [2], Virginia Hernández-Santana [3], Fernando Ferrera Gil [1], Amelia García Escalera [1], José Juan Amores-Agüera [1] and Iván Francisco García-Tejero [1]

[1] Instituto Andaluz de Investigación y Formación Agraria y Pesquera (IFAPA), Centro "Las Torres", Carretera Sevilla-Cazalla Km 12.2, 41200 Sevilla, Alcalá del Río, Spain; fernandoferreragil@icloud.com (F.F.G.); ameliagaes7@gmail.com (A.G.E.); josej.amores@juntadeandalucia.es (J.J.A.-A.); ivanf.garcia@juntadeandalucia.es (I.F.G.-T.)

[2] Instituto Andaluz de Investigación y Formación Agraria y Pesquera (IFAPA), Centro "Camino de Purchil", Camino de Purchil s/n, 18004 Granada, Spain; victorh.duran@juntadeandalucia.es

[3] Instituto de Recursos Naturales y Agrobiología de Sevilla, Consejo Superior de Investigaciones Científicas, Avda. Reina Mercedes 10, 41012 Sevilla, Spain; virginiahsa@irnas.csic.es

* Correspondence: saray.gutierrez@juntadeandalucia.es; Tel.: +34-671532852

Received: 9 April 2020; Accepted: 18 May 2020; Published: 20 May 2020

Abstract: Almond (*Prunus dulcis* Mill. (D.A. Webb)) plantations in irrigated semi-arid areas need to successfully face the new scenarios of climate change combining sustainable irrigation strategies and tolerant cultivars to water stress. This work examines the response of young almond (*cvs*. Guara, Marta, and Lauranne) subjected to different irrigation doses under semi-arid conditions (South-West Spain). The trial was conducted during two seasons (2018–2019) with three irrigation strategies: A full-irrigated treatment (FI), which received 100% of the irrigation requirements (IR), and two sustained-deficit irrigation strategies that received 75% (SDI_{75}) and 65% (SDI_{65}) of IR. Crop water status was assessed by leaf water potential (Ψ_{leaf}) and stomatal conductance (g_s) measurements, determining the yield response at the end of each season. Different physiological responses for the studied cultivars were observed, especially considering the Ψ_{leaf} measurements. In this way, *cv*. Marta behaved more tolerant, while *cvs*. Guara and Lauranne maintained higher g_s rates in response to water stress. These differences were also observed in terms of yield. The *cv*. Lauranne did not reflect yield losses, and the opposite trend was observed for *cv*. Guara, in which reductions on fruit numbers per tree were detected. On overall, effective irrigation water savings ($\approx$2100 $m^3 \cdot ha^{-1}$ in SDI_{65}) could be feasible, although these responses are going to be substantially different, depending on the used cultivar.

Keywords: almond cultivars; crop physiological response; irrigation water productivity; nut yield

---

1. Introduction

Today, deficit irrigation (DI) strategies cannot be considered as a novelty, and their effects and consequences have been widely studied in order to improve water resources management under water shortage conditions. With them, irrigation can be reduced until certain levels, with the main aim of maximizing the water savings keeping the yields within an acceptable range, close to those obtained under non-water restrictions [1]. Several authors have reported positive results when these strategies have been applied in different woody crops, many times with promising results of water savings, low yield reductions [2], and even improvements on fruit quality [3–5]. Although almond is a well-known drought-tolerant crop [6], water availability is the main restricting factor in semi-arid environments, determining the nut yield and its components (kernel unit weight and fruit numbers per tree) [7]. Moreover, yield reductions are not exclusively affected by the water stress but also the DI practice imposed; among them; regulated deficit irrigation (RDI), sustained-deficit irrigation (SDI), partial root-zone (PRD), or low-frequency deficit irrigation (LFDI) [8]. RDI and SDI are the most studied strategies; the first one, characterized by applying a smaller amount of water in that period in which the crop is less sensitive to this withholding of water [9], whereas SDI consists of applying a sustained water reduction throughout the growth cycle [10]. Very promising results have been obtained when LFDI is applied during the kernel-filling period, without yield losses even under long-term experiences [11]. This DI strategy consists of applying irrigation-restriction cycles which are derived by means of physiological threshold values previously defined. Thus, its application has the disadvantage of requiring proper crop water monitoring, which many times results in being very difficult for farmers and technicians. Finally, in PRD strategies, part of the root system is exposed to drying soil while the remaining part is irrigated normally [12], and in the case of almond, its application has not improved the obtained results in comparison to RDI or SDI treatments that had received similar irrigation amounts [13].

Currently, the RDI is the most-studied strategy for the case of almond and many authors have described its response under different water stress levels [9,14–16], but there are few works about the response of almond trees to SDI. In this sense, Egea et al. [17] compared the response of almond (*cv.* Marta) productivity to different irrigation strategies, reporting yield reductions of 12% and water savings of 37% with SDI strategy. Phogat et al. [18] evaluated the response of soil plant system in *cvs.* Nonpareil, Carmel and Ne Plus Ultra under SDI, obtaining that water uptake efficiency was substantially higher under SDI, compared to normal application conditions. Other authors [19] have evaluated the financial viability of applying SDI or RDI, concluding that SDI was the strategy that allowed greater savings in financial terms.

Not only the DI strategy is determinant, but also the almond cultivar, because the response to water stress could be different among them [20,21]. In this sense, Gómez-Laranjo et al. [22] studied the differences in terms of physiological response of five almond cultivars (*cvs.* Lauranne, Masbovera, Ferragnès, Francoli, and Glorieta) under full-irrigated and rainfed situations. This author concluded that *cv.* Lauranne and Francoli were the less sensitive to irrigation withholdings. Moreover, Oliveira et al. [23] studied leaf anatomy changes and water relations in five traditional cultivars (*cvs.* Bonita, Casanova, Parada, Pegarinhos, and Verdeal) and two commercial (*cvs.* Ferragnès and Glorieta), highlighting *cv.* Ferragnès as most sensitive cultivar to drought conditions. This fact was associated with low values for cuticle and for the ratio between palisade and spongy parenchyma; as well as higher values of vulnerability index, conducting vessels area, or xylem area. Yadollahi et al. [24] apprised that six almond genotypes had different reactions to water stress, displaying the ability to tolerate moderate and severe water stress conditions. In addition, Barzegar et al. [25] studied the responses of six almond cultivars (*cvs.* Azar, Marcona, Mission, Nonpareil, Sahand, and Supernova) to water stress, reporting as less sensitive *cvs.* Supernova and Azar in contrast to *cvs.* Marcona and Sahand. More recently, Gutiérrez-Gordillo et al. [26] analyzed the yield of three almond cultivars (*cvs.* Guara, Marta, and Lauranne) under RDI and over irrigated conditions, revealing that, although no significant reductions were observed between full irrigation and RDI; *cv.* Marta offered significant improvements in terms of yield and physiological response when this cultivar received irrigation doses around 150% of crop evapotranspiration, this response not being observed in the remaining cultivars.

In summary, plants' reactions to water stress are complex, implicating adaptive changes and/or detrimental consequences, and the different responses are regulated by the innate plant features as well as the duration and intensity of the imposed stress. In this line, we hypothesize that under water stress almond cultivars are able to exhibit different tolerance levels, and the physiological reactions could be divergent in response to SDI strategies, which ultimately effects on yield and its components. Thus, the aim of this study was to assess the nut yield and physiological response of three commercial almond cultivars namely Guara, Marta, and Lauranne subjected to sustained-deficit irrigation strategies, elucidating the tolerant cultivar under these strategies in a semi-arid Mediterranean environment.

2. Materials and Methods

2.1. Experimental Site

The trial was conducted during two consecutive years (2018–2019) in a commercial almond orchard (*Prunus dulcis* Mill. (D.A. Webb) *cvs.* Guara, Marta, and Lauranne (Figure 1); grafted onto GN15 rootstock), and located in the Guadalquivir river basin (SW Spain, 37°30′27.4″ N, 5°55′48.7″ W). Trees were planted in 2013, spaced 8 × 6 m, and drip irrigated using two pipelines with emitters of 2.3 L·h^{-1}, at 0.75 m intervals. Canopy volumes were very similar within each cultivar. That is, for *cv.* Marta, canopy volumes ranged from 64 to 65 m^3; for *cv.* Guara between 65 and 66 m^3; and for the case of *cv.* Lauranne, some higher than the previous cultivars, with canopy volumes between 72 and 74 m^3.

Figure 1. Experimental plot. Flowering (**A**) and vegetative (**B**) phenological period.

The soil is a silty loam typical Fluvisol, more than 2 m deep, with organic matter <1.5%. Roots were located predominately in the first 50 cm soil depth, corresponding to the intended wetting depth. Soil water content values at field capacity (−0.033 MPa) and permanent wilting point (−1.5 MPa) were close to 0.40 and 0.15 m^3·m^{-3}, respectively. The climatic classification of the study area is attenuated meso-Mediterranean, with an annual ET$_0$ rate of 1400 mm and accumulated rainfall of 540 mm (average data corresponding to the last 15 years; obtained from the Andalusian Weather information Network [27]).

Regarding to the experimental conditions registered in the irrigation period (from March to October) during the monitored years, in 2018 the average temperature was 20.5 °C with minimum and maximum average temperatures of 7.3 and 37.7 °C, respectively. Similarly, during the second year (2019), the average temperature was 25 °C with minimum and maximum average values of 7.2 and 36.5 °C, respectively. In relation to the average vapor pressure deficit (VPD) in 2018 it ranged from 1.20 to 3.36 KPa, while during 2019 it was between 1.10 and 2.49 KPa. Reference evapotranspiration (ET$_0$) and rainfall during the irrigation period (March–October) amounted to 1102 and 326 mm for 2018, meanwhile, in the second year ET$_0$ and rainfall were 1221 and 85 mm, respectively. Crop evapotranspiration (ET$_C$) rates for both seasons averaged 860 mm, which was in line with previous reports by López-López et al. [28] or García-Tejero et al. [11] who estimated water requirements of 800 and 900 mm for mature almond trees in Guadalquivir river basin.

2.2. Irrigation Treatments

Three irrigation treatments were designed; (i) a full-irrigated treatment (FI), which received 100% of irrigation requirements (IR) during the irrigation period, (ii) a sustained-deficit irrigation (SDI$_{75}$) treatment, which received 75% IR, and (iii) a sustained-deficit irrigation (SDI$_{65}$) treatment, which received 65% IR.

In both seasons, irrigation was applied from the middle of March to the end of October, the IR being estimated according to the methodology proposed by Allen et al. [29] (Equations (1) and (2)), obtaining the values of reference evapotranspiration (ET$_0$) by using a weather station installed in the same experimental orchard (Davis Advance Pro2, Davis Instruments, Valencia, Spain).

$$ET_C = K_C \times K_r \times ET_0 \tag{1}$$

$$IR \ (mm) = (ET_C - Rainfall) \tag{2}$$

where ET$_C$ is the crop evapotranspiration; K$_C$ is the single-crop coefficient; K$_r$ is the crop reduction coefficient, which depends on the percentage of shaded area cast by the tree canopy; ET$_0$ is the reference evapotranspiration; and IR is the irrigation requirements.

The local crop coefficients used during the experimental period ranged from 0.4 to 1.2 according to García-Tejero et al. [30]. Additionally, the IR was reduced for SDI$_{75}$ and SDI$_{65}$ by multiplying it by 0.75 and 0.65, respectively.

According the climatic conditions the irrigation doses received for FI, SDI$_{75}$ and SDI$_{65}$ during 2018 were 4974, 3713 and 3342 m^3·ha^{-1}; and during 2019; 7700, 5744 and 5159 m^3· ha^{-1}, respectively.

2.3. Plant Measurements

During the maximum evapotranspirative demand period, coinciding with the kernel-filling and pre-harvest stages (from early June to mid-August; 160–225 DOY in 2018; and 160–219 DOY in 2019), crop-water monitoring was done by means of measurements of leaf water potential (Ψ_{leaf}) and the stomatal conductance to water vapor (g_s); these readings being taken between 12:00 and 13:30 GTM, and with a periodicity of 10–15 days. The Ψ_{leaf} was measured using a pressure chamber (Soil Moisture Equipment Corp., Sta. Barbara, CA, USA), monitoring 8 trees per irrigation treatment and two leaves per tree, located in the north side of the tree and being totally mature, fresh and shaded; at 1.5 m of height, approximately and NW exposed. In these same trees, we measured the g_s, using a porometer SC-1 (Decagon Devices, INC, Pullman, WA, USA), these measurements being done in two leaves per monitored tree, completely exposed to the sun, at 1.5 m of height, and preferably with south-eastern facing.

Additionally, with the aim of quantifying total water stress supported by the crop, the stress integral (SI) was calculated from Ψ_{leaf} and g_s data, following the methodology proposed by Myers et al. [31] (Equations (3) and (4)):

$$SI_{leaf} = \left| \sum (\Psi_{leaf}^{av} - \Psi_{leaf}^{max}) \times n \right| \tag{3}$$

$$SI_{gs} = \sum (g_s^{av} - g_s^{min}) \times n \tag{4}$$

where SI_{leaf} is the stress integral in terms of leaf-water potential values, Ψ_{leaf}^{av} is the average leaf water potential for any interval; Ψ_{leaf}^{max} is the maximum value of leaf-water potential registered during the experimental period; SI_{gs} is the stress integral in terms of stomatal conductance values, g_s^{av} is the average stomatal conductance for any interval; g_s^{min} is the minimum value of stomatal conductance during the experimental period; and n is the days numbers within each interval.

According to these indexes, higher water stress gathered by the crop would be related to higher values of SI_{leaf} and lesser values of SI_{gs}. In this sense, these indexes provide information to quantify the crop water stress accumulated along a period, allowing quantifying the effect of water restriction beyond its temporal distribution, integrating the global stress supported by the crop in comparison to the punctual measurements.

At the end of each season it was monitored the almond yield in terms of nut and kernel weight. Harvesting was carried out by using a mechanical vibrator with a mechanical peeling to remove the hull. Once cleaned, almonds were left to air dry and weighed once they reached a humidity content of around 6%. Finally, almonds were processed with shelling machine, obtaining the kernel yield for each irrigation treatment and cultivar.

In relation to the harvesting dates these were different for each cultivar and season. In 2018, the almond harvest labors were done at 232 DOY for *cv.* Guara and 239 DOY for *cvs.* Marta and Lauranne; meanwhile in 2019, these were done at 219, 221, and 235 DOY for Guara, Marta, and Lauranne, respectively.

Taking into consideration that two components determines the yield (the almond size and number of almonds per tree); the first one was obtained by weighing 100 almonds per monitored tree ($n = 8$); obtaining the kernel unit weight and ratio between kernel and nut (kernel + shell). After this, the second component (number of almonds per tree) was estimated by dividing "the kernel yield per monitored tree" by "the kernel unit weight", tracing the most affected yield component by water stress imposed in each cultivar.

Finally, and considering the total irrigation applied in each treatment, it was estimated the irrigation water productivity (IWP; $kg \cdot m^{-3}$), defined as the ratio between kernel yield and irrigation, water applied.

2.4. Experimental Design and Stadistical Analysis

The experimental design was of randomized blocks, with four replications per irrigation treatment and cultivar. Each replication had 12 trees (3 rows and 4 trees per row), and the two central trees for each replication were monitored. Thus, eight trees per treatment of irrigation strategy were monitored ($n = 8$).

Statistical analysis was developed by using the Sigma Plot statistical software (version 12.5, Systat Software, Inc., San Jose, CA, USA) and the SPSS software (SPSS Inc., 15.0 Statistical packages; Chicago, IL, USA). Year-to-year, an exploratory descriptive analysis of the whole of physiological measurements (Ψ_{leaf} and g_s) for each treatment and cultivar was done; applying a Levene's test to check the variance homogeneity of the variables studied. After this, for each cultivar, an ANOVA for repeated measures was developed (three treatments and 2 freedom degrees), applying a Bonferroni test to compare pairs of treatments when significant differences in the ANOVA were detected. Moreover, and with the aim of identifying those days in which differences between irrigation treatments were detected, a one-way analysis of variance (ANOVA) and a Tukey's test were done for each measurement day.

Additionally, considering the results provided by SI_{gs} and SI_{leaf}; for each irrigation treatment and cultivar, a two-way ANOVA was developed, followed by a Tukey's multiple range test. These results allow development of a simpler statistical analysis (considering that all measurements were transformed into an stress index that summarizes the total water stress accumulated by the crop) and to find a single value to evaluate the water stress supported by each cultivar and irrigation treatment and to analyze the hypothetical differences between cultivars and treatments in physiological terms with ease. Moreover, the linear regressions between the average values of SI_{leaf} and SI_{gs} registered for each treatment in both years were obtained ($n = 6$), applying an analysis of covariance (ANCOVA) to evaluate the differences in the interception points and slopes of these regressions. These analyses would allow the determination of whether these relationships are cultivar dependent and if this dependence is accompanied by similar yield responses.

Finally, and year-to-year, the kernel yield and its components were analyzed (kernel unit weight, the ratio between kernel weight vs. almond weight (kernel + shell), and fruits number per tree); by applying a Levene's test to check the variance homogeneity and ANOVA with a Tukey's test, considering as factors, the irrigation treatment, the cultivar, and their interactions.

3. Results and Discussion

3.1. Physiological Response to Water Stress

In the course of the first experimental season (2018) the ANOVA for repeated measures did not show significant differences between irrigation treatments within each cultivar. That is, on overall as Ψ_{leaf} as g_s evidenced similar results for the studied treatments; although afterwards, the analysis done independently for each monitoring day reflected significant differences between treatments. These differences between treatments were especially noticeable for the case Ψ_{leaf} (Figure 2), in comparison to the higher similarities reported by g_s measurements (Figure 3). When confronting the Ψ_{leaf} values registered in each cultivar, these ranged from −0.9 to −2.3 MPa for *cvs.* Guara and Lauranne, meanwhile for *cv.* Marta these ranged between −0.8 and −2.0 MPa.

In relation to g_s, the obtained values between treatments and cultivars were very similar, which was very evident by the absence of significant differences. Comparing the g_s values for each cultivar, *cv.* Guara showed values between 80 and 175 mmol·m^{-2}·s^{-1}, *cv.* Marta between 75 and 180 mmol·m^{-2}·s^{-1}, and *cv.* Lauranne registered g_s rates between 90 and 175 mmol·m^{-2}·s^{-1}.

As happened in 2018, during 2019 the differences among treatments were higher in Ψ_{leaf} than in g_s. (Figure 2). Again, the ANOVA for repeated measures did not reflect significant differences between irrigation treatments. For *cv.* Guara, Ψ_{leaf} values ranged from −1.4 to −1.9 MPa in FI, whereas in SDI treatments these values oscillated from −1.7 to −2.5 MPa. Somewhat lower were the values for *cv.* Lauranne with Ψ_{leaf} between −1.1 and −1.7 MPa in FI, and for both SDI$_{75}$ and SDI$_{65}$ between −1.5 and −2.0 MPa. Also, some differences for *cv.* Marta were observed with Ψ_{leaf} with values between −1.1 and −1.6 MPa in FI; and for both SDI treatments between −1.3 and −1.9 MPa. In relation to g_s in 2019, no significant differences among treatments were observed for any studied cultivar (Figure 3), very similar to that reported in the previous season. In particular, for *cv.* Guara, the g_s rates were between 163 and 314 mmol·m^{-2}·s^{-1} whereas *cv.* Lauranne displayed similar rates between 161 and 275 mmol·m^{-2}·s^{-1}. Finally, *cv.* Marta registered g_s values that ranged from 150 to 310 mmol·m^{-2}·s^{-1}, these being similar in all studied irrigation treatments.

In summary and considering the observed values during the two studied years, *cv.* Marta showed a higher capability to register higher values for Ψ_{leaf} with respect to the other remaining studied cultivars, especially under SDI conditions.

More perceptible findings were achieved in relation to the water stress integral (Table 1). During the first experimental season (2018) significant effects in relation to cultivar ($p < 0.05$) were found in $SI_{\Psi leaf}$ and SI_{gs}. In this sense, *cvs.* Guara and Lauranne showed similar values of $SI_{\Psi leaf}$ and SI_{gs}, and significant ($p < 0.05$) higher than those calculated in *cv.* Marta. Particularly noticeable were the values of SI_{gs} for *cv.* Marta, which were, globally, 38% and 30% lower than those registered in *cvs.* Lauranne and Guara, respectively. Respect to the irrigation doses effects, no significant differences were observed for $SI_{\Psi leaf}$ and SI_{gs}, this being in agreement with the absence of differences reported by the ANOVA for repeated measured previously discussed.

During 2019, the highest differences were observed again among cultivars, as for $SI_{\Psi leaf}$ as for SI_{gs}. Thus, the lowest and highest values of $SI_{\Psi leaf}$ were reached by *cvs.* Marta and Guara, respectively, very similar to the response observed the previous season. Moreover, the highest differences in 2019 were determined in SI_{gs} among cultivars, with values for *cvs.* Guara and Lauranne higher that those obtained for *cv.* Marta ($p < 0.01$). That is, irrigation doses did not promote differences in $SI_{\Psi leaf}$ (as the previous season), different to the observed in SI_{gs} with values for SDI$_{65}$ 20% higher than those detected in FI.

Taking into account the interaction between both considered factors (irrigation × cultivar), these were detected only the second studied year, as for $SI_{\Psi leaf}$ as for SI_{gs}, the cultivar being the main factor of this interaction. More interesting were the relationships between the SI_{leaf} and SI_{gs}, which were estimated for each cultivar (Figure 4). Whereas *cvs.* Guara and Lauranne showed similar relationships, and the ANCOVA did not reveal differences for the intercepts and slopes between both cultivars;

significant differences were registered with the interception point obtained in *cv.* Marta, although the slopes were very similar.

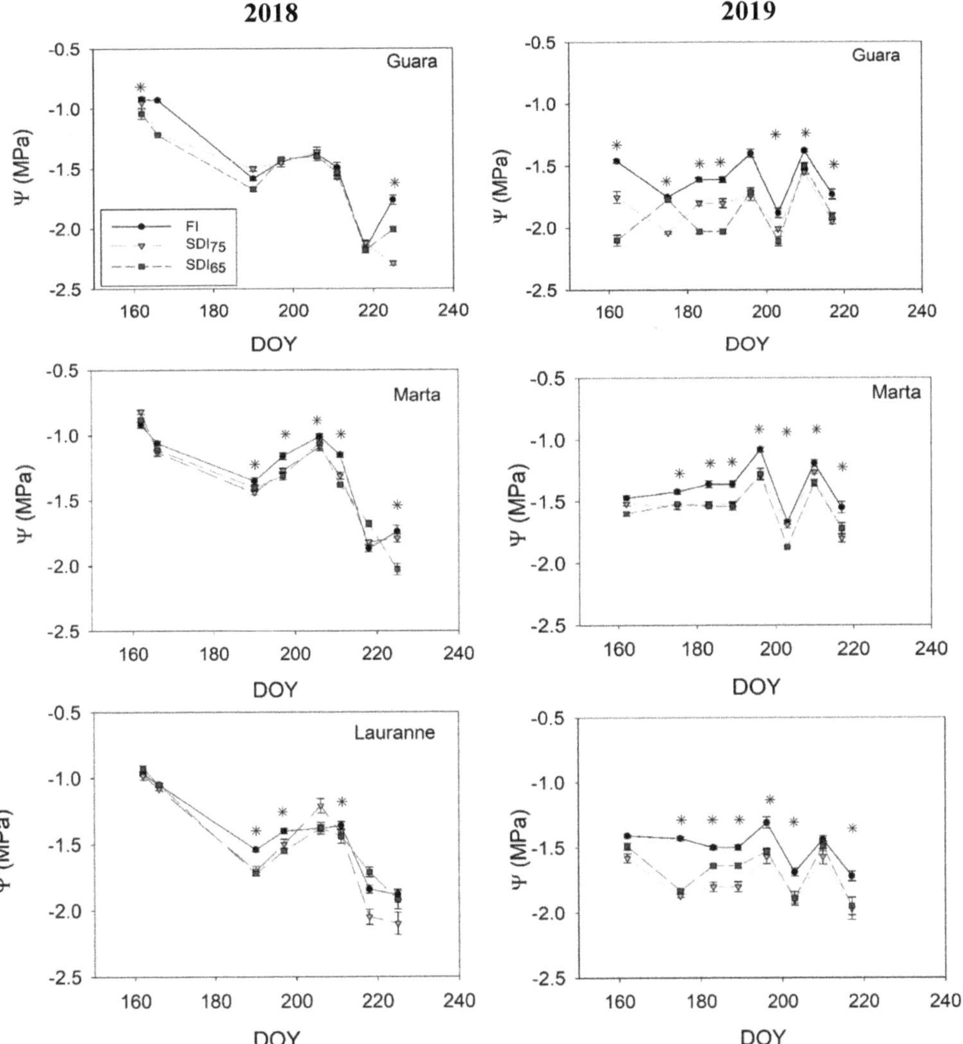

Figure 2. Seasonal dynamics of leaf water potential (Ψ_{leaf}) during 2018 and 2019. FI, Full irrigated treatment; SDI_{75}, sustained-deficit irrigation at 75% IR; SDI_{65}, sustained-deficit irrigation at 65% IR, DOY; Day of the year. Vertical bars are standard deviation. Asterisks show the intervals with significant differences between FI and SDI treatments ($p < 0.05$).

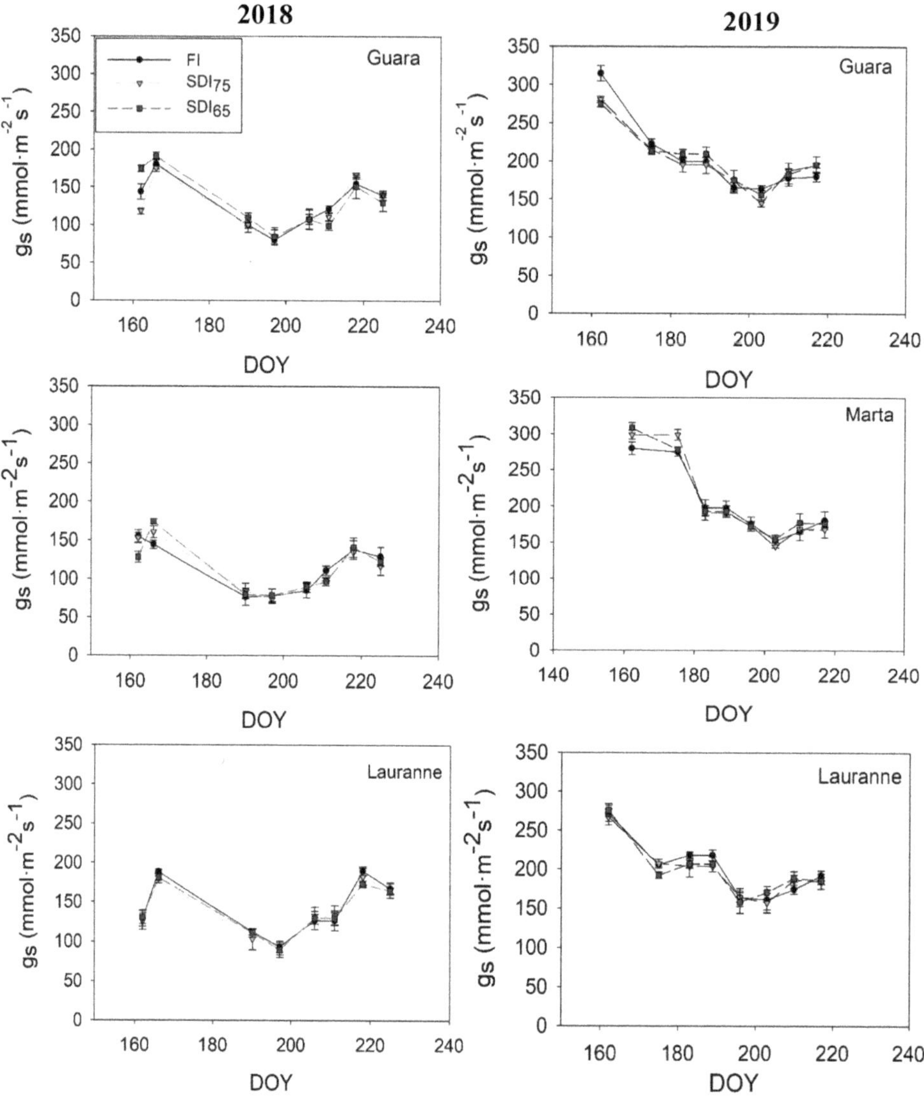

Figure 3. Seasonal dynamics of stomatal conductance (g_s) during 2018 and 2019. FI, Full irrigated treatment; SDI75, sustained-deficit irrigation at 75% IR; SDI65, sustained-deficit irrigation at 65% IR, DOY; Day of the year. Vertical bars are standard deviation. Asterisks show the intervals with significant differences between FI and SDI treatments ($p < 0.05$).

Table 1. Effect of deficit irrigation treatment and cultivar on water stress integrals for SI_{leaf} and SI_{gs}.

	First Year (2018)		Second Year (2019)	
	SI_{leaf}	SI_{gs}	SI_{leaf}	SI_{gs}
ANOVA	(MPa·day)	(mmol m^{-2} s^{-1}·day)	(MPa·day)	(mmol m^{-2} s^{-1}·day)
Irrigation	ns	ns	ns	*
Cultivar	*	*	*	**
Irrigation × cultivar	ns	ns	*	*
Tukey multiple range test				
Irrigation				
FI	198.7a	3069a	188.7a	3136a
SDI$_{75}$	199.4a	3071a	197.9a	3426ab
SDI$_{65}$	197.9a	3152a	199.6a	3758b
Cultivar				
Marta	189.6b	2301b	187.8b	2651b
Lauranne	201.6a	3685a	194.3ab	4127a
Guara	204.7a	3306a	204.1a	3532a
Irrigation × cultivar				
cv. Marta				
FI	190.9a	2174a	178.8d	3005b
SDI$_{75}$	189.8a	2306a	187.6c	2874b
SDI$_{65}$	188.2a	2422a	197.2ab	2075c
cv. Lauranne				
FI	201.9b	3784b	193.9bc	4151a
SDI$_{75}$	202.9b	3633b	197.7ab	4023a
SDI$_{65}$	200.2b	3639b	191.3c	4207a
cv. Guara				
FI	203.3b	3249b	193.5bc	4089a
SDI$_{75}$	205.5b	3276b	208.5a	3381b
SDI$_{65}$	205.4b	3395b	210.2a	3127b

FI, Full irrigated treatment; SDI$_{75}$, sustained-deficit irrigation at 75% IR; SDI$_{65}$, sustained-deficit irrigation at 65% IR. ns, not significant; * and **, significant at $p < 0.05$ and $p < 0.01$, respectively. Values followed by the same letter within the same column and factor are not significantly different ($p > 0.05$) by Tukey's test.

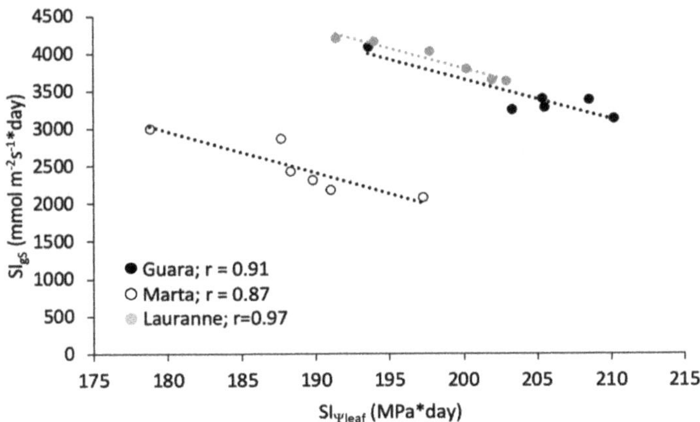

Figure 4. Linear relationships between the stress integral of leaf water potential (SI_{leaf}) and stomatal conductance (SI_{gs}) for cvs. Guara, Marta, and Lauranne. Each point is defined by the average values for each irrigation treatment, season, and cultivar.

According to these results, when physiological response was analyzed by using SI_{leaf} and SI_{gs}, these differences were more evident, this being mostly relevant in comparing the type of cultivars. Considering previous results reported by Gutiérrez-Gordillo et al. [26], *cv.* Marta was more sensitive to water stress in physiological terms, evidencing a stronger stomatal control under RDI strategy. On the contrary, *cv.* Guara triggered physiological mechanisms that were able to maximize the gas-exchange rates by increasing g_s and decreasing $_{leaf}$ values. By contrasting these findings with those outlined by García-Tejero et al. [32], declines of $_{leaf}$ encourage to lessen the carbon assimilation rate, disclosing the almond capability in maintaining high g_s values even when $_{leaf}$ is close to −2.5 MPa [33]. Moreover, before reaching important exhaustion in g_s rates, reductions in $_{leaf}$ are not accompanied by relevant depletions in g_s [32]. This might happen because almond is competent at holding optimum g_s and carbon assimilation rates, improving the water-use efficiency, at least under moderate water-stress situations. Therefore, almond reaction to water stress would demand different regulation mechanisms to counteract the adverse impact of water stress as was corroborated with the results of the present study. In addition, according to Fernández et al. [34] and Fu et al. [35], the almond has a lower capability to regulate the stoma under mild water stress situations. Moreover, these findings corroborate with those revealed by García-Tejero et al. [11] who defined two threshold values for Ψ_{leaf}; the first about −1.4 MPa without reductions for g_s, and a second of −2.0 MPa, when significant depletions in g_s are observed. Also, other authors reported that under moderate water stress conditions the almond tree decreased the $_{leaf}$ rates much more than g_s, and only when a certain threshold value is reached, significant depletions on photosynthesis rates are detected [36–38].

3.2. Nut Yield and Irrigation-Water Productivity (IWP)

Yield and its related components showed results that agreed with the physiological responses observed in the three studied cultivars (Table 2). During the first experimental season, significant differences on kernel yield were observed in *cv.* Guara, with yield reductions around 13% in SDI$_{65}$ in comparison to FI treatment. These differences were exclusively reflected in fruits number per tree, with a fruit number depletion similar to that reported on total yield. Something similar happened with *cv.* Marta, with yield reductions around 11% in SDI$_{65}$ comparing to FI treatment, as consequence of fruits number reductions of 13.5% on average. However, these depletions were partially corrected by increasing the kernel unit weight, especially in SDI$_{75}$. Moreover, no differences on kernel yield were observed for *cv.* Lauranne, with similar results among treatments in all the yield components. FI, Full irrigated treatment; SDI$_{75}$, Sustained-deficit irrigation at 75% RI; SDI$_{65}$, Sustained-deficit irrigation at 65% RI. Values followed by the same letter within the same row and factor were not significantly different ($p < 0.05$) by Tukey's test.

During the second year, *cv.* Guara showed again significant differences between SDI$_{65}$ and FI conditions, although in this case SDI$_{75}$ offered similar productions to those detected under FI; with yield reductions around 8% and 17% on SDI$_{75}$ and SDI$_{65}$, respectively. These depletions were associated with fruit number reductions roughly 9% on SDI$_{75}$ and 32% on SDI$_{65}$. It is noticeable that these reductions in the fruits number were partially corrected because of a significant increasing of kernel unit weight on SDI$_{65}$ (23% higher than the obtained value of FI). Something different were the obtained values for *cv.* Marta during the second year, which did not evidence relevant yield losses by effect of SDI strategies. Finally, as it was previously discussed for the first studied season, *cv.* Lauranne registered the best response to SDI strategies, without relevant differences in the studied yield components.

Analysing these results regarding to the irrigation water productivity (IWP, kg·m^{-3}) (Table 3), in the first studied season and within each treatment, *cvs.* Guara and Marta evidenced similar results, these being 20%, 40%, and 31% higher in *cv.* Lauranne for FI, SDI$_{75}$, and SDI$_{65}$; respectively. Regarding to the effects of water stress, all treatments fixed significant improvements in comparison to the results detected under FI. Thus, comparing the IWP obtained in SDI$_{65}$ with that registered under FI, the average values for both studied seasons offered improvements on IWP of 31%, 36%, and 43% for Guara, Marta, and Lauranne; respectively. These results were confirmed during the second season. All cultivars offered similar values of IWP for FI and SDI$_{75}$, meanwhile, for the case of SDI$_{65}$ relevant improvements were detected in *cvs.* Marta and Lauranne in comparison to *cv.* Guara. Regarding to the effects of water stress, all cultivars registered significant increasing trend under

SDI strategies, *cvs*. Lauranne and Marta offering the best response versus *cv*. Guara when sustained water withholdings of 35% were imposed. These values would be comparable to those reported by Egea et al. [39] for *cv*. Marta (0.25–0.40 kg·m^{-3}) or Phogat et al. [18] who highlighted that water productivity increased substantially respect to full irrigated trees when SDI strategies were applied.

Table 2. Impact of irrigation strategies on yield components during the studied years.

Cultivar	First Year (2018)			Second Year (2019)		
	FI	SDI$_{75}$	SDI$_{65}$	FI	SDI$_{75}$	SDI$_{65}$
	Kernel yield (kg ha^{-1})					
Guara	1929a	1659b	1704b	2254a	2081ab	1871b
Marta	1933a	1677b	1775b	2218a	2209a	2243a
Lauranne	2349a	2343a	2241a	2326a	2105a	2196a
	Ratio (kernel/nut)					
Guara	0.41a	0.41a	0.41a	0.34a	0.39b	0.37b
Marta	0.32a	0.31a	0.33a	0.34a	0.34a	0.34a
Lauranne	0.36a	0.36a	0.36a	0.33a	0.33a	0.32a
	Kernel unit weight (g)					
Guara	1.40a	1.40a	1.41a	0.99b	1.00b	1.22a
Marta	1.31a	1.37b	1.33a	1.21a	1.18a	1.18a
Lauranne	1.12a	1.13a	1.13a	1.03a	1.05a	1.08a
	Fruits number tree^{-1}					
Guara	6611a	5688b	5801b	10930a	9989ab	7363b
Marta	7083a	5875b	6407ab	8799a	8984a	9125a
Lauranne	10068a	9952a	9519a	10828a	9621b	9758b

FI, Full irrigated treatment; SDI$_{75}$, Sustained-deficit irrigation at 75% RI; SDI$_{65}$, Sustained-deficit irrigation at 65% RI. Values followed by the same letter within the same row and factor were not significantly different ($p < 0.05$) by Tukey's test.

Table 3. Irrigation-water productivity (IWP, kg·m^{-3}) in each cultivar, treatment and studied year.

Cultivar	First Year (2018)			Second Year (2019)		
	(kg·m^{-3})					
	FI	SDI$_{75}$	SDI$_{65}$	FI	SDI$_{75}$	SDI$_{65}$
Guara	0.39c	0.45b	0.51a	0.29b	0.36a	0.36a
Marta	0.39c	0.45b	0.53a	0.29b	0.34b	0.44a
Lauranne	0.47b	0.63a	0.67a	0.30b	0.37ab	0.43a

FI, Full irrigated treatment; SDI$_{75}$, Sustained-deficit irrigation at 75% IR; SDI$_{65}$, Sustained-deficit irrigation at 65% IR. Values followed by the same letter within the same raw and year were not significantly different ($p < 0.05$) by Tukey's test.

Taking into account the whole of data, *cv*. Guara was the most sensitive cultivar to water stress under sustained-deficit irrigation strategies. This fact is especially noticeable, taking into account that this cultivar evidenced a very positive response when water withholding was applied under RDI strategies during the kernel-filling period [26,28,32]. This contrary response would evidence that the final response to DI strategies would be determined by the added effect of three factors: The cultivar, the water stress, and the irrigation strategy. However, taking into consideration the findings in the present work, when this water stress was applied during the whole irrigation period, it was detected a significant fruit dropping (during the fruit setting period) (Table 2), and ultimately, this determined the final yield with significant reductions linked to less fruit numbers per tree. These results agree with those reported by other authors. In this sense, for a sustained-deficit irrigation study (2500 m^3·ha^{-1}), Alegre et al. [40] in Catalonia (North-Eastern Spain) reported productions for seven-year-old almond plantations of *cvs*. Guara and Lauranne of 1.65 and 2.02 t·ha^{-1}, respectively. The absolute difference with respect to our findings could mainly be ascribed to the amount of irrigation water applied in water-stressed treatments (4250 and 4730 m^3·ha^{-1}) and the climatic conditions of South-Western Spain. Likewise, these results would confirm the better response of *cv*. Lauranne in comparison to *cv*. Guara under SDI strategies. This is in line with Miarnau et al. [41] who outlined that kernel yield of new almond plantations could be ranged between 1.50 and 2.0 t·ha^{-1} with water allocations of 2000–3000 and 6000 m^3·ha^{-1}, respectively. In general, the yield potential of almond is highly related to the irrigation amount provided as it was revealed by Miarnau et al. [42]. These authors pointed out that under a SDI strategy with water applications around 2000 m^3·ha^{-1}, the nut yield for *cvs.* Guara and Marta amounted to 1.20 and 1.85 t·ha^{-1}, whereas under FI conditions ($\approx$7500 m^3·ha^{-1}) these values were 2.80 and 3.55 t·ha^{-1}, respectively. Consequently, this fact would suggest that *cv.* Marta would be able to activate a physiological prevention mechanism to mitigate the water stress, yielding more than *cv.* Guara, as was observed in the present work. Moreover, similar agronomical and physiological responses to water stress of *cvs.* Guara and Lauranne were highlighted by Girona et al. [16].

According to effects of SDI in the fruit unit weight, Alegre et al. [40] reported similar values than those found in the present experiment with kernel unit weights of 1.5 and 1.2 g for *cv.* Guara and Lauranne. In addition, according to Miarnau et al. [43] kernel unit weight FI conditions for *cvs.* Guara, Lauranne, and Marta would be around 1.50, 1.20, and 1.50 g, respectively, implying that water stress provoked weight reduction as it was found in the present study. Likewise, it is worth mentioning the improvements in terms of fruit unit weight observed in *cv.* Marta under SDI$_{75}$ during the first experimental season; and something similar in *cv.* Guara under SDI$_{65}$, during the second year of this experiment. These results would reinforce the possibility of improving the fruit size when SDI is imposed, this being an added value in relation to fruit marketability and consumer acceptance [44].

4. Conclusions

Combining the type of almond cultivars with water stress through deficit irrigation will be vital to reach an equilibrium between water allocations and sustainable nut yields under climate change scenarios. In the framework of the present experiment, the almond response to SDI strategies was cultivar-dependent, and hence, this fact should be considered before designing a proper DI strategy.

The findings allow for a conclusion on the importance of the cultivar when a DI strategy is being applied because different physiological behaviors will promote different responses in terms of yield and its components. In this way, the *cv.* Marta exhibited the most conservative behavior to water stress in physiological terms, which allowed it to obtain very similar productions than those registered under FI conditions. Furthermore, *cv.* Lauranne, despite showing a physiological behavior similar to *cv.* Guara, it was able to reach the best yield values when a moderate-to-severe SDI was applied. Also, according to our findings, *cv.* Guara registered the lesser promising results, with significant yield reductions ($\approx$14%) when water restrictions around 35% of irrigation requirements were applied; these being particularly promoted by depletions in the fruit number per tree. That is, SDI$_{65}$ would be suitable strategies to *cvs.* Lauranne and Marta, whereas for the case of *cv.* Guara we should select a more moderate SDI strategy (as SDI$_{75}$) or re-consider the application of other more appropriate treatments such as RDI during the kernel-filling period.

Finally, taking into consideration the absence of differences in *cvs.* Lauranne and Marta, and the differing results observed in *cv.* Guara; long-term experiment could be advisable; in order to get a deeper knowledge respect to cumulative effects of more severe water stress strategies imposed during several consecutive seasons.

Author Contributions: Writing: S.G.-G.; Conception or design: S.G.-G. and I.F.G.-T.; Acquisition, analysis, and interpretation of data: S.G.-G., I.F.G.-T., A.G.E., J.J.A.-A., and F.F.G.; Critical revision of the manuscript for important intellectual content: S.G.-G., I.F.G.-T., V.H.D.Z., and V.H.-S.; Statistical analysis: S.G.-G.; Reagents/materials/analysis tools contribution: S.G.-G., I.F.G.-T., A.G.E., and F.F.G. All authors have read and agreed to the published version of the manuscript.

Funding: Part of this work was sponsored by the research project "Impact of climate change and adaptation measures (INNOVA-Climate)" (AVA.AVA2019.051), Research and Technological Innovation Projects for period 2019–2021 Andalusia moves with Europe.

Acknowledgments: The author S. Gutiérrez-Gordillo has a contract co-financed by National Institute of Agrarian and Food Research and technology (FPI-INIA 2016) and European Social Fund (ESF) "The European Social Fund invests in your future". The authors particularly thank Paco Campos and Juan Luis Viveros from Nutalia SL Comp. for their capable research assistance.

Conflicts of Interest: The authors declare no conflict of interest.

References

1. Fereres, E.; Soriano, M.A. Deficit irrigation for reducing agricultural water use. *J. Exp. Bot.* **2007**, *58*, 147–159. [CrossRef] [PubMed]
2. Sepaskhah, A.R.; Ahmadi, S.H. A review on partial root-zone drying irrigation. *Int. J. Plant Prod.* **2012**, *4*, 241–258.
3. Expósito, A.; Berbel, J. Sustainability implications of deficit irrigation in a mature water economy: A case study in Southern Spain. *Sustainability* **2017**, *9*, 1144. [CrossRef]
4. Kaya, S.; Evren, S.; Dasci, E.; Adiguzel, M.C.; Yilmaz, H. Effects of different irrigation regimes on vegetative growth, fruit yield and quality of drip-irrigated apricot trees. *Afr. J. Biotechnol.* **2010**, *9*, 5902–5907.
5. Sánchez-Rodríguez, L.; Corell, M.; Hernández, F.; Sendra, E.; Moriana, A.; Carbonell-Barrachina, Á.A. Effect of Spanish-style processing on the quality attributes of HydroSOStainable green olives. *J. Sci. Food Agric.* **2019**, *99*, 1804–1811. [CrossRef] [PubMed]
6. Torrecillas, A.; Ruiz-Sanchez, M.C.; Del Amor, F.; Leon, A. Seasonal variations on water relations of Amygdalus communis L. under drip irrigated and non irrigated conditions. *Plant Soil* **1988**, *106*, 215–220. [CrossRef]
7. Naor, A.; Birger, R.; Peres, M.; Gal, Y.; Elhadi, F.A.; Haklay, A.; Assouline, S.; Schwartz, A. The effect of irrigation level in the kernel dry matter accumulation period on almond yield, kernel dry weight, fruit count, and canopy size. *Irrig. Sci.* **2018**, *36*, 1–8. [CrossRef]
8. García Tejero, I.F.; Moriana, A.; Rodríguez Pleguezuelo, C.R.; Durán Zuazo, V.H.; Egea, G. Chapter 12-Sustainable Deficit-Irrigation Management in Almonds (*Prunus dulcis* L.): Different Strategies to Assess the Crop Water Status. In *Water Scarcity and Sustainable Agriculture in Semiarid Environment*; García Tejero, I.F., Durán Zuazo, V.H., Eds.; Academic Press: Cambridge, MA, USA, 2018; pp. 271–298.
9. Stewart, W.L.; Fulton, A.E.; Krueger, W.H.; Lampinen, B.D.; Shackel, K.A. Regulated deficit irrigation reduces water use of almonds without affecting yield. *Calif. Agric.* **2011**, *65*, 90–95. [CrossRef]
10. Monks, D.P.; Taylor, C.; Sommer, K.; Treeby, M.T. Deficit irrigation of almond trees did not decrease yield. *Acta Hortic.* **2017**, 251–260. [CrossRef]
11. García-Tejero, I.F.; Gutiérrez Gordillo, S.; Souza, L.; Cuadros-Tavira, S.; Durán Zuazo, V.H. Fostering sustainable water use in almond (*Prunus dulcis* Mill.) orchards in a semiarid Mediterranean environment. *Arch. Agron. Soil Sci.* **2019**, *65*, 164–181. [CrossRef]
12. Kang, S.; Zhang, J. Controlled alternate partial root-zone irrigation: Its physiological consequences and impact on water use efficiency. *J. Exp. Bot.* **2004**, *55*, 2437–2446. [CrossRef] [PubMed]
13. Egea, G.; González-Real, M.M.; Baille, A.; Nortes, P.A.; Sánchez-Bel, P.; Domingo, R. The effects of contrasted deficit irrigation strategies on the fruit growth and kernel quality of mature almond trees. *Irrig. Sci.* **2009**, *36*, 1–8. [CrossRef]
14. Goldhamer, D.A.; Viveros, M.; Salinas, M. Regulated deficit irrigation in almonds: Effects of variations in applied water and stress timing on yield and yield components. *Irrig. Sci.* **2006**, *24*, 101–114. [CrossRef]
15. Romero, P.; Botia, P.; Garcia, F. Effects of regulated deficit irrigation under subsurface drip irrigation conditions on vegetative development and yield of mature almond trees. *Plant Soil* **2004**, *260*, 169–181. [CrossRef]

16. Girona, J.; Mata, M.; Marsal, J. Regulated deficit irrigation during the kernel-filling period and optimal irrigation rates in almond. *Agric. Water Manag.* **2005**, *75*, 152–167. [CrossRef]

17. Egea, G.; Nortes, P.A.; Domingo, R.; Baille, A.; Pérez-Pastor, A.; González-Real, M.M. Almond agronomic response to long-term deficit irrigation applied since orchard establishment. *Irrig. Sci.* **2013**, *31*, 445–454. [CrossRef]

18. Phogat, V.; Skewes, M.A.; Mahadevan, M.; Cox, J.W. Evaluation of soil plant system response to pulsed drip irrigation of an almond tree under sustained stress conditions. *Agric. Water Manag.* **2013**, *118*, 1–11. [CrossRef]

19. Alcon, F.; Egea, G.; Nortes, P.A. Financial feasibility of implementing regulated and sustained deficit irrigation in almond orchards. *Irrig. Sci.* **2013**, *31*, 931–941. [CrossRef]

20. Karimi, S.; Yadollahi, A.; Arzani, K.; Imani, A.; Aghaalikhani, M. Gas-exchange response of almond genotypes to water stress. *Photosynthetica* **2015**, *53*, 29–34. [CrossRef]

21. Palasciano, M.; Logoluso, V.; Lipari, E. Differences in drought tolerance in almond cultivars grown in apulia region (Southeast Italy). *Acta Hortic.* **2014**, 319–324. [CrossRef]

22. Gomes-Laranjo, J.; Coutinho, J.P.; Galhano, V.; Cordeiro, V. Responses of five almond cultivars to irrigation: Photosynthesis and leaf water potential. *Agric. Water Manag.* **2006**, *83*, 261–265. [CrossRef]

23. Oliveira, I.; Meyer, A.; Afonso, S.; Gonçalves, B. Compared leaf anatomy and water relations of commercial and traditional *Prunus dulcis* (Mill.) cultivars under rain-fed conditions. *Sci. Hortic.* **2018**, *229*, 226–232. [CrossRef]

24. Yadollahi, A.; Arzani, K.; Ebadi, A.; Wirthensohn, M.; Karimi, S. The response of different almond genotypes to moderate and severe water stress in order to screen for drought tolerance. *Sci. Hortic.* **2011**, *129*, 403–413. [CrossRef]

25. Barzegar, K.; Yadollahi, A.; Imani, A.; Ahmadi, N. Influences of severe water stress on photosynthesis, water use efficiency and proline content of almond cultivars. *J. Appl. Hortic.* **2012**, *14*, 33–39. [CrossRef]

26. Gutiérrez-Gordillo, S.; Durán-Zuazo, V.H.; García-Tejero, I. Response of three almond cultivars subjected to different irrigation regimes in Guadalquivir river basin. *Agric. Water Manag.* **2019**, *222*, 72–81. [CrossRef]

27. RIA. Red de Información Agroclimática de Andalucía. Instituto Andaluz de Investigación y Formación Agraria y Pesquera. Consejería de Agricultura, Ganadería, Pesca y Desarrollo Sostenible. Junta de Andalucía. Available online: https://ifapa.junta-andalucia.es/agriculturaypesca/ifapa/ria/servlet/FrontController?action= Static&url=coordenadas.jsp&c_provincia=41&c_estacion=19 (accessed on 27 March 2020).

28. López-López, M.; Espadador, M.; Testi, L.; Lorite, I.J.; Orgaz, F.; Fereres, E. Water use of irrigated almond trees when subjected to water deficits. *Agric. Water Manag.* **2018**, *195*, 84–93. [CrossRef]

29. Allen, R.G.; Pereira, L.S.; Raes, D.; Smith, M.; W, B. Crop evapotranspiration-Guidelines for computing crop water requirements-FAO Irrigation and drainage paper 56. *Irrig. Drain.* **1998**, 1–15.

30. García-Tejero, I.F.; Hernandez, A.; Rodriguez, V.M.; Ponce, J.R.; Ramos, V.; Muriel, J.L.; Zuazo, V.H.D. Estimating Almond Crop Coefficients and Physiological Response to Water Stress in Semiarid Environments (SW Spain). *J. Agric. Sci. Technol.* **2015**, *17*, 1255–1266.

31. Myers, B.J. Water stress integral—a link between short-term stress and long-term growth. *Tree Physiol.* **1988**, *4*, 315–323. [CrossRef]

32. García-Tejero, I.F.; Gutiérrez-Gordillo, S.; Ortega-Arévalo, C.; Iglesias-Contreras, M.; Moreno, J.M.; Souza-Ferreira, L.; Durán-Zuazo, V.H. Thermal imaging to monitor the crop-water status in almonds by using the non-water stress baselines. *Sci. Hortic.* **2018**, *238*, 91–97. [CrossRef]

33. Hernandez-Santana, V.; Fernández, J.E.; Rodriguez-Dominguez, C.M.; Romero, R.; Diaz-Espejo, A. The dynamics of radial sap flux density reflects changes in stomatal conductance in response to soil and air water deficit. *Agric. For. Meteorol.* **2016**, *218–219*, 92–101. [CrossRef]

34. Fernández, J.E. Understanding olive adaptation to abiotic stresses as a tool to increase crop performance. *Environ. Exp. Bot.* **2014**, *103*, 158–179. [CrossRef]

35. Fu, X.; Meinzer, F.C. Metrics and proxies for stringency of regulation of plant water status (iso/anisohydry): A global data set reveals coordination and trade-offs among water transport traits. *Tree Physiol.* **2018**, *39*, 122–134. [CrossRef] [PubMed]

36. Hernandez-Santana, V.; Rodriguez-Dominguez, C.M.; Fernández, J.E.; Diaz-Espejo, A. Role of leaf hydraulic conductance in the regulation of stomatal conductance in almond and olive in response to water stress. *Tree Physiol.* **2016**, *36*, 725–735. [CrossRef] [PubMed]

37. Rodriguez-Dominguez, C.M.; Buckley, T.N.; Egea, G.; de Cires, A.; Hernandez-Santana, V.; Martorell, S.; Diaz-Espejo, A. Most stomatal closure in woody species under moderate drought can be explained by stomatal responses to leaf turgor. *Plant Cell Environ.* **2016**, *39*, 2014–2026. [CrossRef]

38. Espadafor, M.; Orgaz, F.; Testi, L.; Lorite, I.J.; González-Dugo, V.; Fereres, E. Responses of transpiration and transpiration efficiency of almond trees to moderate water deficits. *Sci. Hortic. (Amst.)* **2017**, *225*, 6–14. [CrossRef]

39. Egea, G.; Nortes, P.A.; González-Real, M.M.; Baille, A.; Domingo, R. Agronomic response and water productivity of almond trees under contrasted deficit irrigation regimes. *Agric. Water Manag.* **2010**, *97*, 171–181. [CrossRef]

40. Alegre Castellví, S.; Miarnau i Prim, X.; Romero Romero, M.; Vargas García, F. Potencial productivo de seis variedades de almendro. *Frutic. Prof.* **2007**, *169*, 23–29.

41. Miarnau, X.; Torguet, L.; Batlle, I.; Romero, A.; Rovira, M.; Alegre, S. Comportamiento agronómico y productivo de las nuevas variedades de almendro. *Fruticultura* **2016**, *49*, 42–59.

42. Miarnau, X.; Torguet, L.; Batlle, I.; Romero, A.; Rovira, M.; Alegre, S. Resultados agronómicos y productivos de los ensayos IRTA de nuevas variedades de almendro. *Irta Jorn. Almendro Ix Ed.* **2017**, 43–63.

43. Miarnau, X.; Torguet, L.; Batlle, I.; Romero, A.; Rovira, M.; Alegre, S. La revolución del almendro: Nuevas variedades y nuevos modelos productivos. *Simp. Nac. Almendro Otros Frutos Secos* **2015**, 6–27.

44. Lipan, L.; Cano-Lamadrid, M.; Corell, M.; Sendra, E.; Hernández, F.; Stan, L.; Vodnar, D.C.; Vázquez-Araújo, L.; Carbonell-Barrachina, Á.A. Sensory profile and acceptability of hydrosostainable almonds. *Foods* **2019**, *8*, 64. [CrossRef] [PubMed]

 *agronomy*

Article

Deficit Irrigation and Its Implications for HydroSOStainable Almond Production

Iván Francisco García-Tejero [1,*,†], Leontina Lipan [2,†], Saray Gutiérrez-Gordillo [1], Víctor Hugo Durán Zuazo [3], I. Jančo [4], F. Hernández [5], Belén Cárceles Rodríguez [3] and Ángel Antonio Carbonell-Barrachina [2]

[1] Centro IFAPA "Las Torres". Ctra. Sevilla-Cazalla, km 12.2, Alcalá del Río, 41200 Sevilla, Spain; saray.gutierrez@juntadeandalucia.es

[2] Department of Agro-Food Technology, University Miguel Hernández-EPSO, Carretera de Beniel, km 3.2, Orihuela, 03312 Alicante, Spain; leontina.lipan@goumh.umh.es (L.L.); angel.carbonell@umh.es (Á.A.C.-B.)

[3] Centro IFAPA "Camino de Purchil", Camino de Purchil s/n, 18004 Granada, Spain; victorh.duran@juntadeandalucia.es (V.H.D.Z.); belen.carceles@juntadeandalucia.es (B.C.R.)

[4] Department of Chemistry, Faculty of Biotechnology and Food Sciences, Slovak University of Agriculture in Nitra, Tr.A. Hlinku 2, 949 76 Nitra, Slovakia; ivonajanco@gmail.com

[5] Department of Plant Science and Microbiology, University Miguel Hernández-EPSO, Carretera de Beniel, km 3.2, Orihuela, 03312 Alicante, Spain; francisca.hernandez@umh.es

* Correspondence: ivanf.garcia@juntadeandalucia.es

† These authors had an equally contribution.

Received: 1 October 2020; Accepted: 21 October 2020; Published: 23 October 2020

Abstract: Deficit irrigation (DI) strategies are considered essential in many arid and semi-arid areas of Mediterranean countries for proper water management under drought conditions. This fact is even more necessary in crops such as almond (*Prunus dulcis* Mill.), which in the last recent years has been progressively introduced in irrigated areas. An essential aspect to be considered would be the ability to improve fruit-quality parameters when DI strategies are imposed, which can boost the final almond price and ensure the sustainability and competitiveness of this crop. This work examines the effects of sustained deficit irrigation (SDI) on three almond cultivars (Marta, Guara, and Lauranne) on parameters related to almond functionality, aroma and sensory profile, which consequently influence its marketability and consumers acceptance. SDI strategies allowed the improvement of physical parameters such as unit weight, kernel length, kernel thickness or color. Moreover, higher total phenolic compounds, organic acids and sugars were found in SDI almonds. Finally, the highest concentrations of volatile compounds were obtained under SDI, this being a clear advantage in relation to almond flavor. Thus, moderate SDI strategy offered relevant improvements in parameters regarding the marketability, by enhancing the final added value of hydroSOStainable almonds with respect to those cultivated under full irrigation conditions.

Keywords: almond quality; sustainability; marketability; semiarid Mediterranean environment; water stress

---

1. Introduction

Water is the most limiting natural resource for sustainable agricultural development in arid and semi-arid areas of Mediterranean and more specifically under climate change scenarios [1]. In this regard, several works have reported the impact of climate change on agriculture [2,3], concluding that crop water demand will increase substantially due to higher evapotranspiration rates, temporal variability of rainfall or heat weaves events [4].

Within these scenarios, implementing drought-tolerant crops in irrigated zones and the application of deficit irrigation (DI) strategies are being considered, especially in the last few years [5,6]. Developing water-saving strategies in Mediterranean woody crops involves many considerations relative to environmental constraints, sustainable yields, and product marketability. However, these must be deeply studied, establishing the most profitable strategies to maximize the fruit production, minimizing the irrigation water consumption and maintaining (or even improving) the fruit quality.

Almond (*Prunus dulcis* Mill.) is the largest tree nut crop, and the world surface dedicated to its cultivation during 2018 amounted to 2,071,884 ha, with Spain being the country with the largest area devoted to this crop, with 657,768 ha, followed by the USA, with 441,107 ha [7]. However, in production terms, relevant differences are found, with 1,872,500 and 339,033 t for the USA and Spain, respectively. Moreover, for the period 2010–2018 in the USA (3500–5608 kg ha^{-1}) and Spain (279–515 kg ha^{-1}), the production per area were highly different. In 2018, the Mediterranean basin (Spain, Italy, Greece, Syria, Tunisia, Argelia and Morocco) produced roughly 29% of the world almond production, with most plantations under rainfed conditions and located in marginal areas. In Spain, the almond cultivation is mainly concentrated in Andalusia (S Spain) (31% of total area), and since approximately 85% of almond crops are rainfed, this provokes important fluctuations in productivity [8].

Today, the increase in almond prices from 2014 to 2019, when it reached an average price of 5€ per kg [9], has resulted in an increase of surface devoted to almond cultivation with different techniques [10], explicitly in irrigated areas that were traditionally occupied by other crops (cereals, cotton and sunflower, among others) [11]. This increase in cultivated area under irrigation from 2014 to 2018 amounted to 25% by using new cultivars [12], which allows to obtain higher yields (>1500 kg ha^{-1}) than those from traditional rainfed plantations (150–500 kg ha^{-1}) [13].

By taking into consideration the advantage of the positive adaptation of almond to drought scenarios [14,15], and its sharp phenology, many authors have reported the positive responses and opportunities of DI for almond cultivation, obtaining competitive yields under moderate-to-severe water stress situations [16–19].

Recently, a novelty research line, focused on food production under hydro-sustainable strategies (hydroSOStainable products), has been successfully developed [20–22]. This also showed advantages of different Mediterranean crops, with significant improvements in the fruit quality, sensory profile and consumer acceptance in crops such as pistachios [23], olives [24] and almonds [25]. In this context, there is in an interest in those characteristics or parameters of raw almonds related to their marketability that could be affected by DI strategies. In other words, the main limitation of DI implementation is ultimately the yield reduction (in comparison to the potential rate when almond is grown under full-irrigated conditions), affecting the plantation viability and its competitiveness. In this regard, Lipan et al. [25,26] reported relevant results, concluding that some fruit-quality parameters could be improved (or at least not affected) when DI strategies are imposed. If that is the case, these kinds of strategies would encourage the product marketability and the consumer acceptance, allowing a recovering in terms of final price, minimizing the losses when these are analyzed in monetary terms. In addition, many aspects are still not clear, such as if these effects would be similar for the new high-yielding cultivars or the dependence in relation to the irrigation strategy imposed or crop physiological status during the water stress period.

On the other hand, consumer appreciation, and hence, almond marketability, can be determined by a wide number of variables that could be classified into physical and chemical parameters, all of them determining the sensory appreciation and the almond appeal. Raw almonds are mainly composed by fats (44–61%), proteins (16–23%) and dietary fiber (11–14%) and high concentrations of vitamin E [27,28]. Despite these being the main compounds of raw almonds, their influence in taste receptors is negligible. In this line, other compounds, more related to flavor properties and sensory and chemical characteristics, can be found. According to Civille et al. [29] the main taste properties of raw almonds are mainly defined by astringency and sweetness degree, and to a lesser extent, the tactile dimensions (almond texture) [30]; this is the highest variability in almond flavor related to odor-active volatiles compounds. That is, volatile compounds are responsible of characteristic flavor properties of raw and processed almonds and contribute to their high consumer acceptance [25,31]. In particular, benzaldehyde is one of the main volatiles in bitter almonds, but its presence in sweet almonds is very variable (highly cultivar dependent), and, overall, it is found in very low concentrations [32]. Despite these low concentrations, its presence is responsible of the typical almond flavor and derivates such as marzipan [33].

Considering that the hydroSOStainable almond production could be a key factor for sustainable development in a semiarid Mediterranean environment; the objective of this study was to evaluate the effects of sustained deficit irrigation strategies and almond cultivars on the main physico-chemical parameters involved in nut sensory profile and improvements in marketability.

2. Material and Methods

2.1. Plant Material, Growing Conditions and Experimental Design

The trial was conducted during 2019 in a commercial orchard of almonds (*Prunus dulcis* Mill., *cvs* Guara, Marta and Lauranne), grafted onto GN15 rootstock, and located in the Guadalquivir river basin (37°30′27.4″ N; 5°55′48.7″ O) (Seville, SW Spain) (Figure 1). The plantation contained seven-year-old almond trees, 8 × 6 m spaced and drip irrigated by using two pipelines with emitters of 2.3 L h^{-1}. The soil is a silty loam typical Fluvisol [34], more than 2 m deep, fertile and with an organic matter content of 15.0 g kg^{-1}. Roots were located predominately in the first 50 cm of soil, corresponding to the intended wetting depth, although these exceed more than one meter in depth. The climatology in the study area is attenuated meso-Mediterranean, with an annual reference evapotranspiration rate (ET$_0$) of 1400 mm and an annual rainfall of 540 mm, mainly distributed from October to April. More details about the experimental site can be found in Gutiérrez-Gordillo et al. [35].

Figure 1. Experimental almond orchard (**A**) and *cvs*. Marta (**B**), Guara (**C**) and Lauranne (**D**).

Three irrigation treatments were designed: (i) a full-irrigated treatment (FI), which received 100% of irrigation requirements (IR) during the irrigation period, and two sustained-deficit irrigation treatments, which received 75% (SDI$_{75}$) and 65% (SDI$_{65}$) of IR. Irrigation was applied from April to

October, the IR being estimated according to the methodology proposed by Allen et al. [36], obtaining the values of ET_0 by using a weather station installed in the same experimental orchard (Davis Advance Pro2, Davis Instruments, Valencia, Spain). The local crop coefficients used during the experimental period ranged from 0.4 to 1.2, according to the results obtained by García-Tejero et al. [37].

2.2. Field Measurements

Physiological response to different irrigation doses was evaluated throughout measurements of leaf water potential (Ψ_{leaf}) in shaded leaves, these readings being taken between 12:00 and 13:30 GTM, and on a weekly basis. Measurements of Ψ_{leaf} were developed by using a pressure chamber (Soil Moisture Equipment Corp., Sta. Barbara, CA, USA), monitoring eight trees per irrigation treatment (two leaves per tree), located in the north side of the tree and being totally mature, fresh and shaded, at 1.5 m of height. These readings were used to quantify the water stress supported by the crop for each week and the whole kernel-filling period by means of the stress integral (Ψ_{Int}), following the methodology proposed by Myers [38] (Equation (1)). This index allows to quantify the effect of water stress provided by the water restriction beyond its temporal distribution, integrating the global stress supported by the crop in comparison to the punctual measurements:

$$\psi_{Int} = \left| \sum \left(\psi_{leaf}^{av} - \left(\psi_{leaf}^{max} \right) \right) \cdot n \right|. \tag{1}$$

where Ψ_{Int} is the stress integral in terms of Ψ_{leaf} values, ψ_{leaf}^{av} is the average leaf water potential for any interval (in our case, for each week), ψ_{leaf}^{max} is the maximum value of Ψ_{leaf} weekly registered, during the experimental period and n is the days numbers within each interval, in our case $n = 7$.

At the end of each season, monitored trees (eight per cultivar and irrigation strategy) were harvested. This process was carried out by using a mechanic vibrator to throw the almond on the ground (previously covered with a plastic mesh). Collected almonds were processed with a mechanic peeling to remove the hull. Finally, once cleaned, almonds were left to air dry and weighed once reached a humidity content around 6%. Around 3 kg of in-shell almonds were sent to Miguel Hernández University for quality and sensory analysis, where the main morphological, physical and chemical parameters were analyzed (Figure 2).

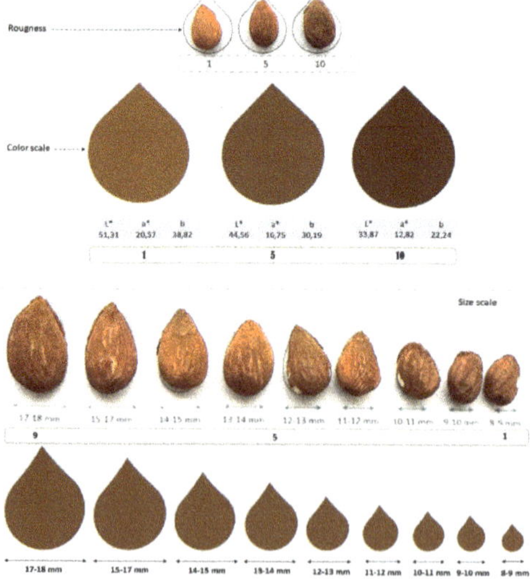

Figure 2. Scales reference used by trained panel to evaluate the almond appearance in this study.

2.3. Morphological and Physical Parameters

The ratio between the mass of in-shell almonds and kernel was calculated from ~1 kg of nuts per cultivar and irrigation treatment. Additionally, 225 almonds (25 samples × 3 varieties × 3 treatments) were randomly selected and analyzed by measuring the weight and size (length, width and thickness) of almonds (both in-shell and kernel) using a digital caliper (Mitutoyo 500-197-20, Kawasaki, Japan) and a scale (Mettler Toledo model AG204, Barcelona, Spain), respectively.

A Minolta Colorimeter CR-300 (Minolta, Osaka, Japan) was used to perform the color measurements in 75 kernels per each variety. This colorimeter uses a D_{65} illuminant and a $10°$ observer as references. The color was provided as CIEL*a*b* coordinates defining the color in a three-dimensional space and it was expressed in three numerical values, which includes L* for the lightness (L* = 0 black; L* = 100 white), a* for the green-red (a* = red; $-a$* = green) and b* for the blue-yellow components (b* = yellow; $-b$* = blue).

2.4. Chemical Composition

2.4.1. Total Sugars

Sugars were determined using a high-performance liquid chromatography (HPLC) equipment. The extraction consisted of 1 g of grinded almond in a Moulinex grinder (AR110830) for 10 s, homogenized with 5 mL of phosphate buffer with an homogenizer (Ultra Turrax T18 Basic) over 2 min at 11.3 rpm, while the tube was maintained in an ice bath and after it was centrifuged for 20 min at 15,000 rpm and 4 °C (Sigma 3–18 K; Sigma Laborzentrifugen, Osterode and Harz, Germany) followed by filtration and injection in the HPLC equipment. Sugar content was determined by using a Supelcogel TM C-610H column (30 cm × 7.8 mm) with a pre-column (Supelguard 5 cm × 4.6 mm; Supelco, Bellefonte, PA, USA) and it was detected by a refractive index detector (RID). Organic acid absorbance was measured at 210 nm in the same HPLC condition using a diode-array detector (DAD). Analyses were triplicated and results were expressed as g kg^{-1} dry weight.

2.4.2. Volatile Compounds

For the extraction of the volatile compounds, headspace solid phase microextraction (HS-SPME) was used. Ground almond (1 g) was added to a hermetic vial with polypropylene cap and PTFE (polytetrafluoroethylene)/silicone septa, together with 500 µL salty water (12.5% NaCl) and 2.5 µL of 2-acethylthiazole (1000 mg L^{-1}) internal standard, needed for the semi-quantification of the volatile compounds. To simulate the mouth temperature, the vial was heated in a laboratory hot plate up to 50 °C. When the temperature was reached and was stable, a 50/30 µm Divinylbenzene/Carboxen/Polydimethylsiloxane (DVB/CAR/PDMS) fiber was introduced in the headspace of the vial for 35 min. A gas chromatograph Shimadzu GC-17A (Shimadzu Corporation, Kyoto, Japan) coupled with mass spectrometer (MS) detector Shimadzu QP-5050A were used for isolation and identification of the volatile compounds. The Gas Chromatography—Mass Spectrometry (GC-MS) was equipped with a SLB-5ms Fused Silica Capillary Column of 30 m × 0.25 mm × 0.25 µm film thickness, 5% diphenyl and 95% dimethyl siloxane (Supelco Analytical). Helium was used as gas carrier at a flow rate of 0.9 mL min^{-1} in a split ratio of 1:5. The oven program was: (a) initial temperature 50 °C, (b) rate of 4.0 °C min^{-1} to 130 °C, (c) rate of 10 °C min^{-1} from 130 °C to 180 °C, (d) rate of 20 °C from 180 °C to 280 °C. The injector and the detector were held at 250 °C. The identification of the volatile compounds was performed using three methods: (a) retention indices, (b) GC-MS retention times of authentic chemicals and (c) mass spectra compounds were extracted using HS-SPME.

Simultaneously, the quantification of the volatile compounds was done on a gas chromatograph, Shimadzu 2010, with a flame ionization detector (FID). The column and chromatographic conditions were those previously reported for the GC-MS analysis. The injector temperature was 200 °C and nitrogen was used as carrier gas (1 mL min^{-1}). The quantification was obtained from electronic integration measurements using flame ionization detection (FID). 2-Acethylthiazole (2.5 µL of 1000 mg L^{-1}) was used as internal standard.

2.5. Descriptive Sensory Analysis

The descriptive sensory analysis was held by a trained panel with a 10 highly qualified panelists from the Food Quality and Safety Group (Miguel Hernández University of Elche, Orihuela, Alicante, Spain). The descriptive sensory analysis was performed to estimate if the significant differences among treatments were found. Although the panelists were highly trained, having more than 600 h of experience with different types of food products, three orientation sessions were done prior to almond tasting, where the panelists were trained with reference products for each attribute according to the lexicon previously described by Lipan et al. [25]. The samples were served in odor-free 30 mL covered plastic cup and randomly coded with three digits. To clean the palate between samples, water and unsalted crackers were served. The descriptive test was performed in a special tasting room with individual booths (controlled temperature of 21 ± 1 °C and combined natural/artificial light), and to collect panelists' evaluations, ballot charts were used. The samples were presented based on a randomized block design to avoid biases. Numerical scale from 0 to 10 was used by the panelists to quantify the intensity of the almond attributes, where 0 represents no intensity and 10 extremely strong with a 0.5 increment (Figure 2).

2.6. Statistical Analysis

The stress integral of Ψ_{leaf} and yield were analyzed by Sigma Plot statistical software (version 12.5, Systat Software, Inc., San Jose, CA, USA). Initially, a descriptive analysis for each treatment and cultivar was done, applying a Levene's test to check the variance homogeneity of the whole of data. Once completed, a one-way analysis of variance (ANOVA) was performed to determine whether there were statistical differences ($p < 0.05$) between irrigation treatments and within each cultivar, applying a Tukey's test to find the differences among them.

Relating to the quality and sensorial parameters, a two-way analysis of variance (ANOVA) was performed, with the cultivar and irrigation being the two factors. Moreover, a Tukey's multiple range test was carried out to establish the means that were significantly different from each other. XLSTAT Premium 2016 (Addinsoft, New York, NY, USA) was used to perform statistically significant differences, with a significant level $p < 0.05$.

3. Results and Discussion

3.1. Irrigation Doses, Crop Physiological and Yield Response

Table 1 summarizes the climatic conditions throughout the experiment with cumulative rainfall and crop evapotranspiration (ET$_C$) of 85 and 840 mm, respectively. According to the registered data, the water irrigation amount for FI, SDI$_{75}$ and SDI$_{65}$ was 770, 574 and 516 mm, respectively.

The irrigation doses imposed different physiological responses and yield reductions were observed in the SDI strategies with respect to FI (Table 2). The *cv.* Marta reported higher values of Ψ_{Int} in SDI$_{65}$ (197 MPa) compared to that registered in FI (179 MPa), with intermediate values for SDI$_{75}$ (188 MPa). These differences were even more pronounced for *cv.* Guara with Ψ_{Int} values in SDI treatments (~210 MPa) significantly higher than in FI (194 MPa). Finally, *cv.* Lauranne did not register variations among treatments for Ψ_{Int} with values of 194 MPa.

Table 1. Monthly average values of weather parameters and irrigation doses during the study period.

Parameters	April	May	June	July	August	September	October
T_{max} (°C)	22.2	30.4	31.3	34.5	36.5	32.4	27.6
T_{min} (°C)	7.2	12.2	17.5	17.9	17.9	16.3	11.7
T_{av} (°C)	19.8	21.5	22.7	25.8	26.9	23.8	18.9
RH_{max} (%)	97.8	85.2	83.2	84.0	77.2	81.9	90.7
RH_{min} (%)	39.8	23.3	23.4	25.3	18.7	27.6	32.9
RH_{av} (%)	72.2	52.3	51.4	55.4	45.9	54.4	63.2
Rad (MJ m^{-2})	1.9	2.1	2.1	2.9	0.8	0.9	0.7
Rainfall (mm)	71.2	0.0	0.0	0.0	0.0	3.4	10.4
ET_0 (mm)	111.0	198.0	202.9	238.7	170.1	121.0	76.4
ET_C (mm)	44	119	135	215	179	97	46
				Irrigation (mm)			
FI	25	115	140	210	170	80	30
SDI_{75}	18	85	104	157	127	61	22
SDI_{65}	16	77	95	141	114	53	20

T_{max}, T_{min}, T_{av}, maximum, minimum and average air temperature; RH_{max}, RH_{min}, RH_{av}, maximum, minimum and average relative humidity; Rad, solar radiation; ET_0, reference evapotranspiration; ET_C, crop evapotranspiration rate; FI, SDI_{75}, SDI_{65}, full-irrigated and sustained-deficit irrigation at 75 and 65% of irrigation requirements, respectively.

Table 2. Integral stress (Ψ_{Int}) values and almond yield for the different cultivars and irrigation treatments.

	FI	SDI_{75}	SDI_{65}
Cultivar		Ψ_{Int} (MPa day)	
Marta	179b	188ab	197a
Guara	194b	209a	210a
Lauranne	194a	198a	191a
		Almond yield (kg ha^{-1})	
Marta	2218a	2208a	2243a
Guara	2254a	2081ab	1872b
Lauranne	2325a	2104a	2195a

Values (average of eight replications; n = 8) within a same row, and followed by different letters, show significant differences between treatments and within each cultivar, according to Tukey's test ($p < 0.05$). FI, SDI_{75} and SDI_{65}, full-irrigated and sustained-deficit irrigation at 75 and 65% of irrigation requirements, respectively.

In line with the physiological pattern previously described, the studied cultivars also presented yield reductions. In this regard, *cvs.* such as Marta and Lauranne did not showed significant variations among treatments, while *cv.* Guara was reduced about 8 and 17% in SDI_{75} and SDI_{65}, respectively (Table 2).

Taking into consideration the obtained results, the water stress promoted different physiological and yield responses depending on the studied cultivar. In this regard, Gomes-Laranjo et al. [39] also reported different physiological responses of almond cultivars when they were subjected to deficit-irrigation strategies, concluding that *cv.* Lauranne would be less sensitive to irrigation restrictions than other cultivars. More recently, Gutiérrez-Gordillo et al. [17] revealed that *cv.* Marta evidenced a stronger stomatal control as compared to *cvs.* Guara and Lauranne, when subjected to regulated deficit-irrigation strategies. On the contrary, *cv.* Guara would show a minor conservative behavior, being able to maximize the gas-exchange rates when subjected to water restriction.

However, as observed in the present study, Guara was the most sensitive cultivar growth under SDI conditions. This point is especially remarkable considering that this cultivar presented very positive responses when water restrictions were applied during the kernel-filling period [40,41]. This means that the final response to water stress would be determined by the effect of water restriction, cultivar and deficit-irrigation strategy. Similar results were also reported by Alegre et al. [42], who obtained higher yield reductions in *cv.* Guara as compared to *cv.* Lauranne, when they were subjected to severe SDI (~2500 m^3 ha^{-1}). Moreover, Miarnau et al. [43] suggested that under SDI strategies, with irrigation applications around 2000 m^3 ha^{-1}, *cv.* Marta would be able to reach higher productions (1850 kg ha^{-1})

than those obtained by *cv.* Guara (1200 kg ha^{-1}) when using similar irrigation doses. Thus, these results would reinforce the statement that *cv.* Marta would be able to activate a physiological prevention mechanism to mitigate the water stress, leading to a higher yield than *cv.* Guara.

3.2. Morphological and Physical Parameters

Table 3 displays the main results related to the effects of irrigation doses and cultivar on the physical and morphological properties of raw almonds. Both irrigation treatments and cultivars offered significant differences on the weight, size and physical parameters. In relation to the almond weight, *cv.* Marta stood out, comparing to *cvs.* Guara and Lauranne. More evident were the improvements fixed in the almond weight from SDI$_{65}$ trees regarding to SDI$_{75}$ and FI treatments. These differences were also found in the morphological parameters, with higher values in SDI$_{65}$ for kernel length, the whole thickness and kernel thickness. As was expected, these morphological differences were even more pronounced between cultivars. Thus, *cvs.* Marta and Lauranne offered a more lengthened morphology in comparison to *cv.* Guara. In relation to the kernel color coordinates, significant differences between cultivars and treatments were observed. Thus, SDI$_{75}$ and SDI$_{65}$ registered higher values of L^*, a^* and b^* that evidenced lighter, redder and yellower almonds than FI and with even greater values of chroma, which means a higher color intensity of samples perceived by humans. Instrumental color was also affected by the cultivar, with Guara being the cultivar with the highest values of L^*, a^*, b^* and *chroma*, whereas *cvs.* Lauranne and Marta showed a higher similarity between them for these parameters.

In relation to the interaction between irrigation dose × cultivar, all the studied parameters reported significant differences. For *cv.* Marta, the most notable effects related to the irrigation doses were found for the almond size, with higher values of kernel thickness and length. Moreover, this cultivar registered lower values of L^*, a^*, b^* and *chro*ma for SDI$_{65}$, while SDI$_{75}$ was mainly similar to FI almonds. More interesting were the irrigation effects in *cv.* Guara with significant improvements in the almond and kernel weight for SDI$_{65}$ compared to SDI$_{75}$ and FI. Within this cultivar, SDI$_{65}$ presented higher values of L^*, b^* and *chroma*, while SDI$_{75}$ generated almonds with a greater hardness and crispiness. Finally, regarding *cv.* Lauranne, higher values of almond weight and color on SDI$_{65}$ were observed, although the weight improvements were more pronounced in the almond shell.

Table 3. Morphology and instrumental color of raw almonds as affected by deficit treatment and almond cultivar.

	Weight (g)			Size (mm)						Kernel Color Coordinates				
	Whole	Kernel	Shell	WL	KL	WW	KW	WT	KT	L*	a*	b*	C	Hue
Irrigation	**	**	**	NS	***	NS	NS	***	***	***	***	***	***	NS
Cultivar	**	**	**	**	***	***	***	***	***	***	***	***	***	NS
Irrigation × Cultivar	**	**	**	**	***	***	***	***	***	***	***	***	***	NS
Tukey Multiple Range Test [‡]														
Irrigation														
FI	3.19ab	1.08b	2.11ab	30.7	22.2b	21.4	13.3	15.0b	8.30a	45.9b	19.1a	29.9b	35.6b	60.4
SDI$_{75}$	3.07b	1.08b	1.98b	30.0	22.2b	21.5	13.1	14.9b	8.03b	48.7a	18.9ab	34.9a	39.7a	61.3
SDI$_{65}$	3.38a	1.16a	2.22a	30.1	22.7a	21.5	13.3	15.4a	8.44a	48.7a	18.2b	34.5a	39.0a	61.8
Cultivar														
cv. Marta	3.49a	1.19a	2.30a	30.2ab	23.1a	20.3c	12.8b	14.8b	8.34a	47.3b	17.9b	32.1b	36.8b	60.4
cv. Guara	2.92b	1.07b	1.85c	29.8b	21.6c	22.7a	13.5a	15.9a	8.43a	49.4a	19.2a	35.0a	40.1a	64.1
cv. Lauranne	3.22ab	1.05b	2.17b	30.9a	22.4b	21.5b	13.4a	14.6b	7.97b	46.6b	19.1a	32.2b	37.5b	58.9
Irrigation × Cultivar														
cv. Marta														
FI	3.55a	1.21a	2.34a	29.7b	22.5abc	20.0d	12.7c	14.3c	8.13abcd	48.5abc	18.2bc	32.6bcd	37.4bcd	60.6
SDI$_{75}$	3.47a	1.18a	2.29a	30.2ab	23.2ab	20.4cd	13.0abc	14.9bc	8.49abc	48.2abc	18.5abc	34.1abc	38.9abc	61.1
SDI$_{65}$	3.46a	1.18a	2.28a	30.6ab	23.7a	20.5cd	12.8bc	15.2b	8.49abc	45.1cd	17.0c	29.6de	34.2de	59.6
cv. Guara														
FI	2.91d	0.99b	1.92bc	30.4ab	21.6cd	22.7ab	13.7a	16.0a	8.72a	47.2bc	20.0a	30.2cde	36.5cde	66.1
SDI$_{75}$	2.56d	1.00b	1.56c	29.5b	21.2d	22.7ab	13.2abc	15.4ab	7.96cd	49.6ab	19.1ab	36.6ab	41.3ab	62.3
SDI$_{65}$	3.30b	1.22a	2.08b	29.6b	22.0bcd	22.8a	13.6ab	16.2a	8.6ab	51.5a	18.6abc	38.2a	42.5a	63.9
cv. Lauranne														
FI	3.12c	1.02b	2.09b	32.1a	22.4bcd	21.7abc	13.4abc	14.7bc	8.05bcd	42.0d	19.0ab	26.9e	33.0e	54.5
SDI$_{75}$	3.18c	1.05b	2.13b	30.3ab	22.1bcd	21.5abc	13.3abc	14.3b	7.63d	48.2abc	19.2ab	34.0abc	39.0abc	60.6
SDI$_{65}$	3.38b	1.08b	2.30a	30.2ab	22.5abc	21.3bcd	13.4abc	14.9bc	8.22abcd	49.5ab	19.0ab	35.7ab	40.5abc	61.8

NS, not significant at $p < 0.05$; ** and *** significant at $p < 0.01$, and 0.001, respectively. [‡] Values (average of 25 replication) followed by the same letter, within the same column and factor, were not significantly different ($p < 0.05$), according to Tukey's least significant difference test. WL, Whole Length; KL, Kernel Length; WW, Whole Width; KW, Kernel Width; WT, Whole Thickness; KT, Kernel Thickness; L*, a*, b*, Color coordinates; C, Chroma. FI, SDI$_{75}$, SDI$_{65}$, full-irrigated and sustained-deficit irrigation at 75 and 65% of irrigation requirements, respectively.

3.3. Total Sugars and Phenolic Content

The total sugars (TSC) and total phenolic (TPC) contents are shown in Table 4. High significant effects were observed in response to the studied cultivars and irrigation dose imposed. Regarding the TSC, the highest values were reached in almonds under water stress conditions, with SDI_{75} and SDI_{65} having a TSC of 62.9 and 62.2 g kg^{-1}, respectively, which increased ~19% with regard to almonds growth under full-irrigated conditions. The *cvs.* Guara and Lauranne registered the highest values of sugars with ~1.4-fold higher than *cv.* Marta. Comparing all cultivars and irrigation treatments the highest values were reached by *cv.* Guara under SDI_{75} (76.1 g kg^{-1}), followed by *cv.* Guara under SDI_{65} (68.4 g kg^{-1}) and *cv.* Lauranne under SDI_{75} and SDI_{65} (65.1 and 64.7 g kg^{-1}, respectively). Thus, SDI conditions led to higher total contents of sugar in all cultivars as compared to fully irrigated trees. As previously reported, raw almonds contain a variable amount of sugars, highlighting sucrose, glucose and fructose [27], whose concentrations are significantly affected by both water stress and cultivar [26]; and hence, their concentrations can vary depending on the water management applied during the fruit development. The increase of sugars in the fruits under stress circumstances is mainly related to the osmotic adjustment, initiated to adapt the plant to dry and saline stress by accumulation of solutes rich in hydroxyl (-OH) groups (sugars, proline etc.) in the cytoplasm [44] and to the induction of the growth inhibitor abscisic acid, inducing the accumulation of osmotically active compounds, which help to protect the cells from harm [45]. Sugars are key compounds in the basic sweet taste of almonds, this fact being important for consumer acceptance [25,46] and essential in the aroma profile of toasted almonds, because these are precursors of aroma compounds formation during thermal processing [47].

Table 4. Impact of deficit irrigation on total phenolic (TPC) and sugars contents (TSC).

	TSC (g kg^{-1})	TPC (g GAE kg^{-1})
	ANOVA [†]	
Irrigation	***	***
Cultivar	***	***
Irrigation × Cultivar	***	***
	Tukey Multiple Range Test [‡]	
	Irrigation	
FI	52.6b	2.97b
SDI_{75}	62.9a	3.81a
SDI_{65}	62.2a	3.80a
	Cultivar	
Marta	48.5b	3.40b
Guara	65.5a	3.50b
Lauranne	63.7a	3.68a
	Irrigation × Cultivar	
	cv. Marta	
FI	44.4f	2.79de
SDI_{75}	47.6ef	3.44cd
SDI_{65}	53.6d	3.98bc
	cv. Guara	
FI	52.1de	2.29e
SDI_{75}	76.1a	3.14cde
SDI_{65}	68.4b	5.06a
	cv. Lauranne	
FI	61.3c	3.82c
SDI_{75}	65.1bc	4.86ab
SDI_{65}	64.7bc	2.37e

[†]—Analysis of variance test (ANOVA), *** significant at $p < 0.001$; GAE, Gallic Acid Equivalent; [‡] Values (average of three replications) followed by the same letter, within the same column and factor, were not significantly different ($p > 0.05$), according to Tukey's least significant difference test. FI, SDI_{75}, SDI_{65}, full-irrigated and sustained-deficit irrigation at 75 and 65% of irrigation requirements, respectively.

On the other hand, the total phenolic content (TPC) expressed as gallic acid equivalent (GAE) was significantly raised by the SDI treatments with an increase of 28% regarding the FI almonds. Additionally, *cv.* Lauranne registered the highest values (3.68 g GAE kg^{-1}), followed by *cv.* Guara (3.50 g GAE kg^{-1}) and *cv.* Marta (3.40 g GAE kg^{-1}), proving the cultivar effect for the TPC. Focusing on the interaction between the irrigation dose and cultivar, the highest TPC values were reached by *cv.* Guara in SDI$_{65}$ and *cv.* Lauranne in SDI$_{75}$ conditions (5.06 g GAE kg^{-1} and 4.86 g GAE kg^{-1}, respectively), being the lowest values obtained by *cvs.* Marta and Guara under FI (2.79 and 2.29 g GAE kg^{-1}) and *cv.* Lauranne under SDI$_{65}$ conditions (2.37 g GAE kg^{-1}). These results agreed with the study by Lipan et al., who found a positive correlation between TPC and water stress in almonds [31]. Moreover, Horner [48] reported that water stress in trees generates an increase in phenolic compounds precursors (free phenylalanine) and their synthesis could be more sensitive in moderate water stress circumstances.

Overall, the water stress in plants decrease the turgor pressure, increase the ion toxicity and inhibits the photosynthesis [49], which leads to the activation of the antioxidant defense system to deal with reactive oxygen species (ROS). The trigger of many defense mechanisms, including the increase in antioxidants to enhance plant tolerance to water stress is mainly done by plant phytohormones [49]. Almond polyphenols are mostly found in skin and are responsible of the kernel color and astringency [50]. In this line, Monagas et al. [51] identified flavonol monomers as well as oligomers up to seven units as the most abundant type of flavonoids in almond skin; moreover, the intensity of the astringency depended on the polymerization degree [52].

3.4. Volatile Compounds

Using NIST libraries and Kovats Index values (REF), a total of 35 volatile compounds were identified and presented in Table 5, together with their retention times, retention indices, and their odor descriptors. These include 10 alcohols, 13 alkanes, five terpenes, four aldehydes, one ketone, one acid and one ester. Significant differences ($p < 0.001$) were promoted by the effect of both irrigation and cultivar (Table 6) factors, with the highest values for V2 and V16 to V34 under SDI$_{65}$ treatment, whereas V4–V9, V12, V14, V15, V17 and V35 reached the highest values under SDI$_{75}$ treatment. According to this and attending to the sum (total volatile compounds content), SDI$_{75}$ was able to increase the volatile compounds. By contrast, the SDI$_{65}$ strategy reflected a reduction in the whole amount, which is significantly lower than that obtained under moderate SDI$_{75}$, which confirms the theory about the quadratic equation of Horner [48], who reported a reduction in fruit chemical compounds when the stress threshold is exceeded. Additionally, differences among cultivars were also found, with Marta and Guara being the cultivars that registered the highest values of the total volatiles. However, focusing the attention in the most abundant compounds (V15 and V17), the highest amounts of benzaldehyde (V15) were registered for SDI$_{75}$ (in terms of irrigation treatment) and Guara (in terms of cultivar) (Table 6). Regarding to pentamethyl heptane, which might be a degradation product of fatty acids, the highest amounts were found for the SDI strategies and *cv.* Marta.

Regarding the interaction irrigation $\times$ cultivar the highest contents of benzaldehyde were reached by *cvs.* Marta and Guara both under RDI$_{75}$. For the case of pentamethyl heptane the highest values were found for *cv.* Marta under SDI$_{75}$, *cv.* Guara under FI, and *cv.* Lauranne under SDI$_{65}$. Finally, and considering the total volatiles, the highest values were registered by *cv.* Marta under RDI$_{75}$, followed by *cv.* Guara under FI and *cv.* Lauranne under SDI$_{65}$.

Alcohols were the most abundant volatile compounds found in the present experiment. In this line, Kwak et al. [32] reported that alcohols are the main volatiles in raw almonds (*cv.* Nonpareil) and are released by enzymatic reactions. These compounds might contribute to the typical raw sweet aroma of almonds and an increase in its concentration may improve consumer acceptance [31,32]. In the present study, the highest content of alcohols was reached by the SDI$_{75}$ treatment (296 µg kg^{-1}), followed by FI (288 µg kg^{-1}) and it was reduced by SDI$_{65}$ (243 µg kg^{-1}). Regarding the cultivar effect, *cv.* Lauranne recorded the highest values of alcohols followed by *cvs.* Marta and Guara with 310, 296 and 220 µg kg^{-1}, respectively.

Table 5. Volatile compounds profile in raw studied almonds cultivars, retention index and main odor and aroma descriptors.

Code	Compound	Chemical Family	RT (min)	Retention Index [†]		Odor Descriptor
				Experimental	Literature [‡]	
V1	3-Methyl-2-butanol	Alcohol	1.65	699	700	Musty, alcoholic, vegetable, cider, cocoa, cheesy [1]
V2	Acetoin	Ketone	1.765	702	707	Sweet, buttery, creamy, dairy, milky, fatty [1]
V3	Acetic acid [¥]	Acid	2.083	717	630	Pungent, acidic, cheesy, vinegar[1]
V4	3-Hexenol	Alcohol	2.548	739	746	Green, leafy, floral, petal, oily, earthy [1]
V5	1-Pentanol	Alcohol	3.053	764	762	Pungent, fermented, bready, yeasty, winey, solvent-like [1]
V6	2-Pentenol	Alcohol	3.097	767	767	Fermented, ripe banana, apple [1]
V7	α-Octene	Alkene	3.677	794	788	
V8	Octane	Alkane	3.805	800	800	
V9	Hexanal	Aldehyde	3.905	805	804	Fresh green fatty aldehydic grassy leafy fruity sweaty [1]
V10	(2E)-2-Octene	Alkene	4.057	812	815	
V11	(2E)-2-Hexenal	Aldehyde	4.457	831	825	Green, banana, aldehydic, fatty, cheesy [1]
V12	Nonane	Alkane	5.903	900	900	Gasoline [1]
V13	α-Pinene	Terpene	6.924	934	933	Sharp, warm, resinous, fresh, pine [1]
V14	Citronellene	Terpene	7.241	945	945	Citronellol, herbal, citrus, terpenic [1]
V15	Benzaldehyde	Aldehyde	7.963	970	967	Almond, fruity, powdery, nutty, cherry, sweet, bitter [1]
V16	Heptanol	Alcohol	8.435	986	977	Musty, leafy green, fruity, apple, banana and nutty and fatty notes [1]
V17	2,2,4,6,6-Pentamethyl heptane [¥]	Alkane	8.627	991	997	
V18	Decane	Alkane	8.87	1000	1000	
V19	2-Octanol	Alcohol	9.192	1010	1010	Fresh, spicy, green, woody, herbal earthy [1]
V20	Limonene	Terpene	9.97	1032	1034	Citrus, orange, sweet, fresh, peely [1]
V21	2-Ethyl-hexanol	Alcohol	10.044	1034	1030	Citrus, fresh, floral oily sweet [1]
V22	3,5,5-Trimethyl-hexanol	Alcohol	10.513	1048	1048	Green, floral, camphoreous, woody, melon, berry.
V23	Butanoate	Ester	10.839	1058	1054	Fruity, pineapple odor [1]
V24	Undecane	Alkane	12.333	1100	1100	Waxy, fruity, creamy, fatty, orris, floral, pineapple [1]
V25	Linalool	Terpene	12.525	1106	1106	Citrus, orange, floral, terpy, rose [1]
V26	Nonanal	Aldehyde	12.862	1115	1107	Waxy, aldehydic, citrus, green lemon peel, orange peel [1]
V27	Octyl-formate	Alkane	13.683	1136	1128	Fruity, rose, orange, waxy, cumber [1]
V28	1-Nonanol	Alcohol	15.425	1185	1181	Fresh, clean, fatty, floral, rose, orange, dusty, wet, oily [1]
V29	(2Z)-2-Dodecene	Alkene	15.701	1193	1193	Pleasant odor [2]
V30	Dodecane	Alkane	15.979	1200	1200	
V31	3,7-Dimethyl-1-octanol [¥]	Alcohol	16.283	1209	1190	Aldehydic citrus, rosy and green woody notes [1]
V32	Tridecane	Alkane	19.525	1301	1300	
V33	Tetradecane	Alkane	22.5	1401	1400	Mild waxy [1]
V34	Pentadecanol	Alcohol	27.993	1770	1772	Mild alcohol odor [2]
V35	Geranyl linalool	Terpene	29.933	2039	2034	Mild floral rose balsam [1]

[¥] tentatively identified (identification only based on spectral database); [†] RT, retention time; [‡] NIST [53]; [1] Company [54]; [2] NCBI [55].

Table 6. Volatile compounds are based on the use of 2-acethyl thiazole as internal standard in raw almonds.

| | ANOVA [†] | | | Irrigation | | | Cultivar | | | Irrigation × Cultivar | | | | | | | | |
| | | | | | | | | | | *cv.* Marta | | | *cv.* Guara | | | *cv.* Lauranne | | |
Code	Irrigation	Cultivar	Irrigation × Cultivar	F_I	RDI_{75}	RDI_{65}	Marta	Guara	Lauranne	FI	SDI_{75}	SDI_{65}	FI	SDI_{75}	SDI_{65}	FI	SDI_{75}	SDI_{65}
									(μg kg^{-1})									
V1	***	***	***	17.2a	6.07b	5.19b	3.25c	8.43b	16.7a	4.08d	4.14d	1.54d	21.2b	2.04d	2.01d	26.2a	12.0c	12.0c
V2	***	***	***	6.61c	9.71b	11.2a	9.17b	4.34c	14.0a	7.53cd	14.3b	5.71de	3.86e	5.94cde	3.23e	8.44cd	8.91c	24.7a
V3	***	***	***	15.4a	4.11b	3.18b	3.75b	4.79b	14.2a	2.32c	5.33bc	3.60c	8.35b	3.58c	2.43c	35.6a	3.43c	3.53c
V4	***	***	***	1.89b	3.24a	1.84b	3.36a	1.29c	2.32b	2.45a	5.76a	1.88d	1.54d	1.35c	0.97d	1.67ab	2.61bc	2.67abc
V5	***	***	***	109a	119a	73.2b	117a	59.5b	124a	149a	151a	52.1d	40.7d	94.9c	42.8d	136ab	111bc	125abc
V6	***	***	***	35.4b	51.4a	22.9c	51.4a	22.7c	35.6b	35.1b	103a	16.2c	31.4b	19.0c	17.6c	39.7b	32.2b	35.0b
V7	***	***	***	83.2b	100a	61.3c	84.4a	68.9b	91.3a	110ab	86.4bc	57.1de	79.2cd	87.7bc	40.0e	60.6d	126a	87.0bc
V8	***	***	***	168b	223a	141c	227a	152b	153b	228b	305a	147de	170cb	203bc	83.4f	107ef	161cd	191bcd
V9	***	***	***	54.6a	50.1a	43.2b	56.2a	50.4a	41.2b	66.6a	48.3bc	53.7ac	67.9a	58.4ab	24.9e	29.2de	43.6cd	50.9bc
V10	***	***	***	16.2b	16.4b	30.2a	36.2a	13.3b	13.3b	28.3b	12.8cde	67.5a	12.0de	19.0c	8.83e	8.31e	17.2de	14.4cd
V11	***	***	***	46.1a	38.9b	31.2c	50.8a	38.9b	26.5c	57.0a	52.3ab	43.2bc	62.5a	37.7cd	16.4e	18.8e	26.7de	34.0cd
V12	***	***	***	30.8a	30.5a	27.2b	35.1a	24.2c	29.2b	39.1ab	45.0a	21.3c	33.7b	23.0c	15.8c	19.6c	23.6c	44.4a
V13	***	***	***	7.85a	4.80b	3.62c	3.82c	7.59a	4.86b	4.88c	5.46bc	1.13e	16.1a	4.11cd	2.54de	2.54de	4.84c	7.20b
V14	***	***	***	8.42b	9.14a	8.78b	8.99a	7.84b	9.51a	7.91c	12.7ab	6.42c	10.6b	7.62c	5.31c	6.76c	7.16c	14.6a
V15	***	***	***	292b	465a	235c	419b	542a	31.3c	260b	686a	310b	587a	686a	353b	28.8c	21.9c	43.3c
V16	***	***	***	24.5a	21.2b	24.0ab	20.7b	28.8a	20.3b	20.5cde	24.1cd	17.4def	38.2a	26.7bc	21.6cde	14.9ef	12.9f	33.0ab
V17	***	***	***	1423b	1482a	1461ab	1534a	1413b	1418b	1328b	2090a	1185b	1809a	1323b	1108b	1131b	1033b	2091a
V18	***	***	***	30.4b	28.1b	37.2a	27.5b	34.1a	34.2a	24.5de	34.9bc	23.1e	40.7b	28.0cde	33.6bc	26.1cde	21.5b	55.1a
V19	***	***	***	11.3ab	11.0b	11.8a	12.1a	10.7c	11.2b	10.2b	17.3a	8.84b	16.7a	8.25b	7.17b	6.99b	7.37b	19.3a
V20	***	***	***	22.9ab	21.5b	24.5a	20.6b	21.3b	27.1a	22.5abc	21.5b	17.7c	17.7c	17.6c	28.6a	28.6a	25.5ab	27.2ab
V21	***	***	***	73.5b	71.4b	83.4a	75.4b	73.0c	79.9a	66.9c	95.4b	63.7c	96.9b	63.6c	58.5c	56.6c	55.3c	128a
V22	***	***	***	56.0b	51.2b	70.5a	52.3b	56.4b	69.0a	47.0c	67.1b	42.7c	74.3b	48.2c	46.7c	46.6c	38.3c	122a
V23	***	***	***	9.08b	9.18ab	9.59a	9.07b	8.36b	10.4a	7.44de	11.9bc	7.88de	12.8b	9.83cd	2.47f	7.03de	5.81e	18.4a
V24	***	***	***	12.5a	3.95b	11.1a	2.53b	12.4a	12.7a	3.27cd	2.61cd	1.70d	7.67b	5.66bc	23.7a	26.7a	3.56cd	7.89bc
V25	***	***	***	3.99b	2.54c	11.9a	3.10c	10.7a	4.61b	2.41de	3.93cd	2.97cde	4.34cd	2.41de	25.4a	5.21bc	1.29e	7.32b
V26	***	***	***	30.1b	23.6c	37.8a	25.3c	30.3b	35.9a	23.1cde	31.4c	21.5de	40.9b	22.0cde	27.8cd	26.3cde	17.3e	63.9a
V27	***	***	***	4.16b	3.86b	5.48a	3.73c	4.51b	5.27a	2.33d	5.75b	3.12cd	6.17b	3.35cd	4.02c	4.00c	2.49d	9.31a
V28	***	***	***	4.89b	3.89c	6.35a	4.04c	4.97b	6.12a	3.63cd	5.07c	3.41d	6.75b	3.60cd	4.54cd	4.28cd	2.98d	11.1a
V29	***	***	***	3.50b	3.64ab	3.84c	2.80b	4.02a	4.16a	1.42e	4.96b	2.01de	5.67ab	3.48c	2.92cd	3.41c	2.48cde	6.58a
V30	***	***	***	8.79a	5.39b	9.49a	5.10c	8.74b	9.83a	4.67cd	6.95bc	3.68d	8.85b	4.74cd	12.6a	12.8a	4.48cd	12.2a
V31	***	***	***	8.60b	6.73c	11.6a	6.77c	8.79b	11.4a	6.10cd	9.06b	5.14d	11.0b	6.12cd	9.28b	8.70bc	5.00d	20.4a
V32	***	***	***	9.91a	7.69b	10.2a	6.32c	8.99b	12.4a	5.29d	9.48c	4.21e	7.84cd	6.17de	13.0b	16.6a	7.42cd	13.3b
V33	***	***	***	2.94ab	2.68b	3.28a	3.20a	2.40b	3.31a	3.12b	4.26a	2.22cd	3.10b	1.53d	2.57bc	2.60bc	2.27bcd	5.06a
V34	***	***	***	1.69b	2.38a	2.65a	2.11b	1.99b	2.61a	1.93c	3.23b	1.17d	1.80cd	2.69b	1.49cd	1.32cd	1.22cd	5.29a
V35	***	***	***	1.09b	1.42a	0.97b	1.21a	1.10b	1.18ab	1.09b	1.79a	0.75c	1.02bc	1.19b	1.07bc	1.16b	1.27b	1.10b
Σ	***	***	***	2,636ab	2,894a	2,536b	2,928a	2,751a	2,387b	2,588bcd	3,989a	2,206cd	3,358ab	2,842bc	2,054d	1,961d	1,853d	3,347ab

[†]—Analysis of variance test (ANOVA), ***, significant at $p < 0.001$; Values (mean of three replications) followed by the same letter within the same row were not significantly different ($p < 0.05$), according to Tukey's least significant difference test. FI, SDI_{75} and SDI_{65}, full-irrigated and sustained-deficit irrigation at 75 and 65% of irrigation requirements, respectively.

The presence of alkanes, alkenes, acids and aldehydes are mainly related to the oxidative decomposition of the triglyceride and fatty acid components [56]. The oxidation of polyunsaturated fatty acids generates monohydroperoxides, which are precursors of volatile aldehydes such as hexanal, octanal, nonanal and decanal [57]. In the present study, only hexanal, hexenal and nonanal were identified, with values between 112 and 131 µg kg^{-1} with the lowest content corresponding to the SDI$_{65}$ and SDI$_{75}$ treatments and the highest one to the FI treatment. In this way, Yang et al. [54] when studying the roasted almond shelf life concluded that hexanal and nonanal concentrations should be less than 2140 and 5970 µg kg^{-1}, respectively, at the endpoint of shelf life to be suitable for their consumption. Thus, the relatively low experimental contents of hexanal and nonanal are indicative of the freshness of the samples under study.

Benzaldehyde is released from amygdalin, its precursor, by enzymatic hydrolysis, which is a cyanogenic glycoside naturally produced in almond [32]. Moreover, benzaldehyde is a characteristic aroma compound of wild/bitter almonds with a low odor threshold and it is found in a lower concentration in sweet almonds, but it is cultivar dependent [47,58]. In this context, the concentration of benzaldehyde in *cvs*. Vairo and Nonpareil was reported to be below that needed to affect the aroma of the almonds [31,47]. In the present study, the benzaldehyde content for *cv*. Lauranne (31 µg kg^{-1}) was similar to that reported for *cv*. Vairo [31]. However, a greater content of this compound was registered by *cv*. Marta (419 µg kg^{-1}) and even higher by *cv*. Guara (542 µg kg^{-1}). Thus, DI strategies significantly affected the benzaldehyde concentration, which was increased by SDI$_{75}$ (465 µg kg^{-1}), but decreased in more severe conditions of water stress such as SDI$_{65}$ (235 µg kg^{-1}) when compared to FI almonds (292 µg kg^{-1}). This fact suggests that the benzaldehyde was cultivar and irrigation treatment dependent, which convert it in an alleged marker for cultivar and hydroSOStainable identification.

3.5. Descriptive Sensory Profile

Descriptive sensory analysis was conducted to quantify the hypothetical effects of cultivars and irrigation doses on the almond sensory profiles. In this sense, 15 attributes were considered, and in general, significant differences both affected by cultivar and irrigation were found (Table 7). Regarding the DI treatments, panelists found that FI and SDI$_{75}$ almonds had an intense red-brown color, which agreed with instrumental data, which also showed the highest values for the *a** coordinate (FI = 19.1a; SDI$_{75}$ = 18.9ab; SDI$_{65}$ = 18.2b), indicating that almonds from FI and SDI$_{75}$ were more reddish than those from SDI$_{65}$. Regarding the size, even though the instrumental measurements were statistically significant the trained panel did not detect significant differences for these parameters among irrigation treatments. Similar findings were revealed by Lipan et al. [46] and Carbonell-Barrachina et al. [23] on hydroSOStainable almonds and pistachios, respectively, where no significant differences on sensory size were detected.

Regarding the flavor attributes, higher intensity of sweetness, aromatics reminiscent of almond (almond ID) and benzaldehyde-like notes were found for SDI$_{65}$ almonds; these results proved that these particular almonds are those having the most intense, typical almond flavor. As shown, the benzaldehyde perception by human was in the contrast with the volatile compound concentration, which was higher in the SDI$_{75}$ in comparison to FI. However, the human perception regarding the sweetness was in agreement with the results of total sugar (Table 4), showing a higher sweetness and sugar content in almonds cultivated under deficit irrigation conditions.

Table 7. Descriptive sensory analysis of raw almonds as affected by deficit irrigation. Scale used ranged from 0 = no intensity to 10 = extremely strong intensity.

	Outer Color	Size	Roughness	Sweetness	Bitterness	Astringency	Overall Nuts	Almond ID	Benzaldehyde Like	Woody
					ANOVA Test [†]					
Irrigation	***	NS	***	***	NS	NS	NS	***	***	NS
Cultivar	***	***	NS	***	NS	NS	***	***	***	NS
Irrigation × Cultivar	***	***	***	***	NS	NS	***	***	***	NS
					Tukey's Multiple Range Test [‡]					
					Irrigation					
FI	3.9a	2.8	4.2a	3.7b	1.2	0.5	4.5	3.4b	2.0a	1.9
SDI_{75}	1.8b	2.7	4.5a	3.7b	1.1	0.6	4.4	3.9ab	1.8b	1.5
SDI_{65}	2.0b	2.8	2.4b	4.2a	1.0	0.7	4.5	4.4a	2.1a	2.2
					Cultivar					
Marta	2.2b	3.2a	3.7	1.9c	1.0	0.6	2.8c	1.9b	1.4b	2.2
Guara	2.5ab	2.4b	3.8	4.4b	1.5	0.7	4.7b	4.8a	2.9a	1.7
Lauranne	3.0a	2.7b	3.7	5.3a	0.9	0.6	5.7 a	5.2a	1.7b	1.6
					Irrigation × Cultivar					
					cv. Marta					
FI	2.1bc	3.1ab	4.6ab	2.1c	1.4	0.5	3.4bc	2.3cd	1.7bc	1.9
SDI_{75}	2.0bc	3.5a	4.4ab	1.4c	0.6	0.4	2.6c	1.4d	1.1c	1.7
SDI_{65}	2.4bc	3.0abc	2.0c	2.3c	0.6	0.8	2.5c	1.9cd	1.3c	3.1
					cv. Guara					
FI	4.9a	2.8abc	4.1ab	4.0b	1.2	0.5	4.4ab	3.6bc	2.4 abc	2.1
SDI_{75}	1.5bc	2.1c	5.1a	4.1b	1.8	0.9	4.9ab	4.8ab	2.9 ab	1.3
SDI_{65}	2.16b	2.1c	2.1c	5.0ab	1.6	0.8	4.8ab	5.8a	3.5a	1.6
					cv. Lauranne					
FI	4.8a	2.5bc	4.0ab	4.9ab	1.0	0.5	5.5a	4.4ab	2.1bc	1.7
SDI_{75}	1.8bc	2.3bc	4.0ab	5.7a	0.9	0.6	5.6a	5.4ab	1.5c	1.5
SDI_{65}	2.66b	3.1ab	3.1bc	5.3ab	0.6	0.6	6.0a	5.6ab	1.5c	1.7

[†] NS, not significant at $p < 0.05$; *** significant at $p < 0.001$; [‡] Values (mean of 10 trained panelists) followed by the same letter, within the same column, were not significantly different ($p > 0.05$), according to Tukey's least significant difference test. FI, SDI_{75} and SDI_{65}, full-irrigated and sustained-deficit irrigation at 75 and 65% of irrigation requirements, respectively. Almond ID, aromatics reminiscent of almond.

Although not many affective studies have been conducted using almond, Lipan et al. [25] concluded that both Spanish and Romanian consumers considered the almond ID (aromatics reminiscent of almond) and sweetness as the main attributes that control the consumer preferences. Moreover, sweetness, flavor, texture and price were the most relevant parameters in the CATA questionnaire when consumers were asked about their buying drivers. Taking into consideration the obtained results in this work, almond ID and sweetness were parameters that reached significant improvements when SDI was imposed, and it would reinforce the statement that water savings strategies in almond crop would help to obtain a final product with a higher acceptance by consumers. Thus, hydroSOStainable almonds with a final added value would allow to recover the economic losses caused by yield reductions, offering a product with a higher competitiveness and marketability (Figure 3), as has been corroborated by authors such as Lipan et al. [46], who concluded that consumers were willing to pay an extra amount of money for the hydroSOStainable almonds.

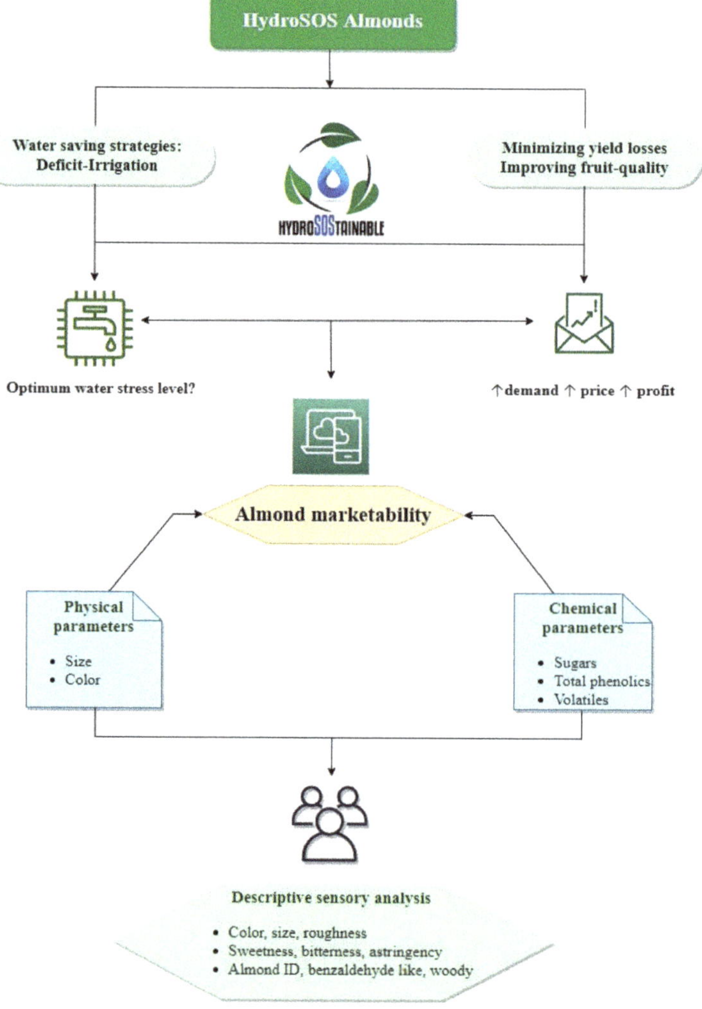

Figure 3. HydroSOStainable almonds: towards an equilibrium among water savings, optimum yields and quality parameters supported by marketability and sensory profile. ↑, increase.

4. Conclusions

This work highlights the main effects of irrigation in three almond cultivars in terms of the morphological, physicochemical and sensory parameters when this crop is subjected to sustained-deficit irrigation treatments. The findings allow us to conclude that almonds subjected to moderate sustained water stress improved substantially the most important features (sugars, total phenolic content and volatiles) related to the sensory profile and, probably, consumer acceptance. These results supported that all the monitored parameters besides water irrigation amounts are also cultivar-dependent, which determines the need of characterization of each cultivar growth under deficit irrigation conditions. Moreover, this study displayed the advantages of these strategies and opened the possibility of showcasing those hydroSOStainable products that have been obtained within a framework of water scarcity and sustainable use of natural resources. Thus, the findings prove the importance of considering the cultivar effect when these strategies are being imposed, not only in terms of final yield, but also from a nut quality perspective.

Author Contributions: Writing: I.F.G.-T., L.L. and S.G.-G.; Conception or design: I.F.G.-T., L.L. and S.G.-G. Acquisition, analysis, and interpretation of data: I.F.G.-T, L.L., S.G.-G., I.J., F.H. and á.A.C.-B.; Critical revision of the manuscript for important intellectual content: á.A.C.-B., V.H.D.Z. and B.C.R.; Statistical analysis: L.L. and I.F.G.-T.; Reagents/materials/analysis tools contribution: L.L., I.J., F.H., B.C.R. and á.A.C.-B. All authors have read and agree to the published version of the manuscript.

Funding: This work has been partially sponsored by the research project "Impact of climate change and adaptation measures (INNOVA-Climate)" (AVA.AVA2019.051) both co-financed by the European Regional Development Fund (ERDF) within the Operational Programme Andalusia 2014–2020 "Andalucía is moving with Europe" and the Spanish Ministry of Economy, Industry and Competitiveness; through the research project (*hydroSOS* mark) including the Universidad Miguel Hernández de Elche (AGL2016-75794-C4-1-R, *hydroSOS* foods) and the Universidad de Sevilla (AGL2016-75794-C4-4-R). The author S. Gutiérrez-Gordillo has a contract co-financed by the National Institute of Agrarian and Food Research and technology (FPI-INIA 2016) and European Social Fund (ESF) "The European Social Fund invests in your future."

Conflicts of Interest: The authors declare no conflict of interest.

References

1. García-Tejero, I.F.; Durán, Z.V.H.; Rodríguez, P.C.R.; Muriel, F.J.L. Water and Sustainable Agriculture. In *Springerbriefs Agriculture*; Springer: Dordrecht, The Netherlands, 2011; p. 94.

2. Garrote, L.; Granados, A.; Iglesias, A. Strategies to reduce water stress in Mediterranean river basins. *Sci. Total Environ.* **2015**, *543*, 997–1009. [CrossRef] [PubMed]

3. Iglesias, A.; Garrote, L. Adaptation strategies for agricultural water management under climate change in Europe. *Agric. Water Manag.* **2015**, *155*, 113–124. [CrossRef]

4. Lorite, I.J.; Ruiz-Ramos, M.; Gabaldón-Leal, C.; Cruz-Blanco, M.; Porras, R.; Santos, C. Water Management and Climate Change in Semi-Arid Environments. In *Water Scarcity and Sustainable Agriculture in Semi-Arid Environments*; García-Tejero, I.F., Durán Zuazo, V.H., Eds.; Elsevier–AP: Cambridge, MA, USA, 2018; pp. 3–40.

5. García-Tejero, I.F.; Durán, Z.V.H.; Muriel, F.J.L. Towards sustainable irrigated Mediterranean agriculture: Implications for water conservation in semi-arid environments. *Water Int.* **2014**, *39*, 635–648. [CrossRef]

6. García Tejero, I.F.; Moriana, A.; Rodríguez, P.C.R.; Durán, Z.V.H.; Egea, G. Deficit-Irrigation Management in Almonds (Prunus dulcis L.): Different Strategies to Assess the Crop Water Status. In *Water Scarcity and Sustainable Agriculture in Semiarid Environment*; García-Tejero, I.F., Durán-Zuazo, V.H., Eds.; Elsevier–AP: Cambridge, MA, USA, 2018; pp. 271–298.

7. FAOSTAT. Food and Agriculture Organization of the United Nations. 2018. Available online: http://www.fao.org/faostat/en/#data/QC (accessed on 1 March 2020).

8. MAPA. Ministerio de Agricultura, Pesca y Alimentación. Superficies y Producciones Anuales de Cultivos. 2019. Available online: https://www.mapa.gob.es/es/estadistica/temas/estadisticas-agrarias/agricultura/superficies-producciones-anuales-cultivos/ (accessed on 8 April 2020).

9. OPM. Observatorio de Precios y Mercados. Consejería de Agricultura, Ganadería, Pesca, y Desarrollo Sostenible. Junta de Andalucía. Available online: https://www.cap.junta-andalucia. es/agriculturaypesca/observatorio/servlet/FrontController?action=Static&subsector=35&producto= 35000&url=generadorInformesOR.jsp (accessed on 10 June 2020).

10. USDA. U.S. Department of Agriculture. Tree Nuts Annual EU-28. Glob. Agric. Inf. Netw. Available online: https://gain.fas.usda.gov/RecentGAINPublications/TreeNutsAnnual_Madrid_EU-28_9-18-2017.pdf (accessed on 23 October 2020).

11. CAPDR. Caracterización del sector de la almendra en Andalucía. Secretaría General de Agricultura y Alimentación, Consejería de Agricultura, Pesca y Desarrollo Rural. Junta de Andalucía. Available online: http://www.juntadeandalucia.es/agriculturaypesca/observatorio/servlet/FrontController?action= RecordContent&table=12030&element=1654785. (accessed on 3 March 2020).

12. MAPA. Ministerio de Agricultura, Pesca y Alimentación. Anuario de estadística agraria 2019. Available online: https://www.mapa.gob.es/es/estadistica/temas/publicaciones/anuario-de-estadistica/2019/default. aspx?parte=3&capitulo=07&grupo=10&seccion=1 (accessed on 28 March 2020).

13. Arquero, O. Manual del almendro. Consejería de Agricultura, Pesca y Desarrollo Rural, Junta de Andalucia Sevilla, España. 2013. Available online: https://www.juntadeandalucia.es/servicios/publicaciones/detalle/ 77668.html (accessed on 16 May 2020).

14. Girona, J.; Marsal, J.; Cohen, M.; Mata, M.; Miravete, C. Physiological, growth and yield responses of almond (Prunus dulcis L.) to deficit irrigation regimes. *Acta Hortic.* **1993**, *335*, 389–398. [CrossRef]

15. Girona, J.; Mata, M.; Marsal, J. Regulated deficit irrigation during the kernel-filling period and optimal irrigation rates in almond. *Agric. Water Manag.* **2005**, *75*, 152–167. [CrossRef]

16. Egea, G.; Nortes, P.A.; González, R.M.M.; Baille, A.; Domingo, R. Agronomic response and water productivity of almond trees under contrasted deficit irrigation regimes. *Agr. Water Manag.* **2010**, *97*, 171–181. [CrossRef]

17. Gutiérrez-Gordillo, S.; Durán, Z.V.H.; García, T.I.F. Response of three almond cultivars subjected to different irrigation regimes in Guadalquivir river basin. *Agric. Water Manag.* **2019**, *222*, 72–81. [CrossRef]

18. López-López, M.; Espadafor, M.; Testi, L.; Lorite, I.J.; Orgaz, F.; Fereres, E. Water requirements of mature almond trees in response to atmospheric demand. *Irrig. Sci.* **2018**, *36*, 271–280. [CrossRef]

19. Romero, P.; García, J.; Botía, P. Cost–benefit analysis of a regulated deficit-irrigated almond orchard under subsurface drip irrigation conditions in South-eastern Spain. *Irrig. Sci.* **2006**, *24*, 175–184. [CrossRef]

20. Cano-Lamadrid, M.; Girón, I.F.; Pleite, R.; Burló, F.; Corell, M.; Moriana, A.; Carbonell-Barrachina, A.A. Quality attributes of table olives as affected by regulated deficit irrigation. *LWT Food Sci. Technol.* **2015**, *62*, 19–26. [CrossRef]

21. Lipan, L.; Sánchez, R.L.; Collado, G.J.; Sendra, E.; Burló, F.; Hernández, F.; Vodnar, D.C.; Carbonell-Barrachina, A.A. Sustainability of the legal endowments of water in almond trees and a new generation of high quality hydrosustainable almonds—A review. *Bull. USAMV. Food Sci. Technol.* **2018**, *75*, 98–108. [CrossRef]

22. Noguera-Artiaga, L.; Lipan, L.; Vázquez, A.L.; Barber, X.; Pérez, L.D.; Carbonell-Barrachina, A.A. Opinion of Spanish consumers on hydrosustainable pistachios. *J. Food Sci.* **2016**, *81*, S2559–S2565. [CrossRef]

23. Carbonell-Barrachina, A.A.; Memmi, H.; Noguera, A.L.; Gijón, L.M.D.; Ciapa, R.; Pérez, L.D. Quality attributes of pistachio nuts as affected by rootstock and deficit irrigation. *J. Sci. Food Agric.* **2015**, *95*, 2866–2873. [CrossRef] [PubMed]

24. Cano-Lamadrid, M.; Hernández, F.; Corell, M.; Burlo, F.; Legua, P.; Moriana, A.; Carbonell-Barrachina, A.A. Antioxidant capacity, fatty acids profile, and descriptive sensory analysis of table olives as affected by deficit irrigation. *J. Sci. Food Agric.* **2017**, *97*, 444–451. [CrossRef] [PubMed]

25. Lipan, L.; Martín, P.M.J.; Sánchez, R.L.; Cano-Lamadrid, M.; Sendra, E.; Hernández, F.; Burló, F.; Vázquez-Araújo, L.; Andreu, L.; Carbonell-Barrachina, á.A. Almond fruit quality can be improved by means of deficit irrigation strategies. *Agric. Water Manage.* **2019**, *217*, 236–242. [CrossRef]

26. Lipan, L.; García-Tejero, I.F.; Gutiérrez, G.S.; Demirbaş, N.; Sendra, E.; Hernández, F.; Durán, Z.V.H.; Carbonell-Barrachina, A.A. Enhancing nut quality parameters and sensory profiles in three almond cultivars by different irrigation regimes. *J. Agric. Food Chem.* **2020**, *68*, 2316–2328. [CrossRef]

27. USDA. U.S. Department of Agriculture, Agricultural Research Service, Nutrient Data Laboratory. Basic Report: 12061, Nuts, almonds. In *USDA National Nutrient Database for Standard Reference*; National Agricultural Library: Washington, DC, USA, 2018.

28. Yada, S.; Lapsley, K.; Huang, G. A review of composition studies of cultivated almonds: Macronutrients and micronutrients. *J. Food Anal.* **2011**, *24*, 469–480. [CrossRef]

29. Civille, G.V.; Lapsley, K.; Huang, G.; Yada, S.; Seltsam, J. Development of an almond lexicon to assess the sensory properties of almond varieties. *J. Sens. Stud.* **2010**, *25*, 146–162. [CrossRef]

30. Vickers, Z.; Peck, A.; Labuza, T.; Huang, G. Impact of almond form and moisture content on texture attributes and acceptability. *J. Food Sci.* **2014**, *79*, S1399–S1406. [CrossRef]

31. Lipan, L.; Moriana, A.; López Lluch, D.B.; Cano-Lamadrid, M.; Sendra, E.; Hernández, F.; Carbonell-Barrachina, A.A. Nutrition quality parameters of almonds as affected by deficit irrigation strategies. *Molecules* **2019**, *24*, 2646. [CrossRef]

32. Kwak, J.; Faranda, A.; Henkin, J.M.; Gallagher, M.; Preti, G.; McGovern, P. Volatile organic compounds released by enzymatic reactions in raw nonpareil almond kernel. *Eur. Food Res. Technol.* **2015**, *241*, 441–446. [CrossRef]

33. Xiao, L.; Lee, J.; Zhang, G.; Ebeler, S.E.; Wickramasinghe, N.; Seiber, J.; Mitchell, A.E. HS-SPME GC/MS characterization of volatiles in raw and dry-roasted almonds (Prunus dulcis). *Food Chem.* **2014**, *151*, 31–39. [CrossRef] [PubMed]

34. USDA. Keys to Soil Taxonomy. In *United States Department of Agriculture Natural Resources Conservation Service*, 11th ed.; National Agricultural Library: Washington, DC, USA, 2010.

35. Gutiérrez-Gordillo, S.; Durán, Z.V.H.; Hernández, S.V.; Ferrera, G.F.; García, E.A.; Amores, A.J.J.; García-Tejero, I.F. Cultivar dependent impact on yield and its components of young almond trees under sustained-deficit irrigation in semi-arid environments. *Agronomy* **2020**, *10*, 733. [CrossRef]

36. Allen, R.G.; Pereira, L.S.; Raes, D.; Smith, M.W.B. Crop evapotranspiration–Guidelines for computing crop water requirements–FAO Irrigation and drainage paper 56. *Irrig. Drain.* **1998**, *300*, 1–15.

37. García-Tejero, I.F.; Hernández, A.; Rodríguez, V.M.; Ponce, J.R.; Ramos, V.; Muriel, J.L.; Zuazo, V.H.D. Estimating Almond Crop Coefficients and Physiological Response to Water Stress in Semiarid Environments (SW Spain). *J. Agric. Sci. Technol.* **2015**, *17*, 1255–1266.

38. Myers, B.J. Water stress integral—A link between short-term stress and long-term growth. *Tree Physiol.* **1988**, *4*, 315–323. [CrossRef] [PubMed]

39. Gomes-Laranjo, J.; Coutinho, J.P.; Galhano, V.; Cordeiro, V. Responses of five almond cultivars to irrigation: Photosynthesis and leaf water potential. *Agric. Water Manag.* **2006**, *83*, 261–265. [CrossRef]

40. García-Tejero, I.F.; Gutiérrez, G.S.; Souza, L.; Cuadros, T.S.; Durán, Z.V.H. Fostering sustainable water use in almond (Prunus dulcis Mill.) orchards in a semiarid Mediterranean environment. *Arch. Agron. Soil Sci.* **2018**, *65*, 164–181. [CrossRef]

41. López-López, M.; Espadafor, M.; Testi, L.; Lorite, I.J.; Orgaz, F.; Fereres, E. Yield response of almond trees to transpiration deficits. *Irrig. Sci.* **2018**, *36*, 111–120. [CrossRef]

42. Alegre Castellví, S.; Miarnau i Prim, X.; Romero, R.M.; Vargas, G.F. Potencial productivo de seis variedades de almendro. *Frutic. Prof.* **2007**, *169*, 23–29.

43. Miarnau, X.; Torguet, L.; Batlle, I.; Romero, A.; Rovira, M.; Alegre, S. La revolución del almendro: Nuevas variedades y nuevos modelos productivos. In Proceedings of the Simp. Nac. Almendro Otros Frutos Secos, Lérida, Spain, 24 September 2015; pp. 6–27.

44. Sanders, G.J.; Arndt, S.K. *Osmotic Adjustment under Drought Conditions Plant Responses to Drought Stress*; Springer: Berlin/Heidelberg, Germany, 2012.

45. Ahanger, M.A.; Morad-Talab, N.; Abd-Allah, E.F.; Ahmad, P.; Hajiboland, R. Plant growth under drought stress: Significance of mineral nutrients. In *Water Stress and Crop Plants: A Sustainable Approach*; Parvaiz, A., Ed.; John Wiley & Sons, Ltd.: Hoboken, NJ, USA, 2016; Volume 2-2, pp. 649–668.

46. Lipan, L.; Cano-Lamadrid, M.; Corell, M.; Sendra, E.; Hernandez, F.; Stan, L.; Vodnar, D.C.; Vázquez, A.L.; Carbonell-Barrachina, A.A. Sensory Profile and Acceptability of HydroSOStainable Almonds. *Foods* **2019**, *8*, 64. [CrossRef]

47. Erten, E.S.; Cadwallader, K.R. Identification of predominant aroma components of raw, dry roasted and oil roasted almonds. *Food Chem.* **2017**, *217*, 244–253. [CrossRef] [PubMed]

48. Horner, J.D. Nonlinear effects of water deficits on foliar tannin concentration. *Biochem. Syst. Ecol.* **1990**, *18*, 211–213. [CrossRef]

49. Ali, M.S.; Baek, K.H. Jasmonic Acid Signaling Pathway in Response to Abiotic Stresses in Plants. *Int. J. Mol. Sci.* **2020**, *21*, 621. [CrossRef]

50. Bolling, B.W. Almond polyphenols: Method of analysis, contribution to food quality and health promotion. *Compr. Rev. Food Sci. Food Saf.* **2017**, *16*, 346–368. [CrossRef]

51. Monagas, M.; Garrido, I.; Lebrón-Aguilar, R.; Bartolomé, B.; Gómez-Cordovés, C. Almond (Prunus dulcis (Mill.) D.A. Web) skins as a potential source of bioactive polyphenols. *J. Agric. Food Chem.* **2007**, *55*, 8498–8507. [CrossRef] [PubMed]

52. Freitas, V.d.; Mateus, N. Protein/polyphenol interactions: Past and present contributions. Mechanisms of astringency perception. *Curr. Org. Chem.* **2012**, *16*, 724–746. [CrossRef]

53. NIST. National Institute of Standards and Technology (Libro del Web de Química del NIST. SRD69. In, U.S. Secretary of Commerce on behalf of the United States of America). Available online: https://webbook.nist.gov/chemistry/ (accessed on 30 June 2019).

54. Company, T.G.S. The Good Scents Company Information System. 2018. Available online: http://www.thegoodscentscompany.com (accessed on 30 June 2019).

55. NCBI. National Center for Biotechnology Information, U.S.N.L.o.M. PubChem open Chemistry Database. Available online: https://pubchem.ncbi.nlm.nih.gov/ (accessed on 30 June 2019).

56. Charalambous, G. *Food Flavors: Generation, Analysis and Process Influence*, 1st ed.; Elsevier Science: Amsterdam, The Netherlands, 1995; Volume 37B, p. 1360.

57. Yang, J.; Pan, Z.; Takeoka, G.; MacKey, B.; Bingol, G.; Brandl, M.T.; Wang, H. Shelf-life of infrared dry-roasted almonds. *Food Chem.* **2013**, *138*, 671–678. [CrossRef]

58. Hojjati, M.; Lipan, L.; Carbonell-Barrachina, A.A. Effect of roasting on physicochemical properties of wild almonds (Amygdalus scoparia). *J. Am. Oil Chem. Soc.* **2016**, *93*, 1211–1220. [CrossRef]

Publisher's Note: MDPI stays neutral with regard to jurisdictional claims in published maps and institutional affiliations.

Article

Long-Term Correlation between Water Deficit and Quality Markers in HydroSOStainable Almonds

Leontina Lipan [1], Marina Cano-Lamadrid [1], Francisca Hernández [2], Esther Sendra [1], Mireia Corell [3,4], Laura Vázquez-Araújo [5,6], Alfonso Moriana [3,4] and Ángel A. Carbonell-Barrachina [1,*]

[1]   Department of Agro-Food Technology, Research Group "Food Quality and Safety, CSA", Universidad Miguel Hernández de Elche (UMH), Escuela Politécnica Superior de Orihuela (EPSO), Carretera de Beniel, km 3.2, 03312 Orihuela, Spain; leontina.lipan@goumh.umh.es (L.L.); marina.cano.umh@gmail.com (M.C.-L.); esther.sendra@umh.es (E.S.)
[2]   Department of Plant Science and Microbiology, Research Group "Plant Production and Technology", UMH, EPSO, Carretera de Beniel, km 3.2, 03312 Orihuela, Spain; francisca.hernandez@umh.es
[3]   Departamento de Ciencias Agroforestales, ETSIA, Universidad de Sevilla, Carretera de Utrera, km 1, 41013 Sevilla, Spain; mcorell@us.es (M.C.); amoriana@us.es (A.M.)
[4]   Unidad Asociada al CSIC de Uso sostenible del suelo y el agua en la agricultura (US-IRNAS), Crta de Utrera, km 1, 41013 Sevilla, Spain
[5]   BCC Innovation, Technological Center in Gastronomy, Juan Avelino Barriola 101, 20009 Donostia-San Sebastián, Spain; lvazquez@bculinary.com
[6]   Basque Culinary Center, Mondragon Unibersitatea, Juan Avelino Barriola 101, 20009 Donostia-San Sebastián, Spain
*   Correspondence: angel.carbonell@umh.es; Tel.: +34-966-749-754

Received: 22 August 2020; Accepted: 23 September 2020; Published: 25 September 2020

Abstract: Global warming enhances the rainfall and temperature irregularity, producing a collapse in water resources and generating an urgent need for hydro-sustainable thinking in agriculture. The aim of this study was to evaluate the correlation between the water stress of almond trees and quality parameters of fruits, after 3 years of experiments, with the objective of establishing quality markers necessary in the certification process of hydroSOStainable almonds. The results showed positive correlations among the stress integral (SI) and dry weight, color coordinates (L^*, a^* and b^*), minerals (K, Fe, and Zn), organic acids (citric acid), sugars (sucrose, fructose, and total sugars), antioxidant activity, and fatty acids [linoleic acid, polyunsaturated (PUFA)/monounsaturated (MUFA) ratio, PUFA and SFA, among others]. As well as negative correlations of SI with water activity, weight (almond, kernel, and shell), kernel size, minerals (Ca and Mg), fatty acids (oleic acid, oleic/linoleic ratio, MUFA, and PUFA/SFA ratio), and sensory attributes (size, bitterness, astringency, benzaldehyde, and woody). Finally, this research helped to prove key quality parameters that can be used as makers of hydroSOStainable almonds. In addition, it was demonstrated that controlling water stress in almond trees by using deficit irrigation strategies can lead to appropriate yields, improve the product quality, and consequently, lead to a final added value.

Keywords: *Prunus dulcis*; Vairo; water stress; regulated deficit irrigation; sustained deficit irrigation; quality markers

---

1. Introduction

Almonds [*Prunus dulcis* (Mill.) D.A. Webb] are an economically and nutritionally important agricultural good, widely consumed in the Mediterranean diet either as a snack or as an ingredient for confectionery (*turrón)* and baking [1]. Almond consumption increased by 1.9% at the end of 2018 in Spain, indicating that consumer appreciation for this nut is high and constantly increasing due to its nutrition values, pleasant flavor, and healthy properties [2,3]. Moreover, raising the number of health-conscious consumers, together with environmental and animals care, lactose intolerance, and hypercholesterolemia in consumers, plant-based milk, yogurt, and cheese has grown over the last decade [4]. For instance, the global almond milk market is predicted to expand 14.3% by 2025; this product being considered a dairy alternative rich in vitamin E and omega 3 and 6 fatty acids. After all, this might be also an important reason that led to almond consumption growth [4–6].

Almond is the third largest crop in terms of surface and the most cultivated tree nut in Spain [7]. Besides, Spain is the main European almond producer and the second-largest in the world (339,033 t in-shell almonds), after the United States of America (1,872,500 t) [8]; Andalusia (111,877 t), Aragon (63,235 t), Castilla La-Mancha (53,201 t), and Valencian Community (40,875 t) are the main producing regions [9]. However, almond production in Spain is relatively low because this crop has been mainly grown in marginal areas where it has traditionally cultivated under restrictive conditions [7]. The almond tree is a drought-tolerant species, but due to the low yield in rainfed conditions (380 kg ha^{-1}), irrigation water is necessary to increase its productivity (1842 kg ha^{-1}) [9].

The Mediterranean regions are the most affected areas by water stress due to the scarcity and irregularity of rainfall. Moreover, the highest crop water needs are found in areas that are hot, dry, windy, and sunny due to the growth needs of the plant (foliage expansion, vegetative growth, and fruit yield) [10]. The water scarcity crisis is considered the biggest global risk for the world economy and it is affecting every continent [11]. Regarding agricultural sector, there is a consensus about the inadequate management of water resources and the need of achieving an equilibrium between rural development, food security, and environment protection [12]. The population growth leads to an expansion in intensive food production that alters the environment due to greenhouse gas emissions, soil deterioration, and water stress [13]. The main impact produced by climate change includes significant alteration in the average temperature and the rainfall irregularity [14], which leads to a substantially increase in irrigation water demand. In almond farming, climate change can provoke phenological variations on fruit, which may affect the final yield, quality, and marketability [15]. Consequently, all these changes might lead to a reduction in the productivity of agro-ecosystems, a progressive decline of rural areas, and even, land abandonment [16].

The implementation of sustainable irrigation strategies is an important tool to attenuate these negative aspects. However, these strategies must fulfill two important requirements: (i) causing minimal production losses and (ii) ensuring the final quality of the fruits. Regulated deficit irrigation (RDI) is one of these strategies meant to increase the water productivity with minimal yield losses and consists of reducing the amount of water during the kernel-filling stage in almond orchards [17]. Sustained deficit irrigation (SDI) is another strategy, which consists of applying a uniform and reduced amount of water during the whole growing cycle, creating a progressive stress in plants throughout the season [18].

Recently, new research lines focused on water resources sustainability have been developed for different crops (almonds, pistachios, olives, etc.), and the foodstuffs produced under controlled water stress conditions are called hydroSOStainable foods [19–21]. However, a variability in crop responses to water stress was reported for these products with the quality parameters being cultivar-, crop-, and year-dependent. Therefore, long-term research to decide which quality parameters are really affected by the waters stress conditions is needed.

Consequently, the aim of this study was to correlate water deficit response with quality parameters after 3 years of experiments (2017, 2018, 2019) to identify those parameters that behave in the same way throughout the trials. These results are essential to establish the future hydroSOStainable markers.

2. Materials and Methods

2.1. Plant and Experimental Conditions

The experiment was performed during 3 growing cycles (2017, 2018, and 2019) in a commercial orchard "La Florida" (37.23° N, −5.91 W, Dos Hermanas, Seville, Spain). The almond [*P. dulcis* (Mill.) D.A. Webb cv. Vairo] orchard was 7 years old at the beginning of the experiment. The tree spacing was an 8 m × 6 m square pattern, while the irrigation system used a drip irrigation line (3.8 L h^{-1}) with drippers separated at 0.4 m distance.

The weather data for each season were obtained from the "Instituto de Investigación y Formación Agraria (IFAPA) Los Palacios" station in the Andalusian weather stations network (Figure 1) located about 6 km away from the experimental orchard.

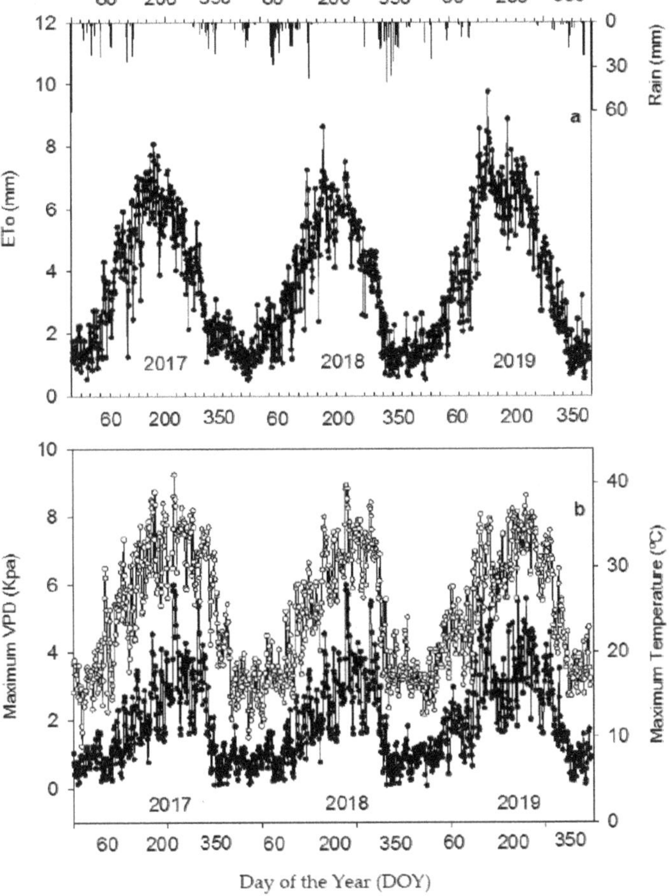

Figure 1. Climatic conditions during the three experimental seasons (2017–2019). (**a**) Seasonal daily reference evapotranspiration (circles) and rain (bars). (**b**) Seasonal daily maximum air temperature (white circles) and maximum vapor pressure deficit (VPD) (black circles). Vertical dots lines indicated from right to left each season, the beginning of pit hardening, early recovery, and regular recovery. DOY: day of the year.

The data for all 3 seasons were typical of Mediterranean zones, with null rainfall during the summer period and warm winters. The threshold values of midday stem water potential (SWP) were measured weekly, most of the dates, or every ten days using a pressure chamber (PMS Instrument Company, Albany, OR, USA). These values were used for the irrigation schedule by evaluating the stress level in the plant with the methodology proposed by Myers [22] according to the following expression Equation (1):

$$SI = |\sum (\psi_{stem} - (-0.2)) \times n \tag{1}$$

where *SI* was the stress integral, ψ_{stem} is the average midday stem water potential for any interval, and *n* is the number of days in the interval. Most of the measurements were weekly.

2.2. Irrigation Treatments

Four irrigation treatments were applied to the experimental plots. Each treatment represents different strategies of farmers in conditions of water scarcity. Moderate RDI is a controlled deficit irrigation in which applied water is lower than full irrigation but restricted considering an accurate water management. Severe RDI (was considered due to the low water availability) represents concentrated irrigation mainly during postharvest. Finally, SDI, is a strategy that was not considered in the phenological stages, and then, postharvest irrigation was very limited. These treatments are described in detail below:

- Full irrigation (T1): irrigated to assure the crop needs. Irrigation was daily and irrigation scheduling was performed every week. Water needs were estimated with the crop evapotranspiration (ETc) approach according to Steduto et al. [23] using reduction coefficients (Kr) around 0.6. In addition, water status was evaluated using midday stem water potential and compared to the McCutchan and Shackel [24] baseline. When water status was more negative than expected, irrigation was increased by 150% ETc.
- Moderate RDI (T2): the water stress was imposed during the kernel-filling period; almond trees were irrigated when SWP was below −1.5 MPa, and for the rest of the time, trees were irrigated to keep an SWP as the baseline proposed by McCutchan and Shackel [24]. Equation (2) estimated optimum midday stem water potential in relation with vapor pressure deficit (VPD):

$$SWP = (-0.41) \times (-0.12VPD) \tag{2}$$

  where: *SWP* is optimum midday stem water potential (MPa) and *VPD* is vapor pressure deficit (KPa).
- Severe RDI (T3): the same as T2, except that trees were irrigated when SWP was below −2.0 MPa during kernel filling and maximum seasonal water was considered (120 mm, around 20% ETc). Therefore, after harvest, when total applied water was reached, irrigation stopped.
- SDI (T4): the same as T3, but tree water status was not considered. Irrigation was applied in a constant daily rate around 1–2 mm per day. The main differences between both strategies (T3 and T4) was that T4 limited postharvest irrigation more than T3.

Harvesting was done with a self-propelled trunk shaker with collector in the mid of August (28 weeks after blossom). The treatments were separately harvested, and almonds were sun-dried until a moisture content lower than 5% was achieved. Later, in-shell almonds were delivered to Miguel Hernández University (Orihuela, Alicante, Spain) for analysis.

2.3. Physical Parameters

2.3.1. Kernel Ratio

The ratio between the mass of in-shell almonds and kernel was calculated from 12 kg of whole fruit per treatment and year.

2.3.2. Dry Weight and Water Activity

For the dry weight content (%) analysis, 2 g of ground almonds (Moulinex grinder AR110830, Alençon, France) were added to an aluminum tray and dried in an oven at 60 °C until a constant weight was reached, while water activity (a_w) was measured by placing the cups with almond (2 g) into an a_w meter (Novasina aw-Sprint TH500; Pfaffikon, Zurich, Switzerland) and reading the value. The experiments were done in quadruplicate.

2.3.3. Weight and Size

For the morphological parameters, 100 almonds per treatment (25 almonds × 4 trees × treatment × year) were randomly selected and measured in terms of weight and size (length, width, thickness) of both in-shell almond and kernel using a digital caliper (Mitutoyo 500-197-20, Kawasaki, Japan) and a precision scale (Mettler Toledo model AG204, Barcelona, Spain), respectively.

2.3.4. Instrumental Color

Color measurements were performed at 25 ± 1 °C using a Minolta Colorimeter CR-300 (Osaka, Japan). Outside color was directly measured on the skin of 100 individual almond kernels per treatment each year. Results were presented as international commission on illumination (CIE) L^*, a^* and b^* color coordinates describing the color in a three-dimensional space as following: L^* for the lightness ($L^* = 0$ black; $L^* = 100$ white), a^* for the green-red ($a^* =$ red; $-a^* =$ green), and b^* for the blue-yellow components ($b^* =$ yellow; $-b^* =$ blue).

2.3.5. Instrumental Texture

The texture of 100 almonds per treatment and year was measured using a texture analyzer (Stable Micro Systems, model TA-XT2i, Godalming, UK) with a 30 kg load cell and a probe Volodkevich Bite Jaw (HDP/VB) as following: trigger was placed at 15 g, test speed was 1 mm s^{-1} over a specific distance of 3 mm. Fracturability (mm), hardness (N), work done to shear (Ns), average force (N), and number of fractures (peaks count) were the parameters analyzed.

2.4. Chemical and Functional Analysis/Parameters

2.4.1. Mineral Content Determination

The digestion of 0.5 g of sample with 8 mL of concentrated HNO_3 and 2 mL H_2O_2 (30%) using a START D Medium Microwave Digestion (SK-10) was first carried out [25]. Followed by the determination of macro-nutrients (Ca, Mg, and K) and micro-nutrients (Fe, Cu, Mn and Zn) with a Unicam Solaar 969 atomic absorption–emission spectrometer (Unicam Ltd., Cambridge, UK). Calcium, Mg, Fe, Cu, Mn, and Zn was determined by atomic absorption and K by atomic emission.

2.4.2. Organic Acids and Sugars

High-performance liquid chromatography (HPLC) was used for organic acids and sugars identification and quantification, as previously described [26]. For this, 1 g of ground almond was homogenized (Ultra Turrax T18 Basic, IKA®-Werke GmbH & Co. KG Janke & Kunkel, Staufen, Germany) with 5 mL of 50 mM phosphate buffer (pH = 7.8) for 2 min at 11,300 rpm, centrifuged (Sigma 3−18 K; Sigma Laborzentrifugen, Osterode and Harz, Germany) at 4 °C and 15,000 rpm for 20 min and filtered (0.45 μm Millipore membrane filter, Billerica, MA, USA). The supernatant was injected (10 μL) into a Hewlett Packard (Wilmington, DE, USA) series 1100 (HPLC) using as mobile phase 0.1% orthophosphoric acid elution buffer. Sugars were analyzed using a Supelcogel TM C-610H column (30 cm × 7.8 mm) with a precolumn (Supelguard 5 cm × 4.6 mm; 219 Supelco, Bellefonte, PA, USA) and detected with a refractive index detector (RID). Organic acids were separated as sugars using a diode-array detector (DAD) at 210 nm for the absorbance measurements. Analyses were run in quadruplicate, and results were expressed as g kg^{-1} dry weight (dw).

2.4.3. Antioxidant Activity and Total Phenolic Content

The antioxidant activity and total phenolic content was carried out both for whole kernel and its blanched skin. For the extraction 0.5 g of finely ground almond were sonicated with 10 mL of extractant [MeOH/H_2O_2 (80:20, *v/v*) + 1% HCl at 20 °C] for 15 min and stored at 4 °C overnight. The mixture was sonicated again under the same conditions and centrifuged at 10,000 rpm for 10 min. The antioxidant activity of the obtained extract was measured using 3 methods: $ABTS^{\bullet+}$ [2,2-azinobis-(3-ethylbenzothiazoline-6-sulfonic acid)], $DPPH^{\bullet}$ (2,2-diphenyl-1-picrylhydrazyl), and FRAP (ferric reducing antioxidant power), as previously described by Brand-Williams et al. [27]. The results were calculated according to the Trolox calibration curve and were expressed as mmol Trolox kg^{-1}.

For total phenolic content (TPC) 100 μL of supernatant was mixed with 200 μL Folin-Ciocâlteu reagent and 2 mL of H_2O_2 and was stored at 22 °C for 3 min. Then, 1 mL of 20% Na_2CO_3 was added, followed by 1 h of incubation at room temperature. The results were calculated with the gallic acid calibration curve and expressed as gallic acid equivalents (GAE), g GAE kg^{-1}. All measurements were performed in an ultraviolet-visible (UV-vis) spectrophotometer (Helios Gamma model, UVG 1002E; Helios, Cambridge, UK).

2.4.4. Fatty Acids

Ground almond (40 mg) was saponified with 100 μL of dichloromethane (Cl_2CH_2) and 1 mL of sodium methoxide solution and refluxed for 10 min at 90 °C. Later, 1 mL of BF_3 methanolic was added followed by 30 min rest in dark for reaction [25]. The fatty acids methyl esters (FAMEs) were separated in a Shimadzu GC17A gas chromatography coupled with a flame ionization detector and a DB-23 capillary column (30 m length, 0.25 mm internal diameter, 0.25 μm film thickness) J&W Scientific, Agilent Technologies using the same conditions, as previously described by Lipan et al. [25]. The identification of FAMEs peaks was done by comparing the retention times of the FAME Supelco MIX-37 standards. Analysis were carried out in quadruplicate, and the results were expressed as g kg^{-1} concentration, using methyl nonadecanoate (C19:0) as internal standard.

2.5. Descriptive Sensory Analysis

Descriptive sensory evaluation was performed following the steps previously published in literature using a trained panel [19]. Ten highly trained panelists from the Food Quality and Safety Group (Miguel Hernández University of Elche, Orihuela, Alicante, Spain) with ages between 25–62 years (5 women and 5 men) conducted the descriptive analysis. Once the orientation sessions were finished (4), the panel was asked to evaluate the 4 samples corresponding to the irrigation treatments in terms of appearance, basic tastes, and flavor intensities of almond. For this, a structured scale from 0 to 10 (0.5 increments) was used to quantify the intensity of the almond attributes, where 0 represents no intensity and 10 extremely strong. The samples were presented using a randomized block design to avoid biases in individual tasting booths (controlled temperature of 21 ± 2 °C and combined natural/artificial light) equipped with water and unsalted crackers for palate cleaning among samples. The analysis was run in triplicate.

2.6. Statistical Analysis

Two-way analysis of variance (ANOVA), using "irrigation treatment" and "year" as factors, followed by Tukey's multiple range test were carried out in order to decide the parameters to be used for the correlations. Two supplementary tables were added with the mean values of 3 years (Tables S1 and S2). Only those parameters significantly different among treatments were considered for Pearson's correlations. All statistical analyses were performed using XLSTAT Premium 2016, while Sigma Plot 11 software was used for figures preparation. Statistically differences were considered significant when $p < 0.05$.

3. Results and Discussion

3.1. Agronomic Parameters

Tables S1 and S2 contains supplementary information about the mean values of 3 years study for all the parameters. As observed, a lower amount of irrigation water was received by the trees' growth under deficit irrigation strategies, being T3 and T4 the treatments, which received the less amount of irrigation water; this rebound in the plant status can be observed from the stress integral values. Almond trees from T3 and T4 were the most stressed, although the latter (T4) was statistically significant to T2. During the first season, T2 and T3 were the most stressed treatments followed by T4, which was statistically similar to T2. In 2018, lower values of SI were shown for all treatments, which means that the stress in plant was less severe than the other seasons. T3 and T4 were the most stressed treatments, followed by T2, which was statistically correlated with the control and to the other deficit irrigation treatments. Finally, 2019 was the season in which almonds trees met the highest values of water stress, with T3 and T4 having the highest values, followed by T2. As observed, the SI behaved different yearly, however in the last two seasons, T3 and T4 were similar in terms of water stress in plant. The difference between these treatments is that in the former (T3) the stress was applied in the kernel-filling period, while in the latter (T4), the stress was imposed throughout the whole growing cycle, creating a progressively stress in plant rather than in a single phenological phase (kernel filling). Supplementary data also showed the mean values of kernel yield of all 3 seasons to check how the previous parameters (SI and applied water) influenced the fruit yield. A reduction of this parameter was observed in all the treatments growth under DI conditions with no significant differences among them. If each year production is analyzed, no differences among control and DI treatment was registered in the first season. However, a decrease of 2.4-fold was found for these treatments (T2, T3, T4) regarding the control in the second season (with no significant differences among them) and in the third season. A reduction in kernel yield in deficit irrigation treatments was also observed in 2019 season (1.2-fold in T2 and 1.5-fold in T3 and T4), although this time was lower than in 2018 and T2 was significantly similar to the control. Is important to highlight that in 2018, even though T3 and T4 received a lower amount of water than T2, the kernel yield was similar among them, and that in 2019, although T2 received lower amount of water than the control (T1), the kernel yield was significantly similar between them.

Pearson's correlation coefficients (R) among SWP and SI with agronomical and physical parameters is shown in Figure 2. A negative and significant correlation was found between SI and (i) SWP (R = −0.67; $p < 0.001$) and (ii) water activity (R = −0.39; $p < 0.01$); this means that at higher waters stress values, lower SWP and a_w values are obtained. On the other hand, a positive and significant correlation was observed between SI and dry weight (R = 0.53; $p < 0.001$) as well as between kernel ratio and applied water (R = 0.36; $p < 0.05$). Regarding the water stress effect on kernel yield, the results showed that in the first year it was not affected (R = 0.11; $p > 0.05$); however, kernel yield was reduced with water stress during 2018 and 2019 seasons demonstrated by the negative correlations of R = −0.50; $p < 0.05$ * and R = −0.79; $p < 0.001$ ***, respectively.

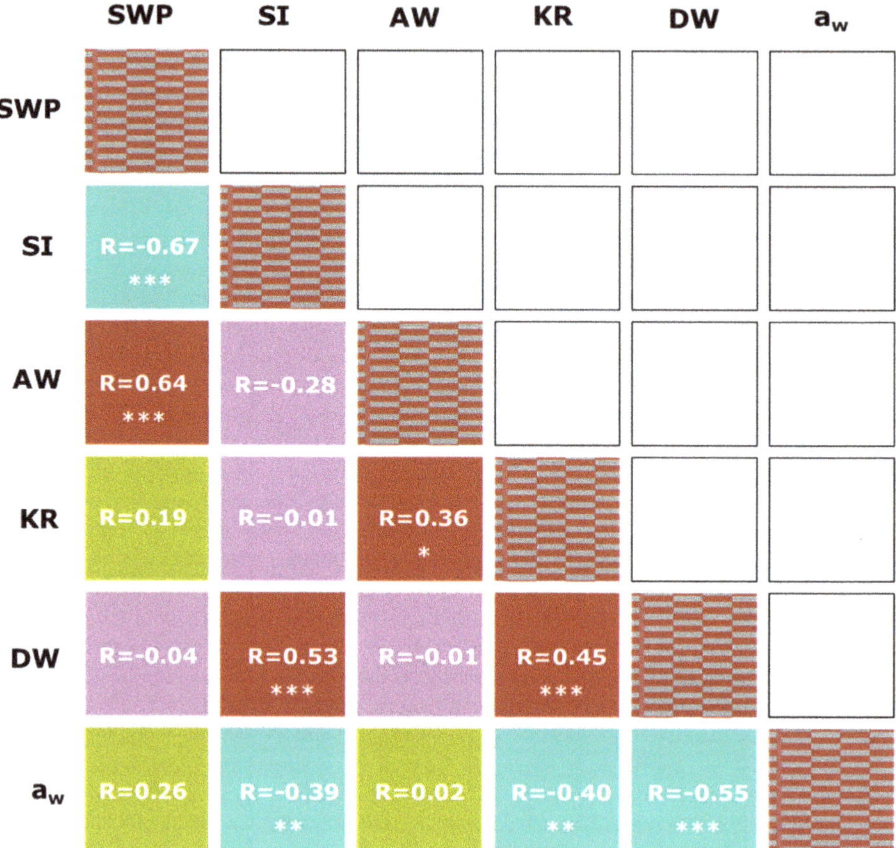

Figure 2. Heat map of correlation matrix of agronomical parameters. Each square indicates Pearson's correlation coefficient for a pair of data and the color represents the positive or negative correlation as: R = 1.00; significant ($p < 0.05$) positive correlation; significant ($p < 0.05$) negative correlation; positive but not correlated; negative but not correlated. *, **, ***, significant at $p < 0.05$, 0.01, and 0.001, respectively. SWP = minimum stem water potential; SI = stress integral; AW = applied water; KR = kernel ratio; DW = dry weight; aw = water activity.

Overall, these results showed that after long-term experiment (3 years), the water stress in almond trees negatively affected the yield but enhanced dry weight. These results suggest that yields differences were related to the number of nuts. Such response could be associated with a postharvest water stress in deficit treatments [18]. However, it was observed that, depending on the treatment, it can lead to yields statistically similar to the control, which might be a good alternative when water restrictions are below the crop needs. In this way, Moderate RDI (T2) could balance water stress effects, because this treatment could secure enough crown volume, which is very important in the tree yield capacity [10] and postharvest recovery, which is according to the current data, the most important effect. On the contrary, T3 and T4 results suggest that water status in postharvest would be better. Then, in conditions of very low water availability, irrigation in this period should be preferential and greater than the ones of T3. Besides, the reduction in the moisture content and water activity with water stress are also important outcomes for food industry, because lower values of these parameters help to maintain at minimum the biological reactions, which essential to increase the almonds shelf life [28]. The obtained results indicate that the use of deficit irrigation in almond trees water management can improve yield and reduce water use. Thus, it contributes to reduce water consumption for irrigation purposes.

3.2. Morphological Parameters

Table 1 shows the Pearson's correlation coefficients (R) between SWP and SI with morphological parameters. The SI was negatively correlated with almond, kernel, and shell weight, with kernel length and width, and with almond thickness. This means that weight and size were reduced with the water stress in plant. The conclusions regarding the effect of deficit irrigation on the morphological parameters are widely spread throughout the literature. For instance, similar results were obtained in almond cultivar (cv.) Nonpareil [18], and no differences were reported for almond cultivars Marta, Guara, Lauranne, Ferragnes, and Texas [29,30]. Additionally, no differences on the morphological parameters were also reported for other crops such as pistachio cv. Kerman and olives cv. Manzanilla if the stress was applied during shell and pit hardening, respectively [21,31]. Finally, an increase in weight and equatorial diameter but a decrease in longitudinal diameter were observed for olives cv. Manzanilla growth under the following RDI conditions: (i) stage I, trees irrigated under non-limited conditions; (ii) stage II, trees under moderate water deficit conditions, they were no irrigated during this period; and (iii) stage III, water applied in order to provide a water status similar to a full irrigated treatment [32].

A significant positive correlation was observed between the SI and color parameters, showing that a higher stress level leads to higher values of L^*, a^*, and b^* coordinates. This means that hydroSOStainable almonds have a lighter color with reddish and yellowish notes (more intense brown color). As the almond color skin is given by the polyphenol profile, which is unique for each cultivar [33], the increase in color coordinates under water stress conditions might be related to a potential increase in polyphenols. For instance, almond flavonoids have been extensively studied in different plants, and it was concluded that they are decisive pigment in color plants [34]. The brown almond skin pigment is largely concentrated in the high-molecular weight fraction such as proanthocyanidins, which are the main polyphenols found in almonds that can impart color formation [35]. A positive correlation between SI and proanthocyanidins ($R = 0.73$; $p = 0.001$) was previously reported in almonds cv. Vairo after one season of experiment. Besides, a^* values were also reported to be higher in almonds cv. Vairo, Marta, Guara, Lauranne growth under deficit irrigation conditions after one season [25,29]. Finally, a^* was reported to be positively correlated with the contents of nine individual flavonols, total kaempferols, and total flavonols in a study about the relationship between rose petals' (*Rosa* spp.) color and polyphenols content [36].

Table 1. Pearson's correlation coefficients (R) among stem water potential and stress integral with morphological parameters.

	SWP	SI	AWe	KWe	SHWe	AL	KL	AWi	KWi	ATi	KTi	L^*	a^*	b^*	Hue	C	H
SWP	1.00																
SI	−0.67 ***	1.00															
AWe	0.17	−0.39 **	1.00														
KWe	0.35 *	−0.56 ***	0.85 ***	1.00													
SHWe	0.11	−0.32 *	0.99 ***	0.75 ***	1.00												
AL	0.05	0.08	0.75 ***	0.43 **	0.80 ***	1.00											
KL	0.26	−0.57 ***	0.84 ***	0.91 ***	0.77 ***	0.40 **	1.00										
AWi	0.08	−0.17	0.90 ***	0.64 ***	0.92 ***	0.89 ***	0.60 ***	1.00									
KWi	0.18	−0.51 ***	0.92 ***	0.88 ***	0.87 ***	0.55 ***	0.88 ***	0.79 ***	1.00								
ATi	0.33 *	−0.41 **	0.80 ***	0.63 ***	0.80 ***	0.68 ***	0.61 ***	0.81 ***	0.70 ***	1.00							
KTi	0.32 *	0.07	−0.36 *	−0.04	−0.44 **	−0.30 *	−0.16	−0.41 **	−0.30 *	−0.15	1.00						
L^*	−0.20	0.61 ***	−0.65 ***	−0.55 ***	−0.64 ***	−0.38 **	−0.61 ***	−0.58 ***	−0.64 ***	−0.64 ***	0.47 ***	1.00					
a^*	−0.30 *	0.80 ***	−0.41 **	−0.57 ***	−0.34 *	0.14	−0.69 ***	−0.13	−0.55 ***	−0.29 *	0.19	0.65 ***	1.00				
b^*	−0.21	0.72 ***	−0.53 ***	−0.55 ***	−0.50 ***	−0.10	−0.67 ***	−0.36 *	−0.61 ***	−0.49 ***	0.37 *	0.87 ***	0.90 ***	1.00			
Hue	0.07	0.02	0.32 ***	−0.04	0.40 **	0.59 ***	−0.06	0.55 ***	0.16	0.48 ***	−0.50 ***	−0.42 **	0.27	−0.05	1.00		
C	−0.09	0.10	−0.40 *	−0.06	−0.48 ***	−0.59 ***	−0.06	−0.60 ***	−0.26	−0.56 ***	0.55 ***	0.56 ***	−0.12	0.23	−0.98 ***	1.00	
H	0.06	−0.40	0.61 ***	0.60 ***	0.58 ***	0.31 *	0.67 ***	0.48 ***	0.63 ***	0.52 ***	−0.32 *	−0.56 ***	−0.56 ***	−0.63 ***	−0.04	−0.08	1.00

*, **, ***, significant at $p < 0.05$, 0.01, and 0.001, respectively. SWP = minimum stem water potential; SI = stress integral; AWe = almond weight; KWe = kernel weight; SHWe = shell weight; AL = almond length; KL = kernel length; AWi = almond width; KWi = kernel width; ATi = almond thickness; KTi = kernel thickness; L^*, a^*, b^* = color coordinates; C = Chroma; H = hardness.

3.3. Mineral, Organic Acids, and Sugars Content

The minerals contained in plant tissue are taken by plants from soil and from the water received in production [34]. For this reason, environmental factors, agronomical practices (location, soil composition, water source, irrigation, and fertilizer) and cultivar are responsible for the final mineral content in kernel. Potassium, Ca, Mg, Fe, P, S, and N are the main elements found in plants mainly accumulated during fruit growing and ripening [37]. It was reported that drought conditions reduces the mineral content transport from root to shoot; however, there are plants with a better water use efficiency (WUE) and consequently with greater drought tolerance [38].

In order to analyze the relationship between SWP and SI with minerals, organic acids, and sugars, Pearson's correlation coefficients (R) were calculated and are displayed in Table 2. Calcium (R = −0.60; $p < 0.001$) and Mg (R = −0.35; $p < 0.01$) showed significant negative correlations with the SI, and the latter was also positively correlated with SWP (R = 0.71; $p < 0.01$). However, if each year is considered both minerals presented significant difference in only one season. Magnesium is a macro element essential component of the chlorophyll molecule, which is necessary in the photosynthesis process [38]. Besides, Mg plays a role in energy preservation and protein synthesis being a cofactor for many enzymes associated with de-phosphorylation, hydrolysis, and in stabilizing the structure of nucleotides and sugar accumulation.

Potassium (R = 0.60; $p < 0.001$), Fe (R = 0.64; $p < 0.001$), and Zn (R = 0.44; $p < 0.01$) were the elements positively correlated with the water stress. Similar results were also reported by other researchers in almonds in which a higher amount of K was reported in moderate RDI attributed to the relationship between water availability and minerals absorption [25]. The authors explain that the excess of water might be the responsible for mineral leaching and also that drought stress could contribute to the saturation of minerals in the rootzone. Potassium is the most important element, after N and P, helping to maintain the plant water status being involved in physiological and molecular mechanisms needed to increase the plant tolerance to stress [38]. Potassium has been reported to be the major mineral cation in almonds kernels (717 mg/100 g in cv. Vairo); in this way, almonds are considered a food high/rich in K because its content is above the minimum threshold (600 mg K/100 g) established in the Regulation (EU) No 1169/2011 of the European Parliament and of the Council [25,39].

Iron and Zn also presented a positive correlation with water stress and these results agreed with other authors reporting that this microelement helps to improve the WUE and the crop yield [38]. Usually, drought induces Fe deficiency with a negative effect on plant tree, causing chlorosis due to low levels of chlorophyll. This microelement is also necessary for an effective function of the antioxidant enzymes because a Fe deficiency reduces the enzymes activity, enhances the ROS production, and reduces the bioactive compounds biosynthesis [38]. The present results might reveal that this controlled stress was below the limit needed to reduce the microelements production; in fact, an opposite phenomenon was observed. Studies in wheat growth in fields under water stress conditions also reported a higher Zn content in grains growth under water stress conditions but not that grown in greenhouses [40].

Positive correlation was shown for citric acid and SI (R = 0.65; $p < 0.001$), which was confirmed by studies in almonds of cv. Marta, Guara, and Lauranne growth under RDI versus full irrigated and over irrigated conditions [29] and other crops such as thyme [41]. However, in studies of cv. Guara under non-irrigated almonds versus drip-irrigated, the citric acid was higher in almonds growth in drip-irrigated conditions [42]. In addition, no differences were reported in cv. Vairo [25] and cv. Marta [43]. These differences could be attributed to the irrigation strategies and the levels of stress created in each experiment. The increase in citric acid in response to drought may result from the larger inhibition of the citrate degrading system relative to citrate synthesis as previously reported in CAM plant (*Aptenia cordifolia*), although the capacity for citric acid oxidation and the citrate synthetase activity decreased during drought [44].

Table 2. Pearson's correlation coefficients (R) among stem water potential and stress integral with minerals, organic acids, and sugars.

	SWP	SI	Ca	Mg	K	Fe	Mn	Zn	Cit	Tar	Mal	ΣOA	Suc	Glu	Fru	ΣS
SWP	1.00															
SI	−0.67 ***	1.00														
Ca	0.14	−0.60 ***	1.00													
Mg	0.13	−0.35 **	−0.12	1.00												
K	−0.06	0.60 ***	−0.79 ***	0.39 **	1.00											
Fe	−0.18	0.64 ***	0.79 ***	−0.20	−0.65 ***	1.00										
Mn	0.13	−0.05	−0.51 ***	−0.46 ***	0.44 **	−0.40 **	1.00									
Zn	−0.41 **	0.44 **	−0.26	−0.15	0.40 **	0.11	0.29 *	1.00								
Cit	−0.19	0.65 ***	−0.61 ***	0.34 *	0.58 ***	−0.58 ***	0.10	0.02	1.00							
Tar	−0.07	0.35	0.07	0.75 ***	0.04	−0.05	−0.80	−0.24	0.41 **	1.00						
Mal	−0.01	−0.17	−0.23	−0.56 ***	0.07	−0.01	0.68 ***	0.22	−0.22	−0.74 ***	1.00					
ΣOA	−0.04	−0.09	−0.29*	−0.49 ***	0.13	−0.07	0.67 ***	0.21	−0.10	−0.66 ***	0.99 ***	1.00				
Suc	−0.42 **	0.71 ***	−0.62 ***	0.06	0.52 ***	−0.43 **	0.17 ***	0.19	0.71 ***	0.20	−0.05	0.02	1.00			
Glu	−0.02	0.26	0.03	0.57 ***	0.12	−0.07	−0.51 ***	−0.02	0.13	0.63 ***	−0.53 ***	−0.50 ***	0.03	1.00		
Fru	−0.36 *	0.30 *	0.21	0.28 *	−0.24	0.28	−0.55 ***	−0.14	0.16	0.51 ***	−0.38 **	−0.34 *	0.19	0.24	1.00	
ΣS	−0.39 **	0.70 ***	−0.36 *	0.44 **	0.37 *	−0.27	−0.31 *	0.09	0.60 ***	0.62 ***	−0.43 **	−0.35 *	0.75 ***	0.64 ***	0.50 ***	1.00

*, **, ***, significant at *p* < 0.05, 0.01, and 0.001, respectively. SWP = minimum stem water potential; SI = stress integral; Cit = citric; Tar = tartaric; Mal = malic; ΣOA = total organic acids; Suc = sucrose; Glu = glucose; Fru = fructose; ΣS = total sugars.

Finally, sucrose, fructose, and total sugars were also positively correlated with the water SI sucrose (R = 0.71; $p < 0.001$), fructose (R = 0.30; $p < 0.05$), and total sugars (R = 0.70; $p < 0.001$), and these results agreed with those of other authors in almonds cv. Marta, Guara, and Lauranne [29], with almonds being grown under RDI circumstances. Authors working with cv. Vairo under RDI and SDI conditions reported no differences for total sugars and sucrose in the first year of water deficit; however, a reduction in glucose was reported for the most stressed treatments [25,26]. Lower amounts of sucrose and glucose were reported in cv. Guara growth under non-irrigated conditions when compared to drip-irrigated, while fructose was not affected [42]. Lower values of sucrose with no differences in glucose, fructose, and the total sugars were also reported for cv. Marta under RDI and partial rootzone drying (PRD) in different levels [43]. Although, sucrose started to increase for the most severe treatment of PRD. Moreover, sucrose was reported to increase in non-irrigated conditions for cv. Ferragnes in early harvest, and the opposite was observed for the same cv. in late harvest, while non-irrigation decreased this sugar in cv. Texas in both situations, with the total sugars not affected [30]. If other crops are considered, sugars were also increased in tomatoes under water stress conditions [17], thyme (*Thymus vulgaris* as drought-tolerant and *T. kotschyanus* as drought-tolerant species) under drought stress [41], and peaches in which experiment it was demonstrated that deficit irrigation can enhance both total and individual sugars, if proper water stress is established for each cultivar [45]. A different behavior was reported for each peach cultivar; therefore, it is essential to establish specific conditions not only for each plant species but for each cultivar.

The sugars' enhancement under stress conditions was related to the osmotic adjustment, activated by accumulation of solutes rich in hydroxyl ($^{-}$OH) groups (sugars, proline, etc.) in the cytoplasm [26]. Osmotic adjustment is a biochemical mechanism that helps plants to adapt to dry and saline conditions by protecting the cellular membrane, protein, and enzymes against dehydration [26]; thus, it enhances the capacity to maintain positive turgor, increasing the sugars and organic acid. Another reason of the sugars accumulation during stress might be the induction of the growth inhibitor abscisic acid (ABA) by plants under stress conditions, which activates the sugar accumulation as an adaptation to stress [46]. Drought increases the biosynthesis and accumulation of ABA, which is considered the main regulator of drought stress response inducing the accumulation of osmotically active compounds, which protect cells from damage [38]. Under stress conditions, this phyto-hormone reduces plant growth and enhances desiccation tolerance by inducing de accumulation of stress-associated transcripts such as low-molecular-weight soluble sugars(sucrose) [47].

In summary, K, Fe, Zn, sucrose, fructose, and total sugars can be considered as good quality markers for hydroSOStainable almonds.

3.4. Antioxidant Activity (AA) and Total Phenolic Compounds (TPC)

The antioxidants are important compounds necessary to inhibit the process of oxidation acting like radical scavengers and converting these pro-oxidants to less reactive species. Antioxidants have attracted considerable consumer interest due to their potential preserving, nutritional, and therapeutic effects. For these reasons, the correlations among SWP and SI with antioxidant activity of almond kernel and kernel skin are important and were evaluated within this study along 3 seasons (Table 3). ABTS$^{\bullet +}$ (R = 0.79; $p < 0.001$ and R = 0.44; $p < 0.01$) and FRAP (R = 0.34; $p < 0.05$ and R = 0.41; $p < 0.01$) in both whole kernel and kernel skin showed a significant and positive correlation with the SI, and only ABTS$^{\bullet +}$ (R = −0.44; $p < 0.01$ and R = −0.30; $p < 0.05$) in both matrixes was negatively correlated with the SWP. This showed that the induced water stress led to almonds with a higher antioxidant activity.

Table 3. Pearson's correlation coefficients (R) among stem water potential and stress integral with antioxidant activity of almond kernel and kernel skin.

	SWP	SI	ABTS$^{•+}$K	DPPH$^{•}$K	FRAP K	TPC K	ABTS$^{•+}$S	DPPH$^{•}$S	FRAP S	TPC S
SWP	1.00									
SI	−0.67 ***	1.00								
ABTS$^{•+}$K	−0.44 **	0.79 ***	1.00							
DPPH$^{•}$K	−0.09	−0.25	−0.46 ***	1.00						
FRAP K	−0.07	0.34 *	0.57 ***	−0.54 ***	1.00					
TPC K	0.05	−0.14	0.21	−0.75 ***	0.38 **	1.00				
ABTS$^{•+}$S	−0.30 *	0.44 **	0.12	0.50 ***	−0.20	−0.82 ***	1.00			
DPPH$^{•}$S	−0.07	0.10	−0.20	0.80 ***	−0.46 ***	−0.95 ***	0.82 ***	1.00		
FRAP S	−0.19	0.41 **	0.14	0.24	−0.17	−0.63 ***	0.87 ***	0.62 ***	1.00	
TPC S	−0.11	0.16	−0.15	0.72 ***	−0.43**	−0.92 ***	0.88 ***	0.94 ***	0.73 ***	1.00

*, **, ***, significant at $p < 0.05$, 0.01, and 0.001, respectively. SWP = minimum stem water potential; SI = stress integral; ABTS$^{•+}$ =2.2-azino-bis (3-ethylbenzothiazoline-6-sulfonic acid); DPPH$^{•}$ = (2,2-diphenil-1-picrylhydrazyl); FRAP = ferric reducing ability of plasma; TPC = total phenolic content; K = kernel; S = skin.

An increase in AA under water stress conditions was previously reported in many crops, including (i) almonds cv. Vairo, in which the stress was imposed in the kernel-filling phase using RDI and SDI strategies, (ii) olives cv. Manzanilla, when the stress was created just before harvest without re-hydration [48], and (iii) pistachios cv. Kerman, where the stress was imposed at stage II, which corresponds to shell hardening [49]. During the water stress, the turgor pressure is decreased, the ion toxicity is increased, and the photosynthesis is inhibited [50]. This increase in AA during stress conditions can be related to the antioxidant defense system used by plants to cope with reactive oxygen species (ROS) and also to the phytohormones accumulation by plants in water stress conditions. Phytohormones, as above mentioned, are responsible for the initiation of many defense mechanisms, including the increase in antioxidants to enhance plant tolerance to water stress. Jasmonate (JA), which is a phytohormone involved in sensing and signaling during the stress response, helps with the alleviation of plant to drought stress by increasing total carbohydrates, polysaccharides and soluble sugars by activating the enzymatic and non-enzymatic antioxidative system [50].

However, no correlation was found between the TPC and SI after 3 seasons, being in contrast with results previously reported by Lipan et al. (2019) [26] for almonds grown in the same conditions but in the first year of study. Several studies are reporting an increase in TPC values in plants grown under stress conditions, due to their role as plant molecules in response to biotic and abiotic stress, because when the carbohydrates exceed the amount used for growth needs, the excess of CO_2 assimilated in stress conditions is used for the biosynthesis of carbon secondary metabolites. Thus, not finding a correlation between SI and TPC may happen due to the level of stress applied, which perhaps was not strong enough to affect the TPC accumulation.

On the other hand, the positive correlation between SI and AA, and the no correlation or negative correlation of TPC with ABTS$^{\bullet+}$ (R = -0.15; $p > 0.05$) and FRAP (R = -0.43; $p < 0.01$) shows that other compounds with antioxidant effect might be responsible for the AA increase observed under water stress rather than only polyphenols. For instance, besides polyphenols, vitamins C and E and carotenoids have been thought to be responsible for most of the AA in foods [41]. Authors reported that, almonds are a valuable source of dietary lipids and have been suggested as a potential source of dietary antioxidants [51]. The same authors in their study about the AA and TPC in 100 different products reported that products with high AA tended to have a higher AA/TPC ratio; thus, this increase may result from compounds with AA that are not phenolic, or some phenolic compounds were more effective than others or with a greater reactivity with peroxyl free radicals (the AA method was an Oxygen Radical Absorbance Capacity (ORAC) Assay on a Plate Reader). Almonds are high/rich in vitamin E (25.6 mg/100 g) [52] because its content is above the minimum threshold (3.6 mg/100 g) established by the European Parliament and Council [39]; thus, this might be a compound contributing to the AA enhancement.

For instance, authors working with almonds cv. Nonpareil under water stress conditions reported higher values of tocopherols when RDI and SDI strategies were applied [53], as well as in sunflower seeds (cvs. Gulshan-98 and Suncross), particularly if the water stress was imposed at the reproductive stage [54]. These results led us to the conclusion that after long term study (3 years), antioxidant activity can be considered an important marker in hydroSOStainable almonds detection, while TPC is not a good indicator, presenting no correlation with water stress.

3.5. Fatty Acids

Pearson's correlation coefficients (R) between SWP and SI integral with fatty acids is showed in Table 4. Polyunsaturated/saturated fatty acids ratio (PUFA/SFA), oleic acid, and consequently, oleic/linoleic ratio (O/L) and monounsaturated (MUFA) fatty acids were significantly negatively correlated with the SI. On the other hand, myristic, palmitic, palmitoleic, margaric, *cis*-heptadecenoic, stearic, *cis*-vaccenic, linoleic, saturated (SFA), polyunsaturated (PUFA) fatty acids, and PUFA:MUFA ratio were positively correlated with the SI. Only linoleic, SFA, and PUFA fatty acids were also correlated in a negative way with the SWP, which helped to confirm the statement that these compounds increased with the water stress in almond trees.

Table 4. Pearson's correlation coefficients (R) among stem water potential and stress integral with fatty acids.

	SWP	SI	C14:0	C16:0	C16:1	C17:0	C17:1	C18:0	C18:1n9	C18:1n7	C18:2	O/L	SFA	MUFA	PUFA	PUFA/SFA	PUFA/MUFA
SWP	1.00																
SI	−0.67 ***	1.00															
C14:0	−0.14	0.73 ***	1.00														
C16:0	−0.20	0.81 ***	0.95 ***	1.00													
C16:1	−0.13	0.75 ***	0.92 ***	0.96 ***	1.00												
C17:0	−0.14	0.69 ***	0.88 ***	0.87 ***	0.86 ***	1.00											
C17:1	−0.06	0.65 ***	0.90 ***	0.90 ***	0.85 ***	0.84 ***	1.00										
C18:0	−0.20	0.80 ***	0.91 ***	0.97 ***	0.95 ***	0.86 ***	0.86 ***	1.00									
C18:1n9	0.22	−0.82 ***	−0.94 ***	−0.99 ***	−0.95 ***	−0.86 ***	−0.90 ***	−0.98 ***	1.00								
C18:1n7	−0.17	0.48 ***	0.51 ***	0.55 ***	0.40**	0.47 ***	0.62 ***	0.57 ***	−0.60 ***	1.00							
C18:2	−0.32 *	0.82 ***	0.86 ***	0.91 ***	0.86 ***	0.76 ***	0.75 ***	0.88 ***	−0.93 ***	0.46 ***	1.00						
O/L	0.24	−0.82 ***	−0.93 ***	−0.97 ***	−0.93 ***	−0.84 ***	−0.84 ***	−0.95 ***	0.98 ***	−0.52 ***	−0.98 ***	1.00					
SFA	−0.21	0.81 ***	0.94 ***	0.99 ***	0.96 ***	0.88 ***	0.89 ***	0.99 ***	−0.99 ***	0.55 ***	0.90 ***	−0.97 ***	1.00				
MUFA	0.27	−0.84 ***	−0.92 ***	−0.97 ***	−0.93 ***	−0.84 ***	−0.83 ***	−0.95 ***	0.98 ***	−0.49 ***	−0.97 ***	0.99 ***	−0.97 ***	1.00			
PUFA	−0.32 *	0.83 ***	0.86 ***	0.91 ***	0.86 ***	0.76 ***	0.75 ***	0.88 ***	−0.93 ***	0.46 ***	1.00 ***	−0.98 ***	0.90 ***	−0.97 ***	1.00		
PUFA/SFA	0.05	−0.65 ***	−0.84 ***	−0.89 ***	−0.90 ***	−0.83 ***	−0.86 ***	−0.91 ***	0.86 ***	−0.46 ***	−0.64 ***	0.78 ***	−0.90 ***	0.79 ***	−0.64 ***	1.00	
PUFA/MUFA	−0.32 *	0.85 ***	0.89 ***	0.94 ***	0.88 ***	0.80 ***	0.79 ***	0.91 ***	−0.96 ***	0.50 ***	0.99 ***	−0.99 ***	0.94 ***	−0.99 **	0.99 ***	−0.70 ***	1.00

*, **, ***, significant at $p < 0.05$, 0.01, and 0.001, respectively. SWP = minimum stem water potential; SI = stress integral; C16:0 (palmitic); C16:1 (palmitoleic); C17:0 (margaric); C17:1 cis (heptadecenoic); C18:0 (stearic); C18:1n9 (oleic); C18:1n7 (cis-vaccenic); C18:2n6 c9,12 (linoleic); O/L (oleic/linoleic); SFA (saturated fatty acids); MUFA (monounsaturated fatty acids); PUFA (polyunsaturated fatty acids).

A reduction in oleic, MUFA, and O/L ratio and an increase in linoleic, PUFA, and PUFA/MUFA ratio was also reported in almond cv. Marta, Guara, Lauranne, Ferragnes, Texas; olives cv. Mazanilla; pistachio cv. Kerman; and sunflower cv. Suncross [29,30,42,49,54,55].

The decrease in oleic and increase in linoleic in drought conditions was reported in many studies in different crops, although sometimes was cultivar dependent [54]. This effect of water stress on these two fatty acids was attributed to the enzyme Δ12 desaturase, which is responsible for the conversion of oleic acid in linoleic under water stress conditions [56].

As observed, PUFA is increased under water stress and similar results were reported in olives cv. Manzanilla [57] after two years of deficit irrigation. The authors reported that the higher the stress applied during stage III in olives, the greater the linoleic acid concentration and, consequently, (PUFA + MUFA)/SFA ratio, with a correlation of $R^2 = 0.71$ and $R^2 = 0.84$, respectively. An increase in linoleic acid may play an important role in the death of cardiac cells and is an essential fatty acid, which cannot be synthesized by human body [29]. It was reported that consuming 50 g of almonds under RDI conditions can cover approximately 33% of the daily intake of linoleic acid recommended by the European Food Safety Authority [29]. The present study showed an increase in PUFA and a decrease in MUFA with water stress, and this led to a low O/L rate. It is well known that a low O/L rate means almonds are more susceptible to oxidation, because this is initiated in the double bonds of PUFA [58]. However, it was also observed that water stress also enhances compounds with antioxidant activity (polyphenols, α-tocopherol, phytoprostanes, phytofurans, jasmonates, abscisic acid, etc., that are also enhanced by water stress) that might help in maintaining the PUFA in a cell's membrane, preserving its bioactivity [59].

Saturated fatty acids were also observed to increase in almonds under water stress conditions, and the American Heart Association (AHA) encourages people to replace SFA with MUFA for a healthy lifestyle and low-density lipoprotein (LDL) cholesterol levels reductions. Thus, controlling the stress in almond trees might help to reduce the SFA content, because other studies reported that moderate deficiency did not negatively affected SFA content in almond [25,29,53]. Moreover, the levels of SFA in almonds are so low that almonds as well as other nuts fits well in AHA guidelines [60]. In fact, Food and Drug Administration (FDA) implemented a healthy claim regarding the almonds and other nuts consumption, stating that diets containing ~42.5 g of almonds per day as part of a diet low in saturated fat, and cholesterol may reduce the risk cardiovascular diseases [61].

To conclude this section, the fatty acids were significantly affected by water stress and are good markers of the hydroSOStainable almonds.

3.6. Descriptive Sensory Analysis

Table 5 shows the Pearson's correlation coefficients (R) between SI and SWP with sensory attributes. These results highlighted strong negative correlations for the size, bitterness, astringency, benzaldehyde, and woody flavors. Previous studies reported that water stress might enhance the sweetness, nutty, almond ID, and crispiness in almonds cv. Lauranne and pistachio cv. Kerman [29,31]. Thus, an increase in sugars and a decrease in bitterness and astringency with water stress conditions as shown in this study might lead to sweeter almonds.

As previously described by Lipan et al. (2019) [19] and Carbonell-Barrachina et al. (2015) [31], the purchase choice of international consumers was based on sweetness, almond ID, pistachio ID, and crispiness. These findings together with those that consumers were willing to pay more for hydroSOStainable almonds [19], pistachios [20], and table olives [62], and the functional properties of the bioactive compounds described here encourage the almond farming sector to bet on deficit irrigation strategy to reduce irrigation water and simultaneously increase the functional and sensorial quality of almonds.

Table 5. Pearson's correlation coefficients (R) among stem water potential and stress integral with sensory analysis parameters.

	SWP	SI	Color	Size	Sweet	Bitter	Astr	Nutty	Al ID	Benz	Woody	Hardness	Crispiness	Aftertaste
SWP	1.00													
SI	−0.69 *	1.00												
Color	0.24	0.04	1.00											
Size	0.35	−0.90 ***	−0.09	1.00										
Sweet	−0.43	0.35	−0.68 *	−0.26	1.00									
Bitter	0.12	−0.62 *	0.27	0.78 **	−0.42	1.00								
Astr	0.32	−0.70 *	0.12	0.77 **	−0.29	0.51	1.00							
Nutty	0.04	−0.39	−0.45	0.83 ***	0.21	0.64 *	0.59 *	1.00						
Al ID	−0.09	−0.34	−0.84 *	0.40	0.64 *	0.03	0.27	0.78 **	1.00					
Benz	0.09	−0.60 *	−0.68 *	0.69 *	0.38	0.26	0.43	0.83 ***	0.83 ***	1.00				
Woody	0.16	−0.71*	−0.67 *	0.76 **	0.20	0.39	0.48	0.89 ***	0.82 ***	0.91 ***	1.00			
Hardness	−0.32	0.03	−0.81 ***	0.05	0.53	−0.34	0.10	0.36	0.73 **	0.57	0.59 *	1.00		
Crispiness	−0.17	−0.12	−0.95 ***	0.17	0.64 *	−0.29	0.05	0.48	0.85 ***	0.71 **	0.71 **	0.91 ***	1.00	
Aftertaste	−0.15	−0.29	−0.83 ***	0.36	0.51	−0.11	0.31	0.65 *	0.91 ***	0.82 ***	0.81 ***	0.86 ***	0.89 ***	1.00

*, **, ***, significant at $p < 0.05$, 0.01, and 0.001, respectively. SWP = minimum stem water potential; SI = stress integral; Sweet = sweetness; Bitter = bitterness; Astr = astringency; Al ID = almond ID; Benz = benzaldehyde like.

4. Conclusions

Globally, data presented here showed that water stress affected the functional and sensorial parameters of hydroSOStainable almonds showing positive correlations with dry weight, color coordinates (L^*, a^*, and b^*), minerals (K, Fe, and Zn), organic acids (citric acid), sugars (sucrose, fructose, and total sugars), antioxidant activity, and fatty acids (linoleic, PUFA, SFA, PUFA/MUFA, among others). On the other hand, the water stress in almonds was negatively correlated with kernel yield, water activity, weight (almond, kernel, and shell), size, minerals (Ca and Mg), fatty acids (oleic acids, oleic/linoleic ratio, MUFA, and PUFA/SFA), and sensory attributes (size, bitterness, astringency, benzaldehyde, and woody). Considering that moderate RDI led to kernel yields similar to the control, agricultural sector can save approximately 45% of the irrigation water obtaining high-quality products. The current long-term research helped to demonstrate which quality parameters are really affected by water stress conditions and to clarify which may be essential markers to distinguish hydroSOStainable almonds from other types of almonds. All these findings help the agro-food sector to understand (i) that is possible to increase the water use efficiency generating products with high functional and sensory quality; (ii) the need of controlling the water stress in plants for the best agronomical and quality responses; and (iii) to set up key agronomic and quality markers to control and establish whether the water stress created at the field/orchard significantly affected the quality and functionality of the final edible nuts.

Supplementary Materials: The following are available online at http://www.mdpi.com/2073-4395/10/10/1470/s1, Table S1: Mean values of morphological and chemical parameters of irrigation treatments (T1, T2, T3, and T4) for 3 years (2017, 2018, and 2019) and Table S2: Mean values of functional and sensorial parameters of irrigation treatments (T1, T2, T3, and T4) for 3 years (2017, 2018, and 2019).

Author Contributions: Conceptualization, L.L. and Á.A.C.-B.; methodology, L.L., F.H., M.C. and A.M.; software, L.L. and L.V.-A.; validation, Á.A.C.-B., A.M. and L.V.-A.; formal analysis, L.L. and M.C.-L.; investigation, L.L. and L.V.-A; resources, Á.A.C.-B. and A.M.; data curation, L.L.; writing—original draft preparation, L.L. and M.C.-L.; writing—review and editing, L.L., M.C., F.H. and E.S.; visualization, L.L. and Á.A.C.-B.; supervision, L.V.-A., E.S. and Á.A.C.-B., project administration, Á.A.C.-B. and A.M.; funding acquisition, Á.A.C.-B. and A.M. All authors have read and agreed to the published version of the manuscript.

Funding: This research was funded by the Spanish Government (Ministerio de Ciencia e Innovacion (MCI), Agencia Estatal de Investigacion (AEI), through a coordinated research project (hydroSOS) including the Universidad Miguel Hernández de Elche (AGL2016-75794-C4-1-R, Productos hidroSOStenibles: identificación de debilidades y fortalezas, optimización del procesado, creación de marca propia, y estudio de su aceptación en el mercado europeo, hydroSOS foods) and the Universidad de Sevilla (AGL2016-75794-C4-4-R); these projects have been also funded by Fondo Europeo de Desarrollo Regional (FEDER) "Una manera de hacer Europa", (MCI/AEI/FEDER, UE).

Acknowledgments: The author M. Cano-Lamadrid was funded by the Formación de Profesorado Universitario (FPU) grant from the Spanish Ministry of Education (FPU15/02158).

Conflicts of Interest: The authors declare no conflict of interest.

References

1. Vázquez-Araújo, L.; Verdú, A.; Navarro, P.; Martínez-Sánchez, F.; Carbonell-Barrachina, A.A. Changes in volatile compounds and sensory quality during toasting of Spanish almonds. *Int. J. Food Sci. Technol.* **2009**, *44*, 2225–2233. [CrossRef]

2. Ministerio de Agricultura Pesca y Alimentación MAPA. Informe del Consumo Alimentario en España. Available online: https://www.mapa.gob.esa (accessed on 15 May 2020).

3. Xiao, L.; Lee, J.; Zhang, G.; Ebeler, S.E.; Wickramasinghe, N.; Seiber, J.; Mitchell, A.E. HS-SPME GC/MS characterization of volatiles in raw and dry-roasted almonds (*Prunus dulcis*). *Food Chem.* **2014**, *151*, 31–39. [CrossRef]

4. Grand View Research. Almond Milk Market Size, Share & Trends Analysis Report by Application (Beverages, Personal Care), by Distribution Channel (Hypermarkets & Supermarkets, Convenience Stores, Online), and Segment Forecasts, 2019–2025. Available online: https://www.grandviewresearch.com (accessed on 21 September 2020).

5. Lipan, L.; Rusu, B.; Lajos, E.S.; Sendra, E.; Hernández, F.; Vodnar, D.C.; Corell, M.; Carbonell-Barrachina, Á. Chemical and sensorial characterization of spray dried hydroSOStainable almond milk. *J. Sci. Food Agric.* **2020**, in press. [CrossRef]

6. Lipan, L.; Rusu, B.; Sendra, E.; Hernández, F.; Vázquez-Araújo, L.; Vodnar, D.C.; Carbonell-Barrachina, Á. Spray drying and storage of probiotic-enriched almond milk: Probiotic survival and physicochemical properties. *J. Sci. Food Agric.* **2020**, *100*, 3697–3708. [CrossRef]

7. García-Tejero, I.F.; Gutiérrez Gordillo, S.; Souza, L.; Cuadros-Tavira, S.; Durán Zuazo, V.H. Fostering sustainable water use in almond (*Prunus dulcis Mill.*) orchards in a semiarid Mediterranean environment. *Arch. Agron. Soil Sci.* **2019**, *65*, 164–181. [CrossRef]

8. Food and Agriculture Organization of the United Nations FAO. Producción de Almendra con Cáscara (FAOSTAT). Available online: http://www.fao.org/faostat/es/ (accessed on 12 May 2020).

9. Ministerio de Agricultura Pesca y Alimentación MAPA. Datos Avances de Frutales no Cítricos y Frutales Secos Año. 2019. Available online: https://www.mapa.gob.es (accessed on 15 May 2020).

10. Egea, G.; Nortes, P.A.; Domingo, R.; Baille, A.; Perez-Pastor, A.; Gonzalez-Real, M.M. Almond agronomic response to long-term deficit irrigation applied since orchard establishment. *Irrig. Sci.* **2013**, *31*, 445–454. [CrossRef]

11. World Resource Institute WRI. 17 Countries, Home to One-Quarter of the World's Population, Face Extremely High Water Stress. Available online: https://www.wri.org (accessed on 15 May 2020).

12. Iglesias, A.; Garrote, L. Adaptation strategies for agricultural water management under climate change in Europe. *Agric Water Manag.* **2015**, *155*, 113–124. [CrossRef]

13. Lazzarini, G.A.; Visschers, V.H.M.; Siegrist, M. How to improve consumers' environmental sustainability judgements of foods. *J Clean Prod.* **2018**, *198*, 564–574. [CrossRef]

14. European Environment Agency Report EEA. In *Climate Change Adaptation In The Agriculture Sector In Europe*; EEA Report 4/2019; EEA: Copenhagen, Denmark, 2019; p. 112. Available online: https://www.eea.europa.eu/publications/cc-adaptation-agriculture (accessed on 15 May 2020).

15. De Ollas, C.; Morillón, R.; Fotopoulos, V.; Puértolas, J.; Ollitrault, P.; Gómez-Cadenas, A.; Arbona, V. Facing climate change: Biotechnology of iconic mediterranean woody crops. *Front Plant Sci.* **2019**, *10*. [CrossRef]

16. Intergovernmental Panel on Climate Change IPCC. Global Warming of 1.5 °C. Available online: http://www.ipcc.ch/report/sr15/ (accessed on 15 May 2020).

17. Du, T.; Kang, S.; Zhang, J.; Davies, W.J. Deficit irrigation and sustainable water-resource strategies in agriculture for China's food security. *J. Exp. Bot.* **2015**, *66*, 2253–2269. [CrossRef]

18. Goldhamer, D.A.; Viveros, M.; Salinas, M. Regulated deficit irrigation in almonds: Effects of variations in applied water and stress timing on yield and yield components. *Irrig. Sci.* **2005**, *24*, 101–114. [CrossRef]

19. Lipan, L.; Cano-Lamadrid, M.; Corell, M.; Sendra, E.; Hernández, F.; Stan, L.; Vodnar, D.C.; Vázquez-Araújo, L.; Carbonell-Barrachina, Á.A. Sensory profile and acceptability of hydrosostainable almonds. *Foods* **2019**, *8*, 64. [CrossRef]

20. Noguera-Artiaga, L.; Lipan, L.; Vázquez-Araújo, L.; Barber, X.; Pérez-López, D.; Carbonell-barrachina, Á. Opinion of spanish consumers on hydrosustainable pistachios. *J. Food Sci.* **2016**, *81*, S2559–S2565. [CrossRef]

21. Sánchez-Rodríguez, L.; Corell, M.; Hernández, F.; Sendra, E.; Moriana, A.; Carbonell-Barrachina, Á.A. Effect of Spanish-style processing on the quality attributes of HydroSOStainable green olives. *J. Sci. Food Agric.* **2019**, *99*, 1804–1811. [CrossRef]

22. Myers, B.J. Water stress integral-a link between short-term stress and long-term growth. *Tree Physiol.* **1988**, *4*, 315–323. [CrossRef]

23. Steduto, P.; Hsiao, T.C.; Fereres, E.; Raes, D. Crop yield response to water. In *FAO Irrigation and Drainage Paper*; FAO: Rome, Italy, 2012; p. 66.

24. McCutchan, H.; A Shackel, K. Stem-water potential as a sensitive indicator of water stress in prune trees (*Prunus domestica L.* cv. French). *J. Am. Soc. Hort. Sci.* **1992**, *117*, 607–611. [CrossRef]

25. Lipan, L.; Martín-Palomo, M.J.; Sánchez-Rodríguez, L.; Cano-Lamadrid, M.; Sendra, E.; Hernández, F.; Burló, F.; Vázquez-Araújo, L.; Andreu, L.; Carbonell-Barrachina, Á.A. Almond fruit quality can be improved by means of deficit irrigation strategies. *Agric Water Manag.* **2019**, *217*, 236–242. [CrossRef]

26. Lipan, L.; Moriana, A.; López Lluch, D.B.; Cano-Lamadrid, M.; Sendra, E.; Hernández, F.; Vázquez-Araújo, L.; Corell, M.; Carbonell-Barrachina, Á.A. Nutrition quality parameters of almonds as affected by deficit irrigation strategies. *Molecules* **2019**, *24*, 2646. [CrossRef]

27. Brand-Williams, W.; Cuvelier, M.E.; Berset, C. Use of a free radical method to evaluate antioxidant activity. *LWT-Food Sci. Technol.* **1995**, *28*, 25–30. [CrossRef]

28. Huang, G. Almond Shelf Life Factors. Available at: Technical Summary, Almond Board of California, 2014, 1–4. Available online: https://www.almonds.com/sites/default/files/content/ (accessed on 20 May 2020).

29. Lipan, L.; Garcia-Tejero, I.F.; Gutierrez-Gordillo, S.; Demirbas, N.; Sendra, E.; Hernandez, F.; Duran-Zuazo, V.H.; Carbonell-Barrachina, A.A. Enhancing nut quality parameters and sensory profiles in three almond cultivars by different irrigation regimes. *J. Agric. Food Chem.* **2020**, *68*, 2316–2328. [CrossRef]

30. Nanos, G.; Kazantzis, I.; Kefalas, P.; Petrakis, C.; Stavroulakis, G. Irrigation and harvest time affect almond kernel quality and composition. *Sci. Hort.* **2002**, *96*, 249–256. [CrossRef]

31. Carbonell-Barrachina, A.A.; Memmi, H.; Noguera-Artiaga, L.; Gijon-Lopez, M.D.; Ciapa, R.; Perez-Lopez, D. Quality attributes of pistachio nuts as affected by rootstock and deficit irrigation. *J. Sci. Food Agric.* **2015**, *95*, 2866–2873. [CrossRef] [PubMed]

32. Cano-Lamadrid, M.; Girón, I.F.; Pleite, R.; Burló, F.; Corell, M.; Moriana, A.; Carbonell-Barrachina, A.A. Quality attributes of table olives as affected by regulated deficit irrigation. *LWT-Food Sci. Technol.* **2015**, *62*, 19–26. [CrossRef]

33. Bolling, B.W. Almond Polyphenols: Methods of Analysis, Contribution to Food Quality, and Health Promotion. *Compr. Rev. Food Sci. Food Saf.* **2017**, *16*, 346–368. [CrossRef]

34. Li, X.; Lu, M.; Tang, D.; Shi, Y. Composition of carotenoids and flavonoids in narcissus cultivars and their relationship with flower color. *PLoS ONE* **2015**, *10*. [CrossRef] [PubMed]

35. Rauf, A.; Imran, M.; Abu-Izneid, T.; Iahtisham Ul, H.; Patel, S.; Pan, X.; Naz, S.; Sanches Silva, A.; Saeed, F.; Suleria, H.A.R.; et al. Proanthocyanidins: A comprehensive review. *Biomed. Pharmacother.* **2019**, *116*. [CrossRef]

36. Wan, H.; Yu, C.; Han, Y.; Guo, X.; Luo, L.; Pan, H.; Zheng, T.; Wang, J.; Cheng, T.; Zhang, Q. Determination of flavonoids and carotenoids and their contributions to various colors of rose cultivars (Rosa spp.). *Front Plant Sci.* **2019**, *10*. [CrossRef]

37. Yada, S.; Lapsley, K.; Huang, G. A review of composition studies of cultivated almonds: Macronutrients and micronutrients. *J. Food Compos. Anal.* **2011**, *24*, 469–480. [CrossRef]

38. Ahanger, M.A.; Morad-Talab, N.; Abd-Allah, E.F.; Ahmad, P.; Hajiboland, R. Plant growth under drought stress: Significance of mineral nutrients. In *Water Stress and Crop Plants: A Sustainable Approach*; John Wiley & Sons, Ltd.: Hoboken, NJ, USA, 2016; Volume 2, pp. 649–668.

39. Official Journal of the European Union EU. Regulation (EU) No 1169/2011 of the European Parliament and of the Council. Available online: https://eur-lex.europa.eu (accessed on 22 May 2020).

40. Karim, M.; Zhang, Y.Q.; Zhao, R.R.; Chen, X.P.; Zhang, F.S.; Zou, C.Q. Alleviation of drought stress in winter wheat by late foliar application of zinc, boron, and manganese. *J. Plant. Nutr. Soil. Sci.* **2012**, *175*, 142–151. [CrossRef]

41. Ashrafi, M.; Azimi-Moqadam, M.R.; Moradi, P.; MohseniFard, E.; Shekari, F.; Kompany-Zareh, M. Effect of drought stress on metabolite adjustments in drought tolerant and sensitive thyme. *Plant Physiol. Biochem.* **2018**, *132*, 391–399. [CrossRef]

42. Sanchez-Bel, P.; Egea, I.; Martinez-Madrid, M.C.; Flores, B.; Romojaro, F. Influence of irrigation and organic/inorganic fertilization on chemical quality of almond (*Prunus amygdalus* cv. *Guara*). *J. Agric. Food Chem.* **2008**, *56*, 10056–10062. [CrossRef] [PubMed]

43. Egea, G.; Gonzalez-Real, M.M.; Baille, A.; Nortes, P.A.; Sanchez-Bel, P.; Domingo, R. The effects of contrasted deficit irrigation strategies on the fruit growth and kernel quality of mature almond trees. *Agric. Water Manag.* **2009**, *96*, 1605–1614. [CrossRef]

44. Peckmann, K.; Herppich, W.B. Effects of short-term drought and rewatering on the activity of mitochondrial enzymes and the oxidative capacity of leaf mitochondria from a CAM plant, Aptenia cordifolia. *J. Plant Physiol.* **1998**, *152*, 518–524. [CrossRef]

45. Thakur, A.; Singh, Z. Responses of 'Spring Bright' and 'Summer Bright' nectarines to deficit irrigation: Fruit growth and concentration of sugars and organic acids. *Sci. Hortic.* **2012**, *135*, 112–119. [CrossRef]

46. Kashem, M.A.; Sultana, N.; Ikeda, T.; Hori, H.; Loboda, T.; Mitsui, T. Alteration of starch- sucrose transition in germinating wheat seed under sodium chloride salinity. *J. Plant Biol.* **2000**, *43*, 121–127. [CrossRef]

47. Jahan, A.; Komatsu, K.; Wakida-Sekiya, M.; Hiraide, M.; Tanaka, K.; Ohtake, R.; Umezawa, T.; Toriyama, T.; Shinozawa, A.; Yotsui, I.; et al. Archetypal roles of an abscisic acid receptor in drought and sugar responses in liverworts. *Plant Physiol.* **2019**, *179*, 317–328. [CrossRef]

48. Sánchez-Rodríguez, L.; Cano-Lamadrid, M.; Carbonell-Barrachina, Á.A.; Wojdyło, A.; Sendra, E.; Hernández, F. Polyphenol profile in manzanilla table olives as affected by water deficit during specific phenological stages and spanish-style Processing. *J. Agric. Food Chem.* **2019**, *67*, 661–670. [CrossRef]

49. Noguera-Artiaga, L.; Sánchez-Bravo, P.; Pérez-López, D.; Szumny, A.; Calin-Sánchez, Á.; Burgos-Hernández, A.; Carbonell-Barrachina, Á.A. Volatile, sensory and functional properties of hydrosos pistachios. *Foods* **2020**, *9*, 158. [CrossRef]

50. Ali, M.S.; Baek, K.-H. Jasmonic acid signaling pathway in response to abiotic stresses in plants. *Int. J. Mol. Sci.* **2020**, *21*, 621. [CrossRef]

51. Wu, X.; Beecher, G.R.; Holden, J.M.; Haytowitz, D.B.; Gebhardt, S.E.; Prior, R.L. Lipophilic and hydrophilic antioxidant capacities of common foods in the United States. *J. Agric. Food Chem.* **2004**, *52*, 4026–4037. [CrossRef]

52. United States Department of Agriculture USDA. Food Data Central. 2019. Available online: https://fdc.nal.usda.gov/fdc-app.html (accessed on 15 May 2020).

53. Zhu, Y.; Taylor, C.; Sommer, K.; Wilkinson, K.; Wirthensohn, M. Influence of deficit irrigation strategies on fatty acid and tocopherol concentration of almond (*Prunus dulcis*). *Food Chem.* **2015**, *173*, 821–826. [CrossRef] [PubMed]

54. Ali, Q.; Ashraf, M.; Anwar, F. Physico-chemical attributes of seed oil from drought stressed sunflower (Helianthus annuus L.) plants. *Grasas y Aceites* **2009**, *60*, 475–481. [CrossRef]

55. Noguera-Artiaga, L.; Sánchez-Bravo, P.; Hernández, F.; Burgos-Hernández, A.; Pérez-López, D.; Carbonell-Barrachina, Á.A. Influence of regulated deficit irrigation and rootstock on the functional, nutritional and sensory quality of pistachio nuts. *Sci. Hortic.* **2020**, *261*. [CrossRef]

56. Baldini, M.; Giovanardi, R.; Tahmasebi Enferadi, S.; Vannozzi, G. Effects of water regime on fatty acid accumulation and final fatty acid composition in the oil of standard and high oleic sunflower hybrids. *Ital. J. Agron.* **2002**, *6*, 119–126.

57. Sánchez-Rodríguez, L.; Lipan, L.; Andreu, L.; Martín-Palomo, M.J.; Carbonell-Barrachina, Á.A.; Hernández, F.; Sendra, E. Effect of regulated deficit irrigation on the quality of raw and table olives. *Agric. Water Manag.* **2019**, *221*, 415–421. [CrossRef]

58. Ros, E.; Mataix, J. Fatty acid composition of nuts-Implications for cardiovascular health. *Br. J. Nutr.* **2006**, *96*, S29–S35. [CrossRef]

59. Traber, M.G.; Atkinson, J. Vitamin E, antioxidant and nothing more. *Free Radic. Bio. Med.* **2007**, *43*, 4–15. [CrossRef]

60. Jenkins, D.J.A.; Kendall, C.W.C.; Marchie, A.; Parker, T.L.; Connelly, P.W.; Qian, W.; Haight, J.S.; Faulkner, D.; Vidgen, E.; Lapsley, K.G.; et al. Dose response of almonds on coronary heart disease risk factors: Blood lipids, oxidized low-density lipoproteins, lipoprotein(a), homocysteine, and pulmonary nitric oxide: A randomized, controlled, crossover trial. *Circulation* **2002**, *106*, 1327–1332. [CrossRef]

61. U.S. Food and Drug Administration FDA. Qualified Health Claims: Letter of Enforcement Discretion-Nuts and Coronary Heart Disease (Docket No 02P-0505). Available online: https://www.fda.gov/food/food-labeling-nutrition/qualified-health-claims-letters-enforcement-discretion (accessed on 15 May 2020).

62. Sánchez-Rodríguez, L.; Cano-Lamadrid, M.; Carbonell-Barrachina, Á.A.; Sendra, E.; Hernández, F. Volatile composition, sensory profile and consumer acceptability of hydrosostainable table olives. *Foods* **2019**, *8*, 470. [CrossRef]

Article

Screening of Popcorn Genotypes for Drought Tolerance Using Canonical Correlations

Samuel Henrique Kamphorst [1,*], Gabriel Moreno Bernardo Gonçalves [1,*], Antônio Teixeira do Amaral Júnior [1], Valter Jário de Lima [1], Jhean Torres Leite [1], Kátia Fabiane Medeiros Schmitt [1], Divino Rosa dos Santos Junior [1], Juliana Saltires Santos [1], Fábio Tomaz de Oliveira [1], Caio Cézar Guedes Corrêa [1], Weverton Pereira Rodrigues [1,2] and Eliemar Campostrini [1]

[1] Laboratório de Melhoramento Genético Vegetal, Centro de Ciências e Tecnologias Agropecuárias, Universidade Estadual Norte Fluminense Darcy Ribeiro (UENF), Campos dos Goytacazes, Rio de Janeiro 28013-602, Brazil; amaraljr@uenf.br (A.T.d.A.J.); valter_jario@hotmail.com (V.J.d.L.); torresjhean@gmail.com (J.T.L.); kmedeirosschmitt@gmail.com (K.F.M.S.); juniorifagro@gmail.com (D.R.d.S.J.); julianasaltiresdossantos@yahoo.com.br (J.S.S.); tomaz_oft@yahoo.com.br (F.T.d.O.); caiocagronomo@gmail.com (C.C.G.C.); wevertonuenf@hotmail.com (W.P.R.); campostenator@gmail.com (E.C.)

[2] Centro de Ciências Agrárias, Naturais e Letras, Universidade Estadual da Região Tocantina do Maranhão, Estreito, Maranhão 65975-000, Brazil

* Correspondence: samuelkampho@hotmail.com (S.H.K.); gabriel.agrobio@gmail.com (G.M.B.G.); Tel.: +55-(22)-3248-6025 (S.H.K. & G.M.B.G.)

Received: 27 August 2020; Accepted: 1 October 2020; Published: 6 October 2020

Abstract: Getting around the damage caused by drought is a worldwide challenge, particularly in Brazil, given that economy is based on agricultural activities, including popcorn growing. The purpose of this study was to evaluate popcorn inbred lines under water stressed (WS) and well-watered (WW) conditions regarding agronomic attributes, root morphology, and leaf "greenness" index (SPAD index), besides investigating the viability of indirect selection by canonical correlations (CC) of grain yield (GY) and popping expansion (PE). Seven agronomic, six morphological root traits were evaluated and SPAD index at five different dates during grain filling. The WS (−29% less water than WW) affected significantly the GY (−55%), PE (−28%), increased the brace and crown root density, and more vertically oriented the brace and crown angles. Higher SPAD index is associated with a higher yield, and these measures were the only ones with no significant genotype × water condition interaction, which may render concomitant selection for WS and WW easier. For associating the corrections of the different traits, CC proved to have better potential than simple correlations. Thus, the evaluation of SPAD index at 29 days after the anthesis showed the best CC, and based on the previous results of SPAD index, may be used regardless of the water condition.

Keywords: leaf greenness index; root morphology; water stress

1. Introduction

In Brazil, popcorn is a much appreciated and consumed food, especially associated with leisure time and movie theaters [1–5]. In view of the increased consumption of this type of grain, it will be necessary to expand the growing area of this crop and enhance genetic values to the cultivars to meet the rising demand for the product [2,3,6]. There is still a wide market to be developed for the growing of popcorn in Brazil, as the area planted has increased by 223% in the last five years [7].

Getting around the drought losses, which will be ever more frequent [8] due to climate change resulting from man-made actions [9], is a challenge for world agribusiness. For Brazil, whose economy is strongly based on large-scale agricultural activities, and annual advances towards new areas with unpredictable rain regimes, the negative effects are supposed to be worse. Drought is the abiotic factor that most limits the agricultural productivity of maize [10–14]. In Brazil, agricultural production is concentrated in the second growing season (February to July), after the soybean harvest [15]. This period also happens during the dry seasons and may contain strong variations in rainfall [8]. Consequently, there is urgency in selecting high productive genotypes under drought conditions, as well as identifying efficient traits for genotypic discrimination in environments with water stress.

Induction of leaf senescence is a mechanism of response of plants to water stress conditions [12,14]. Under soil water stress conditions, it is often seen that maize genotypes with delayed leaf senescence—stay green—are the highest productive ones, and that is why breeders select them as germplasm adapted to the soil water stress condition [16–20]. As stated by some authors, stay-green cultivars are the best option for drought-prone environments [21,22]. This characteristic can be easily measured by means of a spectrometric method, using SPAD index, an important tool for diagnosing plant stress [13,23–25]. Choosing the proper phenological stage for "greenness" index measurements is a major bottleneck, which is why the effective application of phenotyping methodologies has to do, in part, with the ability to apply them in critical stages associated with grain production [19].

When there are environmental constraints intrinsic to the soil—either water or nutritional—a good conformation of the root system architecture is essential for greater agricultural yield [26]. A root system adapted to the specific conditions of abiotic soil stress represents an agronomic advantage [27]. Maize genotypes with a deep and branched axial root system can access water in deeper layers, where water availability is greater [28]. According to Gao and Lynch (2016) [29], having fewer crown roots improves the water uptake capacity in maize plants under water stress.

As for the popcorn breeding for drought adaptation, in order to identify promising variables for indirect selection of grain yield and popping expansion, Kamphorst et al. (2019) [30] adopted the path analysis, and using the supertrait expanded popcorn volume per hectare as a dependent variable, they sought to get around the inconvenience of considering only one dependent variable for path analysis [31]. On the other hand, as it is a method of linear correlation between two multidimensional variables, the analysis of canonical correlations (CCA) turns out to be a more adequate statistical tool to uncover the relationships of the independent variables, when the focus is to quantify the association to more than one dependent variable. Moreover, the CCA takes into account the maximum correlation between two variable complexes [32], being a more viable option in comparison with the simple linear correlation, besides its implementation avoiding the obtaining of Type I error results [33].

The present study aimed at investigating the impact of water stress on agronomic traits, leaf greenness index, and root architectural traits of 20 popcorn inbred lines in two water conditions, and investigating the viability of using canonical correlations for the selection of easy-to-measure variables that are efficient predictors of grain yield and popping expansion.

2. Materials and Methods

2.1. Plant Material

It was evaluated a total of 20 popcorn lines (S_7), whose genealogy is derived from germplasm with tropical (L61, L63, L65, L69, L70, and L71—from the "BRS-Angela" population) and temperate/tropical climate adaptation (P1, P5, and P7—from the hybrid "Zélia"; P2 and P3—from the compound "CMS-42"; P4—from South American breeds; P6, P8, and P9—from the hybrid "IAC-112"; L54, L55, and L59—from the population "Beija Flor"; and L75 and L76—from the population "Barão de Viçosa") [34].

2.2. Experimental Design, Cultural Traits, and Water Conditions Applied

The experiments were conducted at the Experimental Station of the Colégio Estadual Agrícola Antônio Sarlo, in Campos dos Goytacazes city, Rio de Janeiro State, Brazil (Latitude 21°42′48″ S, Longitude 41°20′38″ O, 14 m altitude). The soil of the Experimental Station has been classified as dystrophic Ultisol, presenting high levels of clay and silt. Trials were performed in the 2016 crop year, during the dry seasons, which include the period from April to August. Sowing was done on 10 April, and harvesting was carried out on 15 August.

The basal dressing application of fertilizers was made using 30 kg ha^{-1} of N (in the form of urea), 60 kg ha^{-1} of P_2O_5 (triple superphosphate), and 60 kg ha^{-1} of K_2O (potassium chloride). Top-dressing fertilization occurred 30 days after sowing (DAS), and 100 kg ha^{-1} of N (in the form of urea) were applied.

Experimental plots comprised four 4.40 m long rows spaced 0.20 m apart between plants and 0.80 m between rows, totaling a density of 62,500 plants per hectare. The useful area of the plot was defined by 6.72 m^2 of the central rows. An experimental randomized complete block design with three replicates in each water condition; that is, well-watered (WW) and water stressed (WS) was conducted. The analyses of variance were performed considering genotypes effect as fixed and block affect as random.

Irrigation was used for both water regimes by using a drip system. One Katif dripper per plant was installed with a flow rate of 2.3 mm h^{-1}. Water condition 1—well-watered (WW) was irrigated to maintain field capacity (-10 kPa), which was monitored using Decagon MPS-6 (Decagon, USA) tensiometers applied to the soil between plants at a depth of 0.20 m. Water condition 2—water stress (WS) was characterized by the suspension of irrigation at a phenological stage of male pre-anthesis. The soil reached the permanent wilting point (-1500 kPa) 12 days after the male anthesis (data not presented).

In WS condition, irrigation was interrupted 49 days after sowing (DAS), remaining this way until harvest (119 DAS). At 56, 70, 77, 105, 112, and 119 DAS, rainfall was recorded, totaling 92 mm during this period. Plants of WW condition were given an additional 78 mm of water via drip during the same period, totaling 170 mm (Figure 1a).

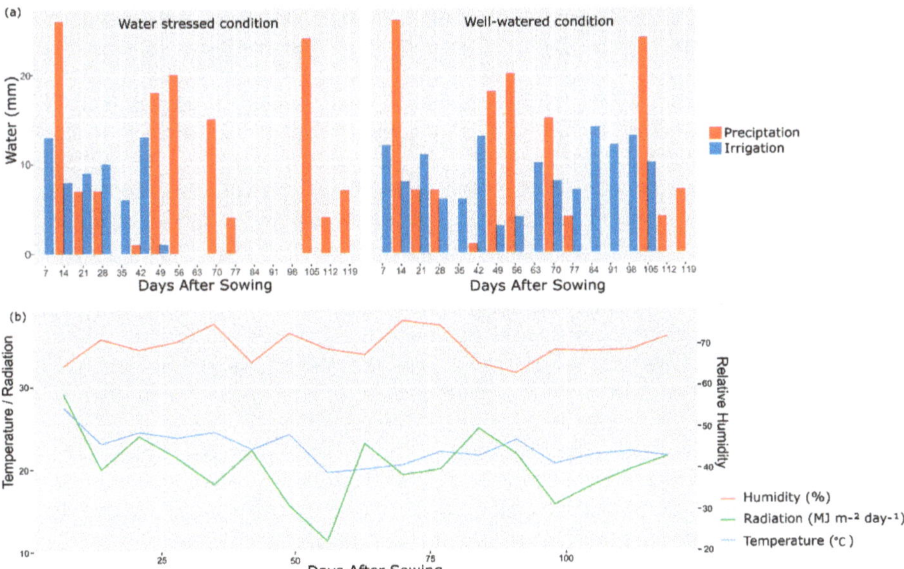

Figure 1. (a) Precipitation occurs during a trial conducted with 20 popcorn lines under well-watered condition and water-stressed condition. (b) Humidity, solar radiation, and temperature values measured in water stressed and well-watered conditions.

Throughout the crop growth and development cycle, temperature and relative humidity ranged from 12 °C to 37 °C and 23% to 97%, respectively, with a mean solar radiation of 20.35 MJ m^{-2} day^{-1} (maximum radiation at the experiment location was 1300 μmol m^{-2} s^{-1} (Figure 1b). Weather conditions were recorded in a weather station of the National Institute of Meteorology (INMET, in Portuguese), installed less than 100 m away from the experimental area.

2.3. Traits Evaluated

Traits were divided into three groups, or trait complex. The reference group, i.e., the so-called agronomic traits (AGRO), is the one related to grain yield and popping expansion. The group of root morphological traits (ROOT) includes the ones measured in the root system, specifically in the brace and crown roots. The third group is related to the intensity of leaf greenness (SPAD index) measured on different dates during the grain filling period.

The agronomic traits (AGRO) evaluated were grain yield (GY), popping expansion (PE), prolificacy (PR), 100-grain mass (HG), ear diameter (ED), ear length (EL), and dry matter weight (DM).

Grain yield (GY), obtained by harvesting all the plants in the useful area of the plot, was corrected to 13% humidity, and expressed in kg ha^{-1}. Popping expansion (PE) was measured for a mass of 30 g of grains, heated by microwaves, in a special paper bag for popping, at a 1000-watt power, for a time of two minutes and fifteen seconds, and the volume of popcorn was measured in a 2000 mL graduated cylinder. Popping expansion (PE) was determined by the quotient of the volume obtained from popcorn and grain mass expressed in mL g^{-1}. Prolificacy (PR) was established by counting the ears and expressed in mean values by the quotient of the number of ears harvested and the total number of plants in the plot. The 100-grain mass (HG) was obtained by weighing (g) three subsamples of 100 seeds each. A random sample of six plants from the plot was used to estimate the ear diameter (ED) and the ear length (EL), quantified by a caliper (cm), and to determine the aerial dry matter (DM) by weighing (g) after drying in a hot air oven at 70 °C for 72 h.

The root system traits evaluated were brace root number (BN), brace root angle (BA), brace root density (BD), crown root number (CN), crown root angle (CA), and crown root density (CD).

The root architectural traits were quantified following the methodology suggested by Traschel et al. (2011) [27], with modifications. After harvesting, the soil of the different water conditions (WW and WS) received 50-mm irrigation, to make it easier to remove the plants mechanically with the use of shovels. The root system of two plants per plot was removed in a 40-cm diameter and 25-cm deep soil cylinder. The roots in the soil cylinders were washed until the soil was completely removed. From the root categories—brace (B) and crown (C), the traits measured were root angle (A) (BA and CA), obtained using a degree protractor and expressed in relation to the soil (°); root number (N) (BN and CN), obtained by counting the structures; and mean density (D) of lateral roots (BD and CD), obtained using a diagrammatic scale proposed by Trachsel et al. (2011) [27]. Root density values vary from 1 to 9, in which higher values indicate higher density.

The trait associated with the intensity of leaf greenness (SPAD index) was estimated by three readings in the middle third of the third leaf counted from the apex and below the flag leaf by means of the portable chlorophyll meter, model SPAD-502 "Soil Plant Analyzer Development" (Minolta, Japan). The SPAD index were measured on five different days after male anthesis (DAA), as follows: 17 DAA (S_1), 22 DAA (S_2), 29 DAA (S_3), 36 DAA (S_4), and 42 DAA (S_5). Measurements were taken when the soil water potential was below the permanent wilting point (>-1500 kPa) (data not provided). SPAD index was measured in a random sample of six useful plants of the experimental plot.

2.4. Analysis of Variance and Statistical—Genetic Parameters

The analysis of variance was individually performed for each water condition (WS and WW) taking into account the linear model: $y_{ij} = \mu + g_i + b_j + e_{ij}$, in which the mean ($\mu$) and genotype effect ($g_i$) were considered to be fixed; and block (b_j) and error (e_{ij}), random.

The statistical-genetic parameters coefficient of variation (*CV*) and heritability (H^2) were estimated from the individual analyses by the estimators: $CV = \sqrt{MSe}/\mu \times 100$, in which *MSe* is the mean square of error; and $H^2 = MSg - MSe/MSg$, in which *MSg* is the mean square of the genotype effect.

To verify the occurrence of genotype interaction with water conditions, a joint analysis was carried out by considering the following linear model: $y_{ijk} = \mu + g_i + w_j + gw_{ij} + b/w_{jk} + e_{ijk}$, in which the mean ($\mu$), the genotype effect ($g_i$), and the water condition effect (w_j) were considered to be fixed, and the block within condition effect (b/w_{jk}) and the error (e_{ijk}), as random.

2.5. Correlation Analysis

For each water condition, a simple linear correlation analysis was applied using the Pearson method, involving all variables between themselves using the following model: $r = \sigma^2_{xy}/\sigma^2_x \times \sigma^2_y$, in which *r* is the phenotypic correlation coefficient between variables *x* and *y*, σ^2_{xy} is the phenotypic covariance between variables *x* and *y*, σ^2_x is the phenotypic variance of *x*, and σ^2_y is the phenotypic variance of *y*.

Existing correlations between trait complex were estimated by means of the canonical correlation analysis, taking into account the following groups: AGRO × ROOT traits and AGRO × SPAD traits.

The canonical correlations were estimated by combining the covariances within and between the traits of the variable complexes of the two groups (S_{11}, S_{22}, and S_{12}) and the weight of the traits of each group: $a' = [a_1\ a_2\ \dots\ a_p]$ is the vector $1 \times p$ of weights of the traits of group I; and $b' = [b_1\ b_2\ \dots\ b_q]$ is the vector $1 \times q$ of weights of the traits of group II.

The first canonical correlation is defined by that which maximizes the relationship between X_1 and Y_1 by means of the expression:

$$r_1 = C\hat{o}v(X_1, Y_1)/\sqrt{\hat{V}(X_1).\ \hat{V}(Y_1)}, \text{ in which } C\hat{o}v(X_1, Y_1) = a'S_{12}b;\ \hat{V} = a'S_{11}a;\ \hat{V} = b'S_{22}b.$$

The simple linear and canonical correlations were separately estimated for the means obtained in water condition (WS and WW), so they could be compared. The multicollinearity diagnostic among the traits in each pair of variable complexes was carried out based on the ratio between the highest and lowest eigenvalue of the inverse matrix and interpreted by the classification of Montgomery and Peck (1981) [35]. All statistical analyses were conducted in the GENES software [36].

3. Results

3.1. Genetic Variability and Effects of Different Water Conditions on SPAD Index and Agronomic (AGRO) and Root (ROOT) Traits

There is genetic variability among genotypes evaluated in both water conditions (WS and WW) for all agronomic traits, for the different times of greenness index measurements (SPAD index) and for traits related to the root system, except BN in WW condition. The experimental coefficient of variation (CV) was below 30.0% for the traits under evaluation. Generally, it was noted that the CV values were lower for the agronomic and root system traits under WS condition and for the different moments of SPAD index under WW condition (Table 1).

Table 1. Summary of analysis of variance, means, coefficients of variation (CV), and coefficient of genotypic determination of agronomic traits (H^2), different dates of measurement of greenness (SPAD index), and traits related to the root system evaluated in popcorn lines under water-stressed (WS) and well-watered (WW) conditions.

Traits	Water Condition (WC)												Mean Squares (MS) of Joint Analysis	
	Water Stressed (MS)						Well Watered (MS)						WC	Genotype × WC
	Block	Genotype	Error	Mean	CV%	H^2	Block	Genotype	Error	Mean	CV%	H^2		
	(DF = 2)	(DF = 19)	(DF = 38)				(DF = 2)	(DF = 19)	(DF = 38)				(DF = 1)	(DF = 19)
Agronomic traits (AGRO)														
GY	137418	55,9525 **	66,624	1139.11	22.65	88.09	60,1287	1,839,610 **	336,794	2548.07	22.77	81.69	5.9×10^7 **	70.2×10^5 *
PE	2.75	36.4 **	3.35	20.91	8.75	90.79	2.52	52.27 **	16.39	29.35	13.79	68.62	2136.4 **	28.35 **
PR	0.00	0.031 *	0.01	0.86	14.08	53.22	0.10	0.060 **	0.02	1.02	12.32	73.64	0.785 *	0.073 **
HG	0.89	5.533 **	0.53	9.69	7.49	90.44	1.33	5.52 **	1.72	12.67	10.36	68.75	265.9 **	1.71 ns
ED	9.04	20.06 **	2.75	27.94	5.94	86.27	10.59	17.70 **	2.39	29.12	5.31	86.46	41.8 **	3.45 **
EL	3.38	6.18 **	1.14	12.62	8.46	81.49	2.97	5.57 **	0.81	12.72	7.10	85.34	0.25 ns	0.79 ns
DM	521.8	6812.3 **	1807.8	313.94	13.54	73.46	21,931.2	10,358.4 **	2557.9	368.31	13.73	75.31	8.8×10^4 *	2614 ns
SPAD index														
S1(17DAA)	33.39	31.77 **	9.35	43.54	7.02	70.55	15.79	53.35 **	4.66	48.83	4.42	91.25	838 **	6.40 ns
S2(22DAA)	17.23	63.53 **	8.45	40.32	7.21	86.69	9.03	38.32 **	7.10	48.96	5.44	81.47	2239 **	9.81 ns
S3(29DAA)	1.52	64.42 **	8.88	39.70	7.50	86.20	24.45	37.26 **	4.59	48.15	4.45	87.67	2145 **	7.68 ns
S4(36DAA)	30.62	100.18 **	22.51	32.50	14.59	77.52	34.45	59.94 **	9.69	45.98	6.77	83.82	5452 **	24.11 ns
S5(42DAA)	39.71	82.73 **	24.20	18.81	26.15	70.74	49.96	108.62 **	13.30	39.35	9.26	87.75	12663 **	45.84 **
Root traits (ROOT)														
BN	0.53	6.472 **	0.50	9.12	7.75	92.27	2.77	3.94 ns	2.79	8.67	19.25	29.27	6.03 ns	5.15 **
BA	14.82	71.01 **	8.12	50.23	5.67	88.55	7.89	243.40 **	31.45	37.20	15.07	87.07	5094 **	48.73 **
BD	0.00	4.002 **	0.16	5.98	6.75	95.91	0.14	4.276 **	0.31	5.26	10.66	92.63	15.95 **	1.43 **
CN	2.60	14.82 **	1.61	16.29	7.78	89.13	1.02	7.16 **	1.52	15.31	8.05	78.74	28.8 ns	6.01 **
CA	4.33	83.81 **	9.08	53.08	5.67	89.16	9.30	166.32 **	11.10	42.63	7.81	93.32	3276 **	38.84 **
CD	0.07	4.737 **	0.14	5.62	6.64	97.05	0.40	2.836 **	0.20	4.12	10.86	92.91	67.45 **	1.76 **

MS: mean squares; WC: water condition; DF: degrees of freedom; GY: grain yield (Kg ha^{-1}); PE: popping expansion (mL g^{-1}); PR: prolificacy; HG: 100-grain mass (g); ED: ear diameter (mm); EL: ear length (cm); DM: dry matter (g); S1, S2, S3, S4 e S5 = SPAD index at 17, 22, 29, 36 and 42 days after male anthesis, respectively; root traits: BN: brace number; BA: brace angle (°); BD: brace density; CN: crown number; CA: crown angle (°) and CD: crown density. Symbols ns, *, and ** represent not significant, significant at 5.0%, and 1.0% probability, by the F test, respectively.

Among the agronomic traits, grain yield (GY), popping expansion (PE), and 100-grain mass (HG) presented higher heritability (H^2) under WS condition, while the other ones—namely, prolificacy (PR), ear diameter (ED), ear length (EL), and dry matter weight (DM)—presented higher H^2 under WW condition. Under the WS condition, the means of GY and PE were reduced by 55.3% and 28.7% compared with WW condition, respectively. Only the EL did not present a significant difference in mean between the different water conditions (WC). The means of PR, ED, and DM were significantly lower under WS condition, in the order of 15.7%, 4.1%, and 14.8%, respectively (Table 1). Significant genotype × water condition (Gen × WC) interactions were observed in GY, PE, PR, and ED (Table 1).

The SPAD index, under WW condition, presented higher values of H^2 in the different measurement dates regarding WS condition. Given the comparison of WS and WW condition, the proportional mean reductions were 10.8% (S_1), 17.6% (S_2), 17.5% (S_3), 29.3% (S_4), and 52.2% (S_5) for the measurements taken 17, 22, 29, 36, and 42 days after the anthesis (DAA). In the joint analysis, a difference was seen between the means of SPAD index between water conditions, with significance in the Gen × WC interaction only in the measurement carried out at 42 days (S_5) (Table 1).

With respect to root traits (ROOT), apart from the crown root angle (CA), H^2 were superior under WS condition. The mean of brace root (BN) and crown (CN) number did not differ in both water conditions; all other root traits, however, expressed means with higher values under WS condition. In the joint analysis, a significant Gen × WC interaction was noted for all root traits (Table 1).

3.2. Phenotypic Correlations within and among AGRO, SPAD Index, and ROOT Trait Groups under WS and WW Conditions

Correlations of greater expression, all in positive direction, among the traits of the AGRO group were seen between GY × ED (0.43), GY × EL (0.40), and ED × DM (0.42), under WS condition, and between GY × ED (0.50), GY × EL (0.52), PE × PR (0.40), ED × DM (0.49), and ED × DM (0.49), under WW condition (Figure 2).

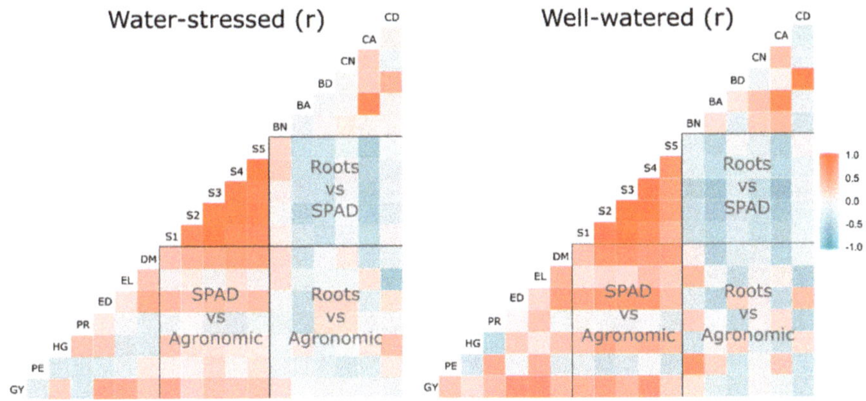

Figure 2. Phenotypical correlation among 18 traits evaluated in 20 popcorn inbred lines under two different water conditions. Agronomic traits—GY: grain yield; PE: popping expansion; PR: prolificacy; HG: 100-grain mass; ED: ear diameter; EL: ear length and DM: dry matter). SPAD index—S_1, S_2, S_3, S_4, and S_5 = SPAD index at 17, 22, 29, 36, and 42 days after male anthesis, respectively; Roots traits—BN: brace number; BA: brace angle; BD: brace density; CN: crown number; CA: crown angle and CD: crown density.

Correlations among the traits of the AGRO and SPAD index groups were more expressive under WW condition. All associations observed between GY × S_4 (0.51), ED × S_2 (0.53), ED × S_3 (0.54), and ED × S_4 (0.53) were positive. Under WS condition, the most intense association estimate occurred between GY × S_1 (0.40).

The correlation of greater expression between the traits of the AGRO and ROOT group, under WW condition, in positive direction, occurred between PE × BN (0.54) and, under WS condition, in negative direction, between EL × CD (−0.46) (Figure 2).

Correlations between the traits of the SPAD index and ROOT group mainly denote negative associations under WW condition. The leaf greenness index, measured 22 DAA (S_2), was moderately correlated with CA (−0.46), and the one measured 29 DAA (S_3) was more intensely associated with BA (−0.48) and CA (−0.53). Under WS condition, the CA trait was correlated with S_2 (22 DAA), S_3 (29 DAA), and S_5 (42 DAA), with values of −0.40, −0.39, and −0.45, respectively (Figure 2).

The traits of the SPAD group showed high correlations between themselves, irrespective of the water condition. Regarding the ROOT group, it was observed a high correlation between BA × CA, under both water conditions, (≈0.72), while the correlation between BD × CD (0.82) was expressive only under WW condition.

3.3. Canonical Correlations among AGRO, SPAD Index, and ROOT Trait Groups in WS and WW Conditions

The multicollinearity classification applied to the phenotypic correlation matrix between the two pairs of variable complexes presented a week multicollinearity for the pair AGRO and ROOT, and moderate for the pair AGRO and SPAD. For that reason, the canonical correlation analysis (CCA) could be carried out with the traits evaluated [37].

There were significant correlations between AGRO and SPAD index groups of traits under both water conditions (WS and WW) (Table 2). The correlation was significant in the first three canonical pairs under WS condition. In the first canonical pair (r = 0.92), the variable DM was associated in the opposite direction with the SPAD index in S_1 (17 DAA) and in the same direction with S_2 (22 DAA). The second canonical pair (r = 0.74) associated higher GY values and lower EL values with higher S_1 values and lower S_3 values. The third canonical pair (r = 0.68) associated PE with all measurement dates of the SPAD index, pointing out that higher values in S_1, S_3, and S_5 and lower S_2 and S_4 were related to the lower PE (Table 2).

Table 2. Correlation and coefficients of canonical pairs estimated between SPAD index and agronomic traits in 20 popcorn inbred lines under two different water conditions.

| | Traits | Canonical Pairs | | | | | | | | | |
| | | Water-Stressed | | | | | Well-Watered | | | | |
		1st	2nd	3rd	4th	5th	1st	2nd	3rd	4th	5th
SPAD index	S1(17DAA)	−1.16	1.74	1.19	−0.13	0.12	1.14	0.82	−0.34	1.43	−0.49
	S2(22DAA)	2.14	−0.07	−0.92	−1.42	2.18	0.01	−1.22	1.41	0.61	−1.63
	S3(29DAA)	−0.33	−1.64	1.05	2.04	−1.19	0.03	0.09	0.79	−1.86	2.41
	S4(36DAA)	0.04	0.45	−1.84	0.50	−0.97	−2.55	−0.34	−0.31	2.46	0.51
	S5(42DAA)	−0.10	0.27	1.20	−0.80	−0.63	1.20	1.48	−0.90	−2.51	−1.00
Agronomic (AGRO)	GY	0.32	1.78	−0.30	−0.76	0.06	−0.14	−0.68	2.10	1.03	1.33
	PE	0.13	−0.45	−0.81	0.25	0.17	0.27	0.02	0.62	−0.12	0.09
	PR	0.33	−0.15	−0.16	0.33	0.94	0.42	1.15	−1.06	0.36	−0.76
	HG	−0.10	0.05	0.39	0.73	0.78	−0.10	−1.16	1.76	0.36	0.07
	ED	−0.30	−0.64	0.04	1.20	−0.70	0.67	0.52	−0.54	−1.13	−0.47
	EL	−0.11	−1.04	−0.34	−0.09	−0.74	0.18	0.47	−1.32	0.04	−1.04
	DM	0.90	−0.23	0.29	−0.04	0.28	−0.94	0.65	−0.08	0.15	−0.40
	CC	0.92 **	0.74 **	0.68 **	0.51 [ns]	0.20 [ns]	0.90 **	0.77 **	0.53 [ns]	0.49 [ns]	0.34 [ns]
	DF	35	24	15	8	3	35	24	15	8	3
	x^2	194.6	93.7	51.5	18.2	2.1	175.9	88.2	39.2	21.2	6.7

CC: canonical correlations; DF: degrees of freedom. AGRO—GY: grain yield; PE: popping expansion; PR: prolificacy; HG: 100-grain mass; ED: ear diameter; EL: ear length and DM: dry matter. SPAD index—S1, S2, S3, S4, and S5 = SPAD readings at 17, 22, 29, 36, and 42 days after male anthesis, respectively. Symbols ns, ** represent not significant, significant at 1.0% probability.

Under WW condition, and considering the AGRO and SPAD index groups of traits, the first canonical pair (r = 0.90) associated the measurements S_1 (17 DAA), S_5 (42 DAA), and S_4 (36 DAA) with the variables ED and DM. The S_1 and S_5 relationships occurred in the same direction of ED and in the opposite direction of DM, while S_4 related itself in the opposite direction of ED and in the same direction of DM. The second canonical pair (r = 0.77) indicates the association of measurements S_1, S_5, and S_2 (22 DAA) with the variables HG and PR. The measurements S_1 and S_5 related themselves in the same direction with PR, but in the opposite direction with HG, while S_2 presented an inverse relationship with the traits PR and HG (Table 2).

In the canonical correlation estimated between the groups of traits, AGRO and ROOT, under WS condition, the first three canonical pairs were significant, with values of r = 0.89, r = 0.84, and r = 0.65, from the first to the third pair, respectively. In this condition, the first canonical pair indicates that the lower the BN, the lower the PE and DM and the higher the ED. The second canonical pair points to lower PR and higher PE concomitant with lower DB and higher CD. The third canonical pair showed inverse correlation between BA and DM (Table 3).

Table 3. Correlation and coefficients of canonical pairs estimated between roots morphology and agronomic traits in 20 popcorn inbred lines under two different water conditions.

Traits		Canonical Pairs											
		Water-Stressed						Well-Watered					
		1st	2nd	3rd	4th	5th	6th	1st	2nd	3rd	4th	5th	6th
Roots	BN	−0.71	0.40	−1.59	0.44	0.01	0.20	−0.26	−0.01	0.86	−0.30	−0.01	0.34
	BA	−0.34	0.48	0.93	0.72	0.92	0.58	−0.49	1.42	0.26	0.38	0.39	−0.88
	BD	0.32	−0.94	0.05	1.27	−1.47	0.20	0.32	0.33	0.38	0.95	0.49	0.03
	CN	0.37	−0.05	−0.25	−0.34	−0.21	0.96	0.71	0.06	0.23	−0.42	0.24	−0.61
	CA	0.07	0.28	0.01	−1.10	−0.96	−1.00	−0.21	−1.47	−0.48	−0.61	0.39	0.90
	CD	0.25	1.59	−0.32	−0.80	1.20	−0.29	0.41	0.39	−0.38	−0.76	−0.04	0.52
Agronomic (AGRO)	GY	−0.42	−0.16	0.10	−1.27	1.04	1.11	0.02	1.21	0.47	1.97	0.18	1.17
	PE	−0.89	0.97	−0.50	0.56	0.37	−0.07	−0.12	0.13	−0.73	0.59	−0.68	0.34
	PR	0.18	−0.37	0.59	0.84	0.40	−0.02	−0.64	−0.28	−0.52	−1.60	−0.40	0.10
	HG	0.22	−1.12	0.34	0.05	0.14	−0.90	0.70	0.27	−0.17	1.68	0.10	0.45
	ED	0.94	0.38	−0.43	0.53	−0.88	−0.43	0.61	−0.70	−0.28	−0.80	0.09	0.07
	EL	0.17	−0.16	0.19	0.55	−1.58	0.13	−0.66	−1.48	−0.27	−1.10	−0.36	−0.22
	DM	−0.63	−0.43	−0.84	0.33	0.64	−0.15	0.09	0.11	0.59	−0.58	−0.75	−0.30
CC		0.89 **	0.84 **	0.65 **	0.52 ns	0.44 ns	0.13 ns	0.88 **	0.82 **	0.68 **	0.55 ns	0.36 ns	0.22 ns
DF		42	30	20	12	6	2	42	30	20	12	6	2
x^2		203.2	121.8	58.0	28.9	12.5	0.9	201.9	121.4	62.2	29.4	10.3	2.6

CC: canonical correlations; DF: degrees of freedom. AGRO—GY: grain yield; PE: popping expansion; PR: prolificacy; HG: 100-grain mass; ED: ear diameter; EL: ear length and DM: dry matter. ROOTS—BN: brace number; BA: brace angle; BD: brace density; CN: crown number; CA: crown angle and CD: crown density. Symbols ns, ** represent not significant, significant at 1.0% probability.

Under WW condition, the canonical correlations estimated between the AGRO and ROOT groups showed significance in the first (r = 0.8), second (r = 0.82), and third (r = 0.68) canonical pair. As per the first canonical pair, the higher the CN, the lower the HG and EL, and the higher the ED and PR. The second canonical pair showed that the higher the BA, the higher will be GY and the lower will be ED and EL, while the third canonical pair suggested a decrease in PE in plants with lower BN (Table 3).

4. Discussion

The soil water limitation strongly affected the traits GY (−55.3%), PE (−28.8%), and HG (−23.5%). Scientific studies assign GY reduction, under drought conditions, to a smaller number of grains yielded per area [19]. The water stress in the soil, throughout anthesis phase, critical period for the crop [38,39], affects the viability of the pollen grain, stigma receptivity, and fertilization [40], resulting in a lower number of grains yielded. In this study, the reduction in GY is explained by the decrease in HG. In WS condition, the insufficiency of photoassimilates for grain filling was pointed out as the cause for the smaller size of these structures [18]. In the evaluation of popcorn hybrids under soil water stress, it was noted, on average, 30% reduction in HG [41]. Prolificacy (−15.7%), ear diameter (−4.1%), ear length

(−0.8%), and dry matter (−14.8%) traits experienced the impact of soil water limitation to a lesser extent. Irrigation was interrupted 15 days before the male anthesis, in which the plants are at the end of their plant development, starting the emission of tassels. Hence, traits like PR, ED, EL, and DM, developed before the anthesis, tend to be less impacted, as they are already partially developed [39]. Durães et al., 2004 [39] described that, in the pre-flowering period, the plants have already reached or are close to the end of the plant development. It should be stated that, in maize, PR presents a high genetic correlation with GY, both under adequate water conditions of growing [42] and under stress conditions [43], being used as a selection trait in breeding programs for WS conditions.

SPAD index (leaf greenness), generated by the evaluation of a portable chlorophyll meter (SPAD), turned out to be a trait of high potential for genotypic discrimination in materials evaluated under different water conditions. At 17 DAA, when the first measurement was taken, differences between genotypes and environments were observed. Moreover, SPAD index decreases faster under WS conditions [13,23,24], because the leaf senescence occurs more sharply under WS in comparison with the WW condition. The foliar senescence is coordinated by reproductive plant stadium, but also strongly influenced by the environment [44]. Thus, SPAD index is an essential tool for diagnosing plant stress [13,23–25]. Actually, spectrometric methods are efficient to evaluate the different water conditions and the performance of cereals [45].

SPAD index measurements suggest a photosynthetic capacity of the leaf tissues, since that this index is correlated with chlorophyll concentration, and enable to obtain better inferences of photosynthesis [25]. Kamphorst et al. (2020) [46] evaluated SPAD values popcorn inbred lines contrasted to grain yield in drought conditions. The authors showed that maximum SPAD values were obtained more later (11 days after anthesis) in WW, when compared with WS (two days after anthesis). Maximum SPAD index values are related to late leaf senescence, due to high chlorophyll concentration and photosynthetic carbon assimilation [46]. In addition, Gregersen (2008) [47], showed that nutrients mobilization from senescent leaves to reproductive organs are important to plant growth in drought conditions. Actually, Killi et al. (2017) [48] evaluated maize varieties with contrasting drought tolerance in water-stressed condition and showed no physiological or morphological traits that may be associated with enhanced yield in the drought tolerant maize variety. As such, the respective yield characteristics of the drought and sensitive maize varieties may be the result of differences in the partitioning of photosynthate between reproduction and vegetative structures [48].

Plants increased the brace root and crown density, displaying the phenotypic root plasticity of adaptive response to WS condition. Maize plants with reduced lateral root branching make greater axial root elongation possible, and the greater depth of rooting promotes water acquisition in deeper layers of dry soils [49]. Because of these characteristics, the metabolic root demand of these genotypes is lower, which may contribute to higher growth, with consequent greater grain yield [50]. The drought tolerant maize varieties exhibited higher investment in root-systems, allowing greater uptake of the available soil water [48]. In this way, lower root density is important for drought tolerance and should be considered as a selection goal in maize breeding programs [49,51].

The brace and crown root angles are directed more vertically towards the soil for the acquisition of water. Trachsel et al. (2011) [27] report that, in water-limited environments, genotypes with larger angles, close to 90°, of the roots in relation to the soil present adaptive advantages to find water. A root ideotype with three characteristics, Steep, Cheap, and Deep, may represent an advantage for water acquisition in deeper soil layers [29]. Steep suggests greater adaptive value to drought for plants with root angles close to 90° in relation to the soil [29].

Agronomic and root traits suffered genotype × water condition interaction, which may make concomitant selection for WS and WW conditions difficult. A number of papers describe the presence of G × WC interaction between common maize traits evaluated with and without drought stress [15,52]. Plant breeders hardly ever select root traits because of the high plasticity in response to soil conditions and measurement difficulty, incurring in low heritability values [50].

Higher SPAD index was associated with higher yield values; in addition, these measurements can be reliable for plant selection considering the absence of genotype × water condition interaction in the present work. Researches in this area have focused on secondary morphological and physiological traits easy to measure under field conditions, which correlate with grain yield [19,20,23,39,53,54]. The use of secondary traits of simple measurement and high adaptive value can increase the efficiency of the selection under stressful condition [55]. A suitable secondary trait is that of high heritability, low cost, easy-to-measure, stable during the measurement period, and, above all, genetically associated with grain yield under stress conditions [56]. Additionally, in recurrent selection programs with maize, it has been said that the mean yield of tropical germplasm grains has increased after eight cycles, when genotypes with delayed leaf senescence were prioritized under intermediate and severe water stress conditions (rain-free winter season in Mexico) [16].

The choice of the appropriate phenological stage for SPAD index measurements is essential for consistent phenotyping of plants under drought conditions. In the view of Cairns et al. (2012) [19], the effective application of phenotyping methodologies relies, partly, on the applicability of these methodologies in the critical stages associated with grain yield. Accordingly, given the results presented herein, the evaluation in S_2 and S_4, under WS, and in S_1, under WW conditions, is recommended. The proper stage of evaluation to identify differences among genotypes by means of spectral indices is the phenological stage of grain filling [10,19,20,23].

On the basis of an agronomic and root characterization and using different dates of measurement of the SPAD index, it was sought, in this research, to identify traits with variability and that do not present genotype × water condition (G × WC) interaction, but that present desirable associations for the traits GY and PE, to favor the indirect selection process of superior genotypes. Firstly, the ideal scenario would be the absence of G × WC interaction in relation to the traits GY and PE. A second promising scenario would be to find a variable associated with GY and PE in both environments. The third promising scenario would be to find a variable associated with GY and PE only in the environment under WS condition, but not presenting G × WC, and which would consequently have the role of ensuring a reliable selection of drought tolerant genotypes in the environment under WW condition. These scenarios were not observed.

Considering the results of the agronomic traits, the G × WC interactions proved to be an obstacle for the selection of high productive genotypes under drought by direct via. In addition to the traits GY and PE have significant G × WC interaction, the traits that did not present G × WC interaction (HG, EL, and DM) expressed either weak or negative correlation with PE (Figure 2). Similar to the agronomic traits, the root traits expressed significant G × WC interaction for all their variables, making their use in indirect selection in an irrigation environment unviable (Table 2), and although, under WS condition, the root traits BN, in the first canonical pair, and BD and CD, in the second canonical pair, are associated with PE, their use is unviable because it is a destructive, indirect and costly measure [27].

The SPAD index did not present significant G × WC interaction in the evaluations from S_1 (17 DAA) to S_4 (36 DAA); hence, they have potential to be used in an environment under WW condition (Table 1). Regarding the root traits, the variables correlated similarly to what was verified by Trachsel et al. [27], which emphasize the association between the two pairs of root traits BA and CA, and DB and CD. When observed the Pearson linear correlation (Figure 2), however, the correlation between the measures of SPAD index and GY under WW condition were positive, while regarding PE, they were negative or weak. This emphasizes that selection for these two traits is a challenge already addressed by many authors, who have reported observing negative correlations between GY and PE [3,6,31,57].

Conversely, the S_3 SPAD index values correlated negatively in the second and third canonical pairs with GY and PE, respectively (Table 2). Taking these two canonical pairs into account, the smaller the S_3 (29 DAA), the greater the GY and PE. Even though the association in the second and third canonical pairs is less relevant in relation to the first one, the results of S_3 association should be considered.

Results of Pearson correlations and canonical correlations (CCA) were in contrast in various situations. However, Pearson linear correlation is a multiple univariate method, whereas CCA considers the interaction between groups of traits. In accordance with Thompson (2005) [33], the CCA better illustrates the reality of traits interacting with each other. For this reason, correlation patterns in CCA are sensitive to changes with the addition or exclusion of a simple variable.

Given that the root and SPAD index traits were well represented, it is assumed that the CCA should be regarded as a selection index for GY and PE increase. In findings from previous analysis, Kamphorst et al. (2019) [30] support this statement using the same genotypes because, as well as in this work, the research cited above concluded that SPAD index measures are associated with simultaneous increase of GY and PE under WS condition, but having as dependent variable the product of GY and PE, that is, volume of popcorn per hectare. Kamphorst et al. (2019) [30] performed a single SPAD index measurement at 35 DAF, equivalent to the S4 measurement, which, in this study, was negatively and with high intensity associated with PE in the third canonical pair and under WS condition. For this reason, CCA meets the need to better understand the relationship among traits and the choice of the best variable for efficient use in indirect selection.

The canonical correlation analysis in popcorn crops has great potential for use in identifying non-destructive and easily quantified traits, which are associated with GY and PE, especially as a result of the negative association between GY and PE proposed by some authors [3,6,31,57]. Even in relation to the path analysis, canonical correlations are of interest, as they allow analyzing the association between different groups of variables simultaneously and not only fixing a variable as dependent. According to the findings of this study and the characteristics of the variable that was highlighted as being optimal for the selection of the high productive genotypes under drought by indirect via, it proved to be possible to identify variables with potential for use in indirect selection or even to compose a selection index for the discrimination of the high productive genotypes under water stress condition based on well-watered condition. The SPAD index, measured at 29 DAA, was significant because it associated grain yield and popping expansion in an environment with limited soil water conditions, as well as the measurements at 22 and 36 DAA, which combined more closely with grain yield.

5. Conclusions

Major guidance has been given regarding the most adequate traits for the development of research correlated to water shortage in popcorn in regard to the determination of the proper phenological stages for measurement. The leaf senescence, estimated by SPAD index, can be applied to identify more productive genotypes under drought conditions, resulting from a non-destructive procedure using a portable chlorophyll meter (SPAD) [14,20,58].

Author Contributions: Conceptualization, S.H.K. and V.J.d.L.; methodology, S.H.K, G.M.B.G., A.T.d.A.J., V.J.d.L., J.T.L., K.F.M.S., and D.R.d.S.J.; software, S.H.K. and G.M.B.G.; validation, S.H.K.; formal analysis, S.H.K, G.M.B.G., and V.J.d.L.; investigation, S.H.K., G.M.B.G., A.T.d.A.J., V.J.d.L., J.T.L., K.F.M.S., D.R.d.S.J., J.S.S., F.T.d.O., C.C.G.C., and W.P.R.; resources, S.H.K.; data curation, S.H.K.; writing—original draft preparation, S.H.K., G.M.B.G., A.T.d.A.J., V.J.d.L., J.S.S., F.T.d.O., W.P.R., and E.C.; writing—review and editing, S.H.K., G.M.B.G., A.T.d.A.J., V.J.d.L., and E.C.; supervision, A.T.d.A.J. and E.C.; project administration, S.H.K.; funding acquisition, S.H.K., A.T.d.A.J., W.P.R., and E.C. All authors have read and agreed to the published version of the manuscript.

Funding: This study was financed in part by the Coordenação de Aperfeiçoamento de Pessoal de Nível Superior – Brasil (CAPES)–Finance Code 001. Funding was also provided by FAPERJ, with the project E26/201.813/2017 to S.H.K., E-26/202.323/2017 to W.P.R., E-26/202.759/2018 to E.C., E26/202.761/2017 to A.T.do A.J and fellowships awarded 303166/2019-3, E.C. (CNPq) and FAPEMA (Fundação de Amparo à Pesquisa e ao Desenvolvimento Científico e Tecnológico do Maranhão) (PVS 00583/20, E.C) and FAPEMA (Fundação de Amparo à Pesquisa e ao Desenvolvimento Científico e Tecnológico do Maranhão) (PVS 00583/20, E.C).

Conflicts of Interest: The authors declare no conflict of interest.

References

1. Freitas, S.P., Jr.; Amaral Juniot, A.T.; Rangel, R.M.; Viana, A.P. Genetic gains in popcorn by full-sib recurrent selection. *Crop. Breed. Appl. Biotechnol.* **2009**, *9*, 1–7. [CrossRef]
2. Amaral Junior, A.T.; Gonçalves, L.S.A.; Freitas, S.D.P., Jr.; Candido, L.S.; Vittorazzi, C.; Pena, G.F.; Ribeiro, R.M.; Silva, T.R.D.C.; Pereira, M.G.; Scapim, C.A.; et al. UENF 14: A new popcorn cultivar. *Crop Breed. Appl. Biotechnol.* **2013**, *13*, 218–220.
3. De Lima, V.J.; do Amaral, A.T., Jr.; Kamphorst, S.H.; Pena, G.F.; Leite, J.T.; Schmitt, K.F.; Vittorazzi, C.; de Almeida Filho, J.E.; Mora, F. Combining ability of S_3 progenies for key agronomic traits in popcorn: Comparison of testers in top-crosses. *Genet. Mol. Res.* **2016**, *15*. [CrossRef] [PubMed]
4. Dos Santos, A.; do Amaral Junior, A.T.; Kamphorst, S.H.; Gonçalves, G.M.B.; Santos, P.H.A.D.; Silva Vivas, J.M.; Sousa Mafra, G.; Khan, S.; Tomaz de Oliveira, F.; Schmitt, K.F.M.; et al. Evaluation of Popcorn Hybrids for Nitrogen Use Efficiency and Responsiveness. *Agronomy* **2020**, *10*, 485. [CrossRef]
5. Mafra, G.S.; do Amaral Junior, A.T.; de Almeida Filho, J.E.; Vivas, M.; Santos, P.H.A.D.; Santos, J.S.; Pena, G.F.; Lima, V.J.d.; Kamphorst, S.H.; de Oliveira, F.T.; et al. SNP-based mixed model association of growth- and yield-related traits in popcorn. *PLoS ONE* **2019**, *14*, e0218552. [CrossRef]
6. Cabral, P.D.S.; do Amaral, A.T., Jr.; de Freitas, I.L.J.; Ribeiro, R.M.; da Silva, T.R.C. Cause and effect of quantitative characteristics on grain expansion capacity in popcorn. *Rev. Ciênc. Agronôm.* **2016**, *47*, 108–117. [CrossRef]
7. Kist, B.B.; de Carvalho, C.; Beling, R.R. *Anuário Brasileiro do Milho 2019*; Beling, R.R., Ed.; Editora Gazeta: Santa Cruz do Sul, Brazil, 2019; ISBN 1808-3439.
8. Awange, J.L.; Mpelasoka, F.; Goncalves, R.M. When every drop counts: Analysis of Droughts in Brazil for the 1901–2013 period. *Sci. Total Environ.* **2016**, *566–567*, 1472–1488. [CrossRef]
9. Van Loon, A.F.; Gleeson, T.; Clark, J.; Van Dijk, A.I.J.M.; Stahl, K.; Hannaford, J.; Di Baldassarre, G.; Teuling, A.J.; Tallaksen, L.M.; Uijlenhoet, R.; et al. Drought in the Anthropocene. *Nat. Geosci.* **2016**, *9*, 89–91. [CrossRef]
10. Adebayo, M.A.; Menkir, A.; Hearne, S.; Kolawole, A.O. Gene action controlling normalized difference vegetation index in crosses of elite maize (*Zea mays* L.) inbred lines. *Cereal Res. Commun.* **2017**, *45*, 675–686. [CrossRef]
11. Wang, W.; Vinocur, B.; Altman, A. Plant responses to drought, salinity and extreme temperatures: Towards genetic engineering for stress tolerance. *Planta* **2003**, *218*, 1–14. [CrossRef]
12. Ghannoum, O. C4 photosynthesis and water stress. *Ann. Bot.* **2008**, *103*, 635–644. [CrossRef]
13. Romano, G.; Zia, S.; Spreer, W.; Sanchez, C.; Cairns, J.; Luis, J.; Müller, J. Use of thermography for high throughput phenotyping of tropical maize adaptation in water stress. *Comput. Electron. Agric.* **2011**, *79*, 67–74. [CrossRef]
14. Araus, J.L.; Kefauver, S.C.; Zaman-Allah, M.; Olsen, M.S.; Cairns, J.E. Translating High-Throughput Phenotyping into Genetic Gain. *Trends Plant Sci.* **2018**, *23*, 451–466. [CrossRef] [PubMed]
15. Dias, K.O.D.G.; Gezan, S.A.; Guimarães, C.T.; Parentoni, S.N.; Guimarães, P.E.D.O.; Carneiro, N.P.; Portugal, A.F.; Bastos, E.A.; Cardoso, M.J.; Anoni, C.D.O.; et al. Estimating Genotype × Environment Interaction for and Genetic Correlations among Drought Tolerance Traits in Maize via Factor Analytic Multiplicative Mixed Models. *Crop Sci.* **2018**, *58*, 72. [CrossRef]
16. Bolaños, J.; Edmeades, G.O. Eight cycles of selection for drought tolerance in lowland tropical maize. I. Responses in grain yield, biomass, and radiation utilization. *Field Crop. Res.* **1993**, *31*, 233–252. [CrossRef]
17. Tollenaar, M.; Ahmadzadeh, A.; Lee, E.A. Physiological Basis of Heterosis for Grain Yield in Maize. *Crop Sci.* **2004**, *44*, 2086. [CrossRef]
18. Araus, J.L.; Sánchez, C.; Cabrera-Bosquet, L. Is heterosis in maize mediated through better water use? *New Phytol.* **2010**, *187*, 392–406. [CrossRef] [PubMed]
19. Cairns, J.E.; Sanchez, C.; Vargas, M.; Ordoñez, R.; Araus, J.L. Dissecting Maize Productivity: Ideotypes Associated with Grain Yield under Drought Stress and Well-watered Conditions. *J. Integr. Plant Biol.* **2012**, *54*, 1007–1020. [CrossRef]
20. Adebayo, M.A.; Menkir, A.; Blay, E.; Gracen, V.; Danquah, E.; Hearne, S. Genetic analysis of drought tolerance in adapted × exotic crosses of maize inbred lines under managed stress conditions. *Euphytica* **2014**, *196*, 261–270. [CrossRef]

21. Joshi, A.K.; Kumari, M.; Singh, V.P.; Reddy, C.M.; Kumar, S.; Rane, J.; Chand, R. Stay green trait: Variation, inheritance and its association with spot blotch resistance in spring wheat (*Triticum aestivum* L.). *Euphytica* **2006**, *153*, 59–71. [CrossRef]

22. Costa, E.F.N.; Santos, M.F.; Moro, G.V.; Alves, G.F.; de Souza Junior, C.L. Herança da senescência retardada em milho. *Pesqui. Agropecu. Bras.* **2008**, *43*, 207–213. [CrossRef]

23. Zia, S.; Romano, G.; Spreer, W.; Sanchez, C.; Cairns, J.; Araus, J.L.; Müller, J. Infrared Thermal Imaging as a Rapid Tool for Identifying Water-Stress Tolerant Maize Genotypes of Different Phenology. *J. Agron. Crop Sci.* **2013**, *199*, 75–84. [CrossRef]

24. Araus, J.L.; Serret, M.D.; Edmeades, G.O. Phenotyping maize for adaptation to drought. *Front. Physiol.* **2012**, *3*, 1–20. [CrossRef] [PubMed]

25. De Castro, F.A.; Campostrini, E.; Netto, A.T.; De Menezes De Assis Gomes, M.; Ferraz, T.M.; Glenn, D.M. Portable chlorophyll meter (PCM-502) values are related to total chlorophyll concentration and photosynthetic capacity in papaya (*Carica papaya* L.). *Theor. Exp. Plant Physiol.* **2014**, *26*, 201–210. [CrossRef]

26. Lynch, J. Root Architecture and Plant Productivity. *Plant Physiol.* **1995**, *109*, 7–13. [CrossRef]

27. Trachsel, S.; Kaeppler, S.M.; Brown, K.M.; Lynch, J.P. Shovelomics: High throughput phenotyping of maize (*Zea mays* L.) root architecture in the field. *Plant Soil* **2011**, *341*, 75–87. [CrossRef]

28. Hund, A.; Ruta, N.; Liedgens, M. Rooting depth and water use efficiency of tropical maize inbred lines, differing in drought tolerance. *Plant Soil* **2009**, *318*, 311–325. [CrossRef]

29. Gao, Y.; Lynch, J.P. Reduced crown root number improves water acquisition under water deficit stress in maize (*Zea mays* L.). *J. Exp. Bot.* **2016**, *67*, 4545–4557. [CrossRef]

30. Kamphorst, S.H.; do Amaral, A.T., Jr.; de Lima, V.J.; Guimarães, L.J.M.; Schmitt, K.F.M.; Leite, J.T.; Santos, P.H.A.D.; Chaves, M.M.; Mafra, G.S.; dos Santos, D.R., Jr.; et al. Can Genetic Progress for Drought Tolerance in Popcorn Be Achieved by Indirect Selection? *Agronomy* **2019**, *9*, 792. [CrossRef]

31. Do Amaral, A.T., Jr.; dos Santos, A.; Gerhardt, I.F.S.; Kurosawa, R.N.F.; Moreira, N.F.; Pereira, M.G.; Gravina, G.A.; Silva, F.H.L. Proposal of a super trait for the optimum selection of popcorn progenies based on path analysis. *Genet. Mol. Res.* **2016**, *15*, 1–9. [CrossRef]

32. Hardoon, D.R.; Szedmak, S.; Shawe-Taylor, J. Canonical Correlation Analysis: An Overview with Application to Learning Methods. *Neural Comput.* **2004**, *16*, 2639–2664. [CrossRef] [PubMed]

33. Thompson, B. Canonical Correlation Analysis. In *Encyclopedia of Statistics in Behavioral Science*; John Wiley & Sons, Ltd.: Chichester, UK, 2005.

34. Vittorazzi, C.; Júnior, A.T.A.; Guimarães, A.G.; Silva, F.H.L.; Pena, G.F.; Daher, R.F.; Gerhardt, I.F.S.; Oliveira, G.H.F.; Santos, P.H.A.D.; Souza, Y.P.; et al. Evaluation of genetic variability to form heterotic groups in popcorn. *Genet. Mol. Res.* **2018**, *17*. [CrossRef]

35. Christensen, H.K.; Montgomery, C.A. Corporate economic performance: Diversification strategy versus market structure. *Strateg. Manag. J.* **1981**, *2*, 327–343. [CrossRef]

36. Cruz, C.D. GENES—A software package for analysis in experimental statistics and quantitative genetics. *Acta Sci. Agron.* **2013**, *35*, 271–276. [CrossRef]

37. Cruz, C.D.; Regazzi, A.J.; Carneiro, P.C.S. *Modelos Biométricos Aplicados ao Melhoramento Genético*; UFV: Viçosa, Brazil, 2012; ISBN 9788572694339.

38. De Sousa Mendes, W.; Drews, T.A.; Medeiros, J.C.; Dalla Rosa, J.; Gualberto, A.V.; Mielezrski, F. Development and productivity of maize in response to spatial arrangement under semiarid condition of Northeastern Brazil. *Aust. J. Crop Sci.* **2017**, *11*, 313–321. [CrossRef]

39. Durães, F.O.M.; dos Santos, M.X.; Gama, E.E.G.; Magalhães, P.C.; Albuquerque, P.E.P.; Guimarães, C.T. Fenotipagem Associada a Tolerância a Seca em Milho para Uso em Melhoramento, Estudos Genômicos e Seleção Assistida por Marcadores. *Embrapa Milho Sorgo Circ. Téc. INFOTECA-E* **2004**, *39*, 18.

40. Zinselmeier, S.A.; Lauer, M.J.; Boyer, J.S. Reversing Drought-Induced Losses in Grain Yield: Sucrose Maintains Embryo Growth in Maize. *Crop Sci.* **1995**, *35*, 1390. [CrossRef]

41. De Lima, V.J.; do Amaral, A.T., Jr.; Kamphorst, S.H.; Bispo, R.B.; Leite, J.T.; de Santos, T.O.; Schmitt, K.F.M.; Chaves, M.M.; de Oliveira, U.A.; Santos, P.H.A.D.; et al. Combined Dominance and Additive Gene Effects in Trait Inheritance of Drought-Stressed and Full Irrigated Popcorn. *Agronomy* **2019**, *9*, 782. [CrossRef]

42. Lima Neto, F.P.; Souza, C.L., Jr. De Number of recombinations and genetic properties of a maize population undergoing recurrent selection. *Sci. Agric.* **2009**, *66*, 52–58. [CrossRef]

43. Li, X.-H.; Liu, X.-D.; Li, M.-S.; Zhang, S.-H. Identification of quantitative trait loci for anthesis-silking interval and yield components under drought stress in maize. *Acta Bot. Sin.* **2003**, *45*, 852–857.

44. Combe, L.; Escobar-Gutiérrez, A.J. Sénescence d'un pied de maïs: Évolution de la floraison à la récolte. *Botany* **2009**, *87*, 1036–1053. [CrossRef]

45. Yousfi, S.; Gracia-Romero, A.; Kellas, N.; Kaddour, M.; Chadouli, A.; Karrou, M.; Araus, J.L.; Serret, M.D. Combined Use of Low-Cost Remote Sensing Techniques and δ13C to Assess Bread Wheat Grain Yield under Different Water and Nitrogen Conditions. *Agronomy* **2019**, *9*, 285. [CrossRef]

46. Kamphorst, S.H.; do Amaral, A.T., Jr.; de Lima, V.J.; Diniz, S.P.H.A.; Rodrigues, W.P.; Silva, V.J.M.; Gonçalves, G.M.B.; Schmitt, K.F.M.; Leite, J.T.; Vivas, M.; et al. Comparison of selection traits for effective popcorn breeding under water limiting conditions. *Front. Plant Sci.* **2020**, *11*, 1–19. [CrossRef] [PubMed]

47. Gregersen, P.L.; Holm, P.B.; Krupinska, K. Leaf senescence and nutrient remobilisation in barley and wheat. *Plant Biol.* **2008**, *10*, 37–49. [CrossRef] [PubMed]

48. Killi, D.; Bussotti, F.; Raschi, A.; Haworth, M. Adaptation to high temperature mitigates the impact of water deficit during combined heat and drought stress in C3 sunflower and C4 maize varieties with contrasting drought tolerance. *Physiol. Plant* **2017**, *159*, 130–147. [CrossRef] [PubMed]

49. Zhan, A.; Schneider, H.; Lynch, J.P. Reduced Lateral Root Branching Density Improves Drought Tolerance in Maize. *Plant Physiol.* **2015**, *168*, 1603–1615. [CrossRef] [PubMed]

50. Lynch, J.P. Root phenes that reduce the metabolic costs of soil exploration: Opportunities for 21st century agriculture. *Plant Cell Environ.* **2015**, *38*, 1775–1784. [CrossRef]

51. Kamphorst, S.H.; de Lima, V.J.; Schimitt, K.F.M.; Leite, J.T.; Azeredo, V.C.; Pena, G.F.; Santos, P.H.A.D.; Júnior, D.R.S.; da Silva Júnior, S.B.; Bispo, R.B.; et al. Water stress adaptation of popcorn roots and association with agronomic traits. *Genet. Mol. Res.* **2018**, *17*, 1–14. [CrossRef]

52. Maazou, A.-R.S.; Tu, J.; Qiu, J.; Liu, Z. Breeding for Drought Tolerance in Maize (*Zea mays* L.). *Am. J. Plant Sci.* **2016**, *7*, 1858–1870. [CrossRef]

53. Santos, M.X.; Andrade, C.L.T.; Oliveira, A.C.; Leite, C.E.P.; Carvalho, H.W.L.; Gama, E.E.G.; Pacheco, C.A.P.; Gumarães, P.E.O.; Parentoni, S.N. Comportamento de Híbridos de Milho Selecionados e não Selecionados para Asi sob Estresse de Água no Florescimento e no Enchimento de Grãos. *Rev. Bras. Milho Sorgo* **2003**, *2*, 71–81. [CrossRef]

54. Câmara, T.M.M.; Bento, D.A.V.; Alves, G.F.; Santos, M.F.; Moreira, J.U.V.; Souza, C.L., Jr. De Parâmetros genéticos de caracteres relacionados à tolerância à deficiência hídrica em milho tropical. *Bragantia* **2007**, *66*, 595–603. [CrossRef]

55. Hallauer, A.R.; Carena, M.J.; Miranda Filho, J.B. *Quantitative Genetics in Maize Breeding*; Springer: New York, NY, USA, 2010.

56. Edmeades, G.O.; Bolaños, J.; Chapman, S.C.; Lafitte, H.R.; Banziger, M. Selection Improves Drought Tolerance in Tropical Maize Populations. *Crop Sci.* **1999**, *39*, 1306. [CrossRef]

57. Pereira, M.G.; Amaral, A.T., Jr. Estimation of Genetic Components in Popcorn Based on the Nested Design. *Crop Breed. Appl. Biotechnol.* **2001**, *1*, 3–10. [CrossRef]

58. Lu, Y.; Hao, Z.; Xie, C.; Crossa, J.; Araus, J.; Gao, S.; Vivek, B.S.; Magorokosho, C.; Mugo, S.; Makumbi, D.; et al. Field Crops Research Large-scale screening for maize drought resistance using multiple selection criteria evaluated under water-stressed and well-watered environments. *Field Crop. Res.* **2011**, *124*, 37–45. [CrossRef]

Article

Maize Hybrid Response to Sustained Moderate Drought Stress Reveals Clues for Improved Management

Samadangla Ao [1], Michael P. Russelle [2], Gary W. Feyereisen [3], Tamás Varga [2] and Jeffrey A. Coulter [4,*]

1   Department of Botany, Kohima Science College, Jotsoma, Kohima, Nagaland 797002, India; aoxxx011@umn.edu
2   Department of Soil, Water, and Climate, University of Minnesota, 439 Borlaug Hall, 1991 Upper Buford Circle, St. Paul, MN 55108, USA; russelle@umn.edu (M.P.R.); tvarga@umn.edu (T.V.)
3   United States Department of Agriculture–Agricultural Research Service Soil and Water Management Research Unit, 439 Borlaug Hall, 1991 Upper Buford Circle, St. Paul, MN 55108, USA; gary.feyereisen@ars.usda.gov
4   Department of Agronomy and Plant Genetics, University of Minnesota, 411 Borlaug Hall, 1991 Upper Buford Circle, St. Paul, MN 55108, USA
*   Correspondence: jeffcoulter@umn.edu

Received: 21 August 2020; Accepted: 10 September 2020; Published: 12 September 2020

Abstract: Crop water productivity (CWP), irrigation water productivity (IWP), actual seasonal basal crop coefficient (K_{ab}), and actual crop evapotranspiration (ET_a) are essential parameters for accurate estimation of crop water requirement to prevent irrigation water waste. These parameters were evaluated by conducting three experiments using a drought-tolerant maize hybrid and a non-drought-tolerant ('standard') maize hybrid receiving 50, 100, and 150% of the recommended optimal nitrogen (N) fertilizer rate and grown under well-watered conditions, drought stress from the 14 leaf collar maize phenological stage (V14) to maize physiological maturity (R6), and drought stress from the blister maize phenological stage (R2) to R6. Across hybrids, ET_a decreased with increased duration of drought stress. The drought-tolerant hybrid had 7 and 8% greater CWP and IWP, respectively, compared to the standard hybrid when drought stress began at V14. Mid-season K_{ab} was 1.08, 0.89, and 0.73 under well-watered conditions and when drought stress began at R2 and V14, respectively. These results reveal that (i) maize achieved more effective physiological acclimation with earlier exposure to drought stress, (ii) grain yield of the drought-tolerant hybrid was unchanged by earlier, compared to later, onset of drought despite a 10% decrease in ET_a, and (iii) two phases of acclimation were identified: Maize K_{ab} declined upon exposure to drought but stabilized as the crop acclimated.

Keywords: crop coefficient; drought stress; evapotranspiration; maize; water productivity

---

1. Introduction

Climate change is associated with increased frequency of drought, which is predicted to be more severe and widespread in the future [1]. According to DeLucia et al. [2], the irrigated area of the U.S. Corn Belt will need to expand at least three-fold into areas where rain usually suffices to maintain U.S. maize production at the same level (10.7 Mg ha^{-1}). Drought adversely affects cereal yields worldwide, with maize (*Zea mays* L.) having greater sensitivity to drought stress than wheat (*Triticum aestivum* L.) [3,4]. Crop yield loss due to drought stress can be mitigated by irrigation, but the amount of water available for irrigation is limited. Therefore, judicious management of irrigation water requires accurate estimation of the factors used to determine water application [5]. There is great opportunity to improve the efficiency of agricultural water management through enhanced knowledge of crop water use or water productivity (WP) [6–9]. Crop water productivity (CWP) and irrigation water productivity (IWP) are measures of grain production per unit of crop transpiration and volume of irrigation, respectively [9–11].

The effects of drought stress on maize have been studied extensively across a wide range of growing environments [12–14]. Much of the research has investigated WP of maize under different water regimes [15–17]. Reduced CWP under dry conditions is often associated with relatively high water uptake but low grain yield [8]. Greater WP can be achieved when there is maximal soil water absorption for transpiration and minimal water loss through soil evaporation [18].

Compared to conventional irrigation, wherein water is applied as needed to reduce yield loss due to drought stress, deficit irrigation can limit unnecessary evaporation, runoff, and leaching [8]. Compared to irrigation at 100% of evapotranspiration requirement, irrigation at 75% of evapotranspiration requirement has been shown to increase CWP of maize with 20% less actual crop evapotranspiration (ET_a) [13]. Djaman and Irmak [14] reported that CWP of maize was maximized at 93 and 100% of ET_a of the fully irrigated treatment, while Panda et al. [15] concluded that deficit irrigation should be based on a 45% maximum allowable depletion of available soil water to achieve high maize grain yield and field water (precipitation plus irrigation) productivity. The phenological stage of crop development affects the amount and timing of water needed for deficit irrigation and thereby influences maize grain yield and WP [19,20]. With drought conditions throughout the reproductive phase of maize, irrigation at the milk and dough stages of kernel development has a greater beneficial effect on grain yield of maize than irrigation at later stages [19].

Maize WP increases with increasing nitrogen (N) supply for both deficit and full irrigation up to the N level required to achieve yield potential at that condition, beyond which an increase in N supply does not provide additional yield benefit [21]. Adequate N supply improves WP in maize in part by increasing the crop canopy, which facilitates greater interception of solar radiation and reduces soil evaporation [22]. Compared to no N fertilization, the application of 120 kg N ha^{-1} to maize was reported to have no effect on WP in water-limited conditions, but it increased WP by increasing the net photosynthetic rate when ample water was available [16].

Accurate estimation of reference evapotranspiration (ET_o), crop evapotranspiration (ET_c), and the crop coefficient (K_c) is vital for calculating the irrigation requirement. Reference ET is evapotranspiration from a grass reference surface under well-watered conditions, and ET_c is water loss from a cropped field as a combination of soil surface evaporation and crop transpiration with agronomic management for full production potential [23]. Dual K_c improves the accuracy of ET_c estimation because it separates the two independent factors representing the basal crop coefficient (K_{cb}) and evaporation coefficient (K_e). The K_{cb} is the ratio of ET_c to ET_o when the surface soil layer is dry, but when the subsurface layer beyond the zone of evaporation has adequate moisture for transpiration to occur at maximum potential [24]. When crops are grown under environmental stress, a stress coefficient (K_s) is used to calculate ET_a [23]. The soil water balance approach, which is an assessment of the influx and efflux of water into the crop root zone during a period of time, can be used to estimate ET_a [6,14,19]; however, ET_a and K_{cb} vary with crop phenological stage, cultivar, plant-available soil water, and environmental conditions [6,7].

Regional and crop-specific ET_a and K_c allow greater precision in crop water management [7,12,25]. Previous estimates of ET_a and WP in maize have been based on full irrigation during the entire growing season, water limitation after the blister stage of maize phenological development (R2), and rainfed conditions, with little emphasis on K_c [14,15,26–28]. There is limited information available on K_c by the phenological development stage for maize under drought stress [25,29], and to our knowledge none exists for drought-tolerant maize hybrids. Globally, drought stress in maize occurs most frequently during the reproductive stages of development, and in some cases begins as early as the late vegetative stages [30–33]. Water requirements of maize are greatest during the late vegetative to early reproductive stages, with drought stress during these stages causing severe yield loss [34,35].

Long-term estimates of maize WP in the western U.S. Corn Belt and China show improvements in CWP over time, which have been attributed to improved cultivars and advances in soil, water, and crop management [36,37]. Improvements in maize grain yield have been associated with enhanced tolerance to drought stress [38,39], and drought-tolerant maize hybrids have potential to alleviate drought-induced yield loss [40–43]. Drought-tolerant hybrids have had greater grain yield and WP than standard hybrids with deficit irrigation at 50 and 75% of evapotranspiration requirements, although no yield difference was observed in the absence of drought or under severe drought [15,44].

The present study was designed to compare ET_a, WP, and K_{cb} of a drought-tolerant maize hybrid and a standard maize hybrid under well-watered conditions and under moderate drought stress imposed during advanced phenological stages that coincide with natural and increasingly likely drought periods encountered in humid and sub-humid maize-producing regions across the globe.

2. Materials and Methods

2.1. Site Description, Experimental Design, and Cultural Practices

In 2013, three independent experiments were conducted at the University of Minnesota Sand Plain Research Farm near Becker, MN (approximately 45°23′17″ N, 93°53′20″ W; 295 m above sea level). The soil at the study sites was a Hubbord-Mosford loamy sand complex (sandy, mixed, frigid Entic Hapludolls and sandy, mixed, frigid Typic Hapludolls). The soil texture was loamy sand for the 0- to 45-cm depth layers and sand for the 45- to 100-cm depth layers. The mean content of sand, silt, and clay, respectively, was 860, 60, and 80 g kg^{-1} for the 0- to 45-cm soil depth and 930, 10, and 60 g kg^{-1} for the 45- to 100-cm soil depth. The soil water content was 0.176 m^3 m^{-3} at field capacity and 0.084 m^3 m^{-3} at permanent wilting point within the 0- to 45-cm depth, and 0.105 m^3 m^{-3} at field capacity and 0.035 m^3 m^{-3} at wilting point within the 45- to 100-cm soil depth [45].

The three experiments were established in close proximity in the same year to make use of spatial and temporal uniformity to increase the accuracy of the results. This enabled precise and timely irrigation application to simulate the drought stress conditions as desired in all three fields without the confounding effects of differences in weather and soil. Each experiment was planted to maize following different previous crops, which were three-year-old alfalfa (*Medicago sativa* L.), soybean [*Glycine max* (L.) Merr.], and winter rye (*Secale cereale* L.) following soybean. Alfalfa and winter rye were terminated with herbicide 10 days before maize planting. All experiments were moldboard plowed seven days before maize planting, then field cultivated and culti-packed on the day of planting. Each experiment evaluated all combinations of three durations of drought stress, two maize hybrids, and three N fertilizer rates using a split-plot arrangement of these 18 treatments in a randomized complete block design with four replications. Main plot treatments were a duration of drought stress and subplots were a factorial arrangement of hybrid and N fertilizer rate. Each main plot was 6.0 m wide by 11.9 m long, and each subplot was 3.0 m wide by 4.0 m long, with maize planted in rows spaced 76 cm apart. The three durations of drought stress were: (i) a well-watered control, restoring soil water content to field capacity at frequent intervals throughout the growing season; (ii) sustained moderate drought stress from the 14 leaf collar maize phenological stage (V14) to maize physiological maturity (R6); and (iii) sustained moderate drought from R2 to R6. Sustained moderate drought stress in this study resulted in maize leaf rolling beginning around mid-day on nearly every day during the drought-stress period, except on days immediately following irrigation or precipitation. The V14 and R2 stages were selected for the onset of drought stress because maize is most sensitive to drought during the late vegetative to early reproductive stages [34,35], and drought commonly begins around these stages in maize growing regions worldwide [30–33]. Two maize hybrids were used: (i) a designated drought-tolerant hybrid, NK Brand N42Z-3011A, reported to have non-transgenic drought tolerance with a relative maturity rating of 99 and maximum yield potential in all growing environments; and (ii) a comparable standard hybrid, NK Brand N36A-3000GT, reported to have a relative maturity rating of 96 and maximum yield in optimum growing conditions.

The three N fertilizer rates were sub-optimal, optimal, and supra-optimal, representing N rates that were 50, 100, and 150%, respectively, of the expected economically optimum N rate for grain yield. Optimum N fertilizer rates for grain yield were 123, 168, and 213 kg N ha^{-1} for maize following alfalfa, soybean, and winter rye, respectively, based on research summarized by Rehm et al. [46,47] and Kaiser et al. [48] for highly productive irrigated sandy soils. Nitrogen was applied as NH_4NO_3, with 45 kg N ha^{-1} broadcast immediately after planting and the remaining amount sidedressed as a surface band 10 cm to the side of each maize row at the six leaf collar maize phenological stage.

Maize was planted on 23 May 2013 in all the experiments. A plant density of 81,500 plants ha^{-1} was achieved by hand thinning at the one leaf collar maize phenological stage. Preemergence and postemergence herbicides were used to control weeds. Additional details on the experimental sites and cultural practices are provided by Ao et al. [49].

2.2. Soil Water Content Measurement and Irrigation Management

The soil water content was measured for the 0- to 20-, 20- to 40-, 40- to 60-, 60- to 80-, and 80- to 100-cm soil layers using a time domain reflectometry soil moisture sensor (TRIME-PICO IPH/T3, IMKO GmbH, Ettlingen, Germany) that was inserted into polyvinylchloride access tubes (Schedule 40, 5.25 cm id). The instrument was calibrated using bulk soil samples collected within the experimental plot area where maize followed soybean [49]. Five days after planting, the access tubes were inserted 1.0 m deep into the soil after boring holes with a reverse-taper bit on a hydraulically-driven soil tube (Giddings Machine Co., Windsor, CO, USA). Within a plot, the tubes were positioned between the center two maize rows at 9.5 and 28.5 cm from one of the rows. Due to time constraints for tube placement and frequent measurement of soil water content throughout the season, the access tubes were placed in the selected treatments that were chosen based on their potential to provide meaningful treatment comparisons. In the experiment where the previous crop was soybean, access tubes were placed in all replications of treatments of both hybrids receiving the optimal N rate and all three durations of drought stress. In the experiments where the previous crop was alfalfa or winter rye, access tubes were placed in all replications of treatments of the drought-tolerant hybrid receiving the optimal N rate and all three durations of drought stress, and in the treatment combination representing the optimal N rate, standard hybrid, and drought stress from V14 to R6.

At the time of maize emergence, three soil cores per block (i.e., replication) in each experiment were collected from the 0- to 1.0-m depth using a hydraulically driven soil tube with an inner diameter of 4.1 cm. Each soil core was separated into 20-cm increments and gravimetric soil water content was determined using the oven-drying method [50]. Gravimetric soil water content was then converted to volumetric soil water content and the sum of the values from the 0- to 1.0-m depth was considered as soil water storage at maize emergence. From the 12 leaf collar maize phenological stage (V12), prior to the application of different irrigation amounts to the drought stress treatments, until R6, soil water content was measured from plots with access tubes once per week unless precipitation occurred, in which case measurements were postponed to the day before the scheduled irrigation (six- to eight-day intervals). Total soil water content in the 0- to 1.0-m soil profile was used to calculate soil water deficit to determine irrigation water application from the 10 leaf collar maize phenological stage until R6, and for estimating ET$_a$ throughout the period of measurement [23,51,52] as described in the following sections. A sampling depth of 1.0 m was used for this study because (i) in these sand-dominated soils, the effective root zone is $\leq$1.0 m; and (ii) sand content in the 80- to 100-cm soil layer was 93% with a field capacity of 68 g kg^{-1} and a wilting point of 17 g kg^{-1}; thus, plant available water is minimal in this soil layer and also in those below it. Actual crop evapotranspiration, K$_{cb}$, and CWP were calculated for the selected treatments containing access tubes for measurement of soil water content, while IWP was determined for all experimental treatments, representing all combinations of three water treatments, three N rates, and two hybrids in each experiment.

Uniform irrigation was applied in all drought stress treatments using a solid-set sprinkler system from the 3 to 10 leaf collar maize phenological stages, and using an on-surface drip irrigation system from V12 to the mid-dent maize phenological stage (R5.5) [53]. The drip irrigation system for each experiment had an automated shut-off valve to apply the precise amount of water in each treatment. The drip tapes were connected to polyethylene header pipes (3.85 cm id) spaced 11.9 m apart and located at the front of each main plot (i.e., drought stress treatment). There was one line of drip tape on each side of every maize row. Each line of drip tape was placed 19 cm from the maize row and had emitters spaced at 15 cm intervals to achieve uniform distribution of water. One maize row was planted between each pair of adjacent main plots to mitigate edge effects resulting from differential water application. These buffer rows were fertilized with the 50% N rate and did not receive drip irrigation. For treatments with drought stress beginning at V14 and R2, irrigation was limited prior to V14 (21 July) and R2 (12 August), respectively, to achieve the desired moderate drought stress at V14 and R2. Compared to the well-watered control, the treatments with drought stress commencing at V14 and R2 received reduced amounts of water through drip irrigation beginning at the 12 and 16 leaf collar maize phenological stages, respectively.

A modified checkbook method [54] was used to determine the irrigation amount to apply twice each week. Daily potential crop evapotranspiration was estimated using reference ET_o [23], weekly volumetric soil moisture data, precipitation, and previous irrigation amounts. At each irrigation event, the amount of water applied in the well-watered treatment brought the soil moisture level back to field capacity, and the amount of water applied in the treatments with drought stress was 60 to 70% of that applied in the well-watered control [49]. Additionally, for every irrigation decision, a visual observation of maize and the weather forecast of the following few days were taken into consideration. Additional details on drip irrigation management for the treatments are described by Ao et al. [49].

2.3. Volumetric Soil Water Content Calculation

Volumetric soil water content was determined for each 20-cm increment within the upper 1.0 m of soil. The equivalent depth of soil water, θ_p (cm), for the entire measured depth of the soil profile was calculated as follows [52]:

$$\theta_p = 20\,(\theta_{20} + \theta_{40} + \theta_{60} + \theta_{80} + \theta_{100}) \tag{1}$$

where θ_{20}, θ_{40}, θ_{60}, θ_{80}, and θ_{100} ($mm^3\ mm^{-3}$) are volumetric soil water content for the 0- to 20-, 20- to 40-, 40- to 60-, 60- to 80-, and 80- to 100-cm depths, respectively. This, along with ET_c, precipitation, and irrigation application, were used to estimate drainage, soil water storage, and ET_a.

2.4. Crop Coefficient, Soil Water Balance, and Evapotranspiration

Daily grass reference evapotranspiration was calculated using the Penman-Monteith equation [23, 51]. Daily air temperature and precipitation were obtained from an onsite weather station, maintained and managed by the University of Minnesota. Solar radiation, wind speed, and relative humidity were obtained from the nearest USDA-Natural Resources Conservation Service Soil Climate Analysis Network (SCAN) weather station [55,56] located 6.4 km away at Crescent Lake, MN (SCAN Site Crescent Lake #1; 45°25′ N, 93°57′ W; 299 m above sea level) [57]. Vegetation at this SCAN weather station site is grass and the soil is classified as Hubbard sandy loam (sandy, mixed, frigid Entic Hapludolls). The soil texture is sandy loam for the 0- to 23-cm depth layer, loamy sand for the 23- to 38-cm depth layer, and sand for the 38- to 203-cm depth layers. The SCAN weather stations receive annual preventative maintenance and sensor repair [55]. Incoming hourly data from SCAN weather stations are automatically validated, and values occurring beyond pre-established limits are identified and subsequently evaluated for accuracy based on plots of data over time and comparisons with data for other weather variables from different sensors [56].

Daily ET_c was estimated as the product of $(K_s K_{cb} + K_e)$ and ET_o [19,23], and was used to calculate deep percolation in the water balance equation. The K_{cb} was calculated using the dual K_c approach, accounting for the water-stress coefficient (K_s). Dual K_c is the sum of K_{cb} and K_e, reduced by water stress with drought conditions:

$$K_c = K_{cb} \times K_s + K_e \qquad (2)$$

where K_{cb} is the basal crop coefficient, K_s is the water stress coefficient, and K_e is the soil evaporation coefficient. All terms are unitless.

The basal crop coefficient was calculated for three phases of maize phenological development, representing initial (from the date of emergence to the three leaf collar stage), mid-season (from the 10 leaf collar stage to R5.5), and late-season (from R5.5 to R6) growth phases ($K_{cb\ ini}$, $K_{cb\ mid}$, and $K_{cb\ end}$, respectively) according to Allen et al. [23]. In all experiments, the maize canopy covered ≥80% of the soil surface at the 10 leaf collar stage of maize phenological development, supporting the use of this crop stage as the beginning of the mid-season growth phase [23]. During rapid canopy development, which in this study was taken from the 3 to 10 leaf collar maize phenological stages, K_{cb} is assumed to increase linearly between $K_{cb\ ini}$ and $K_{cb\ mid}$, at which time the full canopy cover reduces K_e to minimum values [23]. The values for $K_{cb\ ini}$, $K_{cb\ mid}$, and $K_{cb\ end}$ from Allen et al. [23] were used to calculate ET_c. When the daily mean minimum relative humidity was different from 45% or when wind speed (u_2) at 2 m was different from 2.0 m s^{-1}, $K_{cb\ mid}$ and $K_{cb\ end}$ values were adjusted as:

$$K_{cb} = K_{cb\ (Table)} + [0.04\ (u_2 - 2) - 0.004\ (RH_{min} - 45)]\ (h/3)^{0.3} \qquad (3)$$

where $K_{cb\ (Table)}$ is the value for $K_{cb\ mid}$ or $K_{cb\ end}$ (if ≥0.45) from Allen et al. [23], u_2 is the mean value for daily wind speed at a 2-m height over grass during the mid- or late-season growth phase (m s^{-1}) for 1 m s^{-1} ≤ u_2 ≤ 6 m s^{-1}, RH_{min} is the mean daily minimum relative humidity during the mid- or late-season growth phase (%) for 20% ≤ RH_{min} ≤ 80%, and h is the mean height of maize during the mid- or late-season phase (m) for 20% ≤ RH_{min} ≤ 80%. The K_s was estimated according to Allen et al. [23] as:

$$K_s = (TAW - D_r)/[(1 - p)\ TAW] \qquad (4)$$

for $D_r >$ (p)TAW, where K_s is a dimensionless transpiration reduction factor based on soil water content (0–1), TAW is total available soil water in the crop root zone (mm), D_r is root zone depletion (mm), and p is fraction of TAW that a crop can extract from the root zone without suffering water stress. The value of p was assumed to be 0.59, based on the recommendation of Allen et al. [23] for maize on coarse-textured soils.

The soil evaporation coefficient describes the evaporation components of ET_c. When surface soil is wet following rain or irrigation, K_e is maximum; when surface soil is dry, K_e is small or zero. The K_e was determined as [23]:

$$K_e = \min\ (K_r\ (K_{c\ max} - K_{cb}),\ f_{ew}\ K_{c\ max}) \qquad (5)$$

where K_r is the dimensionless evaporation reduction coefficient, $K_{c\ max}$ is the maximum value of K_c following precipitation or irrigation, and f_{ew} is the portion of soil that is exposed to solar radiation and wetted. Additional details regarding these equations used for calculating K_{cb}, K_e, and K_s are provided by Allen et al. [23,51].

Actual ET_a was calculated on six- to eight-day intervals using the water balance equation [23]:

$$ET_a = P + I - R - D + C \pm \Delta SF \pm \Delta S \qquad (6)$$

where P is precipitation, I is irrigation, R is surface runoff, D is deep percolation below the measured crop root zone, C is capillary rise, ΔSF is change in the horizontal subsurface water flux within the root zone, and ΔS is change in soil water storage during the measured time interval. All terms are in millimeters per unit time. The ΔS was determined using gravimetric soil moisture content measured at maize emergence and volumetric soil moisture content measured at regular intervals from V12 to R6 according to Equation (1).

Deep percolation was estimated using daily precipitation, soil water content at the time of planting, dates and amounts of irrigation water supplied, effective crop rooting depth, crop phenology, and soil properties. Deep percolation was calculated as follows [15]:

$$D_j = \max\,(P_j + I_j - R_j - ET_{cj} - \Delta CD_{j-1},\, 0) \tag{7}$$

where D_j is deep percolation on day j, P_j is precipitation on day j, I_j is irrigation on day j, R_j is precipitation and/or irrigation runoff from the soil surface on day j, ET_{cj} is crop evapotranspiration on day j, and ΔCD_{j-1} is cumulative depletion depth in the root zone at the end of day j–1. All terms are in millimeters per unit time. Surface runoff from the experimental sites was estimated using the curve number method, with a curve number of 75 based on land use and soil properties of the sites [58].

It was assumed that subsurface upward water flux and capillary rise were negligible in the sand and loamy sand soil. Therefore, the soil water balance equation for ET_a calculation was reduced to:

$$ET_a = P + I - R - D \pm \Delta S \tag{8}$$

For comparative purposes, actual seasonal basal crop coefficient (K_{ab}) was determined following a modified approach used by previous researchers for K_{cb} calculation under deficit irrigation [59,60]. The K_{ab} was calculated on six- to eight-day intervals using estimated ET_o and evaporation (E) according to the equations from Allen et al. [23], and ET_a was calculated from the field data for the well-watered control and the two treatments with sustained moderate drought stress as:

$$K_{ab} = K_{cb} \times K_s = (ET_a - E)/ET_o \tag{9}$$

where ET_a is actual crop ET (mm d^{-1}) and ET_o is reference ET (mm d^{-1}). Field-based K_s was computed using calculated ET_a and estimated ET_c based on K_{cb} and Allen et al. [23] as:

$$K_s = (ET_a - E)/(ET_c - E) \tag{10}$$

Crop water productivity and IWP of maize were calculated using modified equations from Payero et al. [19]:

$$CWP = GY/ET_a \tag{11}$$

$$IWP = GY/I \tag{12}$$

where CWP is crop water productivity (kg m^{-3}), GY is maize grain yield (g m^{-2}), ET_a is total seasonal actual crop evapotranspiration (mm), IWP is irrigation water productivity (kg m^{-3}), and I is total seasonal irrigation for a given drought stress treatment (mm).

The date of emergence was 31 May for all treatments and the date of R6 was 24, 20, and 18 September for the well-watered control and the treatments with drought stress beginning at V14 and R2, respectively. Reference evapotranspiration, ET_a, average $K_{ab\,mid}$, CWP, and IWP were calculated using a spreadsheet software program from maize emergence to 24 September for all three durations of drought stress. Additionally, we calculated growth-stage-specific K_{ab} using K_{ab} computed at six- to eight-day intervals from the 10 leaf collar maize phenological stage to R6 from the experiment where maize followed soybean. All possible precautions were taken to achieve precision in the measurement of soil water content and calculation of water balance components [61,62]. However, this study used the soil water balance approach to estimate ET_a, which is based on calculations that are reliant upon book values and assumptions. A more direct approach of measuring the influx and efflux of soil water into the crop root zone, such as weight-based lysimetry, may produce more representative estimates of ET_a [61].

2.5. Maize Grain Yield Measurement

At R6, maize ears were harvested from the center 2.4 m of the two center rows of each plot and oven dried at 60 °C until constant mass, followed by shelling and weighing of grain. Maize grain yield was calculated at 155 g kg^{-1} moisture. A detailed summary of the results of maize grain yield and yield components was reported by Ao et al. [49]. The grain yield data used in this manuscript were (i) analyzed for the treatment combinations with calculated ET_a, (ii) used to determine CWP for these same treatment combinations, and (iii) used with the irrigation amount to calculate IWP for all experimental treatments.

2.6. Statistical Analysis

Statistical analysis was performed for three datasets based on the treatments sampled, which represent: (i) all durations of drought stress for both hybrids at the optimal N rate from the experiment where maize followed soybean; (ii) all durations of drought stress for the drought-tolerant hybrid at the optimal N rate from all experiments; and (iii) drought stress from V14 to R6 for both hybrids at the optimal N rate from all experiments. Actual crop evapotranspiration, CWP, IWP, maize grain yield, and average $K_{ab\,mid}$ were analyzed for datasets (i), (ii), and (iii), and growth-stage-specific K_{ab} was analyzed for dataset (i).

Data were analyzed with mixed-effect linear models using the MIXED procedure of SAS [63] at $P \leq 0.05$. Block (nested within experiment), experiment (i.e., previous crop), and interactions with block were considered random effects, duration of drought stress and hybrid were considered fixed effects, and maize phenological stage for K_{ab} was considered a fixed effect with repeated measurement. The UNIVARIATE procedure of SAS and scatterplots of residuals versus predicted values were used to assess the normality and homogeneity of variance [64]. Means from significant fixed effects were compared with pairwise *t*-tests using the PDIFF option of the MIXED procedure of SAS.

3. Results

3.1. Growing Conditions and Water Supply

With the exception of precipitation, weather conditions were within the range for favorable maize growth (Table 1) [65], thereby allowing the imposed drought stress treatments to be evaluated in the absence of other abiotic stresses. Seasonal average air temperature was 19.8 °C and maximum daily average air temperature for July through August was 29.0 °C. Average seasonal relative humidity was 71%, monthly average wind speed did not exceed 2 m s^{-1} except in June, and the average vapor pressure deficit in July through August was 1.11 kPa. Total seasonal ET_0 from maize emergence to R6 was estimated as 469 (Table 2). Compared to ET_0, the estimated total seasonal ET_c was 25 mm greater for the well-watered control and 59 and 21 mm less for the treatments where drought stress began at V14 and R2, respectively, when averaged across experiments. The range of ET_c, 409 to 494 mm, estimated in this study was comparable to ET_c under deficit irrigation estimated by Trout and DeJonge in a more arid climate with lower relative humidity and ET_0 [66].

Total precipitation from maize emergence to R6 averaged 258 mm across all treatments (Table 2). Total precipitation from the 13 leaf collar maize phenological stage to the dough stage of maize was 33 mm, with the highest daily amount of 9 mm near the tasseling stage (Figure 1). This low amount of precipitation during this period of maize growth, coupled with judicious irrigation amounts, enabled sustained moderate drought stress to be achieved in the treatments with drought stress from V14 to R6 and R2 to R6. Irrigation was applied from the three leaf collar stage of maize phenological development until R5.5 (Tables 2 and 3; Figure 1), although soil water deficits may have occurred beyond this point (Figure 1). Averaged across experiments, total irrigation was 345, 224, and 177 mm for the well-watered control and treatments with drought stress beginning at R2 and V14, respectively (Table 2). A total of 105 mm of irrigation was applied from V14 to R6 in the treatment where drought stress began at V14. In the treatment where drought stress began at R2, 67 mm of irrigation was applied from R2 to R6. The approximate deep percolation water loss beyond the effective root zone was 150 mm in the well-watered control, and 35 and 31% less in the treatments where drought stress began at V14 and R2, respectively. This study considered the effective root zone to be 0 to 1.0 m, which is supported by the results of total root length density. Averaged across the experiments where maize followed soybean and winter rye, the well-watered control and the treatment with drought stress from V14 to R6, and both hybrids, total root length density of maize after harvest was 13.48, 10.04, 1.10, 0.44, and 0.10 cm cm^{-3} for the 0- to 15-, 15- to 30-, 30- to 45, 45- to 60-, and 60- to 90 cm soil layers, respectively, and the value of 0.10 cm cm^{-3} from the 60- to 90-cm soil layer was not significantly different from zero ($P = 0.439$) [67]. However, the measurement of soil water content below the 1.0-m depth in this study may have provided a more comprehensive assessment of soil water balance [14].

At the time of maize emergence, the soil was nearly at field capacity (only 9 mm soil water deficit) for all three drought stress treatments (Table 2, Figure 1). Across experiments, the average change in water storage from maize emergence to R6 was −18, −32, and −33 mm for the well-watered control and the treatments where drought stress began at V14 and R6, respectively. During the period of imposed drought stress, the average water storage was 75 and 72 mm, respectively, for the treatments where drought stress began at V14 and R2, and these values were 23 and 19% of total available water (Figure 1). During the milk stage of maize phenological development on 19 August and 26 August, soil water storage for the well-watered treatment was slightly less than the threshold below which evapotranspiration is reduced to less than potential values (90 mm). This occurred near the time when precipitation was absent for about three weeks, and may have resulted in a slight underestimation of irrigation amount to meet the crop evapotranspiration demand. The treatments with imposed drought stress from V14 to R6 and R2 to R6 experienced soil water storage that was near the permanent wilting point (57 mm) during the mid-R2 to early dent stages of maize phenological development from 16 August to 8 September. This may have caused maize in these treatments to experience more water stress near this time compared to other times during the period of imposed drought stress. The estimated cumulative growing season evaporation was 91, 98, and 98 mm for the well-watered control and the treatments where drought stress began at V14 and R2, respectively (Table 2). These values are comparable with the evaporation values reported by others from a semiarid area at the western edge of the central U.S. High Plains [66,68]. Additional details on growing conditions and water supply were reported by Ao et al. [49].

Table 1. Monthly average maximum and minimum air temperature (T_{max} and T_{min}, respectively), maximum and minimum relative humidity (RH_{max} and RH_{min}, respectively), wind speed, solar radiation, vapor pressure deficit, and reference evapotranspiration (ET_o), and monthly total precipitation during the growing season of 2013.

Month	T_{max}	T_{min}	RH_{max}	RH_{min}	Wind Speed	Solar Radiation	Vapor Pressure Deficit	ET_o	Total Precipitation
	°C	°C	%	%	m s^{-1}	MJ m^{-2} d^{-1}	kPa	mm d^{-1}	mm
June	28.8	14.3	94.3	44.1	2.3	18	0.9	4.2	111
July	27.6	13.3	97.5	48.0	1.9	21	1.1	4.5	61
August	29.9	14.4	93.8	47.2	1.3	20	1.1	4.1	23
September	23.0	8.1	95.0	44.3	1.7	14	0.9	3.0	62

Table 2. Means and standard errors of cumulative water balance components from maize emergence to maize physiological maturity (R6) for plots with the optimal nitrogen rate [1].

Previous Crop	Drought Stress [2]	ET_o [3]	ET_c [4]	Precipitation	Irrigation	Surface Runoff	ΔS [5]	Deep Percolation [6]	Evaporation [7]
						mm			
Alfalfa	None	469	494	258	345	8	−26 (7)	158 (9)	91
	R2-R6	469	448	258	224	6	−28 (2)	100 (3)	98
	V14-R6	469	409	258	177	4	−35 (2)	101 (1)	98
Soybean	None	469	494	258	348	8	−8 (2)	144 (2)	91
	R2-R6	469	448	258	221	5	−38 (3)	104 (2)	98
	V14-R6	469	411	258	184	4	−35 (3)	100 (2)	98
Winter rye	None	469	494	258	346	8	−19 (9)	149 (7)	91
	R2-R6	469	449	258	224	6	−33 (5)	108 (7)	98
	V14-R6	469	409	258	177	4	−25 (2)	93 (1)	98

[1] In the experiment where the previous crop was soybean, the means are from all combinations of three drought stress treatments and both hybrids (standard and drought tolerant). In the experiments where the previous crop was alfalfa and winter rye, the means are from three drought stress treatments with the drought-tolerant hybrid, and the treatment with drought stress from the 14 leaf collar maize phenological stage (V14) to maize physiological maturity (R6) with the standard hybrid. Standard errors are within parentheses for those variables with values that differed among replications due to variability in measured soil water content. [2] None: No drought stress; R2-R6: Drought stress from the blister maize phenological stage to R6; V14-R6: Drought stress V14 to R6. [3] Reference evapotranspiration calculated using the Penman-Monteith equation [23]. [4] Estimated crop evapotranspiration according to Allen et al. [23]. [5] Change in soil water storage during the measured time interval. [6] Estimated deep percolation of water below the measured crop root zone. [7] Estimated evaporation from wet soil.

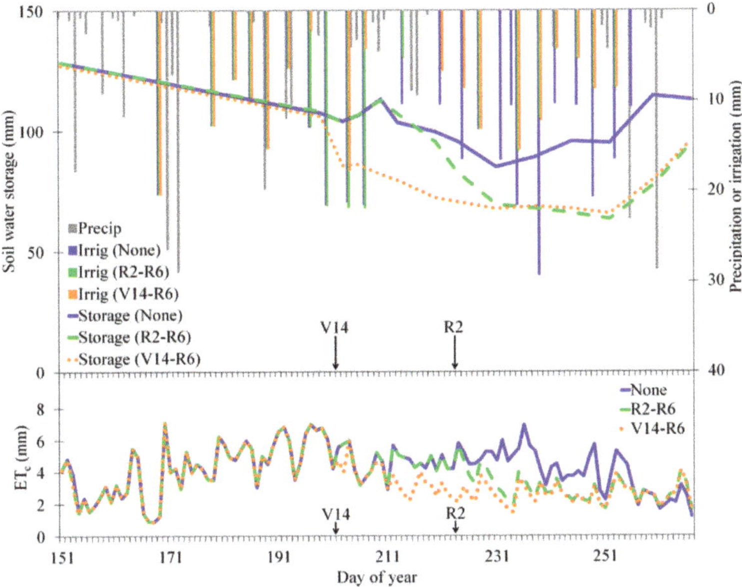

Figure 1. Daily precipitation (Precip), irrigation (Irrig), estimated soil water storage (Storage), and estimated crop evapotranspiration (ET_c) from maize emergence (day 151) to maize physiological maturity (R6, day 267) for treatments with no drought stress (None), drought stress from the 14 leaf collar maize phenological stage (V14) to R6 (V14-R6), and drought stress from the blister maize phenological stage (R2) to R6 (R2-R6) that received the optimal nitrogen rate, averaged across experiments. In the experiment where the previous crop was soybean, the values for soil water storage are from all combinations of three drought stress treatments and both hybrids (standard and drought tolerant). In the experiments where the previous crop was alfalfa and winter rye, the values for soil water storage are from three drought stress treatments with the drought-tolerant hybrid, and the treatment with drought stress from V14 to R6 with the standard hybrid.

Table 3. Total monthly irrigation (mm) for treatments with no drought stress (None), drought stress from the blister maize phenological stage (R2) to maize physiological maturity (R6) (R2-R6), and drought stress from the 14 leaf collar maize phenological stage (V14) to R6 (V14-R6) for experiments where maize followed alfalfa, soybean, and winter rye.

Month	Alfalfa			Soybean			Winter Rye		
	None	R2-R6	V14-R6	None	R2-R6	V14-R6	None	R2-R6	V14-R6
June	33	33	33	33	33	33	33	33	33
July	115	118	61	118	115	68	116	118	62
August	138	50	61	139	50	60	138	50	60
September	58	23	23	59	22	23	58	23	22

3.2. Actual Crop Evapotranspiration

Actual crop evapotranspiration and related CWP, IWP, and K_{ab} were evaluated for treatment combinations where volumetric soil moisture measurement access tubes were placed (Tables 4 and 5). In the experiment where maize followed soybean, seasonal ET_a for the treatment with the optimal N rate was affected only by drought stress (Table 5). Averaged across hybrids in the experiment following soybean, seasonal ET_a declined by 56 and 90 mm when drought stress began at R2 or V14 compared to the well-watered control, respectively (Table 4). Seasonal ET_a for the drought-tolerant hybrid with the optimal N rate was influenced by drought stress (Table 5). Averaged across experiments, seasonal

ET_a of the drought-tolerant hybrid with the optimal N rate declined by 60 and 97 mm when drought stress began at R2 and V14, respectively, compared to the well-watered control. When drought stress commenced at V14 with the optimal N rate, seasonal ET_a did not differ between hybrids and averaged 366 mm (Table 4). Maize grain yield is presented in Tables 4 and 5 to assist in the interpretation of ET_a, CWP, and IWP. A detailed description of grain yield in this study was reported by Ao et al. [49].

Table 4. Maize actual crop evapotranspiration (ET_a), grain yield, crop water productivity (CWP), irrigation water productivity (IWP), and mid-season actual basal crop coefficient ($K_{ab\ mid}$) for treatments with the optimal nitrogen rate, as influenced by the main effects of drought stress treatment and hybrid.

Components of Dataset	Treatment [1]	ET_a	Grain Yield	CWP	IWP	$K_{ab\ mid}$
		mm	kg m^{-2}	kg m^{-3}	kg m^{-3}	
Experiment where maize followed soybean, with	None	462 a [2]	1.374 a	2.97 a	3.94 b	1.08 a
four replications of both hybrids subjected to	R2-R6	406 b	0.932 b	2.29 b	4.22 b	0.89 b
three drought stress treatments	V14-R6	372 c	0.920 b	2.47 b	5.01 a	0.73 c
Three experiments, with four replications of the	None	464 a	1.459 a	3.14 a	4.22 b	1.08 a
drought-tolerant hybrid subjected to three	R2-R6	404 b	0.981 b	2.43 c	4.40 b	0.87 b
drought stress treatments	V14-R6	367 c	0.954 b	2.60 b	5.32 a	0.72 c
Three experiments, with four replications of both	ST	365	0.885 b	2.43 b	4.94 b	
hybrids subjected to drought stress from V14 to R6	DT	367	0.954 a	2.60 a	5.32 a	

[1] DT: Drought-tolerant hybrid; None: No drought stress; R2-R6: Drought stress from the blister maize phenological stage to maize physiological maturity (R6); ST: Standard hybrid; V14-R6: Drought stress from the 14 leaf collar maize phenological stage to R6. [2] Within a column for a given dataset, means followed by the same letter are not significantly different ($P \leq 0.05$).

Table 5. Significance of *F* tests ($P > F$) for fixed sources of variation involving drought stress treatment (D) and hybrid (H) on maize seasonal actual crop evapotranspiration (ET_a), grain yield, crop water productivity (CWP), irrigation water productivity (IWP), and mid-season actual basal crop coefficient ($K_{ab\ mid}$) for treatments with the optimal nitrogen rate.

Components of Dataset	Source of Variation	Dependent Variable				
		ET_a	Grain Yield	CWP	IWP	$K_{ab\ mid}$
Experiment where maize followed soybean, with	D	<0.001	<0.001	0.001	<0.001	<0.001
four replications of both hybrids subjected to	H	0.386	0.509	0.440	0.163	0.897
three drought stress treatments	D × H	0.802	0.152	0.121	0.079	0. 986
Three experiments, with four replications of the drought-tolerant hybrid subjected to three drought stress treatments	D	<0.001	<0.001	<0.001	<0.001	<0.001
Three experiments, with four replications of both hybrids subjected to drought stress from V14 to R6	H	0.216	0.010	0.0171	0.010	0.087

3.3. Water Productivity

Crop WP and IWP were influenced by drought stress (Table 5). Across hybrids in the experiment where maize followed soybean, CWP averaged 25% greater under well-watered conditions compared to when drought stress was imposed, and CWP did not differ between treatments with drought stress from V14 to R6 and R2 to R6 (Table 4). Across hybrids in the experiment where maize followed soybean, IWP averaged 23% greater when maize was exposed to drought stress beginning at V14 compared to the well-watered control and the treatment where drought stress began at R2. Across experiments for the drought-tolerant hybrid, CWP was 21 and 29% greater under well-watered conditions compared to when drought stress began at V14 and R2, respectively, and IWP with drought stress commencing at V14 averaged 23% greater compared to the well-watered control and the treatment with drought stress beginning at R2. Across experiments for the treatment where drought stress began at V14, CWP and IWP were 7 and 8% greater for the drought-tolerant hybrid than the standard hybrid, respectively, as a result of yield differences because hybrid did not affect ET_a.

3.4. Basal Crop Coefficient

In the experiment where maize followed soybean and received the optimal N rate, $K_{ab\,mid}$ was influenced by the drought stress treatment, but not by hybrid or interactions with hybrid (Table 5). Averaged across hybrids in the experiment following soybean, $K_{ab\,mid}$ declined as the duration of drought stress increased (Table 4).

For the drought-tolerant hybrid receiving the optimal N rate, $K_{ab\,mid}$ was affected by the drought stress treatment (Table 5). Across previous crops for the drought-tolerant hybrid with the optimal N rate, $K_{ab\,mid}$ was greater under well-watered conditions than under drought stress (Table 4). Compared to the well-watered control, $K_{ab\,mid}$ declined by 0.36 and 0.21 when drought stress began at V14 and R2, respectively. With drought stress imposed at V14 and the optimal N rate, $K_{ab\,mid}$ did not differ between hybrids.

When maize followed soybean and received the optimal N rate, growth stage-specific K_{ab} was influenced by the interaction between onset of moderate drought stress and phase of maize phenological development ($P < 0.001$). At each stage of maize development from V12 to R6, K_{ab} was greatest in the absence of drought (Table 6). Both K_{ab} and K_s declined below those of the well-watered control at the onset of each drought treatment (Figure 2). The field-based $K_{s\,act}$ was slightly higher than the K_s estimated according to Allen et al. [23] during the period of sustained moderate drought stress. Imposition of drought stress at V14 decreased K_{ab} more than drought stress imposed at R2 for most of the subsequent phenological stages.

Table 6. Actual basal crop coefficient (K_{ab}) by maize phenological stage for treatments with no drought stress (None), drought stress from the blister maize phenological stage (R2) to maize physiological maturity (R6) (R2-R6), and drought stress from the 14 leaf collar maize phenological stage (V14) to R6 (V14-R6), across hybrids in the experiment where maize followed soybean and received the optimal nitrogen rate.

Maize Phenological Stage [1]	None	R2-R6	V14-R6
V10	0.93 dA [2]	0.93 cA	0.93 bA
V12	1.13 abA	1.13 aA	1.02 aB
V16	1.13 abA	1.13 aA	1.06 aB
VT	1.12 abA	1.07 bB	0.81 cC
R1	1.10 bA	1.13 aA	0.74 dB
R2	1.13 abA	0.75 dB	0.61 efC
R3	1.12 abA	0.72 dB	0.44 hC
R4	1.11 bA	0.63 eB	0.57 gC
Early R5	0.84 eA	0.60 fB	0.60 efB
Mid-R5	1.15 aA	0.65 eB	0.59 fgC
Late R5/R6 [3]	0.99 cA	0.57 gC	0.63 eB
R6 [4]	0.64 [5]		

[1] V10: 10 leaf collar; V12: 12 leaf collar; V16: 16 leaf collar; VT: Tasseling; R1: Silking; R3: Milk; R4: Dough; R5: Dent. [2] Within a column, means followed by the same lowercase letter are not significantly different ($P \leq 0.05$). Within a row, treatment means followed by the same uppercase letter are not significantly different. [3] Late R5/R6 represents R5 for the treatment with no drought stress and R6 for the treatments with drought stress. [4] R6 maize phenological stage for the treatment with no drought stress. [5] Not included in the statistical analysis.

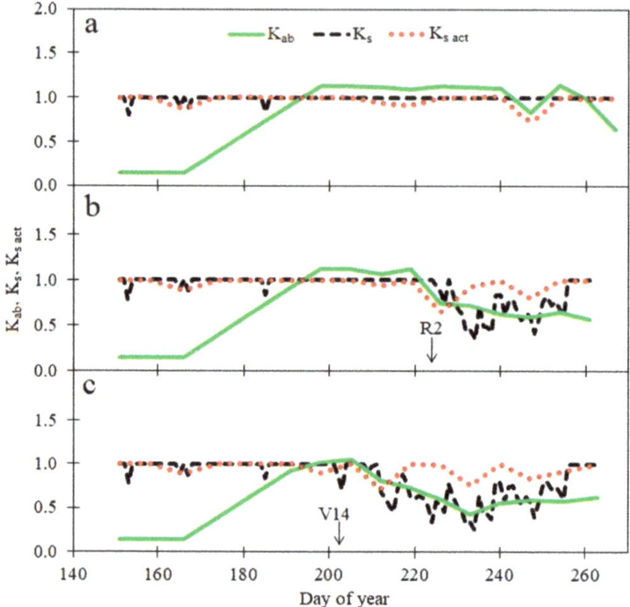

Figure 2. Actual basal crop coefficient (K_{ab}), estimated water stress coefficient (K_s) according to Allen et al. [22], and the field-based calculated water stress coefficient ($K_{s\ act}$) of maize by day of year when there was no drought stress (**a**), drought stress from the blister maize phenological stage (R2) to maize physiological maturity (R6) (**b**), and drought stress from the 14 leaf collar maize phenological stage (V14) to R6 (**c**), across hybrids for the experiment where maize followed soybean and received the optimal nitrogen rate.

4. Discussion

4.1. Maize Evapotranspiration

Results from our research support findings by many others that ET_a is dependent on maize physiological requirements, maize phenological stage at the onset of drought, and environmental conditions. Our results for the well-watered treatment are consistent with those from Dietzel et al. [8], who reported a minimum threshold of 400 to 450 mm of seasonal ET_a for optimum maize growth in the central U.S. Corn Belt, although this varies with seasonal ET_o, which is influenced by climatic variables including season length, air temperature, relative humidity, and solar radiation [69]. The range of ET_a in our experiments is also consistent with the range of 356 to 566 mm reported for a deficit irrigation study of maize in a temperate climate in Serbia [70]. However, in a longer growing season in the western U.S. Corn Belt with nearly twice as much rainfall, lower relative humidity, and higher solar radiation, Djaman and Irmak [14] reported substantially higher ET_a (587 to 627 mm). Both ET_a and maize grain yield were reduced by moderate drought, consistent with other reports [14,19]. However, our experiments showed that the grain yield was similar for the two treatments with imposed drought stress, despite lower ET_a when drought stress began at V14 compared to R2. Thus, our results may indicate that maize achieves more effective physiological acclimation with earlier exposure to drought stress. This was especially apparent for the drought-tolerant hybrid, which produced a similar grain yield with 367 mm of ET_a over the longer period of drought stress compared to 404 mm of ET_a with later onset of stress. This implies that additional mid-season water application may not contribute to increased WP. This result is unexpected since the late vegetative to early reproductive period is often reported as the phase when maize is most sensitive to drought stress [34,35].

4.2. Maize Water Productivity

As found in the experiment where maize followed soybean, greater ET_a in the absence of drought stress contributed to greater grain yield and greater CWP, supporting previous research showing that CWP is greater when maize is grown under well-watered compared to water deficit conditions [71–73]. However, CWP can be maximized under moderate drought stress [14,70,74] when the relative reduction in grain yield is less than the reduction in ET_a under deficit irrigation [15,29,66]. In our experiments, the drought-tolerant maize hybrid had greater yield and thus CWP and IWP than the standard hybrid under drought stress, but only when drought was initiated at V14. When drought was delayed until R2, prolonged vegetative growth increased ET_a and subsequent drought stress impaired grain production to a greater extent [49].

This difference between maize hybrids may reflect better physiological acclimation of the drought-tolerant hybrid to reduced ET_a under water deficit conditions [17]. For the particular drought-tolerant hybrid in the present study, this occurred when extended moderate drought stress began at a late vegetative stage, but not with the later onset of drought. Adaptation of this hybrid to drought stress may have been due to greater stomatal conductance, higher radiation use efficiency, and reduced transpiration rate [70,75], the benefits of which would be enhanced during late vegetative growth [34]. Recognizing that the mechanisms of drought tolerance may differ among improved maize hybrids, we infer that hybrids with similar mechanisms as the one we evaluated may allow farmers to limit water application earlier in maize development than recommended by Payero et al. [19].

4.3. Basal Crop Coefficient

In the present study, $K_{ab\,mid}$ of maize was reduced with the earlier onset (increased duration) of drought stress. Compared to the treatment with drought stress beginning at R2, drought stress beginning at V14 resulted in smaller plants with less leaf area for transpiration [76] and more instances of lower ET_a, which likely contributed to lower $K_{ab\,mid}$. The treatment with later onset of drought stress had lower $K_{ab\,mid}$ compared to the well-watered control, possibly because the exposed leaf area for transpiration was reduced by drought stress-induced leaf rolling that we observed in the field [34]. In the absence of drought stress, $K_{ab\,mid}$ in this study (1.08) was similar to the values reported by previous researchers [23,25,77], slightly higher than the range of 0.96 to 1.02 found by Facchi et al. [78], and 0.18 lower than the grass-based crop coefficient of maize reported by Djaman and Irmak [59]. The lower $K_{ab\,mid}$ in the two drought stress treatments was associated with less total available water in the 0- to 1.0-m soil layer, which limits evapotranspiration and reduces ET_a [23].

Growth stage-specific values of K_{ab} throughout the growing season in this study exhibited greater fluctuation when maize was exposed to drought stress. The smaller K_{ab} at V12 for the treatment with drought stress imposed at V14 compared to the other treatments was due to reduced irrigation beginning at the 11 leaf collar maize stage in order to create moderate drought stress by V14. Similarly, the 6% reduction in K_{ab} at tasseling compared to the 16 leaf collar maize phenological stage in the treatment with drought stress from R2 to R6 reflects judicious management of irrigation in preparation for exposing maize to moderate drought stress at R2. The drop in $K_{s\,act}$ for the well-watered treatment during the late reproductive period was likely caused by soil water storage during the milk stage of maize phenological development that was slightly less than the threshold below which evapotranspiration is reduced to less than potential values, and may have caused some stress in the well-watered treatment. For the treatments with imposed drought stress from V14 to R6 and R2 to R6, $K_{s\,act}$ was only moderately depressed compared to the estimated K_s. This may have been due to the crop being able to physiologically adjust to reduce evapotranspiration [23].

The decline in K_{ab} after imposition of drought stress coincided with the gradual decline in calculated K_s in both drought stress treatments. However, K_{ab} did not continue to decline with numerous, periodic declines in K_s. These results suggest that change in K_{ab} during the drought stress period occurred in two phases that can be categorized as a transitional or intermediate phase and an acclimated phase. During the transitional phase, K_{ab} declined as the crop acclimated to increasing crop water stress (K_s). Once acclimated, little change in K_{ab} occurred after sustained moderate drought stress conditions were established, even though K_s varied widely around a mean. This finding from the field supports the conclusions by Harb et al. [79], based on research conducted with *Arabidopsis* under highly controlled conditions. The initial decline in K_{ab} could have been brought about by the abrupt reduction in soil moisture content that reduced plant water potential, which in turn triggered a physiological mechanism for stomatal control of transpiration and gradual acclimation of K_{ab} to the newly adjusted soil moisture condition [80].

An important determinant of ET_a for making irrigation decisions is K_{ab}, which can vary depending on crop phenological stage and growing environment. Our results contribute to better estimation of K_{ab} for more precise irrigation management, especially when maize experiences prolonged, moderate drought stress.

5. Conclusions

This research was conducted under conditions that may well become more common as climate change alters rainfall amounts and timing in many temperate, humid, and sub-humid areas. Using a soil with relatively limited water holding capacity and a precision drip irrigation system allowed us to control the severity of drought stress. We exposed a standard maize hybrid and a drought-tolerant maize hybrid to extended periods of moderate drought stress, evidenced by midday leaf rolling that was relieved overnight. Imposition of drought stress during late vegetative development improved CWP compared to stress initiated during early reproductive development, even though these treatments produced a similar grain yield. The drought-tolerant hybrid showed greater improvement in IWP with a longer duration of drought stress than the standard hybrid.

We observed two phases of the crop coefficient, K_{ab}, in response to drought: an initial transitional phase represented by declining K_{ab}, followed by an acclimated phase with stable K_{ab}. If drought is prolonged, two subdivisions of K_{ab} may be required to estimate irrigation requirements accurately. In addition, it appears that maize can achieve a more effective physiological acclimation with earlier exposure to drought stress, since grain yield did not differ between the shorter and longer drought periods we imposed.

Author Contributions: Conceptualization, J.A.C. and M.P.R.; methodology, J.A.C., M.P.R., G.W.F., T.V., and S.A.; validation, J.A.C., M.P.R., and G.W.F.; formal analysis, S.A., J.A.C., M.P.R., and G.W.F.; investigation, S.A., J.A.C., T.V., M.P.R., and G.W.F.; resources, J.A.C., M.P.R., and G.W.F.; data curation, S.A. and J.A.C.; writing—original draft preparation, S.A. and J.A.C.; writing—review and editing, J.A.C., M.P.R., G.W.F., T.V., and S.A.; visualization, J.A.C., M.P.R., G.W.F., S.A., and T.V.; supervision, J.A.C., M.P.R., and G.W.F.; project administration, J.A.C.; funding acquisition, J.A.C. All authors have read and agreed to the published version of the manuscript.

Funding: This research was funded by the Minnesota Corn Research and Promotion Council, grant 4099-13SP.

Acknowledgments: The authors appreciate technical assistance from Eric Ristau, Ronald Faber, Todd Schumacher, Peter Boulay, and several others.

Conflicts of Interest: The authors declare no conflict of interest.

References

1. Dai, A. Increasing drought under global warming in observations and models. *Nat. Clim. Chang.* **2013**, *3*, 52–58. [CrossRef]
2. DeLucia, E.H.; Chen, S.; Guan, K.; Peng, B.; Li, Y.; Gomez-Casanovas, N.; Kantola, I.B.; Bernacchi, C.J.; Huang, Y.; Long, S.P.; et al. Are we approaching a water ceiling to maize yields in the United States? *Ecosphere* **2019**, *10*, e02773. [CrossRef]
3. Daryanto, S.; Wang, L.; Jacinthe, P.A. Global synthesis of drought effects on food legume production. *PLoS ONE* **2015**, *10*, e0127401. [CrossRef] [PubMed]
4. Daryanto, S.; Wang, L.; Jacinthe, P.A. Global synthesis of drought effects on maize and wheat production. *PLoS ONE* **2016**, *11*, e0156362. [CrossRef] [PubMed]

5. Nair, S.; Johnson, J.; Wang, C. Efficiency of irrigation water use: A review from the perspectives of multiple disciplines. *Agron. J.* **2013**, *105*, 351–363. [CrossRef]

6. Irmak, S.; Djaman, K.; Sharma, V. Winter wheat (*Triticum aestivum* L.) evapotranspiration and single (normal) and basal crop coefficients. *Trans. Am. Soc. Agric. Eng.* **2015**, *58*, 1047–1066.

7. Kumar, V.; Udeigwe, T.K.; Clawson, E.L.; Rohli, R.V.; Miller, D.K. Crop water use and stage-specific crop coefficients for irrigated cotton in the mid-south, United States. *Agric. Water Manag.* **2015**, *156*, 63–69. [CrossRef]

8. Dietzel, R.; Liebman, M.; Ewing, R.; Helmers, M.; Horton, R.; Jarchow, M.; Archontoulis, S. How efficiently do corn-and soybean-based cropping systems use water? A systems modeling analysis. *Glob. Change Biol.* **2016**, *22*, 666–681. [CrossRef]

9. Molden, D.; Murray-Rust, H.; Sakthivadivel, R.; Makin, I. A Water-Productivity Framework for Understanding and Action. In *Water Productivity in Agriculture: Limits and Opportunities for Improvement*; Kijne, J.W., Barker, R., Molden, D., Eds.; International Water Management Institute: Colombo, Sri Lanka, 2003; pp. 1–18.

10. Geerts, S.; Raes, D. Deficit irrigation as an on-farm strategy to maximize crop water productivity in dry areas. *Agric. Water Manag.* **2009**, *96*, 1275–1284. [CrossRef]

11. Li, X.; Zhang, X.; Niu, J.; Tong, L.; Kang, S.; Du, T.; Li, S.; Ding, R. Irrigation water productivity is more influenced by agronomic practice factors than by climatic factors in Hexi Corridor, Northwest China. *Sci. Rep.* **2016**, *6*, 37971. [CrossRef]

12. Sadler, E.J.; Bauer, P.J.; Busscher, W.J.; Millen, J.A. Site-specific analysis of a droughted corn crop: II. Water use and stress. *Agron. J.* **2000**, *92*, 403–410. [CrossRef]

13. Hao, B.; Xue, Q.; Marek, T.H.; Jessup, K.E.; Becker, J.; Hou, X.; Xu, W.; Bynum, E.D.; Bean, B.W.; Colaizzi, P.D.; et al. Water use and grain yield in drought-tolerant corn in the Texas High Plains. *Agron. J.* **2015**, *107*, 1922–1930. [CrossRef]

14. Djaman, K.; Irmak, S. Soil water extraction patterns and crop, irrigation, and evapotranspiration water use efficiency of maize under full and limited irrigation and rainfed settings. *Trans. Am. Soc. Agric. Eng.* **2012**, *55*, 1223–1238. [CrossRef]

15. Panda, R.K.; Behera, S.K.; Kashyap, P.S. Effective management of irrigation water for maize under stressed conditions. *Agric. Water Manag.* **2004**, *66*, 181–203. [CrossRef]

16. Hernández, M.; Echarte, L.; Della Maggiora, A.; Cambareri, M.; Barbieri, P.; Cerrudo, D. Maize water use efficiency and evapotranspiration response to N supply under contrasting soil water availability. *Field Crops Res.* **2015**, *178*, 8–15. [CrossRef]

17. Tolk, J.A.; Evett, S.R.; Xu, W.; Schwartz, R.C. Constraints on water use efficiency of drought tolerant maize grown in a semi-arid environment. *Field Crops Res.* **2016**, *186*, 66–77. [CrossRef]

18. Blum, A. Effective use of water (EUW) and not water-use efficiency (WUE) is the target of crop yield improvement under drought stress. *Field Crops Res.* **2009**, *112*, 119–123. [CrossRef]

19. Payero, J.O.; Tarkalson, D.D.; Irmak, S.; Davison, D.; Petersen, J.L. Effect of timing of a deficit-irrigation allocation on corn evapotranspiration, yield, water use efficiency and dry mass. *Agric. Water Manag.* **2009**, *96*, 1387–1397. [CrossRef]

20. Bausch, W.; Trout, T.; Buchleiter, G. Evapotranspiration adjustments for deficit irrigated corn using canopy temperature: A concept. *Irrig. Drain.* **2011**, *60*, 682–693. [CrossRef]

21. Rudnick, D.R.; Irmak, S. Impact of water and nitrogen management strategies on maize yield and water productivity indices under linear-move sprinkler irrigation. *Trans. Asabe* **2013**, *56*, 1769–1783.

22. Ogola, J.B.O.; Wheeler, T.R.; Harris, P.M. Effects of nitrogen and irrigation on water use of maize crops. *Field Crops Res.* **2002**, *78*, 105–117. [CrossRef]

23. Allen, R.G.; Pereira, L.S.; Raes, D.; Smith, M. *Crop. Evapotranspiration Guidelines for Computing Crop Water Requirements. Irrig. Drainage Paper 56*; Food Agriculture Organization: Rome, Italy, 1998.

24. Allen, R.G. Using the FAO-56 dual crop coefficient method over an irrigated region as part of an evapotranspiration intercomparison study. *J. Hydrol.* **2000**, *229*, 27–41. [CrossRef]

25. Piccinni, G.; Ko, J.; Marek, T.; Howell, T. Determination of growth-stage-specific crop coefficients (K_c) of maize and sorghum. *Agric. Water Manag.* **2009**, *96*, 1698–1704. [CrossRef]

26. El-Hendawy, S.E.; El-Lattief, E.A.A.; Ahmed, M.S.; Schmidhalter, U. Irrigation rate and plant density effects on yield and water use efficiency of drip-irrigated corn. *Agric. Water Manag.* **2008**, *95*, 836–844. [CrossRef]

27. Lindsey, A.J. Agronomic and Physiological Responses Modern Drought-Tolerant Maize (*Zea mays* L.) hybrids to Agronomic Production Practices. Ph.D. Thesis, Ohio State University, Columbus, OH, USA, 2015.

28. Irmak, S.; Mohammed, A.T.; Kranz, W.L. Grain yield, crop and basal evapotranspiration, production functions, and water productivity response of drought-tolerant and non-drought-tolerant maize hybrids under different irrigation levels, population densities, and environments: Part II. In south-central and northeast Nebraska's transition zone and sub-humid environments. *Appl. Eng. Agric.* **2019**, *35*, 83–102.

29. Kang, S.; Gu, B.; Du, T.; Zhang, J. Crop coefficient and ratio of transpiration to evapotranspiration of winter wheat and maize in a semi-humid region. *Agric. Water Manag.* **2003**, *59*, 239–254. [CrossRef]

30. Aydinsakir, K.; Erdal, S.; Buyuktas, D.; Bastug, R.; Toker, R. The influence of regular deficit irrigation applications on water use, yield, and quality components of two corn (*Zea mays* L.) genotypes. *Agric. Water Manag.* **2013**, *128*, 65–71. [CrossRef]

31. Cairns, J.E.; Hellin, J.; Sonder, K.; Araus, J.L.; MacRobert, J.F.; Thierfelder, C.; Prasanna, B.M. Adapting maize production to climate change in sub-Saharan Africa. *Food Secur.* **2013**, *5*, 345–360. [CrossRef]

32. Lobell, D.B.; Roberts, M.J.; Schlenker, W.; Braun, N.; Little, B.B.; Rejesus, R.M.; Hammer, G.L. Greater sensitivity to drought accompanies maize yield increase in the US Midwest. *Science* **2014**, *344*, 516–519. [CrossRef]

33. National Drought Mitigation Center. U.S. Drought Monitor Map Archive. Univ. of Nebraska, Lincoln. Available online: https://droughtmonitor.unl.edu/maps/maparchive.aspx (accessed on 10 September 2020).

34. Çakir, R. Effect of water stress at different development stages on vegetative and reproductive growth of corn. *Field Crops Res.* **2004**, *89*, 1–16. [CrossRef]

35. Kranz, W.L.; Irmak, S.; Van Donk, S.J.; Yonts, C.D.; Martin, D.L. Irrigation Management for Corn. Nebguide G1850. Univ. of Nebraska, Lincoln. 2008. Available online: http://extensionpublications. unl.edu/assets/html/g1850/build/g1850.htm (accessed on 10 September 2020).

36. Sharma, V.; Irmak, S.; Djaman, K.; Sharma, V. Large-scale spatial and temporal variability in evapotranspiration, crop water-use efficiency, and evapotranspiration water-use efficiency of irrigated and rainfed maize and soybean. *J. Irrig. Drain. Eng.* **2015**, *142*, 04015063. [CrossRef]

37. Lu, Y.; Zhang, X.; Chen, S.; Shao, L.; Sun, H. Changes in water use efficiency and water footprint in grain production over the past 35 years: A case study in the North China Plain. *J. Clean. Prod.* **2016**, *116*, 71–79. [CrossRef]

38. Lopes, M.S.; Araus, J.L.; Van Heerden, P.D.; Foyer, C.H. Enhancing drought tolerance in C_4 crops. *J. Exp. Bot.* **2011**, *62*, 3135–3153. [CrossRef] [PubMed]

39. Edmeades, G.O. *Progress in Achieving and Delivering Drought Tolerance in Maize—An Update*; ISAAA: Ithaca, NY, USA, 2013.

40. Edmeades, G.O.; Chapman, S.C.; Lafitte, H.R. Selection improves drought tolerance in tropical maize populations: I. Gains in biomass, grain yield, and harvest index. *Crop. Sci.* **1999**, *39*, 1306–1315. [CrossRef]

41. Tollenaar, M.; Lee, E.A. Yield potential, yield stability and stress tolerance in maize. *Field Crops Res.* **2002**, *75*, 161–169. [CrossRef]

42. Campos, H.; Cooper, M.; Habben, J.E.; Edmeades, G.O.; Schussler, J.R. Improving drought tolerance in maize: A view from industry. *Field Crops Res.* **2004**, *90*, 19–34. [CrossRef]

43. Campos, H.; Cooper, M.; Edmeades, G.O.; Loffler, C.; Schussler, J.R.; Ibanez, M. Changes in drought tolerance in maize associated with fifty years of breeding for yield in the US Corn Belt. *Maydica* **2006**, *51*, 369–381.

44. Roth, J.A.; Ciampitti, I.A.; Vyn, T.J. Physiological evaluations of recent drought-tolerant maize hybrids at varying stress levels. *Agron. J.* **2013**, *105*, 1129–1141. [CrossRef]

45. USDA-Natural Resources Conservation Service. Web Soil Survey. Available online: http://websoilsurvey. nrcs.usda.gov/ (accessed on 10 September 2020).

46. Rehm, G.W.; Malzer, G.L.; Wright, J.A. *Managing Nitrogen for Corn Production on Iirrigated Sandy Soils*; University of Minnesota Extension: St. Paul, MN, USA, 1989; Available online: http://www.wadenaswcd. org/AG-FO-2392-1.pdf (accessed on 10 September 2020).

47. Rehm, G.W.; Lamb, J.; Rosen, C.; Randall, G. *Best Management Pracrtices for Nitrogen on Coarse Textured Soils*; University of Minnesota Extension: St. Paul, MN, USA, 2008. Available online: https://conservancy.umn. edu/handle/11299/198230 (accessed on 10 September 2020).

48. Kaiser, D.E.; Lamb, J.A.; Eliason, R. *Fertilizer Guidelines for Agronomic Crops in Minnesota*; BU-06240-S; University of Minnesota Extension: St. Paul, MN, USA, 2011. Available online: https://conservancy.umn.edu/bitstream/handle/11299/198924/Fertilizer%20Guidelines%20for%20Agronomic%20Crops%20in%20Minnesota.pdf?sequence=1&isAllowed=y (accessed on 10 September 2020).

49. Ao, S.; Russelle, M.P.; Varga, T.; Feyereisen, G.W.; Coulter, J.A. Drought tolerance in maize is influenced by timing of drought stress initiation. *Crop. Sci* **2020**, *60*, 1591–1606. [CrossRef]

50. Cresswell, H.P.; Hamilton, J.G. Bulk density and pore space relations. In *Soil Physical Measurement and Interpretation for Land Evaluation*; McKenzie, N.J., Coughlan, K.L., Cresswell, H.P., Eds.; CSIRO Publishing: Collingwood, Australia, 2002; pp. 35–58.

51. Allen, R.G.; Pereira, L.S.; Smith, M.; Raes, D.; Wright, J.L. FAO-56 dual crop coefficient method for estimating evaporation from soil and application extensions. *Irrig. Drain. Eng.* **2005**, *131*, 2–13. [CrossRef]

52. Schmidt, J.P.; Sripada, R.P.; Beegle, D.B.; Rotz, C.A.; Hong, N. Within-field variability in optimum nitrogen rate for corn linked to soil moisture availability. *Soil Sci. Soc. Am. J.* **2011**, *75*, 306–316. [CrossRef]

53. Varga, T.; Feyereisen, G.W.; Russelle, M.P.; Ao, S.; Coulter, J.A. Versatile, precise drip irrigation system for agronomic small-plot research. In Proceedings of the ASABE 2018 Annual International Meeting, Detroit, MI, USA, 19 July–1 August 2018. Available online: https://doi.org/10.13031/aim.201801700 (accessed on 10 September 2020).

54. Wright, J. *Irrigation Scheduling Checkbook Method*; University of Minnesota Extension: St. Paul, MN, USA, 2002; Available online: https://extension.umn.edu/irrigation/irrigation-scheduling-checkbook-method (accessed on 10 September 2020).

55. Schaefer, G.L.; Cosh, M.H.; Jackson, T.J. The USDA Natural Resources Conservation Service Soil Climate Analysis Network (SCAN). *J. Atmos. Ocean. Technol.* **2007**, *24*, 2073–2077. [CrossRef]

56. USDA-Natural Resources Conservation Service National Water and Climate Center. Soil Climate Analysis Network (SCAN). Available online: https://www.wcc.nrcs.usda.gov/scan/scan_brochure.pdf (accessed on 10 September 2020).

57. USDA-Natural Resources Conservation Service National Water and Climate Center. Scan Site: Crescent Lake #1. Available online: https://wcc.sc.egov.usda.gov/nwcc/site?sitenum=2002 (accessed on 10 September 2020).

58. USDA-Natural Resources Conservation Service. *Chapter 10: Hydrology. National English Handbook*; USDA: Washington, DC, USA, 2004.

59. Djaman, K.; Irmak, S. Actual crop evapotranspiration and alfalfa- and grass-reference crop coefficients of maize under full and limited irrigation and rainfed conditions. *J. Irrig. Drain. Eng.* **2013**, *139*, 433–446. [CrossRef]

60. Suleiman, A.A.; Soler, C.M.T.; Hoogenboom, G. Evaluation of FAO-56 crop coefficient procedures for deficit irrigation management of cotton in a humid climate. *Agric. Water Manag.* **2007**, *91*, 33–42. [CrossRef]

61. Allen, R.G.; Pereira, L.S.; Howell, T.A.; Jensen, M.E. Evapotranspiration information reporting: I. Factors governing measurement accuracy. *Agric. Water Manag.* **2011**, *98*, 899–920. [CrossRef]

62. Allen, R.G.; Pereira, L.S.; Howell, T.A.; Jensen, M.E. Evapotranspiration information reporting: II. Recommended documentation. *Agric. Water Manag.* **2011**, *98*, 921–929. [CrossRef]

63. SAS Institute. *The SAS system for Windows*; Version 9.3; SAS Institute: Cary, NC, USA, 2011.

64. Kutner, M.H.; Nachtsheim, C.J.; Neter, J. *Applied Linear Regression Models*, 4th ed.; McGraw-Hill: New York, NY, USA, 2004; pp. 107–152.

65. Crafts-Brandner, S.J.; Salvucci, M.E. Sensitivity of photosynthesis in a C_4 plant, maize, to heat stress. *Plant. Physiol.* **2002**, *129*, 1773–1780. [CrossRef]

66. Trout, T.J.; DeJonge, K.C. Water productivity of maize in the US high plains. *Irrig. Sci.* **2017**, *35*, 251–266. [CrossRef]

67. Ao, S. Morpho-Physiological Traits Associated with Drought Tolerance of Maize Hybrids Subjected to Different Water and Nitrogen Supply. Ph.D. Thesis, University of Minnesota, Twin Cities, MN, USA, 2016. Available online: http://hdl.handle.net/11299/185203 (accessed on 10 September 2020).

68. Trout, T.J.; DeJonge, K.C. Crop water use and crop coefficients of maize in the Great Plains. *J. Irrig. Drain. Eng.* **2018**, *144*, 04018009. [CrossRef]

69. Wang, K.; Xu, Q.; Li, T. Does recent climate warming drive spatiotemporal shifts in functioning of high-elevation hydrological systems? *Sci. Total Environ.* **2020**, *719*, 137507. [CrossRef]

70. Kresović, B.; Tapanarova, A.; Tomić, Z.; Životić, L.; Vujović, D.; Sredojević, Z.; Gajić, B. Grain yield and water use efficiency of maize as influenced by different irrigation regimes through sprinkler irrigation under temperate climate. *Agric. Water Manag.* **2016**, *169*, 34–43. [CrossRef]
71. Igbadun, H.E.; Salim, B.A.; Tarimo, A.K.P.R.; Mahoo, H.F. Effects of deficit irrigation scheduling on yields and soil water balance of irrigated maize. *Irrig. Sci.* **2008**, *27*, 11–23. [CrossRef]
72. Howell, T.A.; Schneider, A.D.; Evett, S.R. Subsurface and surface microirrigation of corn —Southern High Plains. *Trans. Am. Soc. Agric. Eng.* **1997**, *40*, 635–641. [CrossRef]
73. Couto, A.; Padín, A.R.; Reinoso, B. Comparative yield and water use efficiency of two maize hybrids differing in maturity under solid set sprinkler and two different lateral spacing drip irrigation systems in León, Spain. *Agric. Water Manag.* **2013**, *124*, 77–84. [CrossRef]
74. Otegui, M.E.; Andrade, F.H.; Suero, E.E. Growth, water use, and kernel abortion of maize subjected to drought at silking. *Field Crop. Res.* **1995**, *40*, 87–94. [CrossRef]
75. Mansouri-Far, C.; Sanavy, S.A.M.M.; Saberali, S.F. Maize yield response to deficit irrigation during low-sensitive growth stages and nitrogen rate under semi-arid climatic conditions. *Agric. Water Manag.* **2010**, *97*, 12–22. [CrossRef]
76. Facchi, A.; Gharsallah, O.; Corbari, C.; Masseroni, D.; Mancini, M.; Gandolfi, C. Determination of maize crop coefficients in humid climate regime using the eddy covariance technique. *Agric. Water Manag.* **2013**, *130*, 131–141. [CrossRef]
77. Lee, E.A.; Tollenaar, M. Physiological basis of successful breeding strategies for maize grain yield. *Crop. Sci.* **2007**, *47*, S-202–S-215. [CrossRef]
78. Escobar-Gutiérrez, A.J.; Combe, L. Senescence in field-grown maize: From flowering to harvest. *Field Crops Res.* **2012**, *134*, 47–58. [CrossRef]
79. Harb, A.; Krishnan, A.; Ambavaram, M.M.R.; Pereira, A. Molecular and physiological analysis of drought stress in Arabidopsis reveals early responses leading to acclimation in plant growth. *Plant. Physiol.* **2010**, *154*, 1254–1271. [CrossRef]
80. Osakabe, Y.; Osakabe, K.; Shinozaki, K.; Tran, L.S. Response of plants to water stress. *Front. Plant. Sci.* **2014**, *5*, 86. [CrossRef] [PubMed]

Article

Designing of an Enterprise Resource Planning for the Optimal Management of Agricultural Plots Regarding Quality and Environmental Requirements

Damián Aguilar Morales [1], Paola Sánchez-Bravo [2], Leontina Lipan [2], Marina Cano-Lamadrid [2], Hanán Issa-Issa [2], Francisco J. del Campo-Gomis [1] and David B. López Lluch [1,*]

[1] Departamento Economía Agroambiental, Escuela Politécnica Superior de Orihuela (EPSO), Universidad Miguel Hernández de Elche (UMH), Carretera de Beniel, km 3.2, 03312 Orihuela, ALC, Spain; damian.aguilarm@gmail.com (D.A.M.); francis.delcampo@umh.es (F.J.d.C.-G.)

[2] Departamento Tecnología Agroalimentaria, Grupo Calidad y Seguridad Alimentaria, Escuela Politécnica Superior de Orihuela (EPSO), Universidad Miguel Hernández de Elche (UMH), Carretera de Beniel, km 3.2, 03312 Orihuela, ALC, Spain; paola.sb94@gmail.com (P.S.-B.); leontina.lipan@goumh.umh.es (L.L.); marina.cano.umh@gmail.com (M.C.-L.); hanan.issa@gmail.com (H.I.-I.)

* Correspondence: david.lopez@umh.es

Received: 12 August 2020; Accepted: 7 September 2020; Published: 8 September 2020

Abstract: One of the main current problems in European quality agricultural production is the lack of objective data for linking quality to origin and to evidence environmental concern (CO_2 uptake and use of water in Spain). The'aim of this study was to develop an agricultural management platform, based on Enterprise Resource Planning (ERP) principles and with the ability to collect geolocated information from different plots related to Protected Designation of Origin (PDO) and Protected Geographical Indication (PGI) wine production. First a survey to farmers, technicians and PDO and PGI managers was carried out to detect the needs of the three groups in relation to ERP platforms; and secondly an ERP platform was developed to collect agronomic information to comply with the Spanish legal requirements. Results showed that the end user completes information database, complies with the legal requirements, and obtains benefits derived from the data analysis. Consequently, the platform (i) solves lack of agricultural data problem; (ii) provides the user with management tools for its agricultural operations; (iii) allows the decision maker to obtain geolocated information in real time; and (iv) sets out the bases for the future development of agricultural systems based on Big Data.

Keywords: ERP; GIS; internet of things; precision agriculture; quality; environment; water; software; platform; web application

---

1. Introduction

European agriculture faces two mayor challenges, quality linked to origin and production systems that must reduce impacts to environment. Both need real and continuous data to show their evidences to society. Currently, to achieve this, a change in the agricultural paradigm is necessary through the incorporation of four new technologies: Big Data, Precision Agriculture, Industry 4.0., and Enterprise Resource Planning (ERP). This is specially required in the case of food quality productions linked to origin such as Protected Geographical Indications (PGI) and Protected Designations of Origin (PDO).

The potential of regional products differentiated for their quality has been recognized by the introduction of legislation governing PGI and PDO. These certifications are intended to facilitate the consumer's recognition of the product and perception of superior quality [1,2]. In addition, this has contributed to quality labelling becoming a source of competitive advantage [2]. Felzensztein [3] analyzed the effect of the country or region of origin of agricultural production as an important source of competitive advantage. They reported that for farm products with a long tradition, such as wine, the region of origin (considered as a more specific area than the country of origin) can provide major market positioning opportunities of creating a sustainable competitive advantage.

Protecting PDO and PGI references is a key aspect of the European Common Agricultural Policy (CAP). A good example is the case of the Tocai/Friulano grape variety traditionally grown in Italy, which came into dispute with the Hungarian wine Tokaji, arguing that Italian wines took advantage of the Tokay Designation of Origin since they sounded similar, and finally the European Union prohibited Italian wines from using the brand [4,5].

In this sense, the name of an area, that is strongly associated with a given product quality, is associated to high quality products even though it might not be, and in many cases these associations drive the consumer's purchasing decision [6]. Terroir underpins the process of demarcation so that the concept fits neatly here and relates to both environmental and cultural factors that together influence the complete production process (in this case, form the grape growing to wine production continuum). The physical factors that influence the process include matching a given agricultural product to its ideal climate along with optimum site characteristics of elevation, slope, aspect, and soil [7].

The quality of an agricultural product is linked to PDO or PGI scheme, however, is the result of the combination of five main factors: the climate, the site or local topography, the nature of the geology and soil, the choice of the variety, and how they are together managed to produce the best crop [7]. To develop a system that collects data from the plot, ensures origin and allows continuous improvement of quality is a driving force for quality agriculture. This is especially relevant in the case of PDO and PGI as they face, at least, three decision levels: farmers, processing companies and PDO and PGI managers.

Furthermore, CAP has three clear environmental goals, each of which are listed in the European Green Deal and Farm to Fork strategy: tackling climate change, protecting natural resources, and enhancing biodiversity. Farmers must be aware about the need to provide evidences that they are ensuring the best practices to get CAP environmental goals. The use of water is a key concern in the case of Spanish agriculture.

All represent a technological challenge for agriculture, but together they can be the definitive tools to achieve optimal and sustainable quality agricultural management. In this sense, Big Data allows interpretation and analysis of the constant flow of information provided by Precision Agriculture through the interactivity between devices given by Industry 4.0., and finally, ERP will manage all this. Big Data is often described as a new frontier within the world of new technologies, providing companies with a competitive advantage [8].

The concept of this new paradigm is based on the management and storage of large amounts of data, which are then analyzed in search of patterns or models and is based on five basic principles [9]: volume, variety, velocity, value, and veracity. The required information for Big Data to be able to improve agricultural management may come from various sources, mainly based on Precision Agriculture. These provide a high and constant flow of information, whose management will be one of the main challenges to overcome. This new situation is going to suppose a revolution in the way in which information is stored and administered and will be able to provide a source of innovation and of added economic value [10]. Big Data faces a similar situation to that which arose at the beginning of the current decade, when it was debated whether cloud computing was an appropriate or necessary strategy and how it should be integrated into companies [11]. The technological challenges related to Big Data will result in the following issues that must be solved by each organization:

- The storage infrastructure. This can reveal the limitations of the company's technical capabilities and requires major changes at different levels of the organization [12].

- The technology to integrate the information into the system. At this point, development of an Application Programming Interface (API) is essential, allowing Internet of Things (IoT) components to communicate and exchange information with the database in a safe way [13].
- The diversity of the formats according to the source. Data collection from such different sources and their translation into useful and structured information are two of the big challenges that Big Data faces [14].
- The calibration system, source verification and data security.
- The speed of data processing. Traditional Relational Database Management Systems (RDBMS) have been the systems mainly used since 1970 [15]. They show great efficiency and integrity in data management; but they are not designed to be systems distributed or balanced depending on the load. This is where NoSQL or non-relational systems can show their full potential, offering functionalities in data analysis that are impossible for traditional RDBMS systems [16]. In this sense, one of the main solutions that companies are adopting is the use of a hybrid data storage system, where the best properties of each of the systems are exploited [17].

Agriculture is an activity traditionally subject to instability and inaccuracy since it depends on biological and climatic factors that add uncertainty to the system. A way to alleviate these problems is incorporating the variability of agricultural activity into the decision-making process by increasing the information points and the flow of data. One of the main tools to do this is Precision Agriculture, which gives the possibility of managing the decision-making on the farm to identify, quantify, and respond to the variability [18].

Therefore, the problem that arises is to establish data flows linked to the farm plot and several tools are available for this:

- Geographic Information System (GIS) which provides information separated into layers. This technological system has been used for decades for territorial and environmental management [19].
- Satellite images. For instance, Normalized Difference Vegetation Index (NDVI) images permit differentiating distinct states of vegetative growth.
- Weather stations at the plot level as climatic variability is an important aspect of uncertainty in agriculture.
- IoT. Allows the development of sensors and automation to obtain plot information in real time.
- External information sources. Access to databases, public or private, that can provide information (climatic, geographic, statistical, economic, etc.).

However, Precision Agriculture cannot provide the most decisive data for model development through Big Data. These data are those obtained from plot management and cover aspects such as water use, pest incidence, fertilizer use, phytosanitary treatments, and costs. These data can only be provided by the farmer. Historically, this information has been collected through field surveys (a slow and expensive process). In addition, many farmers have reservations about sharing their data. In this respect it is worth mentioning that, in October 2014, the American Farm Bureau conducted a survey of a group of farmers and 77.5% of them responded that they were afraid that either the regulators or the government could have access to the data that they were facilitating [20]. This concern, regarding such valuable information, should lead to the implementation of a data collection system that respects farmer privacy.

It is possible to go further through the use of new technologies that allows the setting-up of systems capable of exchanging information among themselves, and even create optimized networks that communicate with each other (via wireless or another system), and finally send the information to the data center where it can be stored and analyzed [21]. This concept of interactivity among devices, creating complex networks that can find solutions to problems, is what is called Industry 4.0 [22]. Industry 4.0 supposes the abolition of traditional separation between the physical world and the virtual world, achieving a fusion of reality and its virtual model [23] using model development that allows the implementation of an effective decision-making system based on objective data. A system based on the concept of agriculture 4.0 should consist of a process based on four phases [24]: data collection, centralization of data, predictive data analysis and visualization of results. Application of Industry 4.0 to agriculture is the next technological advance and is based on a more intelligent and predictive behavior [25], where technology takes a leading role and offers tools for business management in the 21st century.

This implies the need to develop ERP II concept. This term was first used by the Gartner Research Group in 2000, when proposing that software should go beyond the company itself and be open to receive information from customers, suppliers, or business partners [26]. This new approach, together with the use of Cloud-based systems, Information and Communication Technology (ICT), Internet of Services (IoS), IoT, Business to Business (B2B), and Business to Consumer (B2C), will allow the development of a tool that facilitates the decision-making process [27], which will be basic in the future of agriculture and will help to move to Agriculture 4.0. Therefore, automation and the use of new technologies will provide agricultural systems with stability and certainty, which was missing in previous scenarios, and that will allow them to meet the required standards for using ERP systems. As a result, farmers will be able to respond to society requirements for quality food linked to origin and an agriculture that respects the environment providing evidences of their commitment.

In this sense, the objective of this paper was to develop an ERP platform for the management of an agricultural farm, based on Agriculture 4.0., by obtaining agronomic data linked to plots through Precision Agriculture and the management of the data through Big Data, to respond to the needs of the agricultural sector. The ERP platform designed for the farm integral management comply with current legal and administrative requirements, use the new technologies, and develop predictive models based on collected data. This platform was presented for the case of wine production under PDO and PGI schemes.

In Figure 1, interconnections among CAP, PDO/IGP and the coordination of their several aspects through an ERP can be seen.

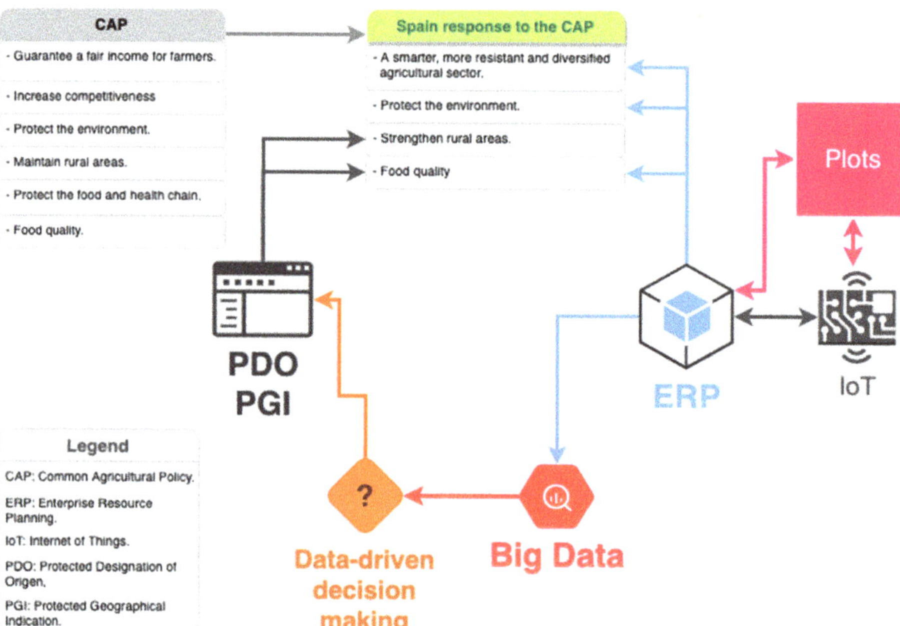

Figure 1. Interconnections among CAP objectives, PDO/IGP and the coordination of their several aspects through an ERP.

2. Materials and Methods

2.1. Survey of Farmers, Technicians, and Quality Managers under PDO and PGI Schemes

To know if there was a need for this type of platform development, a survey and interviews with farmers, winery technical managers, and quality directors of entities that manage wine linked to origin (PDO and PGI) were conducted. This survey was important to understand the platform viability, their needs, and if there are differences in the perception of the utility of the platform.

In the case of vine farmers, a stratified random sampling was carried out around Spain with an allocation proportional to the number of farmers that commercialize their grape production through PDO and PGI schemes. The sample size was 400, which supposes a sampling error of 5% with a 95.5% confidence level. The information was obtained via personal interview after arranging an appointment by telephone.

The final questionnaire, after two pre-tests, included a first part whose purpose was to classify the respondent according to place of birth, age, years in office, level of studies, etc. In the second part, reference was made to the data of the company or entity: number of partners, volume traded, area, number of employees, etc. Finally, the questionnaire (Table 1) focused on obtaining information directly related to the validation of the proposed platform.

Table 1. Questionnaire answered by farmers, technicians, and managers.

1st Block. Analysis of the perception of the ERP applications in agriculture.
P1a. Do you use electronic devices or the internet in your work on a daily basis?
P2a. What is your perception of the internet and computer programs in the management of your company?
Evaluate, from 0 to 10 in order of importance, the following characteristics of a computer application intended to help you in your management:
P2b. It must be simple
P2c. It must help to show the profitability
P2d. It must assist the paperwork
P2e. It must integrate the soil, climate, and production data in each plot
P2f. It is important that it serves as a field notebook
P2g. It must serve to make the daily tasks easier
2nd Block. Analysis of the recording of data in the holdings.
P3a. Do you have a register of the climatic data related to the plots that you manage?
P3b. And a register of the yields?
Evaluate, from 0 to 10 in order of importance, the following modules in a possible application for the management of the plots:
P3c. Climatic data
P3d. Production data
P3e. Incidence of pests and diseases
P3f. Data for the machinery used in the holding
P3g. Register of the cultivation tasks performed
P3h. Management of documents
3th Block. Management and decision-making needs.
P4a. Do you know how each of the plots that you cultivate behaves?
P4b. Do you have to manage data concerning the members of your organization?   P4c. Do you know the sizes of those plots?
P4d. Can you obtain up-to-date information for each of your associates in a reasonable amount of time?
P4e. When was the last time that someone asked you for any information about your plots?
P4f. Can you tell me, with real data, if edaphoclimatic differences exist among the zones that your organization looks after or manages?
P4g. Can you tell me, using real edaphoclimatic data, if there are differences among your distinct plots?
P4h. When was the last time that someone asked you for specific information about one of your associates?
P4i. Do you find it difficult to fill in the documentation that you are required to complete by the current legislation?
P4j. Do you need information grouped by zones in order to make general decisions?
P4k. Do you consider necessary an accounting system linked to the supply of production data?

Table 1. *Cont.*

4th Block. Use of applications and solutions.
P5a Do you use any type of computer program to manage the information relating to your plots?
P5b Do you use any type of computer program to manage the information relating to your different associates?
P5c And to assist you to make decisions concerning production?
P5d And to make decisions at the territorial level?

Note: "P" represents an internal code for the questionnaire used for the farmers, technicians and managers survey.

2.2. Design of an ERP Platform for an Agricultural Holding

Considering the need of ERP system implementation according to farmers, technicians and managers' survey, an ERP platform has been developed following three basic ideas: (i) the use of Open Source (everybody can use, modify and share software) software whenever possible allowing to reduce production and maintenance cost; (ii) the development of a communication system among devices, which will collect data in real time to the platform throw an API; and (iii) the simplicity of the system to avoid complex interfaces for the final user.

Figure 2 shows the entire research process carried out for the development of the ERP application.

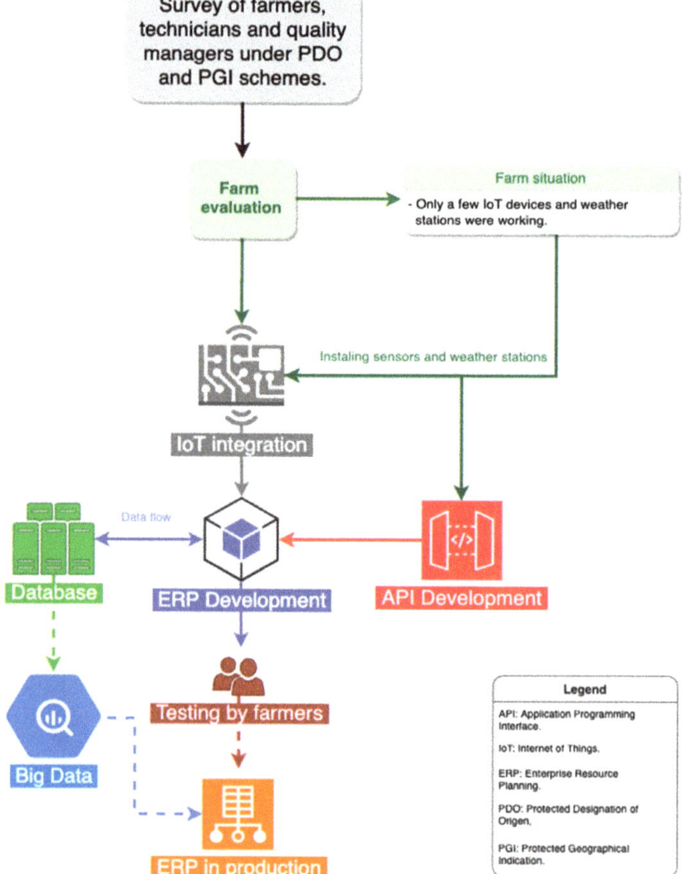

Figure 2. Flowchart of the different steps of the research methodology.

2.2.1. System Architecture Design

Laravel framework was chosen for the ERP platform development. Laravel is a free, open-source PHP (Hypertext Pre-Processor) web framework, created by Taylor Otwell. Frameworks like Laravel, prepackage a collection of third-party components together with custom configuration files, service providers, prescribed directory structures and application bootstrap [28], and is based on the Model, View, and Controller (MVC) architecture (Figure 3).

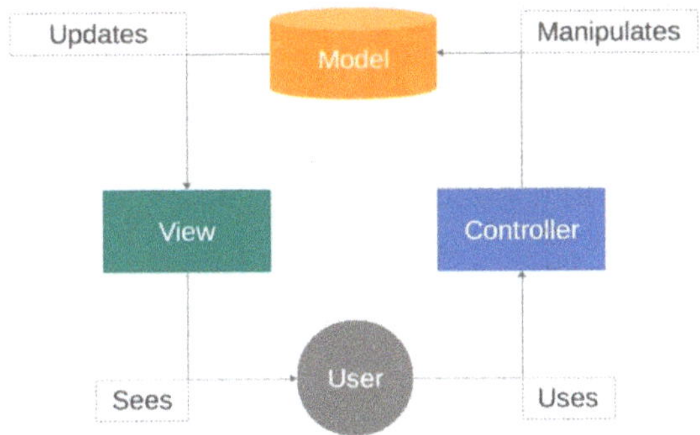

Figure 3. Collaboration between components in an MVC structure.

This structure is designed to divide the software into three blocks (the model, the view, and the controller) to have better control over final product quality [29]. The framework provides a list of functionalities, out of the box, such us authentication, routing, session manager, caching, IoC (Inversion of Control) container, middleware, Eloquent ORM (Object-Relational Mapping), database migration and seed tools, integrated unit testing support, etc. Laravel 5.6 and PHP 7.1 was used for this project development.

When a client sends a HTTP (HyperText Transfer Protocol,) request, the web server executes this request through the PHP engine, and this is where Laravel executes all the procedures that will lead to a final response [30]. This can be returned through the browser, in HTML5 (HyperText Markup Language, version 5) format, or through the platform API (Application Programming Interface), in JSON (JavaScript Object Notation) format.

The Laravel lifecycle, starts with the request that executes the initial framework file, located at public/index.php which initializes the whole process: loading the Kernel, the Service Providers, configuration files, etc. after that will dispatched the routes, the middleware, the controllers, the models, the views and finally the response is sent to the user's browser (Figure 4).

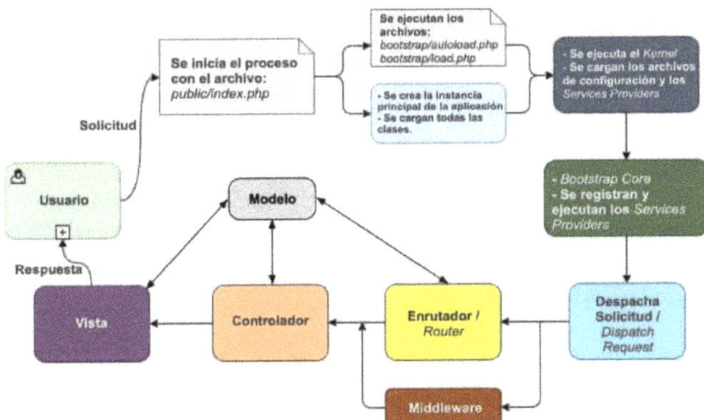

Figure 4. Laravel lifecycle.

The project uses a list of Laravel packages as are mentioned below (the packages bellow are Open Source):

- anhskohbo/no-captcha: wrapper for Google no-captcha, version 3.0 [31].
- askedio/laravel-soft-cascade: used to manage the deletion of entries in the database from related tables, version 5.6 [32].
- barryvdh/laravel-dompdf: used to generate reports in pdf format, version 0.8.2 [33].
- genealabs/laravel-model-caching: for automatic model caching, version 0.2.62 [34].
- intervention/image: powerful tool for image management, version 2.4 [35].
- grimzy/laravel-mysql-spatial: used to store geolocated data in the database, version 1.0 [36].
- maatwebsite/excel: used to import data from excel files, version 2.1.0 [37].
- spatie/laravel-cookie-consent: a configurable way to display the cookie consent, version 2.2 [38].
- spatie/laravel-html: for advanced HTML management, version 2.19 [39].
- spatie/laravel-permission: for roles and permissions, version 2.1 [40].

Finally, for the management of the administration panel, Laravel Nova [41] (a commercial package for Laravel) has been used, being the only resource that is not Open Source, and which entailed the purchase of a user license. This project uses the default middleware from Laravel, in addition to a series of custom middleware.

However, in some cases it was needed to extend its functionalities, and a series of custom middlewares were created, which are described below:

- The Https middleware, which forces the use of secure routes under a secure server.
- The Locale middleware, which investigates user preferences for his default language and determines this language as preselected.
- The IsAdmin middleware, to identify the system administrator.

2.2.2. Databases and Eloquent ORM

For the management of the databases, a hybrid system using MariaDB (version 5.7.31) and MongoDB (version 4.0) was selected. MariaDB is a database based on MySQL an Open Source project, free, with a fast response, very easy to use, and frequently employed by large companies [42]. MongoDB is a powerful, scalable, and flexible database that stores information in files with a JSON structure [43]. This decision was based on the characteristics of each of the platform's needs. Relational systems, such as MariaDB, are more effective for the management of multi-column transactions, such as user management or accounting operations, while non-relational systems, such as MongoDB, are more effective when it comes to dealing with data management in real time, such as climate data or data from sensors [44]. The decision on which parts of the data structure will use each system is complex a priori, thus it will be an issue that might be solved with the use of the application. As a starting point, it was decided that the climatic data and data coming from the IoT-based devices would be stored in the non-relational database, while the rest of the information would be stored in the relational database. This decision will be flexible and must adapt to system needs.

Laravel includes by default an ORM (technique for converting data among incompatible type systems using object-oriented programming languages [45]) called Eloquent, based on active records. and which is intuitive and easy to manage. The operation is simple, each table in the database is related to a PHP class, which includes all the logic necessary to interact with the database [46]. This classes are the Model in the MVC pattern and has also support for relational tables, providing specific classes to perform these operations in Models.

2.2.3. API Connection

Regarding data collection from outside the application, an API was used as recommended [47]. API is defined as a secure connection bridge between external data and the platform, allowing the system to send data from external devices. The APIs are being so standardized that the vast majority of companies are playing the same rules, using the Representational State Transfer (REST) model on the Hypertext Transport Protocol Secure (HTTPS) standard, and using actions such as GET, POST, PUT, and DELETE as if they were web browsers [48]. In this case, a REST model was chosen, using it for API development. API operation consists of the source of information (for example, a sensor) which must connect to the Internet Protocol (IP) address (for security reasons, 127.0.0.1 is used) where the platform is located, and will require access control based on the authorization system that includes Laravel by default. Table 2 explains all the parameters supported by the API and the HTTPS connection address.

Table 2. Description of the API.

Name	Summary	Value
HTTPS Connection	The API HTTPS gateway	https://127.0.0.1/API/
Key	The API identification key. This is a unique value for any device or external user	An alphanumeric value
ItemID	For *get*, *update*, or *deleting*, specific data from the database.	A numeric value
KeyName	This is an optimal field, if we want to add a custom name for the device or user	An alphanumeric value
Date	The current date	Format: DD/MM/YYYY
ValueName	The item name. For example: max-temperature	An alphanumeric value
Value	The item value. For example: 37	An alphanumeric value
ValueNameItem	When we need to send multiple values for different items. This item can be repeated as many times as we need. For example: max-temperature::37.	Format: ValueName::Value
Action	The action type. Not all the types are allowed; by default, the system uses PUT.	GET, PUT, DELETE, UPDATE

The system allows the actions GET, PUT, DELETE, and UPDATE, but by default only enables the PUT action, because the rest of the actions must be authorized by the system administrator. In this case, user is authorized to perform actions only with his/her own data. Therefore, the objective of the API was to collect information from third parties; the basic action done was the PUT, the rest of the actions not being currently relevant. In any case, the GET, DELETE, and UPDATE actions were developed in case they were necessary in the future.

2.2.4. Cloud Computing with PaaS

The application was hosted in the cloud, using the concept Platform as a Service (PaaS), where the computing platform was entirely in the cloud (the operating system, programming languages, databases, web servers, etc.) and this was accessed through an API, a Software Development Kit (SDK), or through services such as Secure Shell (SSH) [49].

For this platform services such as Secure Sockets Layer (SSL) Certificate Management, Manage Queue Workers, Cron Jobs, Load Balancing, Horizontal Scaling, and memory cache systems like Redis1 or Memcached2 were needed. The PaaS systems allowed to manage all these features in a simple way. They gave access and control to the platform as if the application was run on a local server, but without the configuration and maintenance problems that this implies.

For the server, a balanced virtual machine was used which contains: 4 GB RAM, 2-core CPU, 80 GB SSD disk and 4 TB in data transfer. This was the basic configuration for the machine, and depending on the needs of each moment, functionality of the server can be improved in a totally flexible way.

2.3. Statistical Analysis

Statistical analysis was run to process farmers, technicians, and quality managers answers. For the statistical analysis, SPSS 12.0 package for Windows was used to perform the Pearson Chi square, Mann-Whitney U, and Kruskal-Wallis H statistical tests, in order to analyze the behavior of the three groups under study, both jointly and bilaterally.

3. Results and Discussion

3.1. Survey Outcomes

The first question, P1a, (*Do you use electronic devices or the internet in your work on a daily basis?*) (Table 1), produced the first significant differences between the group of farmers and those of technicians and managers (Table 3). Thus, only 23.7% of farmers claimed to use them against 89.3% of technicians and 100% of managers.

The first block (Figure 5) of questions revolves around the analysis of the perception of ERP applications in agriculture. Answers to question P2a (*What is your perception of the internet and computer programs in the management of your company?*) showed that perception was good (1.1 for technicians and 1.2 for managers as average mark), although this perception was worse in the case of farmers (1.6 as average mark) (Table 4). proposed scores were: Good = 1; Moderate = 2; Bad = 3. After the first block questions, respondents were asked to evaluate, from 0 to 10 in order terms of importance, a series of proposed features for a computer application that would be helpful in management (Table 4). The most valued was the option *"Must serve to make the daily tasks easier"* (P2g), scored above 9 within three groups. On the contrary, the least valued feature was *"Must help in the paperwork"* (P2g), scoring around 6 in all groups.

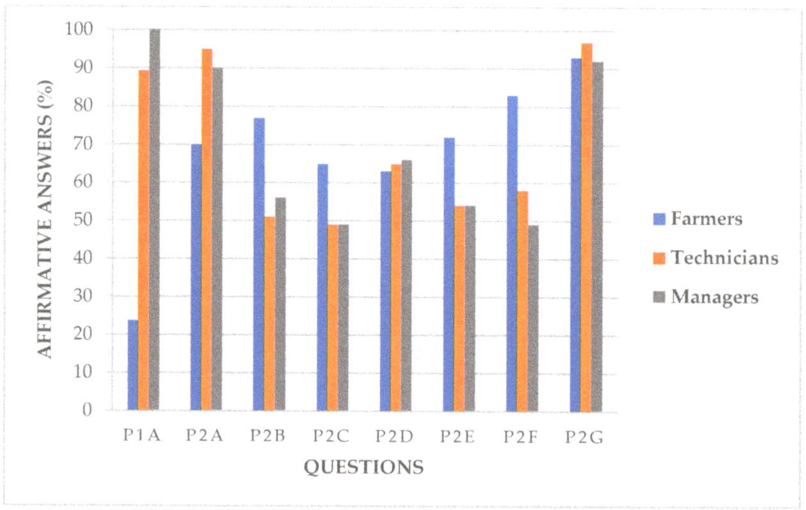

Figure 5. Answers for questions in Block 1 (all groups).

Significant differences appeared in the rest of the characteristics between the group of farmers and those of technicians and managers. *"It is important that it serves as a field notebook"* (P2f), *"It must serve to help show the profitability"* (P2c), *"It must be simple"* (P2c), and *"It must integrate the soil, climate, and production data in each plot"* (P2e) options were higher scored by farmers than by the other groups.

The second block (Figure 6) of questions dealt with the analysis of data collection on farms (Table 3). Both question P3a (*Do you have a register of climatic data related to the plots you manage?*) and P3b (*And a record of the production?*) presented a greater number of affirmative answers in the groups of technicians and managers than in farmers. For the first question, these differences were statistically significant (Table 3). Additionally, another evaluation was requested, from 0 to 10 regarding the importance of a series of modules in a possible application for plot management (Table 4). The most valued module by all groups was *"Management of documents"* (P3h). *"The data modules of machinery used in the holding"* (P3f) and *"registration of cultivation tasks"* (P3g) were more valued by farmers than by technicians and managers, with significant statistical differences between the first group and the other two. The other three modules proposed were also well valued by the three groups, with scores above 7.0.

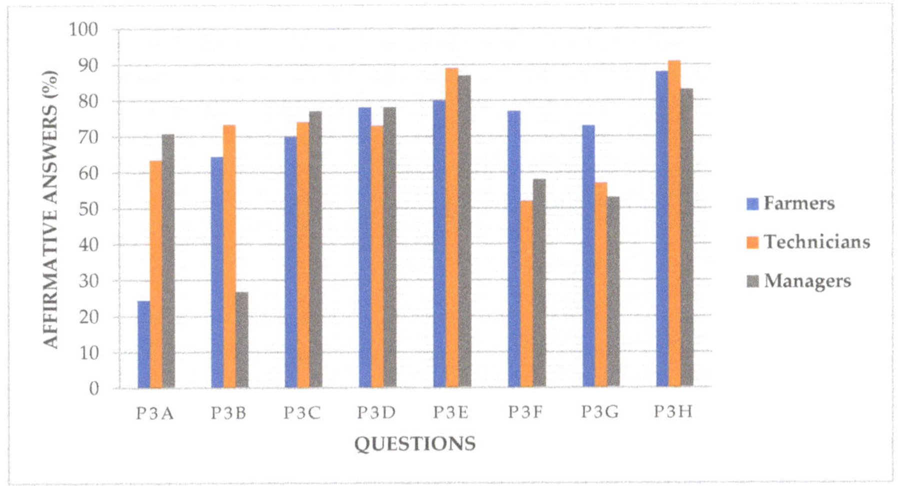

Figure 6. Answers for questions in Block 2 (all groups).

Third block of questions focused on management and decision-making needs (Table 1). Some questions were addressed only to farmers, others only to technicians and managers, and others to the three groups. Regarding farmers (Figure 7) results (Table 3) showed that: (i) 73.5% of respondents, answered that they are aware of how the cultivated plots behave (P4a); (ii) 73.0% answered to know the plot size (P4c); (iii) 65.3% answered that they know when was the last time that someone asked them for any information about their plots (P4e); and (iv) only 31.9% answered that they know (using real edaphoclimatic data) if there are differences among their distinct plots (P4g).

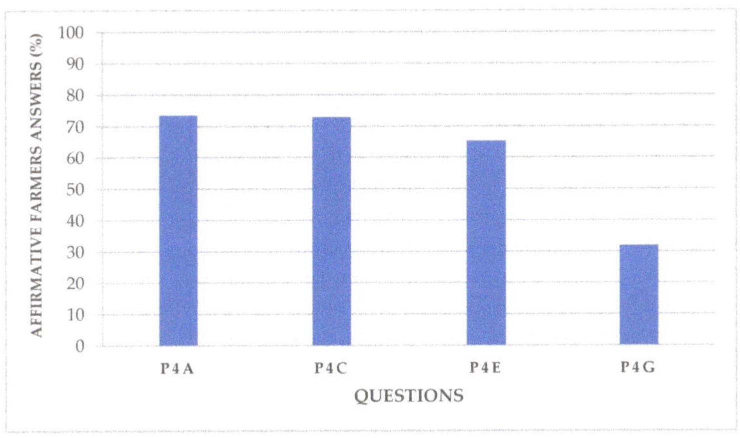

Figure 7. Farmers answers for questions in Block 3.

Regarding technicians and managers answers (Figure 8): (i) 63.6% of technicians and 93.3% of managers answered that they have to manage data concerning the members of their organization (P4b); (ii) 80.1% of technicians and 93.3% of managers were sure about that they are able to obtain updated information for each of their associates in a reasonable amount of time (P4d); (iii) 93.3% of technicians and 81.8% of managers could tell, based on real data, if there are edaphoclimatic differences among the zones that their organization manages (P4f); and (iv) 66.7% of technicians and 63.6% of managers were able to remember when was the last time that someone asked then for a specific information about one of their associates.

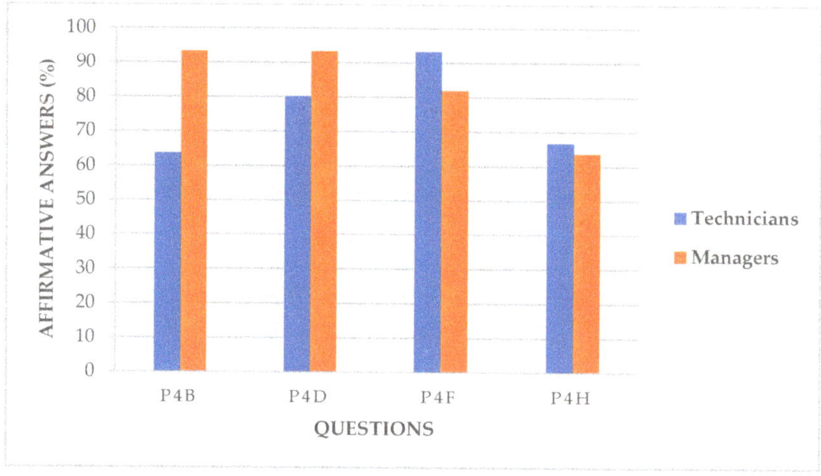

Figure 8. Answers for questions in Block 3 (technicians and managers).

Finally, answers to questions addressed to all groups (Figure 9) were as following: (i) 84.3% of farmers, 70.7% of technicians, and 80.8% of managers answered that they found difficult to fill in required documentation by current legislation (P4i); (ii) 26.5% of farmers, 73.3% of technicians, and 72.7% of managers deemed that they need information grouped by zones in order to make general decisions (P4j) and statistically significant differences were found between farmers and the two other groups; and (iii) 90.3% of farmers, 86.3% of technicians, and 93.3% of managers considered necessary an accounting system linked to yield data supply (P4k).

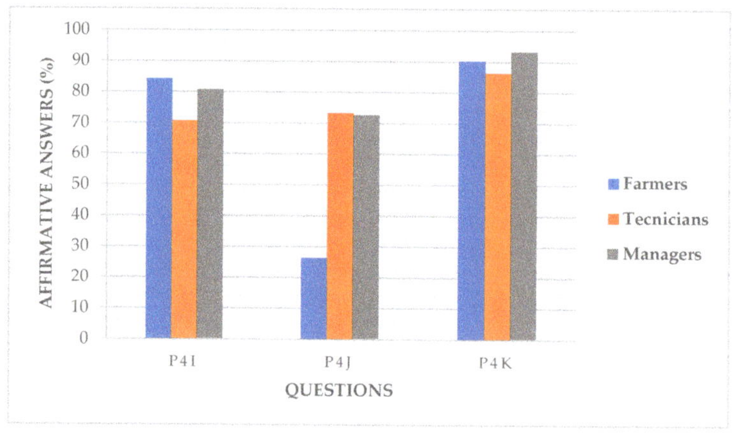

Figure 9. Common answers for questions in Block 3 (all groups).

The fourth and last block (Figure 10) of questions addressed the use of applications and solutions (Table 1). Regarding farmers (Table 3), 90.2% of them answered that they do not use any type of computer program to manage the information concerning their plots and to assist them in decisions about concerning production (P5a and P5c).

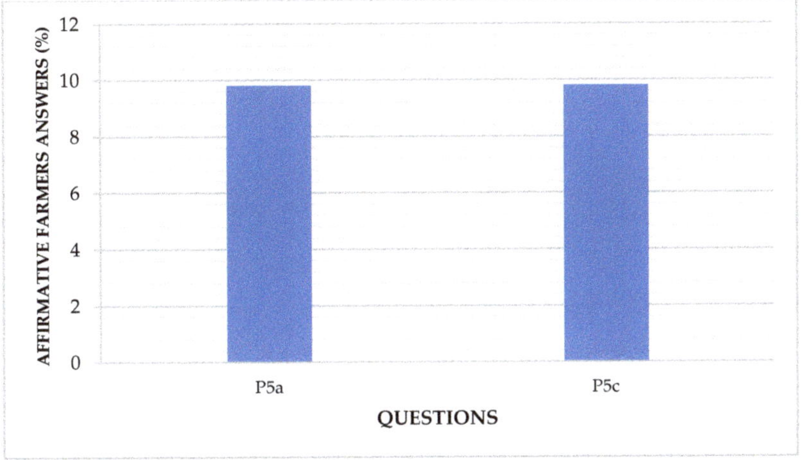

Figure 10. Answers for questions in Block 4 (farmers).

On the other hand, 70.4% of technicians and 80.3% of managers answered (Figure 11) that they use a computer program to manage the information related to their different associates. Also, 66.7% of technicians and 63.6% of managers affirmed to make decisions at the territorial level (P5b and P5d).

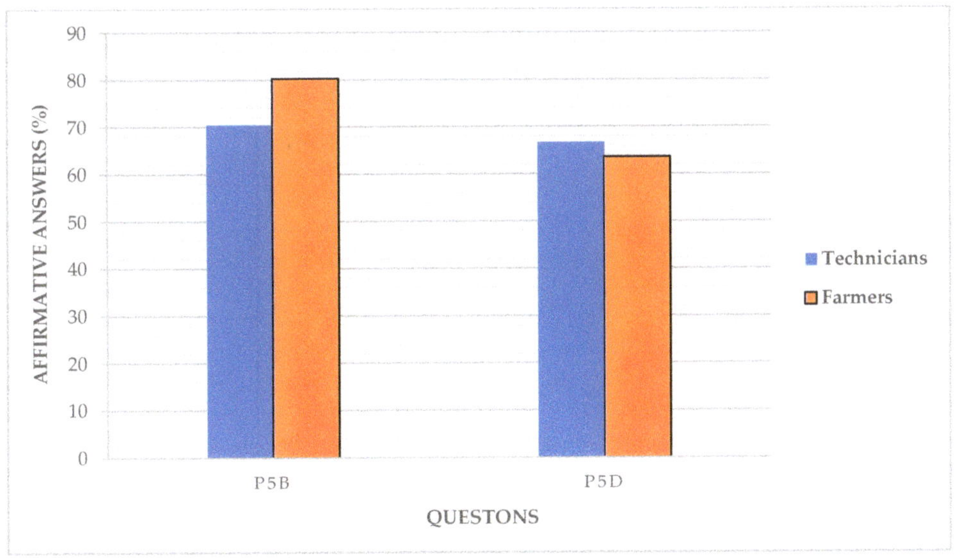

Figure 11. Answers for questions in Block 4 (technicians and managers).

So, it can be said that there is a need for an ERP platform that coordinates the plot production information at the three decision making levels: farmers, technicians, and managers.

Table 3. Analysis of the dichotomous qualitative variables of the survey.

	Farmers (F)		Technicians (T)		Managers (M)		Interactions									
							$F \times T \times M$		$T \times M$		$F \times M$		$F \times T$		$F \times (T + M)$	
							X^2 Pearson									
	%Yes	%No	%Yes	%No	%Yes	%No	X^2	p	X^2	p	X^2	p	X^2	p	X^2	p
P1a	23.7	76.3	89.3	10.7	100	0	44.6	0.00 *[2]	0.1	0.74	26.1	0.00 *	27.3	0.00 *	44.4	0.00 *
P3a	24.5	75.5	63.5	36.5	70.7	29.3	23.1	0.00 *	0.0	0.97	10.9	0.00 *	14.7	0.00 *	23.1	0.00 *
P3b	67.4	42.6	73.3	26.7	72.7	27.3	2.4	0.31	-	-	-	-	-	-	-	-
P4a	73.5	26.5	-	-	-	-	-[1]	-	-	-	-	-	-	-	-	-
P4b	-	-	63.6	36.4	93.3	6.7	-	-	3.6	0.06	-	-	-	-	-	-
P4c	73.0	27.0	-	-	-	-	-	-	-	-	-	-	-	-	-	-
P4d	-	-	80.1	20.9	93.3	6.7	-	-	0.0	0.76	-	-	-	-	-	-
P4e	65.3	34.7	-	-	-	-	-	-	-	-	-	-	-	-	-	-
P4f	-	-	93.3	6.7	81.8	18.2	-	-	0.8	0.36	-	-	-	-	-	-
P4g	31.9	68.1	-	-	-	-	-	-	-	-	-	-	-	-	-	-
P4h	-	-	66.7	33.3	63.6	36.4	-	-	0.0	0.87	-	-	-	-	-	-
P4i	84.3	15.7	70.7	29.3	80.8	19.2	0.1	0.92	-	-	-	-	-	-	-	-
P4j	26.5	73.5	73.3	26.7	72.7	27.3	23.1	0.00 *	0.0	0.97	10.9	0.00 *	14.7	0.00 *	23.1	0.00 *
P4k	90.3	9.7	86.3	13.7	93.3	6.7	2.2	0.32	-	-	-	-	-	-	-	-
P5a	9.8	90.2	-	-	-	-	-	-	-	-	-	-	-	-	-	-
P5b	-	-	70.4	29.6	80.3	19.7	-	-	7.3	0.24	-	-	-	-	-	-
P5c	9.8	90.2	-	-	-	-	-	-	-	-	-	-	-	-	-	-
P5d	-	-	66.7	33.3	63.6	36.4	-	-	0.0	0.87	-	-	-	-	-	-

[1] The symbol "-" indicates that the question was not applied to that group, while [2] the symbol "*" Statistically significant at $p < 0.05$.

Table 4. Analysis of the quantitative variables of the survey.

	Farmers (F)		Technicians (T)		Managers (M)		Interactions									
							F × T × M		T × M		F × M		F × T		F × (T + M)	
							H-Kruskal Wallis				U-Mann-Whitney				U-Mann-Whitney	
	m	sd	m	sd	m	sd	X^2	p	U	p	U	p	U	p	U	p
P2a#	1.6	0.7	1.1	0.5	1.2	0.5	0.8	0.73								
P2b	7.7	2.4	5.1	2.0	5.6	2.5	22.3	0.00 *	57.5	0.19	531.5	0.04 *	551.0	0.00 *	1132.5	0.00 *
P2c	6.5	2.3	4.9	2.7	4.9	3.0	8.2	0.02 *	73.0	0.79	497.5	0.02 *	968.0	0.01 *	1745.5	0.00 *
P2d	6.3	2.7	6.5	2.2	6.6	2.3	0.1	0.89								
P2e	7.2	2.4	5.4	2.3	5.4	3.0	11.6	0.00 *	79.0	0.96	724.0	0.03 *	875.5	0.00 *	1670.5	0.00 *
P2f	8.3	2.4	5.8	2.0	4.9	2.5	26.2	0.00 *	75.4	0.19	582.5	0.04 *	564.0	0.00 *	1134.5	0.00 *
P2g	9.3	4.7	9.2	1.3	9.7	3.6	2.4	0.73								
P3c	7.0	4.4	7.4	1.1	7.7	1.9	1.4	0.50								
P3d	7.8	2.1	7.3	1.5	7.8	2.0	8.1	0.34								
P3e	8.0	2.1	8.9	1.7	8.7	1.6	4.2	0.12								
P3f	7.7	2.4	5.2	2.0	5.8	2.5	24.3	0.00 *	57.5	0.19	571.5	0.04 *	551.0	0.00 *	1122.5	0.00 *
P3g	7.3	3.2	5.7	3.2	5.3	2.3	9.9	0.01 *	80.0	0.89	655.5	0.02 *	1023.0	0.03 *	1678.5	0.00 *
P3h	8.8	4.2	9.1	4.6	8.3	4.2	0.8	0.66								

The letters m = mean; sd = standard deviation. The symbol # = Score (Good = 1; Moderate = 2; Bad = 3). The symbol * = statistically significant at $p < 0.05$.

3.2. Functioning of the ERP Platform for an Agricultural Holding

The operation application of the ERP platform designed to manage an agricultural operation in the Autonomous Community of Valencia (Spain) is explained below. The platform was developed and tested under the supervision of the government department of the Autonomous Community that deals with agriculture (Conselleria de Agricultura, Medio Ambiente, Cambio Climático y Desarrollo Rural). Additionally, a group of farmers has also used the application sending continuous feedback that allowed the optimization and improvement of the platform.

3.2.1. Agronomic Management of the Plots

Spanish legislation (Real Decreto 1311/2012) requires farmers to use the Operational Notebook as a record of the daily activity that takes place in an agricultural plot. Thus, this platform permits automatic generation of the documentation required by the regulations, allowing it to be printed or exported to the following telematic formats: PDF, XLS, and CSV. An example of the plots distribution is presented in Figure 12.

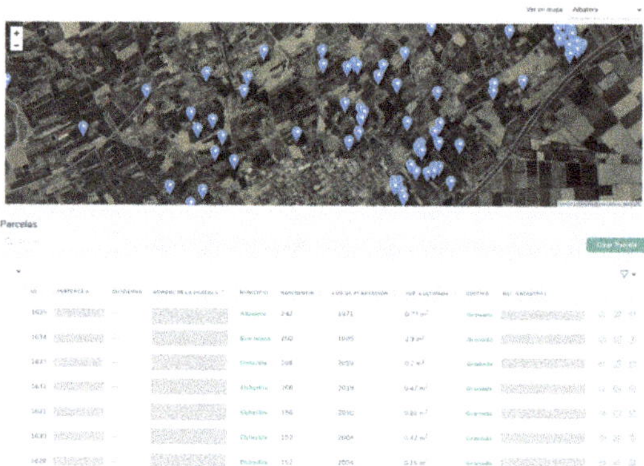

Figure 12. A list of geolocated plots. Personal data has been removed from the image.

First, the plot is geolocated (Figure 13) using WMS technology, and all the information available for the plot is compiled from third-party sources such as the Geographic Information System for Agricultural Parcels (SIGPAC, a tool originally conceived with the purpose of making it easier for farmers to submit applications, with graphic support, as well as to facilitate administrative and on-site controls [50]) or the Catastro (through its cadastral map publication service on the internet [51]). From this moment, all the information added to the plot is automatically geolocated.

Figure 13. Plot geolocation process.

In order to act as a log of the activities in the plot, the platform was able to manage the agronomic information related to phytosanitary treatments, pests, irrigation, cultivation tasks (pruning, fertilizers, amendments, etc.), incidents that have occurred in the agricultural holding, harvesting, workers and machinery manager, and management of seeds.

The system stores the agronomic actions that are added to it by linking them with a date (date of application or completion), thus their history for each plot can be provided. There is also a list of agronomic actions that can be filtered, listed, and exported according to the needs of the user.

Moreover, the ERP platform allows to provide an evidence of sustainable agricultural practices at the plot level and can set the basis for CO_2 uptake. This allows farmers to fulfil CAP and society requirements. Territorial brand managers can also show to society how buying quality food linked to origin contribute to a better environment.

3.2.2. Irrigation Records

The application also permits managing the water inputs to the plots, allowing to keep a historical record of water contributions. Another functionality is to manage all the indicated items required by Spanish legislation, such as bodies of water (Figure 14) and their specific characteristics.

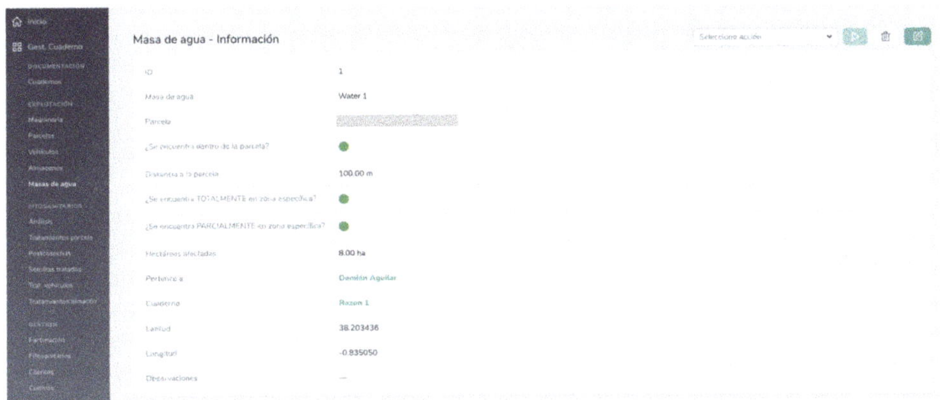

Figure 14. Bodie of water report. Personal data has been removed from the image.

All this data can be displayed on the screen by means of graphs and related to climatic data, such as rainfall on the plot. This supposes an essential part of the project as water use in agriculture is a major concern in Spain. This provides evidences of farmer environmental concern.

3.2.3. Reports of Climatic Data

The platform allows users to manage their own weather data from an on-site meteorological station, connected to the platform through the connection API. If the user does not have this infrastructure, when geolocating the plot, the system searches for the closest AEMET meteorological station and those the data from will be taken. The user must indicate to the system, through a selector in the administration panel, where the data want to be consulted.

In the case of AEMET, historical data since 2011 from the meteorological stations are available, although for some stations they date back 20 years. Users can access the data records (mainly temperature, humidity, and precipitation) for each day, from the different meteorological stations available in the database. Figure 15 shows an example of data received by the platform from AEMET.

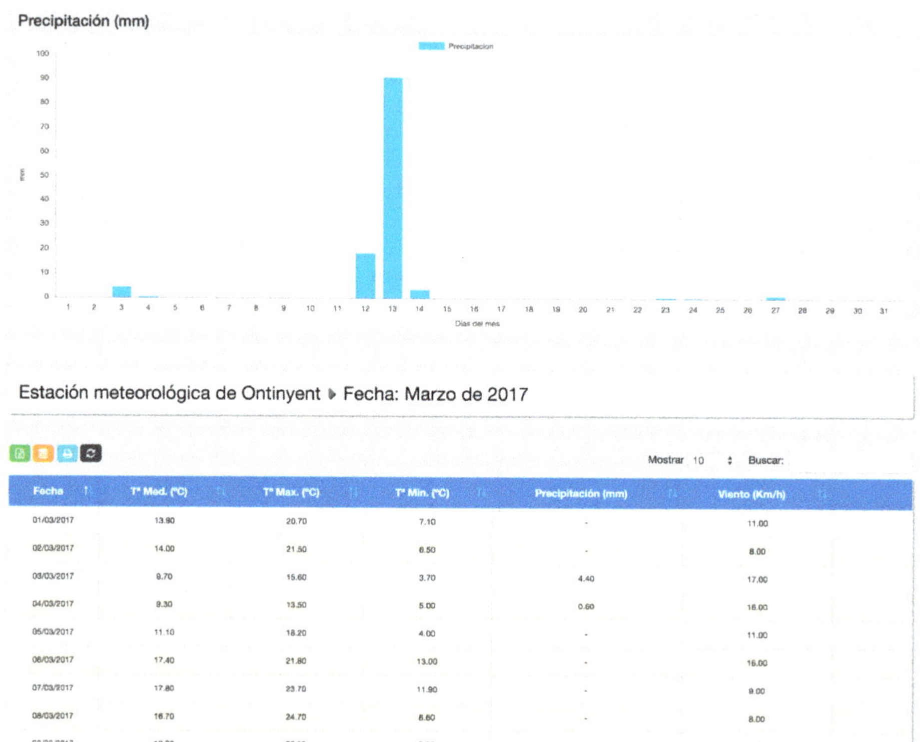

Fecha	Tª Med. (°C)	Tª Max. (°C)	Tª Min. (°C)	Precipitación (mm)	Viento (Km/h)
01/03/2017	13.90	20.70	7.10	-	11.00
02/03/2017	14.00	21.50	6.50	-	8.00
03/03/2017	9.70	15.60	3.70	4.40	17.00
04/03/2017	9.30	13.50	5.00	0.60	16.00
05/03/2017	11.10	18.20	4.00	-	11.00
06/03/2017	17.40	21.80	13.00	-	16.00
07/03/2017	17.80	23.70	11.90	-	9.00
08/03/2017	16.70	24.70	8.60	-	8.00

Figure 15. Example of climatological data from the AEMET climate station, located in Ontinyent (Valencia).

3.2.4. Management System

The platform has a global management system, which allows the administration of roles and permissions which bring access to different parts of the system. The permission system allows different users to access the data based on the parameters established for each of them. The roles and permissions system are managed through the packages developed by Spatie (a software development company), which is called, Laravel permission [40]. This package permits to associate users with roles and permissions. For instance, a farmer can see only their data, while the system administrator can access to all the data. Therefore, a list of system access roles can be established, and each of them with different access permissions. Figure 16 presents users with different accesses and roles on the platform.

This opens the possibility to companies technicians to check where the product is processed, and to quality PDO and PGI managers to access information useful in the decision making process from a territorial point of view. Additionally, gives evidence of production good practices regarding CO_2 uptake and water use, etc.

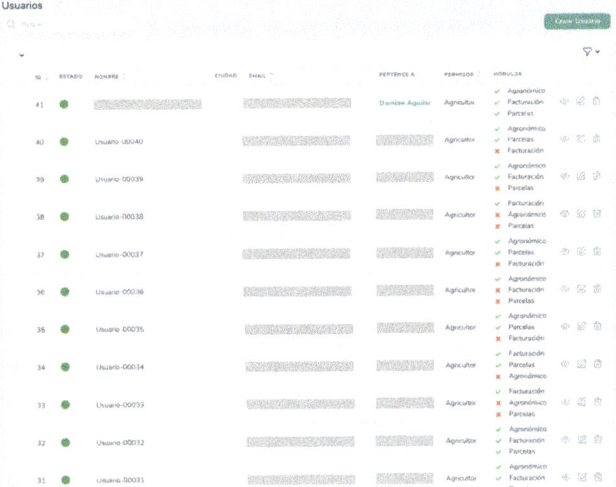

Figure 16. Roles and permission administration. Personal data has been removed from the image.

3.2.5. Other Components

The platform also has a series of components such as (i) machinery (Figure 17) and vehicle management (inventory management, insurance, maintenance, etc.); (ii) personnel records and management (personal information of workers, training and authorization for the handling of phytosanitary products); (iii) agricultural product commercialization records; (iv) generate reports and documentation and (v) warehouse control.

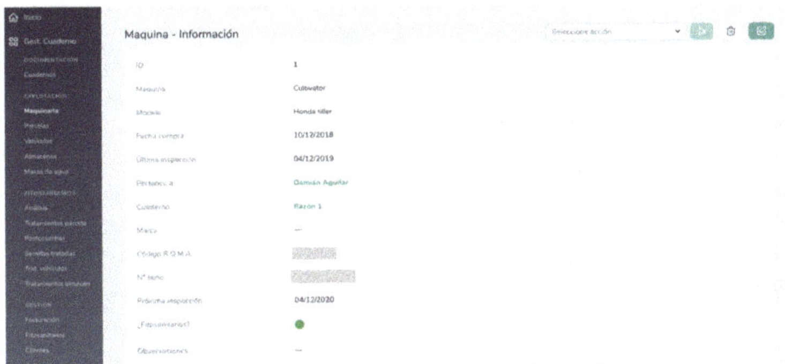

Figure 17. Machinery management detail. Personal data has been removed from the image.

4. Conclusions

This study revealed that farmers, technicians, and managers of territorial quality brands are aware of the changes needed in the agricultural business model to facilitate a new way of performing the internal management of farms. This must produce an increase in food quality and in environmental concern. Therefore, technological solutions are needed to boost the agricultural sector, which nowadays, as far as it concerned, this is not covered by the technology sector. Thus, the present ERP tool can be the global solution to face both mentioned challenges because allows data collection, information management, its subsequent analysis and finally a detailed traceability of the products and the processes. The importance of this platform refers to the allowance of agricultural information collection in real time, which help to increase the knowledge of the farm and agricultural area situation. This platform, based on the use of Precision Agriculture, is being used by more than 1500 farms that are continuously adding information to the system. At the same time, they are evaluating the product, with the aim of converting all that information into predictive models capable of anticipating the most common problems suffered by agricultural operations, such as pests, droughts, and fertilization. Thus, this is an essential tool which allow farmers to optimize the farm management based on this information. In the future, the platform will be able to process and model the data in real time, offering suggestions and improvements on farms, due to the use of Big Data.

Author Contributions: Conceptualization, D.A.M.; D.B.L.L. and M.C.-L.; Methodology, D.A.M., D.B.L.-L. and F.J.d.C.-G.; Software, D.A.M.; Formal Analysis, D.B.L.-L., L.L. and H.I.-I.; Writing—Original Draft Preparation, D.A.M., D.B.L.-L., L.L. and P.S.-B.; Writing—Review and Editing, D.A.M., L.L. and D.B.L.-L.; Project Administration, D.A.M. and D.B.L.-L. All authors have read and agreed to the published version of the manuscript.

Funding: This research was partly funded by Conselleria de Agricultura, Medio Ambiente, Cambio Climático y Desarrollo Rural (Generalitat Valenciana).

Conflicts of Interest: The authors declare no conflict of interest.

References

1. Van Ittersum, K.; Candel, M.J.J.M.; Meulenberg, M.T.G. The influence of the image of a product's region of origin on product evaluation. *J. Bus. Res.* **2003**, *56*, 215–226. [CrossRef]
2. Ruiz, M.P.M.; Zarco, A.I.J. La potenciación del origen en las estrategias de marketing de productos agroalimentarios. *Boletín Económico ICE* **2006**, *2880*, 18.
3. Felzensztein, C.; Hibbert, S.; Vong, G. Is the Country of Origin the Fifth Element in the Marketing Mix of Imported Wine? *J. Food Prod. Mark.* **2004**, *10*, 73–84. [CrossRef]
4. Kur, A.; Cock, S. Nothing but a GI Thing: Geographical Indications under EU Law. *Intellect. Prop. Media Entertain. Law J.* **2007**, *17*, 19.
5. Simon, Z. UE: Campaña en Tokaj para Acabar con las Imitaciones. Available online: http://elmundovino.elmundo.es/elmundovino/noticia.html?vi_seccion=7&vs_fecha=200406&vs_noticia=1086850889 (accessed on 4 September 2020).
6. Skuras, D.; Vakrou, A. Consumers' willingness to pay for origin labelled wine: A greek case study. *Brit. Food J.* **2002**, *104*, 898–912. [CrossRef]
7. Jones, G.V.; Snead, N.; Nelson, P. Geology and Wine 8. Modeling Viticultural Landscapes: A GIS Analysis of the Terroir Potential in the Umpqua Valley of Oregon. *Geosci. Can.* **2004**, *31*, 4.
8. Caesarius, L.M.; Hohenthal, J. Searching for big data: How incumbents explore a possible adoption of big data technologies. *Scand. J. Manag.* **2018**, *34*, 129–140. [CrossRef]
9. Sakr, S.; Gaber, M.M. *Large Scale and Big Data—Processing and Management*; Auerbach Publications: Boca Raton, FL, USA, 2014.
10. Cukier, K.; Mayer-Schönberger, V. *Big Data: La Revolución de los Datos Masivos*, 1st ed.; Turner Publicaciones S.L.: Madrid, Spain, 2013; ISBN 978-84-15427-81-0.
11. Aguilar, L.J. *Big Data, Análisis de Grandes Volúmenes de Datos en Organizaciones*; Alfaomega Grupo Editor: Mexico City, Mexico, 2016; ISBN 978-607-707-757-2.
12. Bhat, W.A. Bridging data-capacity gap in big data storage. *Future Gener. Comput. Syst.* **2018**, *87*, 538–548. [CrossRef]
13. De, B. *API Management: An Architect's Guide to Developing and Managing APIs for Your Organization*; Apress: New York, NY, USA, 2017; ISBN 978-1-4842-1305-6.
14. Nguyen, V.-Q.; Nguyen, S.N.; Kim, K. Design of a Platform for Collecting and Analyzing Agricultural Big Data. *J. Digit. Contents Soc.* **2017**, *18*, 149–158. [CrossRef]
15. Corbellini, A.; Mateos, C.; Zunino, A.; Godoy, D.; Schiaffino, S. Persisting big-data: The NoSQL landscape. *Inf. Syst.* **2017**, *63*, 1–23. [CrossRef]
16. Bicevska, Z.; Oditis, I. Towards NoSQL-based Data Warehouse Solutions. *Procedia Comput. Sci.* **2017**, *104*, 104–111. [CrossRef]
17. Liao, Y.-T.; Zhou, J.; Lu, C.-H.; Chen, S.-C.; Hsu, C.-H.; Chen, W.; Jiang, M.-F.; Chung, Y.-C. Data adapter for querying and transformation between SQL and NoSQL database. *Future Gener. Comput. Syst.* **2016**, *65*, 111–121. [CrossRef]

18. Bongiovani, R.; Chartuni, E.; Best, S.; Roel, Á. *Agricultura de Precisión: Integrando Conocimientos para una Agricultura Moderna y Sustentable*; Procisur/IICA: San José, Costa Rica, 2006; ISBN 978-92-9039-741-0.

19. Pucha-Cofrep, F.; Fries, A.; Cánovas-García, F.; Oñate-Valdivieso, F.; González-Jaramillo, V.; Pucha-Cofrep, D. *Fundamentos de SIG: Aplicaciones con ArcGIS*; Ediloja Cia. Ltda.: Loja, Ecuador, 2017; ISBN 978-9942-28-901-8.

20. Carbonell, I.M. The ethics of big data in big agriculture. *Internet Policy Rev.* **2016**, *5*. [CrossRef]

21. Roblek, V.; Meško, M.; Krapež, A. A Complex View of Industry 4.0. *SAGE Open* **2016**, *6*. [CrossRef]

22. Weltzien, C. Digital agriculture—Or why agriculture 4.0 still offers only modest returns. *Agric. Eng.* **2016**, *71*, 66–68.

23. Braun, A.-T.; Colangelo, E.; Steckel, T. Farming in the Era of Industrie 4.0. *Procedia Cirp* **2018**, *72*, 979–984. [CrossRef]

24. Lee, J.; Lapira, E.; Bagheri, B.; Kao, H. Recent advances and trends in predictive manufacturing systems in big data environment. *Manuf. Lett.* **2013**, *1*, 38–41. [CrossRef]

25. Latorre, M. Historia de las Web, 1.0, 2.0, 3.0 y 4.0. Available online: http://umch.edu.pe/arch/hnomarino/74_Historia%20de%20la%20Web.pdf (accessed on 7 August 2020).

26. Surjit, R.; Rathinamoorthy, R.; Vardhini, K.J.V. *ERP for Textiles and Apparel Industry*; CRC Press: Boca Ratón, FL, USA, 2016; ISBN 978-93-85059-59-9.

27. Elragal, A.; Haddara, M. The Future of ERP Systems: Look backward before moving forward. *Procedia Technol.* **2012**, *5*, 21–30. [CrossRef]

28. Stauffer, M. *Laravel: Up and Running: A Framework for Building Modern PHP Apps*; O'Reilly Media, Inc.: Sebastopol, CA, USA, 2016.

29. Hasan, S.S.; Isaac, R.K. An integrated approach of MAS-CommonKADS, Model–View–Controller and web application optimization strategies for web-based expert system development. *Expert Syst. Appl.* **2011**, *38*, 417–428. [CrossRef]

30. Chen, X.; Ji, Z.; Fan, Y.; Zhan, Y. Restful API Architecture Based on Laravel Framework. *J. Phys. Conf. Ser.* **2017**, *910*, 012016. [CrossRef]

31. Anh, N.V. Anhskohbo/no-captcha. Available online: https://github.com/anhskohbo/no-captcha (accessed on 10 August 2020).

32. Askedio/laravel-soft-cascade: Cascade Delete & Restore When Using Laravel SoftDeletes. Available online: https://github.com/Askedio/laravel-soft-cascade (accessed on 10 August 2020).

33. Heuvel, B. vd barryvdh/laravel-dompdf: A DOMPDF Wrapper for Laravel. Available online: https://github.com/barryvdh/laravel-dompdf (accessed on 10 August 2020).

34. Bronner, M. GeneaLabs/laravel-model-caching: Eloquent Model-Caching Made Easy. Available online: https://github.com/GeneaLabs/laravel-model-caching (accessed on 10 August 2020).

35. Intervention Image—Introduction. Available online: http://image.intervention.io/ (accessed on 10 August 2020).

36. Estefane, J. Grimzy/laravel-mysql-spatial: MySQL Spatial Data Extension Integration with Laravel. Available online: https://github.com/grimzy/laravel-mysql-spatial (accessed on 10 August 2020).

37. Brouwers, P. Maatwebsite/Laravel-Excel: Supercharged Excel Exports and Imports in Laravel. Available online: https://github.com/Maatwebsite/Laravel-Excel (accessed on 10 August 2020).

38. Spatie/laravel-cookie-consent: Make Your Laravel App Comply with the Crazy EU Cookie Law. Available online: https://github.com/spatie/laravel-cookie-consent (accessed on 10 August 2020).

39. Spatie/laravel-html: Painless Html Generation. Available online: https://github.com/spatie/laravel-html (accessed on 10 August 2020).

40. Spatie/laravel-permission: Associate Users with Roles and Permissions. Available online: https://github.com/spatie/laravel-permission (accessed on 10 August 2020).

41. Laravel Nova. Available online: https://nova.laravel.com/ (accessed on 11 August 2020).

42. Forta, B. *MariaDB Crash Course*; Addison-Wesley Professional: Boston, MA, USA, 2011; ISBN 978-0-13-284235-8.

43. Chodorow, K. *MongoDB: The Definitive Guide: Powerful and Scalable Data Storage*; O'Reilly Media, Inc.: Sebastopol, CA, USA, 2013; ISBN 978-1-4493-4482-5.

44. Sarig, M. MongoDB vs. MySQL: The Differences Explained. Available online: https://blog.panoply.io/mongodb-and-mysql (accessed on 10 August 2020).

45. Object-Relational Mapping. Available online: https://en.wikipedia.org/w/index.php?title=Object-relational_mapping&oldid=952535425 (accessed on 10 August 2020).

46. Bean, M. *Laravel 5 Essentials*; Packt Publishing Ltd.: Birmingham, UK, 2015; ISBN 978-1-78528-329-1.

47. Xu, C.; Sun, X.; Li, B.; Lu, X.; Guo, H. MULAPI: Improving API method recommendation with API usage location. *J. Syst. Softw.* **2018**, *142*, 195–205. [CrossRef]

48. Sturm, R.; Pollard, C.; Craig, J. Chapter 11—Application Programming Interfaces and Connected Systems. In *Application Performance Management (APM) in the Digital Enterprise*; Sturm, R., Pollard, C., Craig, J., Eds.; Morgan Kaufmann: Boston, MA, USA, 2017; pp. 137–150.

49. Helmke, M. *Ubuntu Unleashed 2015 Edition: Covering 14.10 and 15.04*; Sams Publishing: Carmel, IN, USA, 2014.

50. Sistema de Información Geográfica de Parcelas Agrícolas (SIGPAC). Available online: https://www.mapa.gob.es/es/agricultura/temas/sistema-de-informacion-geografica-de-parcelas-agricolas-sigpac-/default.aspx (accessed on 10 August 2020).

51. Portal de la Dirección General del Catastro. Available online: http://www.catastro.minhap.gob.es/esp/wms.asp (accessed on 10 August 2020).

 *agronomy*

Article

Absence of Yield Reduction after Controlled Water Stress during Prehaverst Period in Table Olive Trees

María José Martín-Palomo [1,2], **Mireia Corell** [1,2], **Ignacio Girón** [2,3], **Luis Andreu** [1,2], **Alejandro Galindo** [1,2], **Ana Centeno** [4], **David Pérez-López** [4] and **Alfonso Moriana** [1,2,*]

1   Departamento de Ciencias Agroforestales, Universidad de Sevilla, ETSIA, Crta de Utrera Km 1, 41013 Seville, Spain; mjpalomo@us.es (M.J.M.-P.); mcorell@us.es (M.C.); landreu@us.es (L.A.); agegea@us.es (A.G.)
2   Unidad Asociada al CSIC de Uso Sostenible del Suelo y el Agua en la Agricultura (US-IRNAS), Crta de Utrera Km 1, 41013 Seville, Spain; iggi@irnase.csic.es
3   Consejo Superior de Investigaciones Científicas, IRNAS, Avda Reina Mercedes, 10, 41012 Seville, Spain
4   Departamento Producción Agraria, CEIGRAM-Universidad Politécnica de Madrid, Av. Puerta de Hierro, 2, 28040 Madrid, Spain; ana.centeno@upm.es (A.C.); david.perezl@upm.es (D.-P.L.)
*   Correspondence: amoriana@us.es; Tel.: +34-54486456

Received: 10 January 2020; Accepted: 8 February 2020; Published: 11 February 2020

Abstract: Deficit irrigation scheduling is becoming increasingly important under commercial conditions. Water status measurement is a useful tool in these conditions. However, the information about water stress levels for olive trees is scarce. The aim of this experiment was to evaluate the effect on yield of a moderate controlled water stress level at the end of the irrigation season. The experiment was conducted in the experimental farm of La Hampa (Coria del Río, Seville, Spain) during three years. A completely randomized block design was performed using three different irrigation treatments. Deficit irrigation was applied several (4 or 2) weeks before harvest. Irrigation was controlled using the midday stem water potential, with a threshold value of −2 MPa and compared with a full irrigated treatment. This water stress did not reduced gas exchange during the deficit period. The effect on yield was not significant in any of the three seasons. In the high-fruit load season, fruit volume was slightly affected (around 10%), but this was not significant at harvest. Results suggest an early affection of fruit growth with water stress, but with a slow rate of decrease. Moderate water stress could be useful for the management of deficit irrigation in table olive trees.

Keywords: fruit size; Manzanilla; olive; regulated deficit irrigation; water potential; water relation

1. Introduction

Olive orchards grow around the world in semi-arid or arid conditions, with great water scarcity. Deficit irrigations are very common in these areas. Sometimes, growers are not aware that their water management may cause water stress periods. The yield response to water stress conditions is related to the duration, the level and the moment when the plant water status is affected [1]. Although olive trees are considered extremely resistant to drought conditions [2], the full bloom/fruit set period [3] and the oil accumulation [4] have been considered sensitive to irrigation shortages. On the other hand, the pit hardening has been traditionally considered the most drought-resistant phenological stage [5], and irrigation restrictions are commonly scheduled in this period. Irrigation scheduling for table olive cultivars is more complex than for oil ones, especially in the case of green olives. The main limitations of regulated deficit irrigation (RDI) for table cultivars are the importance of the fruit size in the final value of the yield and the short period available for rehydration. In addition, the harvest period for green table olives occurs at the end of the irrigation season, when there is a very small amount of irrigation water available and rains are scarce.

RDI scheduling was defined using the drought resistant and sensitive periods, when irrigation shortages could or could not be performed [6]. Some of these sensitive periods occur at the end of the season, during fruit ripening or in the last fruit growth stages. A moderate water stress before harvest has been linked to a flavour improvement of the fruit (in tomato, Ref. [7]; in peaches, Ref. [8]) although it usually causes a reduction of fruit yield (peaches, Ref. [9]). Reference [10] discussed several irrigation works for peach trees where the RDI strategies results are affected by the water holding capacity of the soils. In the review, similar irrigation restrictions produced different results [10]. On the other hand and also for peach trees, Ref. [8] reported the absence of significant differences in yield between full irrigation and water stress during stage III of fruit development, when the water stress level was controlled. Both works suggested that if the water stress level is controlled during drought-sensitive phenological periods, such as stage III for peaches, a reduction of irrigation could be applied without any yield reduction. This benefit arising from moderate water stress conditions is also reported in olive trees. Water stress conditions increased total amount of phenols in fruit with significant changes in the olive oil flavour [11]. These results would explain the reduction of susceptibility to bruising in cv Manzanilla suggested by the water stress in this period [12]. Moderate water stress conditions have also been linked to an improvement of the oil accumulation rate [3,13]. Therefore, the final growth and ripening phases in olive trees could be not as sensitive as initial works suggested. For table olive trees, Ref. [14] reported that midday stem water potential values higher than −2 MPa did not affect fruit growth in cv Manzanilla. More recently, Ref. [15] also suggested this value as a possible threshold for olive irrigation scheduling. The tree response to this level of water stress (−2 MPa) could be different if the irrigation restriction was performed later in the season, or if no rehydration was considered. The aim of this work is to study the yield response of table olive trees to water restrictions before harvest, using −2 MPa of midday stem water potential as threshold value. In addition to water stress level, the duration of the stress was also considered in the evaluation of treatments.

2. Material and Methods

2.1. Site Description and Experimental Design

Experiments were conducted at La Hampa, the experimental farm of the Instituto de Recursos Naturales y Agrobiología (IRNAS-CSIC), located in Coria del Río near Seville (Spain) (37 17′N, 6 3′W, 30 m altitude). In them, 44-year-old table olive trees (*Olea europaea* L cv Manzanillo) were observed from the 2014 to the 2016 seasons. The tree spacing followed a 7 m × 5 m square pattern. The sandy loam soil (about 2 m deep) of the experimental site was characterised by a volumetric water content of 0.33 m^3 m^{-3} at saturation, 0.21 m^3 m^{-3} at field capacity and 0.1 m^3 m^{-3} at permanent wilting point, and 1.30 (0–10 cm) and 1.50 (10–120 cm) g cm^{-3} bulk density. Pest control, pruning and fertilization practices were those commonly used by growers, and weeds were removed chemically within the orchard, pruning was avoided only in the last season. Drip irrigation was carried out at night, using one lateral pipe per row of trees and five emitters per plant, spaced 1m and delivering 8 L h^{-1} each. There were problems with the irrigation system at the beginning of the experiment (2014 season) that reduced water applied in some plots. However, such reduction did not affect plant water relations.

The experimental design was a completely randomized block experiment with 3 blocks and 3 irrigation treatments. Each treatment was carried out in a plot with two trees aligned in a single row and two adjacent guard rows. The amount of water was measured using a water meter for each plot. Control trees were irrigated in order to obtain the optimum tree water status throughout the season. The midday stem water potential (see below for more details) was used to estimate the water stress level. All treatments were irrigated in the same way as the Control treatment from the end of the spring to the beginning of summer, depending on the water status of the trees. Each repetition was scheduled independently. The optimum values for Control treatments were −1.2 MPa before and −1.4 MPa after the beginning of massive pit hardening [16]. The beginning of the massive pit hardening period was estimated according to [17] and the date was determined when a change in the slope of the longitudinal fruit growth was measured. This period started around mid-June, 17, 10 and 15 June, respectively in 2014, 2015 and 2016. The irrigation restrictions were applied according to the estimated harvest date (around mid-September), starting approximately four weeks before harvest (RDI 2) and two weeks before harvest (RDI 1). The beginning and the end (harvest) of the treatment were changed according to the fruit load and fruit development in each season. Harvest date occurred on 15 (2014 season), 3 (2015 season) and 27 (2016 season) September. Deficit trees were irrigated during this period only if the midday stem water potential measurements were below −2.0 MPa [14]. No irrigation was performed after harvest for any of the treatments.

The irrigation needs of individual plots were changed weekly depending on the distance to the threshold value considered [14]. Three levels of irrigation rate were estimated based on the maximum average daily crop evapotranspiration (ETc) of the orchard (4 mm day^{-1}). This estimation was calculated for the last ten years with the Kc and Kr recommended values [18]. The irrigation rate varied according to Table 1.

Table 1. Applied water was estimated according to the comparison of the measured midday stem water potential and the threshold considered. Irrigation was provided only when measured values were lower than threshold.

% of Decrease from the Threshold	Amount of Irrigation (mm day^{-1})	% of Average Maximum ETc
Less than 15%	1 mm	25%
Between 15% and 30%	2 mm	50%
Greater than 30%	4 mm	100%

2.2. Meteorological Conditions throughout the Experiment

Weather data during the three seasons were obtained from "La Puebla" station in the Andalusian weather stations network (Andalusian weather stations network. SIAR, Spanish Agriculture, Fish and Food Spanish Ministry). This station is around 5 km away from the experimental orchard, and data are available at SIAR web page (http://eportal.mapama.gob.es/websiar/SeleccionParametrosMap. aspx?dst=1). During the three seasons, the patterns of daily potential evapotranspiration (ETo) and rainfall events were similar to the average season (Figure 1). The ETo reached a peak at the end of spring-beginning of summer period with maximum values near 8mm day^{-1}; the average of summer data was around 6mm day^{-1}. Minimum ETo values were measured during winter, with values around 1mm. Rain was almost null from the end of spring until end of summer throughout the experiment. The rainy period concentrated in autumn and winter, as usual in Mediterranean climates, with 80% of the total amount of rain in 2014 and 2015 and around 60% in 2016. There were some rainy episodes a few days before harvest (2 mm from 7 to 12 September in 2014 and harvest was on 15 of September, 11.5 mm on 13 September in 2016 and harvest was on 22 of September). According to the average season (539 mm, Ref. [19]), the year 2015 was very dry, with only 289 mm of rain, and this increased the amount and duration of the irrigation. Conversely, the 2016 season was slightly wetter than average, with 643 mm and an important increase of rain events in Spring.

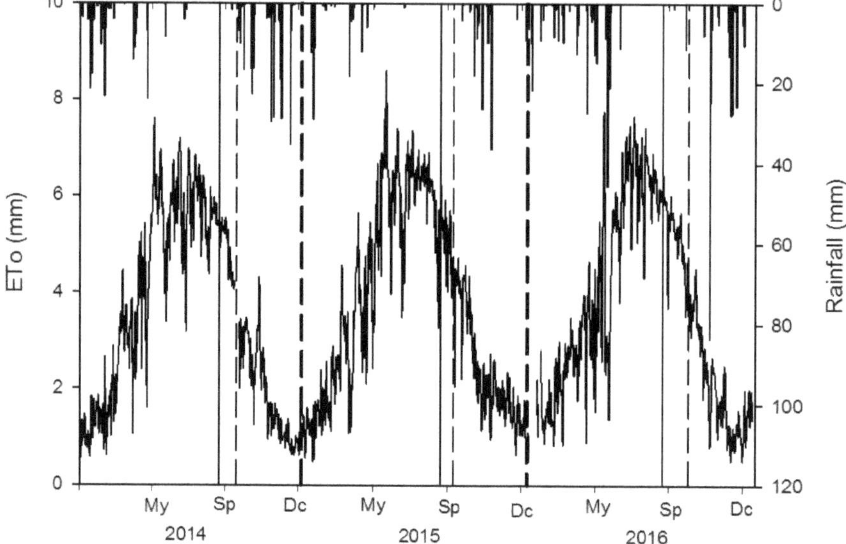

Figure 1. Pattern of daily potential evapotranspiration (ETo, mm) and rainfall (mm) during the three seasons of the experiment (2014 to 2016). Vertical bold dash lines separate each season. Dates between vertical solid and dash lines in each year represent the deficit period. Based on data from "La Puebla" weather station (Andalusian weather stations network). Data download from SIAR, Agriculture, Fish and Food Spanish Ministry (http://eportal.mapama.gob.es/websiar/SeleccionParametrosMap.aspx?dst=1).

2.3. Measurements

Irrigation strategies were characterized using several water relations measurements. A locally calibrated portable FDR (HH2, Delta T, UK) was used to obtain soil moisture measurements. The measurements were made in three plots per treatment. The access tubes for the FDR sensor were placed in the irrigation line, at about 30 cm from the nearest emitter [20]. Data were obtained at 1 m depth and 10 cm intervals. These data were used to estimate the relative extractable water (REW) using the Equation (1).

$$REW = (R - Rmin)/(Rmax - Rmin), \tag{1}$$

Where:

R: Actual soil water content

Rmin: Minimum soil water content measured in the experiment (0.1 m^3 m^{-3})

Rmax: Soil water content at field capacity. Estimated as approximately 0.21 m^3 m^{-3}

The leaf gas exchange was measured using two different methods. Daily cycle of leaf conductance in olive trees presents a maximum value during the morning and decreases until minimum plateau at midday [21]. Maximum values have been reported as earlier indicators for detecting water stress [22]. Then, during the 2014 and 2015 seasons, the abaxial leaf conductance was measured at around 10:00 a.m. in order to estimate the maximum daily value in two fully expanded sunny leaves per tree with a steady state porometer (SC-1, Decagon devices Inc., Pullman, WA, USA). However, differences between treatment were lower than expected. This lack of results could be related with the time of the measurement because the period of maximum values could be shorter tan expected. Then, in the last season, a more sensitive method was used during midday when leaf conductance values were at steady state. During 2016, the midday leaf net photosynthesis was measured with an infrared gas analyser (CI-340, CID BioScience, Camas, USA) in two fully expanded sunny leaves per tree. The water potential was measured weekly at midday in one leaf per tree, using the pressure chamber technique [23]. The leaves near the main trunk were covered in aluminium foil at least one hour before measurements were taken, and a pressure bomb was used for the measurements (PMS model 1000). In order to describe the cumulative effect of the water deficit, the water stress integral (SI) was calculated based on the midday stem water potential data (Equation (2), Ref. [24]) from the beginning of pit hardening until harvest and from the beginning of the irrigation restriction. The maximum value suggested by [24] was changed with the same reference value of −1.4 MPa, which is the threshold value suggested by [16] in fully-irrigated olive trees. Any water potential value higher than the reference was considered as equal to this. The expression used was:

$$SI = |\sum (SWP - (-1.4) * n)|, \tag{2}$$

where SI is the stress integral; SWP is the average midday stem water potential for any interval; n is the number of days in the interval.

At the beginning of each season, ten shoots per tree were selected randomly. The number of fruits were measured periodically in the morning for each shoot. The fruit volume was estimated from a survey of ten fruits per tree. Fruits were randomly selected on each date of measurement. Two measurements were made for each fruit: the longitudinal dimension and the transversal dimension (at the equatorial point) and used both for estimated the volume of an ellipsoid. The pattern of the longitudinal fruit growth was used to estimate the beginning of the pit hardening [17].

The irrigation treatments were also evaluated from the point of view of quantity and quality of yield. For table olives, the quality of the fruit is related to several parameters; some of the most common in the industry are the pulp/stone (PS) ratio, fruit size, fruit colour and fruit firmness. High values of pulp/stone (PS) ratio are considered an indicator of better-quality fruits. The pulp-stone ratio was measured in fresh and dry weight. Pulp and stone of 30 fruits per treatment (10 fruits per plot) were separated and weighed while fresh. Then, these samples were dried at 60 °C until the weight became constant; then, they were weighed again. The final fruit size was estimated in 6 trees per treatment using the number of fruits per kilogram. A sample of around 500 fruits per tree was counted and weighed. Fruit colour is also an important feature for green table olive trees. Spots in the fruit due to the beginning of ripening reduce the yield value. The fruit colour was evaluated using the mature index [25], applied to a sample of 50 fruits that, according to the spots, were ranked with 0 for green fruit, 1 for yellow-green fruit (optimum for green table olives), 2 for purple spots in less than the 50% of the fruit, 3 for purple spots in more than 50% of the fruit and 4 for 100% purple fruit. Each sample was evaluated using the average weight, number of fruits and marks. Fruit firmness was defined as the maximum force required to compress a sample; more specifically, the peak force of the first compression of the fruits [26] was measured using a penetrometer (PCE FM 200, Albacete, Iberica) in 10 fruits per plot. Crop water productivity (CWP) was estimated as the ratio between fruit yield and water consumed in each plot [27]. The water consumed was estimated as the sum of the total water applied and the rainfall during the irrigation period, because the rest of components in the water balance are considered negligible [28]. Only drainage could be considered significant during some periods. Drainage was detected using the comparison of consecutive soil moisture data measurements in the deeper horizon (1m). The data indicating drainage was lower than 10%, and some of them were related to previous rainfall events, only during 2015, when this percentage increased slightly to 15%. Since some works reported olive root growth below 1 m depth (i.e., Ref. [3]), drainage was considered negligible.

Data analyses were performed using ANOVA, and the mean separation was made using Tukey's test with the Statistix (SX) program (8.0). Significant differences were considered when p-level < 0.05 in both tests. Calculations of the p-level were performed considering the F-test of variance equality. When conditions of variance equality were not obtained, a decrease in the degree of freedom and, therefore, more restrictive p-values were calculated. The number of samples measured is specified in the text and figures. The relationship between relative fruit volume and water potential was calculated using all individual data available. The relative fruit volume was calculated considering the maximum value as 100% for each season. The enveloping curve for all data was estimated as the regression of percentile 75% and intervals of 0.5 MPa (SWP) and 5 MPa × day (SI).

3. Results

The water pattern applied during the three years of the experiment is shown in Figure 2. The irrigation season lasted from mid-spring until the end of summer, around mid-September, when trees were harvested. This was the dry period of the season, although rainfall events were measured at the end of the 2014 and 2016 seasons (Figure 1).

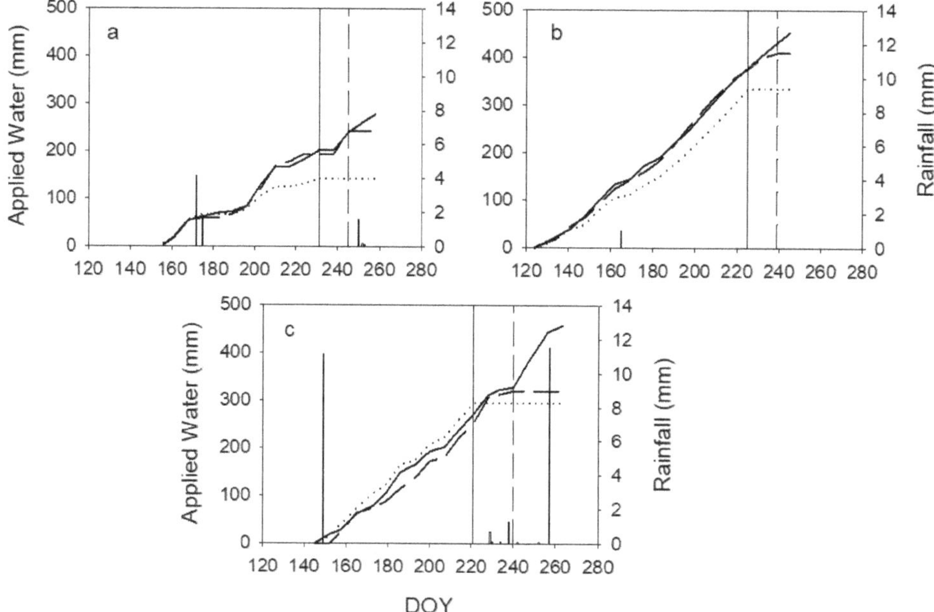

Figure 2. Water applied (lines) and rainfall (bars) during the irrigation season in the three years of the experiment (**a**) 2014, (**b**) 2015, (**c**) 2016. Solid lines, control treatment; dash line, regulated deficit irrigation (RDI) 1 treatment; and dotted line, RDI 2 treatments. Values are the average of three data. The vertical solid line indicates the beginning of the RDI 2 treatment. The vertical dash line indicates the beginning of the RDI 1 treatment.

The pattern of water applied is similar for the different seasons, with almost a linear increase in all treatments until the period of irrigation reduction. The irrigation was scheduled based on water potential measurements and these produced slight changes in some treatments before the irrigation restrictions. The maximum amount of water used corresponded to the 2015 season (Figure 2b), a low fruit load year, because of the scarce rainfall in spring and summer responsible for advancing the beginning of irrigation (around one month before than the rest of the seasons). The moment to start the irrigation restriction was different in each year because the harvest date varied. The shortest water stress period took place in 2015 (20 days in RDI 2 and 6 days in RDI 1, Figure 2b) because the low fruit load brought forward the beginning of the ripening and, therefore, the harvest. Conversely, the longest period happened in 2016 (42 days in RDI 2 and 23 days in RDI 1, Figure 2c) because this season was the one with the highest yield.

The soil moisture pattern is showed in Figure 3 using relative extractable water (REW). During the 2014 season, there were significantly lower values in RDI 2 compared to the rest, before treatments started. Such results were likely related to problems with irrigation during the season. These differences were not found in the following two seasons, when the values of the three treatments were similar before irrigation restrictions.

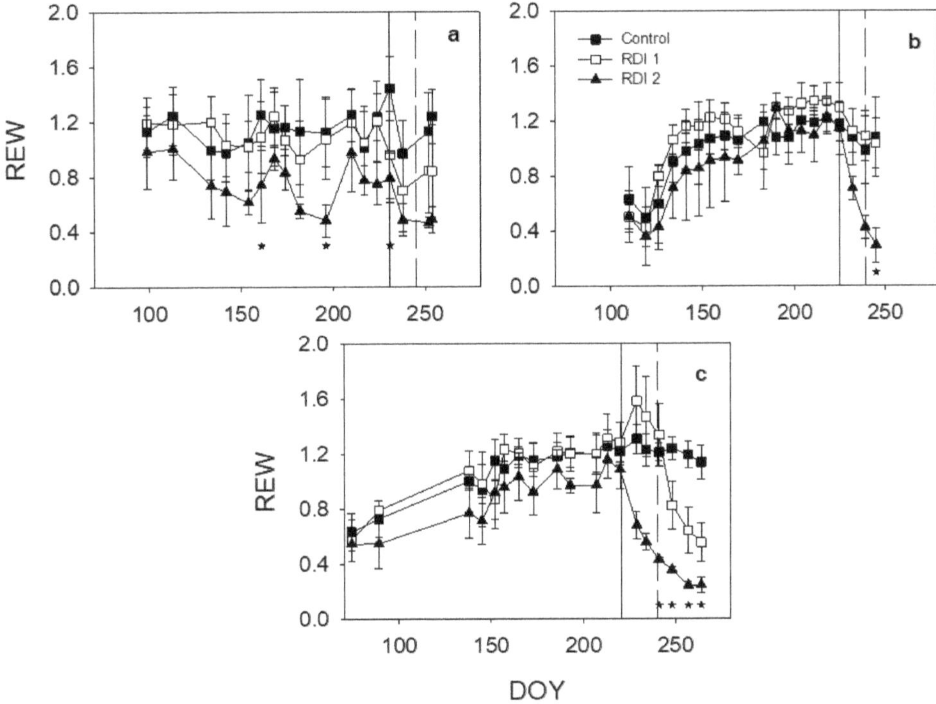

Figure 3. Relative extractable water (REW) at 1 m depth throughout the experiment during the 2014 (**a**), 2015 (**b**) and 2016 (**c**) seasons. Solid squares are Control treatments; empty squares are RDI 1; and triangles are RDI 2 treatments. Values are the average of three data. The vertical solid line indicates the beginning of the RDI 2 treatment. The vertical dash line indicates the beginning of the RDI 1 treatment. Asterisks show when significant differences were found. (Tukey Test, $p < 0.05$).

There were no significant differences in the dry period of the 2014 and 2015 seasons, only in the last data of RDI 2 in 2015. However, soil moisture decreased clearly in RDI 2 trees after the beginning of the treatments. On the contrary, the 2016 season had the longest dry period, and significant differences were found between RDI 2 and Control from DOY 242, and between RDI 1 and Control from DOY 258. RDI 1 and RDI 2 were significantly different only in the period 242–249 in this latter season. Values below 0.4 were measured only in RDI 2, only at the end of the 2015 season and in the last three weeks of 2016. REW data were slightly higher than 1 on most of the dates; only at the beginning of the season in 2015 and 2016 values around 0.4 were found in all treatments.

Irrigation was scheduled based on the midday stem water potential (SWP) in the three seasons (Figure 4).

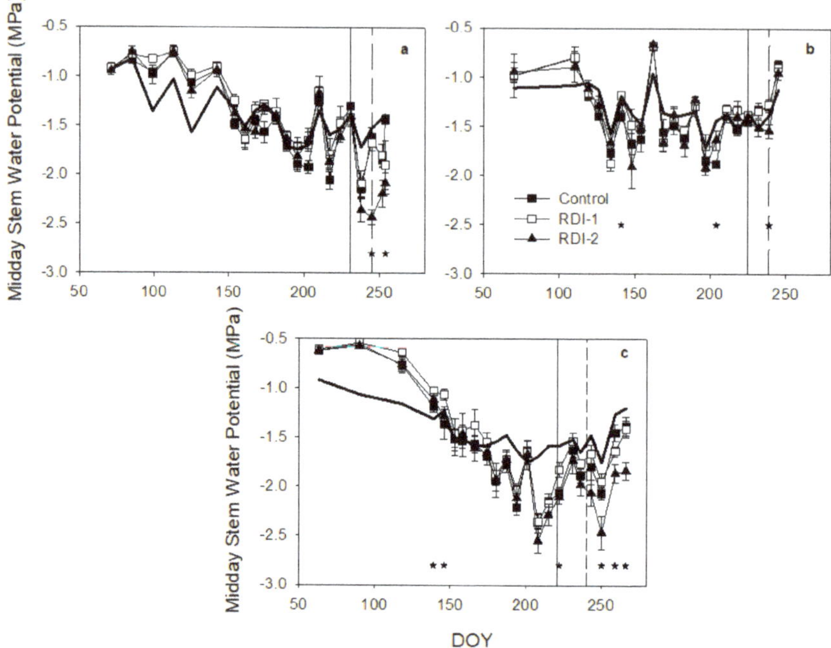

Figure 4. Midday stem water potential throughout the experiment during the 2014 (**a**), 2015 (**b**) and 2016 (**c**) seasons. The solid line represents the baseline calculated according to Corell et al. (2016). Values are the average of six data. The vertical solid line indicates the beginning of the RDI 2 treatment. The vertical dash line indicates the beginning of the RDI 1 treatment. Asterisks show when significant differences were found. (Tukey Test, $p < 0.05$).

Before the beginning of the irrigation restriction, most data were almost equal; the seasonal pattern was also similar between treatments and close to the baseline of [29]. Only during 2015 and 2016, there were two significant differences before the period of restriction. Such differences were small and isolated during the experiment. In 2016, a very high-fruit-load season, all treatments reduced the SWP values. The baseline also predicted this SWP pattern, so the evaporative demand was, in part, related to this response. However, the SWP values in the period 208–222 were even lower than the baseline, and moderate water stress conditions could occur. During the deficit period, RDI 2 SWP decreased around 2 weeks after the irrigation restriction in 2014 and 2016 (Figure 4a,c); only in the low fruit load season, it was almost null (2015, Figure 4b). Such delay was likely related to a not limited capacity of the trees to obtain soil water, probably because the reduction of evaporative demand (Figure 1) increased the percentage of water available in the soil. Only during 2016, there was a long clear period of around three weeks with significant lower SWP in RDI 2 than in Control. In RDI 1, these differences were even lower and not significant. In all the seasons, the recovery for all the treatments was related to rainfall events. Minimum values of SWP, around −2.5 MPa, were measured in RDI 2 during the high fruit load season (2014 and 2016).

The stress integral (SI) was not significantly different between treatments during the low-fruit-load season (2015, Figure 5), when the lowest SI values were calculated. During 2014, no significant differences were found when the overall season was considered. However, in the deficit period, significant differences in SI were found. The RDI 2 value was approximately double (20 MPa·day) that of Control and RDI 1 (approximately 10 MPa·day). In the 2016 season, the SI values from the beginning of pit hardening were the highest, around two-fold higher than the ones from 2014 and 50% higher when the RDI 2 irrigation restriction was considered. RDI 2 was significantly higher than RDI 1 but not higher than Control, from pit hardening in 2016. However, it was significantly higher than the other two treatments when only the period of restriction is considered.

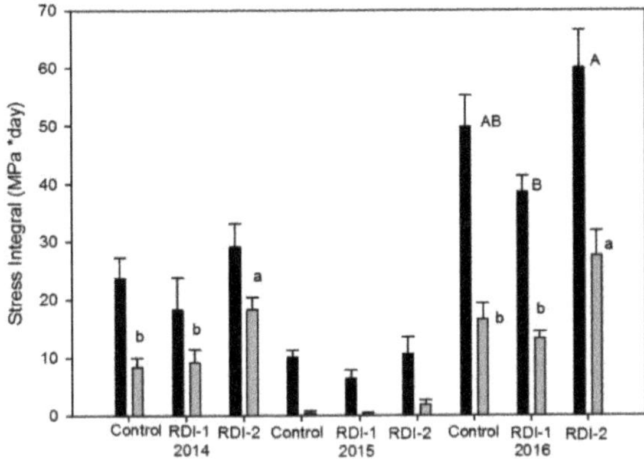

Figure 5. Stress integral (SI) during the three seasons of the experiment. Black bars represent the SI from the pit hardening phase until harvest (2014 DOY 168–254, 2015 DOY 162–245, 2016 DOY 166–266). Grey bars show the SI values only from the RDI 2 irrigation restriction (2014 DOY 231–254, 2015 DOY 225–245, 2016 DOY 221–266). In the three seasons, the reference value to calculate the SI was −1.4 MPa. Each bar is the average of three data. Capital cases indicate significant differences in the season considered from the period of pit hardening. Lower cases show significant differences in the season considered only from the RDI 2 irrigation restriction. (Tukey Test, $p < 0.05$).

Table 2 summarises irrigation, water potential and the stress integral for the three seasons according to three different phenological phases. The first two phases considered (vegetative growth and pit hardening) were the periods when irrigation target was an optimum water status in all treatments. During the vegetative growth period, rains partially covered the water needs and the water status of the trees was typically around the threshold (−1.2 MPa). During pit hardening, the evaporative demand increased and rainfall was almost null, which increased the amount of irrigation in order to maintain the midday water potential around −1.4 MPa. For both periods in all seasons, water potential and stress integral values in all treatments were very similar. In the deficit period, rains were also very scarce but the amount of irrigation was lower because the duration was shorter than in previous periods. The water status of RDI 2 was worse than Control and RDI 1 in all seasons.

Table 2. Summary of irrigation amount (Irr, mm), average midday water potential (SWP, MPa) and stress integral (SI, MPa × day) per season and phenological stage. Each period is characterized using the average daily potential evapotranspiration (ETo, mm) and total rainfall (R, mm).

		2014		
		Control	RDI 1	RDI 2
Vegetative Growth (71–168)	**Irr**	54	56	56
ETo 4.7/R 78.6	**SWP**	−1.12	−1.04	−1.11
	SI	2.3	1.1	1.6
Pit Hardening (169–230)	**Irr**	149	139	87
ETo 5.7/R 5.9	**SWP**	−1.62	−1.49	−1.54
	SI	17.5	11.6	14.2
Deficit Period (231–258)	**Irr**	75	47	0
ETo 4.9/ R 1.9	**SWP**	−1.76	−1.87	−2.27
	SI	6.2	6.6	14.9

Table 2. *Cont.*

		2015		
		Control	**RDI 1**	**RDI 2**
Vegetative Growth (70–161)	Irr	134	124	105
ETo 4.7/R 55.9	SWP	−1.28	−1.22	−1.25
	SI	5.7	3.6	5.1
Pit hardening (162–225)	Irr	244	252	228
ETo 6.0/R 1.6	SWP	−1.46	−1.40	−1.44
	SI	9.9	6.2	9.5
Deficit Period (225–245)	Irr	74	33	0
ETo 4.9 /R 0.0	SWP	−1.16	−1.17	−1.34
	SI	3.2	1.8	10.7
		2016		
		Control	**RDI 1**	**RDI 2**
Vegetative Growth (63–166)	Irr	62	63	73
ETo 4.1/R 253.9	SWP	−1.14	−1.02	−1.12
	SI	3.3	0.9	2.0
Pit hardening (167–221)	Irr	210	190	222
ETo 6.3 /R 0.0	SWP	−1.98	−1.90	−2.01
	SI	36.5	29.0	37.0
Deficit period (222–263)	Irr	185	67	0
ETo 4.8/R 13.9	SWP	−1.70	−1.70	−1.99
	SI	16.5	13.3	27.5

Figure 6 shows the gas exchange data. Significant differences were found in the three seasons, but they were only on isolated dates in the 2014 and 2015. Such results were not in agreement with the soil moisture and water potential, and they were likely not produced for the irrigation management. In the 2014 and 2015 seasons, the maximum leaf conductance was measured (Figure 6a,b). The leaf conductance was increasing throughout the experiment in both seasons but with a similar pattern for the different treatments. No clear trends were found between treatments. The midday net photosynthesis (Pn) was measured throughout 2016 (Figure 6c). Pn values were more similar during the season than those for leaf conductance. The significant differences measured from DOY 194–231 were not likely related to the irrigation treatments because water potential and soil moisture showed the opposite trend and, on all the dates, the Control treatment showed the lowest significant Pn value. Only the decrease in midday stem water potential on DOY 194 (Figure 4c) was coincident with a decrease in Pn (Figure 6c) in the 2016 season.

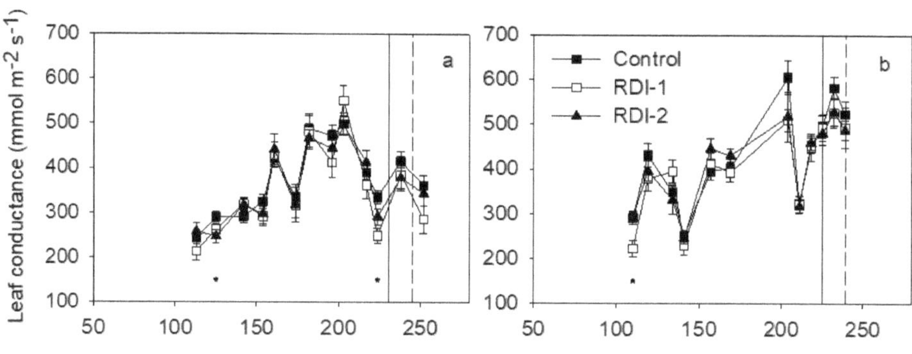

Figure 6. *Cont.*

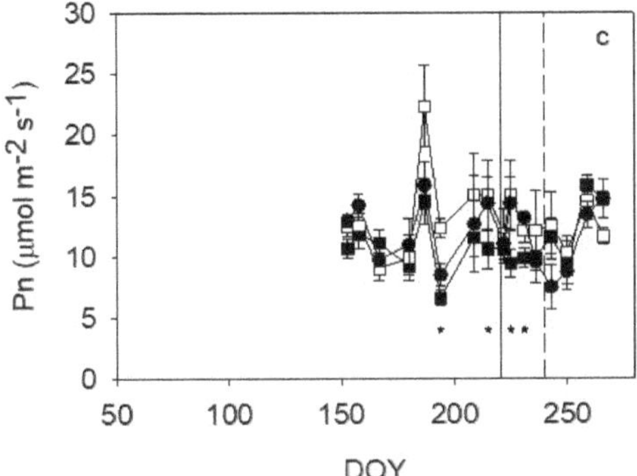

Figure 6. Gas exchange during the three seasons of the experiment. Maximum daily leaf conductance in the 2014 (**a**) and 2015 (**b**) season. Midday net photosynthesis (Pn) in 2016 (**c**). Solid squares are the Control treatments; empty squares are the RDI 1 treatments; and triangles are the RDI 2 treatments. Each point is the average of 6 data. The vertical solid line indicates the beginning of RDI 2 treatment. The vertical dash line indicates the beginning of RDI 1 treatment. Asterisks show when significant differences were found. (Tukey Test, $p < 0.05$).

The data of fruits per shoot presented clear differences between seasons (Figure 7). The fruit load in the 2015 season was almost null in all the treatments, while in 2016 the fruit load was extremely high. This pattern of alternate bearing was the same for the three treatments. Two sampling dates are presented for each season, at the beginning of massive pit hardening and at harvest. There were no significant differences between treatments within the year on the two sampling dates. The difference between both sampling dates (fruit drop) was also not significant.

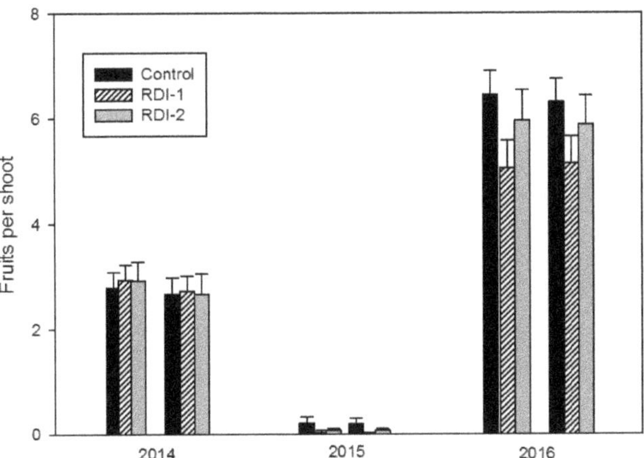

Figure 7. Amount of fruit peer shoot in the three seasons of the experiment. Black bars are the Control, white bars are the RDI 1 and grey bars are the RDI 2. Each bar is the average of 60 data. The vertical bar represents the standard error. For each season, bars on the left include the measurements at the beginning of the pit hardening and on the right the data before harvest. There were no significant differences between treatments in any of the seasons.

The pattern of fruit volume was almost linear before the beginning of the treatments in the three seasons (Figure 8). Although there were significant differences between treatments before these dates, they were small and unclear. The final fruit size was clearly related to the season; maximum values were measured in 2015 (Figure 8b), the low fruit load year, and minimum values in 2016 (Figure 8c), the highest fruit load year. In the 2014 season, the deficit treatments presented a slight reduction in fruit volume during the stress period, but only in RDI 1 this was significantly lower than in Control. Rainfalls before harvest rehydrated all the treatments, and no differences were found in the end. In the 2015 season, the fruit growth was unaffected by the irrigation deficit and showed a linear increase of fruit volume. In 2016, the growth stopped after the deficit period in RDI 1 and RDI 2, while in Control it continued. Such differences were not very high, around 10%, and were reduced by the rainfall events just before harvest.

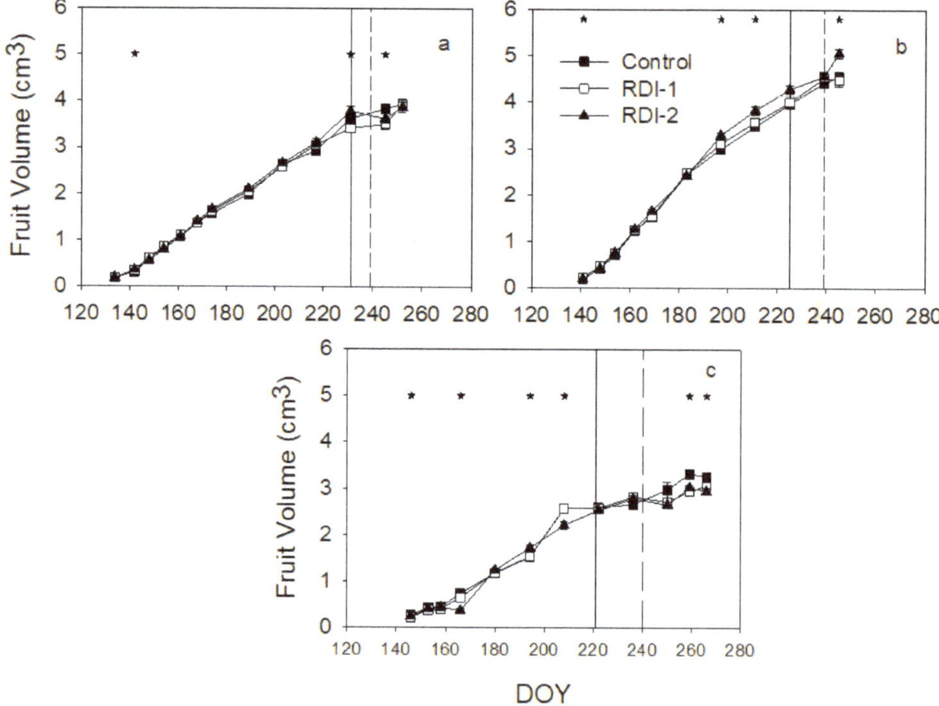

Figure 8. Patterns of fruit volume during the three seasons of the experiment (**a**) 2014, (**b**) 2015 and (**c**) 2016. Full squares are the Control treatments; empty squares are the RDI 1, and triangles are the RDI 2 treatments. Each point is the average of 60 data. The vertical bar represents the standard error. The vertical solid line indicates the beginning of RDI 2 treatment. The vertical dash line indicates the beginning of RDI 1 treatment. Asterisks show when significant differences were found. (Tukey Test, $p < 0.05$).

Figure 9 shows the relationship between relative fruit volume, SWP (a) and SI (b). There was a very high scattering in both indicators. The enveloping curve used the percentile of 75% highest data in several intervals. These curves suggest a slight decrease of fruit size with water stress. In SWP data, values below −1.5 MPa tended to a clear reduction in relative fruit volume, but around −2.5 MPa this was still only 10%. Regarding SI data, the decrease of relative fruit volume started around 10 MPa day, but such decrease is also slow and values around 30 MPa day still reached around 90% of fruit volume.

Table 3 shows the yield data during the three seasons of the experiment. There were no significant differences in yield between treatments in any of the seasons. The highest yield was harvested in 2016, and the minimum was obtained in 2015, when it was almost null. In the 2014 season, the yield was around 15% lower than in 2016, but higher than the orchard average (8 t ha^{-1}). Control trees tended to produce a higher yield than those in deficits treatments in 2014 and 2016. In the 2014 season, the Control trees production was around 18% higher than those in the RDI, which were almost equal. Such trend decreased in 2016, when the maximum differences were 9% between Control and RDI 1 trees, and almost null with RDI 2. Fruit size is one of the most important yield features for table olives. There were no significant differences in the number of fruits per kg between treatments in any of the seasons. The maximum size was measured in the 2015 season and the minimum in 2016 for all treatments. In 2014 and 2016, high fruit load seasons, Control trees tended to produce greater sizes than those with deficit irrigation, which were similar to each other. Such differences were similar in both seasons, between 9% and 12% and close to the values measured in the field (Figure 8). There were not significant differences in fruit load between treatments in any of the season. In 2014 and 2016, high yield season, Control trended to greater values and RDI 1 to the lowest, but such differences were only around 5%. In the 2015 season, RDI 2 presented the highest value of fruit load, almost double than Control, but there were no significant differences because of the high variability within treatments.

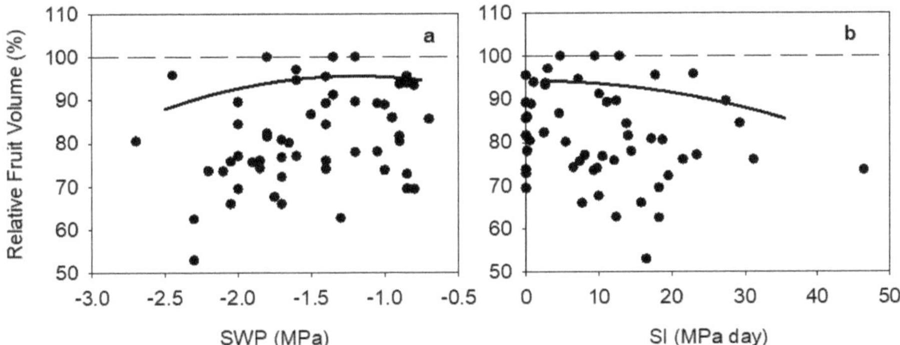

Figure 9. Relationship between relative fruit volume and midday stem water potential (SWP, **a**) and stress integral (SI, **b**). Each point is the data for an individual tree. The relative fruit volume was calculated considering 100% in each season as the greatest fruit volume from all the data available.

The pulp vs stone ratio (PS) is also a very important feature for table olives. There were significant differences in PS fresh weight in 2014 and 2015. In 2016 season, Control trees presented significantly larger PS ratio than the other two treatments. This trend was also measured in 2014 between Control and RDI 2. Conversely, the PS dry weight was significantly larger during 2015 in RDI 1 compared to Control, while in the other two seasons, the PS ratio in dry weight was almost equal for the different treatments. In the two seasons, when texture was measured (2015 and 2016), it was significantly larger in Control than in the two deficit treatments. Also related to this parameter, the mature index (MI) was significantly lower in 2015 in Control trees than in those with deficit treatments. In the 2014 and 2016 season, there were no clear trends, and the MI was below 1 in all treatments. Values of MI higher than 1 indicate a considerable number of fruits with black/purple spots related to the beginning of the ripening. Such spots reduce the yield value, and they were the reason for the early harvesting in 2015. The crop water productivity (CWP) was not significantly different between treatments in any of the seasons, but it was clearly larger in deficit treatments than in Control, mainly in RDI 2, where the minimum differences were 30% higher than in Control. This value was affected mainly by the amount of water used before the deficit period, because irrigation during treatments was higher in Control than in RDI 2 and intermediate in RDI 1, although not always significantly different.

Table 3. Yield quantity and quality features in the three years of the experiment. Yield (Y, t·ha^{-1}), Fruit size (S, fruits·kg^{-1}), Fruit load (FL, fruit tree^{-1}), Mature index (MI), Pulp stone ratio in fresh (PS F) and dry weight (PS D), Fruit texture (T, N), Water Applied per Season (AW, mm), Crop water productivity (CWP, kg·m^{-3}). Different letters in the same season and feature indicate significant differences ($p < 0.05$, Tukey Test).

	2014		
	Control	RDI 1	RDI 2
Yield (T ha^{-1})	14.7±1.6	12.2±2.4	12.0±2.4
Size (Fruit kg^{-1})	244±9	261±21	275±15
FL (Fruit tree^{-1})	12710±1656	11952±2953	11981±2922
MI	0.82±0.03	0.98±0.09	0.78±0.08
Pulp Stone Fresh	4.6±0.1	4.6±0.2	4.4±0.2
Pulp Stone Dry	2.1±0.1	2.2±0.1	2.2±0.1
Fruit texture (N)			
Applied Water (mm)	278±22a	242±54ab	143±13b
CWP (kg m^{-3})	5.1±0.3	5.0±1.0	8.3±1.5
	2015		
	Control	RDI 1	RDI 2
Yield (T ha^{-1})	0.2 ± 0.1	0.4 ± 0.1	0.5 ± 0.4
Size (Fruit kg^{-1})	208 ± 4	192 ± 6	194 ± 4
FL (Fruit tree^{-1})	144 ± 61	277 ± 110	348 ± 307
MI	0.87 ± 0.1b	1.26 ± 0.1ab	1.39 ± 0.1a
Pulp Stone Fresh	5.8 ± 0.1	6.2 ± 0.1	5.7 ± 0.1
Pulp Stone Dry	2.5 ± 0.1b	2.7 ± 0.1a	2.5 ± 0.1ab
Fruit texture (N)	5.9 ± 0.1a	5.2 ± 0.2b	5.2 ± 0.2b
Applied Water (mm)	452 ± 28	409 ± 75	333 ± 33
CWP (kg m^{-3})	0.1 ± 0.0	0.1 ± 0.1	0.1 ± 0.1
	2016		
	Control	RDI 1	RDI 2
Yield (T ha^{-1})	17.6 ± 1.8	16.1 ± 2.3	17.0 ± 2.0
Size (Fruit kg^{-1})	309 ± 17	338 ± 10	331 ± 3
FL (Fruit tree^{-1})	20,083 ± 4198	19,183 ± 4095	19,622 ± 556
MI	0.85 ± 0.1	0.71 ± 0.0	0.87 ± 0.1
Pulp Stone Fresh	4.8 ± 0.3a	4.0 ± 0.2b	3.9 ± 0.1b
Pulp Stone Dry	1.8 ± 0.1	1.8 ± 0.1	1.8 ± 0.1
Fruit texture (N)	6.2 ± 0.2a	5.0 ± 0.1b	5.0 ± 0.1b
Applied Water (mm)	457 ± 30	320 ± 68	295 ± 15
CWP (kg m^{-3})	3.9 ± 0.8	5.5 ± 0.7	5.3 ± 0.3

4. Discussion

Accurate irrigation will need accurate information about water stress conditions. In these regards, threshold values or at least their estimation are very important to optimize limited water resources. Water status measurements are very useful for this purpose. The SWP baseline used in the present work [29] indicated that for most dates, the water status was close to the optimum (Figure 4). The current work validates this baseline in most of the data. But at the beginning of the season and during mid-summer in the high fruit load season, there were clear deviations (Figure 4). Reference [29] reported different equations according to fruit load but concluded that a unique baseline could be used. Decrease of SWP in olive trees like the ones reported in the current work during high fruit load season is reported in the literature [30,31]. Several fruit trees are including the use of water status with similar baselines in their irrigation management [32,33]. However, these baselines are not useful to manage water stress conditions, and threshold values are needed. During the deficit period in the current work, water stress level was not severe because leaf gas exchange was not clearly affected (Figures 4 and 6). Minimum midday stem water potential values around −2 MPa in the literature, similar to the current work, are commonly considered moderate for olive trees, with no effect on yield [14] or even suggesting as no water stress conditions [15]. Such level of water stress could affect the current or the following season's yield. No effects were noticed on the yield of the following season (Table 3), because the vegetative growth occurred before the period of irrigation restriction and the number of fruits per shoot was similar in all treatments during the three seasons considered (Figure 7). Reference [34] reported that the alternate bearing in olive trees is mainly linked to shoot growth in the previous year. Reference [35] suggested in olive trees that floral induction could happen at the end of the summer period, but [14], considering a similar water stress level to the current work, reported no effect on the inflorescence number in the following season. Reference [36] also reported in olive trees the absence of effects on the flowering of the following season with irrigation restrictions in July and August, though unfortunately, no water stress level was reported in this latter work.

Current seasonal yield effects of water stress would affect the amount, size and colour of fruits, which are very important in table olive. There were no significant differences in colour in all seasons, except during the low fruit load one (Table 3). During this season, the main problem was early ripeness, because fruit spotting increased (Table 3). On the other hand, this response was not found in high fruit load seasons (2014 and 2016), although the levels and durations of water stress were clearly higher than in 2015 (Figures 4 and 5). Early fruit ripening has been related with water stress in some works in different olive cultivars (Arbequina, Ref. [37]; Carolea, Ref. [38]). However, this effect was not reported in cv Manzanilla in high fruit load seasons [14]. This reduction of the period of ripening is also described as a drought-avoiding response in plants [39]. Based on the current results, a low fruit load amplified this response in olive trees, even under mild water stress conditions, which did not occur in high fruit load seasons at least in cv Manzanilla. This earlier ripening is a very important factor to be considered in the irrigation for table olive production.

The fruit size is one of the main quality factors for table olives. No significant reductions in fruit size were found in the present work at harvest; however, there were significant differences in the fruit growth pattern and clear reduction trends around 10% (Figures 8 and 9, Table 3). Similar results in fruit size have been reported in other olive cultivars [40,41] but also greater decrease [38,42]. This effect of water deficit would associate with duration or level of water stress. The current work applied a fast and short water stress with minimum water potential around -2.5 MPa and a maximum SI around 30 MPa day (Figures 4 and 5). Figure 9 suggests that size reduction could start at around -1.5 MPa and 10 MPa day, slightly lower values have been suggested in the same and other cultivars, between -1.8 to -2.4 MPa [43–45]. However, this decrease was, sometimes, reversible at harvest. More severe water stress conditions than the current work (predawn or midday, lower than -4 MPa) clearly reduced the fruit size at the deficit period [40–42,45]. At yield, fruit size was near an optimum level or only 10% below in some works [40,41,45] but with a clear reduction after rehydration in others [42,45]. Fruit size recovery in [40,45] was reported even though there were partial rehydrations and SWP before harvest was around -2 MPa all time but not with values lower [45]. This threshold value is similar to the water stress level of the current work. However, this water stress level was not enough for a full recovery of fruit size in other works [41,42]. Such lack in the recovery was associated with a water stress during endocarp growth, which limited fruit growth even though water status was, in some treatments, optimum during recovery [42,46]. Water stress in the current work occurred after maximum endocarp size [17] and confirmed that SWP level around -2 MPa in this period would affect fruit size minimally.

5. Conclusions

Effects of water stress before harvest on yield quantity and quality were very limited. Yield quantity was not significantly affected for the irrigation treatments. Nevertheless, some quality parameters presented changes with water stress conditions. In low fruit load season, water stress level was mild, but it advanced ripening of the deficit treatments. This effect was not observed in high fruit load seasons. Fruit growth was sensitive to water deficit, but fruit size reduction was slow at moderate water stress conditions. Midday stem water potential around −2 MPa was a successful threshold to reduce applied water with a minimum, most of the time no significant, effect on fruit size.

Author Contributions: Conceptualization, M.C., I.G. and A.M.; data curation, M.J.M.-P., M.C., I.G., L.A. and A.M.; formal analysis, M.J.M.-P., M.C., L.A., A.G., A.C., D.P.-L. and A.M.; Writing—Original draft, M.J.M.-P. and M.C.; Writing—Review and editing, A.M. All authors have read and agreed to the published version of the manuscript.

Acknowledgments: This research was supported by the Agencia Española de Investigación (AEI) and the Fondo Europeo de Desarrollo (FEDER) projects AGL2013-45922-C2-1-R and AGL2016-75794-C4-4-R.

Conflicts of Interest: The authors declare no conflict of interest.

References

1. Hsiao, T.C. Measurements of plant water status. In *Irrigation of Agricultural Crops*; Stewart, B.A., Nielsen, D.R., Eds.; American Society of Agronomy: Madison, WI, USA, 1990; pp. 243–279. ISBN 0-89118-102-4.

2. Díaz-Espejo, A.; Fernández, J.E.; Torres-Ruíz, J.M.; Rodríguez-Domínguez, C.M.; Pérez-Martín, A.; Hernández-Santana, V. The olive tree under water stress: Fitting the pieces of response mechanisms in the crop performance puzzle. In *Water Scarcity and Sustainable Agriculture in Semiarid Environment*; García-Tejero, I., Durán-Zuazo, V.H., Eds.; Academic Press: London, UK; Elsevier: Amsterdam, The Netherlands, 2018; pp. 439–480, ISBN 978-0-12-813164-0.

3. Moriana, A.; Orgaz, F.; Fereres, E.; Pastor, M. Yield responses of a mature olive orchard to water deficits. *J. Am. Soc. Hortic. Sci.* **2003**, *128*, 425–431. [CrossRef]

4. Lavee, S.; Wodner, M. Factors affecting the nature of oil acccumulation in fruit of olive (*Olea europea* L.) cultivars. *J. Hortic. Sci. Biotechnol.* **1991**, *66*, 583–591. [CrossRef]

5. Goldhamer, D.A. Regulated deficit irrigation for California canning olives. *Acta Hortic.* **1999**, *474*, 369–372. [CrossRef]

6. Chalmers, D.J.; Canterford, R.L.; Jerie, P.H.; Jones, T.R.; Ugalde, T.D. Photosynthesis in Relation to Growth and Distribution of Fruit in Peach Trees. *Aust. J. Plant Physiol.* **1975**, *2*, 635–645. [CrossRef]

7. Johnstone, P.R.; Hartz, T.K.; LeStrange, M.; Nunez, J.L.; Miyao, E.M. Managing fruit soluble solids with late-season deficit irrigation in drip irrigated processing tomato production. *HortScience* **2005**, *40*, 1857–1861. [CrossRef]

8. Besset, J.; Genard, M.; Girard, T.; Serra, V.; Bussi, C. Effect of water stress applied during the final stage of rapid growth on peach trees (cv. Big-Top). *Sci. Hortic.* **2001**, *91*, 289–303. [CrossRef]

9. Girona, J.; Fereres, E.; Marsal, J.; Goldhamer, D.A.; Naor, A.; Soriano, M.A. Peach. In *Crop Yield Response to Water, Irrigation and Drainage Paper No. 66*; Steduto, P., Hsiao, T.C., Fereres, E., Raes, D., Eds.; FAO: Roma, Italy, 2012; pp. 392–409. ISBN 978-92-5-107274-5.

10. Girona, J. Regulated deficit irrigation in Peach. A global analysis. *Acta Hortic.* **2002**, *592*, 335–342. [CrossRef]

11. D'Amato, R.; Proietti, P.; Onofri, A.; Regni, L.; Esposto, S.; Servili, M.; Businelli, D.; Selvaggini, R. Biofortification (Se): Does it increase the content of phenolic compounds in virgin olive oil (VOO). *PLoS ONE* **2017**. [CrossRef]

12. Casanova, L.; Corell, M.; Suárez, M.P.; Rallo, P.; Martín-Palomo, M.J.; Jiménez, M.R. Bruising susceptibility of Manzanilla de Sevilla table olive cultivar under regulated deficit irrigation. *Agric. Water Manag.* **2017**, *189*, 1–4. [CrossRef]

13. Lavee, S.; Hanoch, E.; Wodner, M.; Abramowitch, H. The effect of predetermined deficit irrigation on the performance of cv. Muhasan olives (*Olea europaea* L.) in the eastern coastal plain of Israel. *Sci. Hortic.* **2007**, *1121*, 156–163. [CrossRef]

14. Girón, I.F.; Corell, M.; Martín-Palomo, M.J.; Galindo, A.; Torrecillas, A.; Moreno, F.; Moriana, A. Feasibility of trunk diameter fluctuations in the scheduling of regulated deficit irrigation for table olive trees without references trees. *Agric. Water Manag.* **2015**, *161*, 114–126. [CrossRef]

15. Marino, G.; Caruso, T.; Ferguson, L.; Marra, F.P. Gas Exchange and stem water potential define stress thresholds for efficient irrigation management in olive (*Olea europaea* L.). *Water* **2018**, *10*, 342. [CrossRef]

16. Moriana, A.; Pérez-López, D.; Prieto, M.H.; Ramírez-Santa-Pau, M.; Pérez-Rodríguez, J.M. Midday stem water potential as a useful tool for estimating irrigation requirements in olive trees. *Agric. Water Manag.* **2012**, *112*, 43–54. [CrossRef]

17. Rapoport, H.F.; Pérez-López, D.; Hammami, S.B.M.; Aguera, J.; Moriana, A. Fruit pit hardening: Physical measurements during olive growth. *Ann. Appl. Biol.* **2013**, *163*, 200–208. [CrossRef]

18. Orgaz, F.; Fereres, E. Riego. In *El Cultivo del Olivo*; Barranco, D., Fernández Escobar, R., Rallo, L., Eds.; Mundiprensa: Madrid, Spain, 1997; pp. 251–271. ISBN 84-7114-657-6.

19. Meteorological Spanish Agency (AEMET). Available online: http://www.aemet.es/es/serviciosclimaticos/datosclimatologicos/valoresclimatologicos (accessed on 8 January 2020).

20. Fernández, J.E.; Moreno, F.; Cabrera, F.; Arrue, J.L.; Martín-Aranda, J. Drip irrigation, soil characteristics and the root distribution and root activity of olive trees. *Plant Soil* **1991**, *133*, 239–251. [CrossRef]

21. Xiloyannis, C.; Pezzarosa, B.; Jorba, J.; Angelini, P. Effects of soil water content on gas exchange in olive trees. *Adv. Hortic. Sci.* **1999**, *2*, 58–63.

22. Moriana, A.; Villalobos, F.J.; Fereres, E. Stomatal and photosynthetic responses of olive (*Olea europaea* L.) leaves to water deficits. *Plant Cell Environ.* **2002**, *25*, 395–405. [CrossRef]

23. Scholander, P.F.; Hammel, H.T.; Bradstreest, E.A.; Hemmingsen, E.A. Sap pressure in vascular plant. *Science* **1965**, *148*, 339–346. [CrossRef]

24. Myers, B.J. Water stress integral a link between short term stress and long term growth. *Tree Physiol.* **1988**, *4*, 315–323. [CrossRef]

25. Hermoso, M.; Uceda, M.; Frías, L.; Beltrán, G. Maduración. In *El Cultivo del Olivo*; Barranco, D., Fernández Escobar, R., Rallo, L., Eds.; Mundiprensa: Madrid, Spain, 1997; pp. 137–153.

26. Szychowski, P.J.; Frutos, M.J.; Burlo, F.; Pérez-López, A.J.; Carbonell-Barrachina, A.A.; Hernández, F. Instrumental and sensory texture attributes of pomegranate arils and seeds as affected by cultivar. *LWT Food Sci. Technol.* **2015**, *60*, 656–663. [CrossRef]

27. Molden, D.; Murray-Rust, H.; Sakthivadivel, R.; Makin, I. A water-productivity framework for understanding and action. In *Water Productivity in Agriculture. Limits and Opportunities for Improvement*; Kijne, J.W., Barker, R., Molden, D., Eds.; CABI Publishing: Oxford, UK, 2003; pp. 1–19. ISBN 0-85199-669-8.

28. Fernández, J.E.; Diaz-Espejo, A.; Cuevas, M.V.; Hernandez-Santana, V. The crop water productivity, that false friend. In Proceedings of the XIV International Plant Water Relations Symposium, Madrid, Spain, 5 October 2018; pp. 119–123.

29. Corell, M.; Pérez-López, D.; Martín-Palomo, M.J.; Centeno, A.; Girón, I.; Galindo, A.; Moreno, M.M.; Moreno, C.; Memmi, H.; Torrecillas, A.; et al. Comparison of the water potential baseline in different locations. Usefulness for irrigation scheduling of olive orchards. *Agric. Water Manag.* **2016**, *177*, 308–316. [CrossRef]

30. Martín-Vertedor, A.; Pérez-Rodriguez, J.M.; Prieto, H.; Fereres, E. Interactive responses to water deficits and crop load in olive (*Olea europaea* L. cv. *Morisca*) I. Growth and water relations. *Agric. Water Manag.* **2011**, *98*, 941–949.

31. Naor, A.; Schneider, D.; Ben-Gal, A.; Zipori, I.; Dag, A.; Kerem, Z.; Birger, R.; Peres, M.; Gal, Y. The effects of crop laod and irrigation rate in the oil accumulation stage on oil yield and water relations of "Koroneiki" olives. *Irrig. Sci.* **2013**, *31*, 781–791. [CrossRef]

32. Shackel, K. A plant-based approach to deficit irrigation in trees and vines. *HortScience* **2011**, *46*, 173–177. [CrossRef]

33. Steduto, P.; Hsiao, T.C.; Fereres, E.; Raes, D. *Crop Yield Response to Water*; FAO Irrigation and Drainage Paper no. 66; FAO: Roma, Italy, 2012.

34. Rallo, L. Fructificación y producción. In *El Cultivo del Olivo*; Barranco, D., Fernández Escobar, R., Rallo, L., Eds.; Mundiprensa: Madrid, Spain, 1997; pp. 107–135. ISBN 84-7114-657-6.

35. Fabbri, A.; Benelli, C. Flower bud induction and differentiation in olive. *J. Hortic. Sci. Biotechnol.* **2000**, *75*, 131–141. [CrossRef]

36. Gómez-del-campo, M. Summer deficit-irrigation strategies in a hedgerow olive orchard cv. Arbequina: Effect on fruit characteristics and yield. *Irrig. Sci.* **2011**. [CrossRef]

37. Motilva, M.J.; Tovar, M.J.; Romero, M.P.; Alegre, S.; Girona, J. Influence of regultaed deficit irrigation strategies applied to olive trees (*Arbequina* cultivar) on oil yield and oil composition during fruit ripening period. *J. Sci. Food Agric.* **2000**, *80*, 2037–2043. [CrossRef]

38. Inglese, P.; Barone, E.; Gullo, G. THe effect of complementary irrigation on fruit growth, ripening pattern and oil characteristics of olive (*Olea europea* L.) cv. Carolea. *J. Hortic. Sci. Biotechnol.* **1996**, *71*, 257–263. [CrossRef]

39. Fereres, E. Variability in adaptive mechanisms to water deficits in annual and perennial crop plants. *Bull. Soc. Bot. Fr.* **1984**, *131*, 17–32. [CrossRef]

40. Cuevas, M.V.; Martín-Palomo, M.J.; Díaz-Espejo, A.; Torres-Ruiz, J.M.; Rodriguez-Dominguez, C.M.; Pérez-Martín, A.; Pino-Mejias, R.; Fernandez, J.E. Assessing water stress in a hedgerow olive orchard from sap flow and trunk diameter measurements. *Irrig. Sci.* **2013**, *31*, 729–746. [CrossRef]

41. Caruso, G.; Gucci, R.; Sifola, M.I.; Selvaggini, R.; Urbani, S.; Esposto, S.; Taticchi, A.; Servilli, M. Irrigation and fruit canopy position modify oil quantity of olive trees (cv. Frantoio). *J. Sci. Food Agri.* **2017**, *97*, 3530–3539. [CrossRef] [PubMed]

42. Gucci, R.; Lodolini, E.; Rapoport, H.F. Productivity of olive trees with different water status and crop load. *J. Hortic. Sci. Biotechnol.* **2007**, *82*, 648–656. [CrossRef]

43. Dell'Amico, J.; Moriana, A.; Corell, M.; Girón, I.F.; Morales, D.; Torrecillas, A.; Moreno, F. Low water stress conditions in table olive trees (*Olea europaea* L.) during pit hardening produced a different response of fruit and leaf water relations. *Agric. Water Manag.* **2012**, *114*, 11–17. [CrossRef]

44. Girón, I.F.; Corell, M.; Galindo, A.; Torrecillas, E.; Morales, D.; Dell'Amico, J.; Torrecillas, A.; Moreno, F.; Moriana, A. Changes in the physiological response between leaves and fruits during a moderate water stress in table olive trees. *Agric. Water Manag.* **2015**, *148*, 280–286. [CrossRef]

45. Hueso, A.; Trentacoste, E.R.; Junquera, P.; Gómez-Miguel, V.; Gómez-del-Campo, M. Differences in stem water potential during oil synthesis determine fruit characteristics and production but not vegetative growth or return blom in an olive hedgerow orchard (cv. Arbequina). *Agric. Water Manag.* **2019**, *223*, 105589. [CrossRef]

46. Gómez-del-Campo, M.; Pérez-Exposito, M.A.; Hammami, S.B.M.; Centeno, A.; Rapoport, H.F. Effect of varied summer déficit irrigation on components of olive fruit growth and development. *Agric. Water Manag.* **2014**, *137*, 84–91. [CrossRef]

Article

Regulation of Agronomic Traits, Nutrient Uptake, Osmolytes and Antioxidants of Maize as Influenced by Exogenous Potassium Silicate under Deficit Irrigation and Semiarid Conditions

Mohamed F. M. Ibrahim [1,*], Gomaa Abd El-Samad [2], Hatem Ashour [1], Ahmed M. El-Sawy [3], Mohamed Hikal [4], Amr Elkelish [5,*], Hany Abd El-Gawad [6,*], Ahmed Abou El-Yazied [6], Wael N. Hozzein [7,8] and Reham Farag [1]

[1] Department of Agricultural Botany, Faculty of Agriculture, Ain Shams University, Cairo 11566, Egypt; Hatem_ashour@agr.asu.edu.eg (H.A.); Reham_hassan@agr.asu.edu.eg (R.F.)

[2] Department of Agronomy, Faculty of Agriculture, Ain Shams University, Cairo 11566, Egypt; Gomaa_abdelsamad@agr.asu.edu.eg

[3] Climate Modification Department, Central Laboratory for Agriculture Climate, Agriculture Research Center, Giza 12411, Egypt; ahmed.elsawy@arc.sci.eg

[4] Department of Biochemistry, Faculty of Agriculture, Ain Shams University, Cairo 11566, Egypt; Mohamed_elsayed4@agr.asu.edu.eg

[5] Botany Department, Faculty of Science, Suez Canal University, Ismailia 41522, Egypt

[6] Department of Horticulture, Faculty of Agriculture, Ain Shams University, Cairo11566, Egypt; ahmed_abdelhafez2@agr.asu.edu.eg

[7] Bioproducts Research Chair, Zoology Department, College of Science, King Saud University, Riyadh 11451, Saudi Arabia; whozzein@ksu.edu.sa

[8] Botany and Microbiology Department, Faculty of Science, Beni-Suef University, Beni-Suef 62511, Egypt;

* Correspondence: Ibrahim_mfm@agr.asu.edu.eg (M.F.M.I.); amr.elkelish@science.suez.edu.eg (A.E.); hany_gamal2005@agr.asu.edu.eg (H.A.E.-G.); Tel.: +2-01123403173 (M.F.M.I.)

Received: 16 July 2020; Accepted: 10 August 2020; Published: 18 August 2020

Abstract: Understanding the link between the protective role of potassium silicate (K_2SiO_3) against water shortage and the eventual grain yield of maize plants is still limited under semiarid conditions. Therefore, in this study, we provide insights into the underlying metabolic responses, mineral nutrients uptake and some nonenzymatic and enzymatic antioxidants that may differ in maize plants as influenced by the foliar application of K_2SiO_3 (0, 1 and 2 mM) under three drip irrigation regimes (100, 75 and 50% of water requirements). Our results indicated that, generally, plants were affected by both moderate and severe deficit irrigation levels. Deficit irrigation decreased shoot dry weight, root dry weight, leaf area index (LAI), relative water content (RWC), N, P, K, Ca, Fe, Zn, carotenoids, grain yield and its parameters, while root/shoot ratio, malondialdehyde (MDA), proline, soluble sugars, ascorbic acid, soluble phenols, peroxidase (POD), catalase (CAT), polyphenol oxidase (PPO), and ascorbate peroxidase (APX) were improved. The foliar applications of K_2SiO_3 relatively alleviated water stress-induced damage. In this respect, the treatment of 2 mM K_2SiO_3 was more effective than others and could be recommended to mitigate the effect of deficit irrigation on maize plants. Moreover, correlation analysis revealed a close link between yield and the most studied traits.

Keywords: drip irrigation; silicon; mineral nutrients; oxidative stress; osmolytes; yield; *Zea mays*

1. Introduction

Semiarid regions are considered a pattern of drylands where the annual precipitation is not sufficient to meet the needs to grow vegetation all year. Generally, the rainfall in these regions ranges from 200 to 750 mm/year. This means the ratio between the total annual rainfall/the potential evapotranspiration reaches approximately 0.20 to 0.50 [1,2]. Currently, water scarcity is predicted to become the most severe environmental challenge that affects the agricultural sector and multiple socio-economic activities in many regions worldwide, especially with frequent climatic changes [3]. The harmful effects of drought stress on plants are usually associated with several events at cellular, biochemical, physiological and molecular levels that may enable the plants to adapt or tolerate such conditions [4–6]. These responses include the rapid generation of reactive oxygen species (ROS) [7], development of an array of complex antioxidant (nonenzymatic and enzymatic) systems [8], regulation of the expression of tolerance related genes [9] and alternation of nutrient uptake [10].

Maize (*Zea mays*) represents the third most important cereal crops cultivated worldwide after wheat and rice [11]. It has a high nutritional value for both human and animals; it contains approximately 72% starch, 10% protein and 4% fat, supplying an energy density of 365 Kcal/100 g [12]. Furthermore, maize provides suitable raw materials for several industries such as starch, fodder, silage and biofuels [13–15]. It is well documented that maize is highly sensitive to drought stress during any period of its growth cycle [16]. Water stress can cause considerable loss in the grain yield ranging from approximately 40–65% according to the genotype, stage of plant growth (the reproductive stage is more sensitive than the vegetative stage) and both the intensity and duration of exposure [17].

Potassium (K) is an essential macronutrient with broad effects on higher plants. In maize, K alleviates the harmful effects of drought stress by different strategies, including the improvement of net carbon assimilation and phloem transport of sugars from leaves to roots [18]. Moreover, K can enhance leaf area, total yield, grain filling and water use efficiency (WUE) in the stressed plants by decreasing leaf evapotranspiration [19]. In addition, K could play a key role in preventing oxidative damage of the maize plants by maintaining ROS homeostasis and enhancing antioxidant capacity [20].

Although silicon (Si) is not considered an essential mineral nutrient, several lines of evidence confirmed its benefits for plants, particularly under biotic and abiotic stresses [21]. It can promote photosynthesis by increasing the concentration of chlorophyll [22], and affect the activities of RuBisCO and PEP-carboxylase that are required for CO_2 fixation [23]. Furthermore, Si regulates antioxidant enzyme systems under diverse stress conditions [21]. Under drought stress, Si deposits in the cell walls of xylem vessels could prevent their compression caused by the high rate of transpiration [24], and it can improve the hydraulic conductivity of the roots in the radial direction leading to enhance uptake of water [25] and several essential nutrients [26]. Moreover, many previous reports indicated that Si could alleviate water deficit stress by improving osmotic adjustment and compatible solutes accumulation, i.e., proline, soluble sugars, free amino acids and polyamines, in several plant species [25,27].

Potassium silicate (K_2SiO_3) is a soluble source of potassium and silicon; it can be used as a fertilizer to maximize the benefits of both elements on plant growth and productivity. In this study, we provide insights into the underlying metabolic changes, uptake of mineral nutrients and some nonenzymatic and enzymatic antioxidants that may differ in maize plants as influenced by the foliar application of K_2SiO_3 under three irrigation regimes. These results may help to understand the link between the protective role of K_2SiO_3 against drought stress and the eventual yield of grains, especially under semiarid conditions.

2. Materials and Methods

2.1. Experimental Layout and Growth Conditions

Two field experiments were carried out during the seasons of 2018 and 2019 on a private farm, Ahmed Orabi Association, Cairo-Ismailia desert road, Qalyubia Governorate, Egypt. To investigate the effect of foliar application of potassium silicate (K_2SiO_3) at 0, 1 and 2 mM on growth, yield and some physiological and biochemical attributes of maize plants grown under three different levels of drip irrigation (100, 75 and 50% water requirements). Before the establishment of the experiments, samples of soil were collected by an Auger T-Handle at depth 30–60 cm for physical and chemical analyses (Table 1). Climatic data were recorded by an agrometeorological station, Ismailia, to monitor the environmental conditions during the experiment (Table 2).

Table 1. Physical and chemical analysis of the experimental soil before cultivation in the seasons of 2018 and 2019.

Season	pH	EC μS cm^{-1}	CaCO$_3$%	Cation meq/L			Anion meq/L		
				Ca^{++}	Mg^{++}	Na$^+$	HCO$_3^-$	CL$^-$	SO$_4^{--2}$
2018	7.84	0.41	2.87	5.52	0.38	1.03	1.59	1.20	1.74
2019	7.61	0.47	3.13	7.04	0.50	0.80	2.14	1.38	1.62

N, P, K	N	P	K	Sand%	Silt%	Clay%	Soil texture
		(ppm)					
2018	2.88	6.38	1.17	91.95	4.81	3.24	Sandy
2019	2.03	6.22	0.91				

EC: Electrical conductivity.

Maize seeds of white single cross hybrid (Hytech 2030) produced by Misr Hytech Seed Int., Egypt was sown on 17th of May 2018 and 2019, respectively. The experiment was arranged in a split plot design with three replicates. A surface drip irrigation system with three levels (100, 75, and 50% of water requirements) was implemented in the main plots, and the foliar applications of K$_2$SiO$_3$ treatments (0, 1, and 2 mM) were randomly distributed in the subplots. The experimental unit area was 60 m^2 (15 m length × 4 m width) consisting of 5 rows with 0.8 m distance between rows. The plant distance was 30 cm apart on one side. Maize plants were irrigated using drippers of 4 L h^{-1} capacity and 0.3 m distance between drippers. A flow meter was installed for each irrigation level treatment, and three rows were left without irrigation as a border between different irrigation levels.

Table 2. Monthly averages of solar radiation, precipitation, wind speed, air temperature and relative humidity during the period of cultivation (May–September) in the season 2018 and 2019.

Date	Solar Radiation Dgt [MJ/m^2]	Precipitation [mm]	Wind Speed [m/s]		Air Temperature [°C]			Relative Humidity [%]
	Average	Sum	Average	Max	Average	Min	Max	Average
2018								
May	671.29	0.0	1.4	8.9	23.9	11.8	38.5	62.5
June	654.76	0.0	1.3	5.6	26.6	13.4	38.2	67.4
July	616.47	0.0	0.9	4.8	27.7	16.9	37.4	75.5
August	542.00	0.0	0.5	3.7	27.5	17.4	38.0	75.0
September	424.92	0.0	0.7	3.8	25.2	14.7	36.2	73.5
2019								
May	689.12	0.0	1.3	6.5	23.9	12.7	36.1	58.1
June	535.47	0.0	1.3	5.2	27.6	15.7	39.6	65.6
July	472.97	0.0	1.1	5.0	27.5	17.8	37.2	71.0
August	415.72	0.0	1.0	4.6	27.3	17.0	37.5	73.4
September	327.05	0.0	1.0	5.3	25.8	14.6	41.2	69.8

2.2. Calculations of Water Regimes

Data of class A pan (Epan) for the experimental site expressed in mm/day were obtained from an agrometeorological station located close to the site. Water requirements (Table 3) for different irrigation levels were calculated for 105 days, and then irrigation was stopped for 11 days before the harvesting date (117 days after sowing). The calculation was made according to Doorenbos [28].

Table 3. Average amounts of the water requirements for the maize plants in the seasons of 2018 and 2019.

Days	Date	Stage	KC*	Irrigation Level (m^3·ha^{-1})		
				100%	75%	50%
10 days	17/5:26/5	initial	0.3	299.52	299.52	299.52
10 days	27/5:5/6		0.6	694.08	694.08	694.08
Starting date of different irrigation regimes						
15 days	6/6:20/6	development	0.9	953.28	714.96	476.64
20 days	21/6:10/7		1.0	1114.56	835.92	557.28
20 days	11/7:30/7	Mid-season	1.2	1319.04	989.28	659.52
20 days	31/7:19/8		1.0	1085.76	814.32	542.88
10 days	20/8:29/8	Last season	0.9	1097.28	822.96	548.64
11 days	30/8:11/9			Not irrigated before harvest Total amount (m^3 ha^{-1})		
116 Days				6563.52	5171.04	3778.56

KC*: Crop coefficient.

2.3. Foliar Application and Sampling

Maize plants were subjected to the foliar application of distilled water as a control and K_2SiO_3 (1 or 2 mM) four times: first at 24 days after sowing (DAS) then the subsequent applications were applied every 15 days. Tween 20 at 0.05 mL L^{-1} was used as a wetting agent for all foliar treatments (K_2SiO_3-treated and control plants). To determine plant growth and physiological and biochemical changes in response to applications, plants samples were collected twice, first after 10 days of the last foliar application. Four plants were randomly collected from the inner rows to determine the vegetative growth (shoots and roots) in each experimental unit. Biochemical analyses were conducted using the 4th fully expanded leaf from the top, which was randomly collected from 3 plants of each experimental plot. In addition, two plants were randomly selected to collect the 4th fully expanded leaf from the top to determine mineral nutrients after drying in an oven at 105 °C. At the end of the experiment (117 DAS), grain yield per plant and its related traits were estimated, while the grain yield per hectare was determined from one inner row that was left for this purpose (12 m^2/experimental unit).

2.4. Studied Parameters

2.4.1. Vegetative Growth

Shoot and root dry weights were determined by drying four plants from each experimental unit in an air-forced ventilated oven at 105 °C. The dry weight ratio of root/shoot ratio was calculated. Leaf area index (LAI) was calculated as described by Iqbal and Hidayat [29].

2.4.2. Leaf Relative Water Content (RWC)

Leaf relative water content was determined according to Ünyayar et al. [30]. Leaf discs (1.8 cm diameter) from 10 fully expanded young leaves (ear leaf) were taken from 6–8 plants at the mid-canopy position before irrigation. Then the discs were weighed (FW) and placed immediately in distilled water for 2 h at 25 °C then their turgid weights (TW) were recorded. The samples were dried in an oven at 110 °C for 24 h (DW). Relative water content (RWC) was calculated using the following formula: RWC = (FW − DW)/(TW − DW) × 100.

2.4.3. Membranes Lipid Peroxidation

Lipid peroxidation was measured by the determination of malondialdehyde (MDA) as described by Heath and Packer [31]. Frozen leaf tissues were homogenized in 0.1% (*w/v*) trichloroacetic acid (TCA).

The absorbance (A) of the supernatant was measured at 535 nm and corrected for nonspecific turbidity at 600 nm using a spectrophotometer (Chrom Tech CT-2200, Taiwan). The MDA concentration (nmol g^{-1} FW) was calculated using ΔOD (A532-A600) and the extinction coefficient (ε =155 mM^{-1} cm^{-1}).

2.4.4. Proline and Soluble Sugars

Proline levels were determined using the method of acid-ninhydrin reagent as described by Bates et al. [32]. Soluble sugars were determined by anthrone-sulfuric acid reagent as described by Plummer [33].

2.4.5. Determination of Mineral Nutrients

Dry leaves were ground and digested using sulfuric acid and hydrogen peroxide. Leaf mineral concentrations of N, P, K, Ca, Fe and Zn were determined according to Cottenie et al. [34]. Nitrogen (N) was determined by the Kjeldahl method (Velp Scientifica, Europe). The colorimetric method by UV/VIS spectrophotometer was used to determine P; potassium (K) was determined by a Flamephotometer (Jenway, UK). Meanwhile, Ca, Fe and Zn were determined by atomic absorption spectrophotometry (AAS-Hitachi, Tokyo, Japan).

2.4.6. Determination of Carotenoids, Ascorbic Acid and Total Soluble Phenols

Carotenoids were determined using the acetone and petroleum ether method as described by de Carvalho et al. [35]. Ascorbic acid (AsA) was determined using the 2, 6-Dichloroindophenol titrimetric method according to Association of Official Analytical Chemists (A.O.A.C) [36]. Total soluble phenols were determined according to the method of Folin-Denis as described by Skalindi and Naczk [37].

2.4.7. Quantification of Antioxidant Enzymes

Leaf tissue of maize plants (0.5 g) was homogenized in 4 mL 0.1 M K-phosphate buffer (pH 7.0) containing 1% (*w/v*) polyvinylpyrrolidone (PVP) and 0.1mM Ethylenediaminetetraacetic acid (EDTA). The homogenate was centrifuged at 10,000 rpm for 15 min and the supernatant was used as a crude enzyme extract. All the preparation steps of the enzyme extract were carried out at 0–4 °C. Total soluble protein was determined according to Bradford [38].

Peroxidase (EC1.11.1.7) activity was quantified by the method of Hammerschmidt et al. [39]. The absorbance was recorded every 30 s for 3 min at 470 nm using a spectrophotometer (Chrom Tech CT-2200). Catalase (CAT) (EC 1.11.1.6) activity was determined according to the method of Cakmak et al. [40]. Polyphenol oxidase (PPO) (EC 1.14.18.1) activity was measured according to Oktay et al. [41]. The reaction mixture consisted of 100 µL crude enzyme, 600 µL catechol and 2.3 mL phosphate buffer (0.1 M, pH 6.5). The absorbance at 420 nm was recorded at zero time and after 1 min. Ascorbate peroxidase (APX) (EC 1.11.1.11) activity was measured according to the method of Nakano and Asada [42] by monitoring the decrease of absorbance at 290 nm following the ascorbate oxidation for 3 min. The reaction was initiated by the addition of H_2O_2. All enzyme activities were expressed as ΔOD min^{-1} mg^{-1} protein.

2.4.8. Determination of Yield Parameters

Maize ears were harvested at 117 DAS and averages of ear length, ear diameter, number of grains·ear^{-1}, weight of grains·ear^{-1}, weight of grains·$plant^{-1}$ were estimated from 10 random plants per each experimental unit. Eventually, total grain yield (t ha^{-1}) was calculated using the average yield of grains/12 m^2 (one inner row was left for this purpose in each experimental unit).

2.4.9. Statistical Analysis

Data of the two seasons were subjected to combined analysis following the two way ANOVA procedure as described by Snedecor and Cochran [43] using MSTAT-C software (Michigan State University, USA). Duncan's test based on a probability of $p \leq 0.05$ was used to determine the significant differences between means.

All data were expressed as means ± standard deviation (SD). The correlation coefficient between the grain yield (t ha^{-1}) and different physiological and biochemical aspects was also estimated.

3. Results

3.1. The Main Effects of the Irrigation Levels and K_2SiO_3 Foliar Applications

Reduction of irrigation (moderate or severe level) caused significant ($p \leq 0.05$) decreases in shoot dry weight and root dry weight. Furthermore, a substantial reduction in the LAI, RWC, N, P, K, Ca, Fe, carotenoids, ear length, ear diameter, number of grains/ear, weight of grains/plant and grain yield (ton/ha) was observed when compared to the well-irrigated plants (Table 4), while, Zn was only decreased when plants were exposed to the irrigation level of 50% WR. In contrast, root/shoot ratio, MDA, proline, soluble sugars, ascorbic acid and soluble phenols, as well as the activities of peroxidase (POD), catalase (CAT), polyphenol oxidase (PPO), and ascorbate peroxidase (APX) were significantly increased (Table 4). The foliar applications of K_2SiO_3 at 1 or 2 mM significantly increased all studied variables except root/shoot ratio, MDA, proline and Zn. The treatment of 2 mM K_2SiO_3 was more effective in enhancing yield and its parameters than the lower concentration (1 mM).

Table 4. Mean comparison shows the main effects of the irrigation levels (100, 75, and 50 % of water requirements) and the foliar applications of K_2SiO_3 (0, 1, and 2 mM). KSi 0: K_2SiO_3 -untreated plants, KSi 1: K_2SiO_3 (1 mM) and KSi 2: K_2SiO_3 (2 mM). On the vegetative growth, water status, lipid peroxidation, osmolytes, mineral nutrients, non-enzymatic antioxidants, antioxidant enzymes, yield and its parameters of maize plants.

Variables	Irrigation Level			Foliar Application		
	100%	75%	50%	KSi 0	KSi 1	KSi 2
Shoot dry weight (g.plant^{-1})	318.4 ^{A}	242.1 ^{B}	208.1 ^{C}	232.6 ^{C}	260.3 ^{B}	275.7 ^{A}
Root dry weight (g.plant^{-1})	45.32 ^{A}	38.20 ^{B}	33.01 ^{C}	36.52 ^{B}	39.43 ^{A}	40.58 ^{A}
Root/shoot ratio	0.144 ^{B}	0.158 ^{A}	0.159 ^{A}	0.157 ^{A}	0.153 ^{A}	0.150 ^{A}
LAI	7.15 ^{A}	5.19 ^{B}	4.08 ^{C}	4.62 ^{C}	5.44 ^{B}	6.35 ^{A}
RWC (%)	88.11 ^{A}	73.13 ^{B}	67.79 ^{C}	75.52 ^{B}	77.17 ^{A}	76.34 ^{AB}
MDA (nmol.g^{-1} FW)	5.82 ^{C}	11.99 ^{B}	13.62^A	11.08 ^{A}	10.27 ^{B}	10.09 ^{B}
Proline (μg.g^{-1} FW)	183.3 ^{C}	314.0 ^{A}	241.6 ^{B}	285.0 ^{A}	241.5 ^{B}	212.2^C
Soluble sugars (mg.g^{-1} DW)	24.50 ^{B}	45.54 ^{A}	44.53 ^{A}	35.13 ^{C}	37.99 ^{B}	41.44 ^{A}
N (mg.g^{-1} DW)	85.29 ^{A}	71.00 ^{B}	62.14 ^{C}	70.39 ^{B}	71.31 ^{B}	76.72 ^{A}
P (mg.g^{-1} DW)	2.24 ^{A}	1.91 ^{B}	1.50 ^{C}	1.77 ^{B}	1.93 ^{A}	1.96 ^{A}
K (mg.g^{-1} DW)	11.74 ^{A}	9.90 ^{B}	8.81 ^{C}	8.02 ^{C}	10.65 ^{B}	11.79 ^{A}
Ca (mg.g^{-1} DW)	7.25 ^{A}	6.52 ^{B}	5.22 ^{C}	6.07 ^{B}	6.31 ^{AB}	6.61 ^{A}
Fe (μg.g^{-1} DW)	203.9^A	180.5 ^{B}	161.7 ^{C}	169.7 ^{C}	180.7 ^{B}	195.6 ^{A}
Zn (μg.g^{-1} DW)	46.3^A	47.0 ^{A}	41.0 ^{B}	48.6 ^{A}	42.9 ^{B}	42.8 ^{B}
Carotenoids (mg.g^{-1} FW)	0.332 ^{A}	0.308 ^{B}	0.288 ^{C}	0.277 ^{C}	0.317 ^{B}	0.334 ^{A}
Ascorbic acid (μmol.g^{-1} FW)	1.36 ^{C}	1.78 ^{A}	1.67 ^{B}	1.57 ^{B}	1.61 ^{AB}	1.64 ^{A}
Soluble phenols (μg.g^{-1} FW)	13.97 ^{C}	16.39 ^{B}	17.29 ^{A}	15.27 ^{C}	15.87 ^{B}	16.49^A
POD (Δ O.D. min^{-1}.mg protein)	14.3 ^{C}	33.6 ^{B}	35.6 ^{A}	25.92 ^{B}	28.90 ^{A}	28.67 ^{A}
CAT (Δ O.D. min^{-1}.mg protein)	2.64 ^{C}	4.17 ^{A}	3.57 ^{B}	3.19 ^{B}	3.58 ^{A}	3.60 ^{A}
PPO (Δ O.D. min^{-1}.mg protein)	6.95 ^{C}	8.59 ^{B}	9.69 ^{A}	7.97 ^{B}	8.57 ^{A}	8.67 ^{A}
APX (Δ O.D. min^{-1}.mg protein)	2.83 ^{C}	4.36 ^{A}	4.01 ^{B}	3.40 ^{B}	3.85 ^{A}	3.94 ^{A}
Ear length (cm)	23.3 ^{A}	19.1 ^{B}	15.7 ^{C}	18.2 ^{C}	19.4 ^{B}	20.5 ^{A}
Ear diameter (cm)	4.7 ^{A}	4.4 ^{B}	3.3 ^{C}	4.00 ^{C}	4.17 ^{B}	4.30 ^{A}
Number of grains/ear	332.4 ^{A}	271.8 ^{B}	226.3 ^{C}	264.7 ^{B}	281.2 ^{A}	284.6 ^{A}
Weight of grains/ear (g)	109.7 ^{A}	84.5 ^{B}	66.2 ^{C}	79.5 ^{C}	88.8 ^{B}	92.2 ^{A}
Weight of grains/plant (g)	172.8 ^{A}	127.6 ^{B}	95.0 ^{C}	118.1 ^{C}	135.0 ^{B}	142.4 ^{A}
Grain yield (ton·ha^{-1})	7.94 ^{A}	5.82 ^{B}	4.39 ^{C}	5.33 ^{C}	6.14 ^{B}	6.68 ^{A}

Data of the two seasons of 2018 and 2019 were subjected to combined analysis with 3 replicates in each season. The different superscript capital letters within a row indicate significantly different values according to Duncan's multiple range tests ($p < 0.05$). LAI, leaf area index; RWC, relative water content; MDA, malondialdehyde; POD, peroxidase; CAT, catalase; PPO, polyphenol oxidase; APX, ascorpate peroxidase.

3.2. Changes in Plant Growth

The progressive reduction in the irrigation level significantly ($p \leq 0.05$) inhibited plant growth in terms of shoot dry weight, root dry weight and LAI. In contrast, root/shoot ratio was not affected compared to the well-irrigated control (Figure 1). When plants were treated with K_2SiO_3 (1 or 2 mM), a significant increase was observed in shoot dry weight and LAI either under nonstressed or stressed conditions. Meanwhile, this trend was just obvious in root dry weight under water shortage conditions. Root/shoot ratio revealed a significant decrease in the K_2SiO_3-treated plants under well-irrigated conditions. Generally, the highest concentration of the K_2SiO_3 treatments (2 mM) was more effective in this respect (Figure 1).

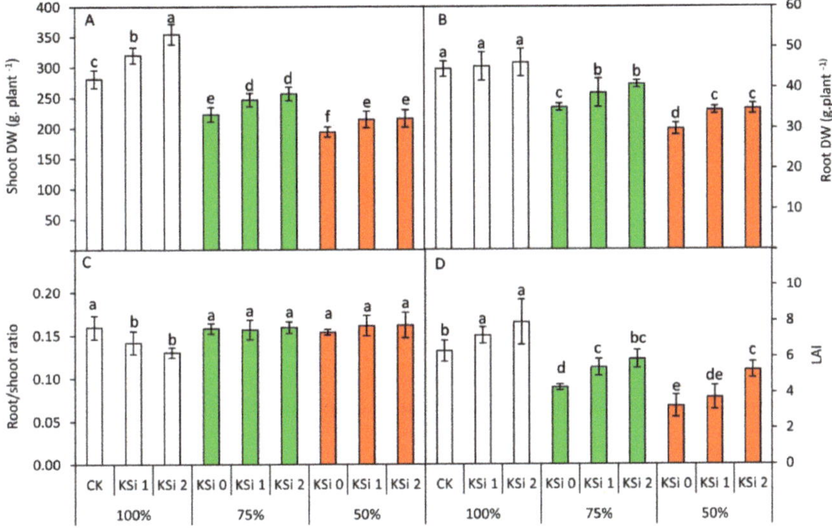

Figure 1. Shoot dry weight (**A**), root dry weight (**B**), root/shoot ratio (**C**) and leaf area index (LAI) (**D**) of the maize plants at 80 days after sowing (DAS) as influenced by the foliar application of K_2SiO_3 (0, 1 and 2 mM) under three irrigation regimes: 100% (white), 75% (green) and 50% (orange) of water requirements. CK: well-watered control, KSi 0: K_2SiO_3-untreated plants, KSi 1: K_2SiO_3 (1 mM) and KSi 2: K_2SiO_3 (2 mM). Data of the two seasons of 2018 and 2019 were subjected to combined analysis. Means were presented ± SD. Different letters are significant differences, according to Duncan's multiple range tests ($p < 0.05$).

3.3. Changes in RWC, MDA, Proline and Soluble Sugars

Plants that were exposed to deficit irrigation demonstrated a significant ($p \leq 0.05$) increase in MDA, proline and soluble sugars, whereas RWC was diminished compared to the well-irrigated conditions (Figure 2). The foliar applications of K_2SiO_3 significantly enhanced RWC under both investigated deficit-irrigation levels, while this tendency was conspicuous in soluble sugars under moderate level of deficit irrigation. Conversely, MDA and proline generally exhibited a significant decrease in K_2SiO_3-treated plants compared to the untreated ones under stressed conditions. Overall, the treatment of 2 mM K_2SiO_3 was more efficient than the other treatments.

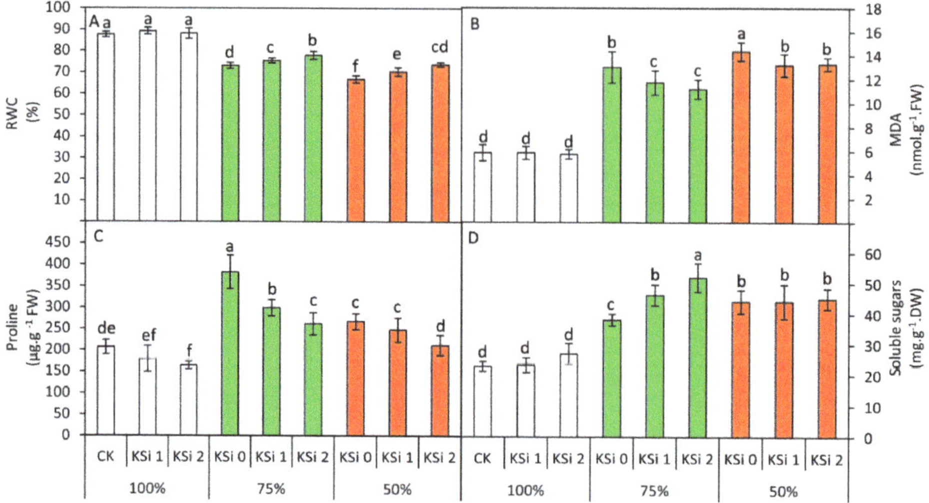

Figure 2. Leaf relative water content (RWC) (**A**), membranes lipid peroxidation as indicated by malondialdehyde (MDA) (**B**), proline (**C**) and soluble sugars (**D**) of t maize plants at 80 DAS as influenced by the foliar application of K_2SiO_3 (0, 1 and 2 mM) under three irrigation regimes: 100% (white), 75% (green) and 50% (orange) of water requirements. CK: well-watered control, KSi 0: K_2SheiO_3-untreated plants, KSi 1: K_2SiO_3 (1 mM) and KSi 2: K_2SiO_3 (2 mM). Data of the two seasons of 2018 and 2019 were subjected to combined analysis. Means were presented ± SD. Different letters are significant differences, according to Duncan's multiple range tests ($p < 0.05$).

3.4. Changes in Mineral Nutrients

To evaluate the nutritional status of plants under continuous deficit irrigation and K_2SiO_3 foliar applications, N, P, K, Ca, Fe and Zn were quantified (Figure 3). The general tendency was that deficit irrigation obviously and significantly ($p \leq 0.05$) decreased N, K, Ca and Fe in K_2SiO_3 nontreated plants under both examined deficit levels of irrigation (75% and 50%). In comparison, P and Zn were only affected under the severe level of deficit irrigation (50%). Applied K_2SiO_3, specifically at 2 mM, significantly improved the concentration of N, P, K, Ca and Fe under unstressed conditions. In contrast, a significant reduction in Zn was manifested in K_2SiO_3-treated plants under well-irrigated conditions. When plants were subjected to continuous deficit irrigation, the treatment of 2 mM K_2SiO_3 exhibited the highest significant increases in N, K and Fe under both investigated levels of deficit irrigation. A similar trend was only observed in P under a moderate level of irrigation.

On the other hand, no significant differences were detected in Ca between K_2SiO_3 nontreated and the treated plants under both deficit irrigation levels (75% and 50%). Meanwhile, Zn was significantly decreased by the treatments of K_2SiO_3 under the moderate level of irrigation. This effect did not occur under the lower level of irrigation (50%).

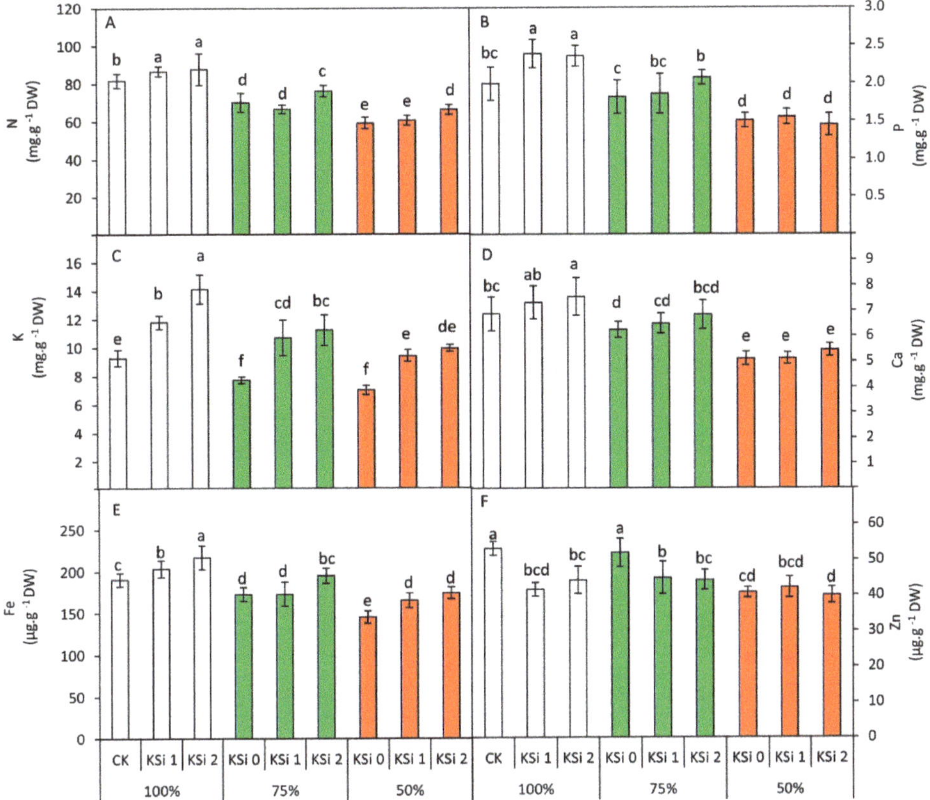

Figure 3. Leaf mineral content including N (**A**), P (**B**), K (**C**) Ca (**D**), Fe (**E**) and Zn (**F**) of the maize plants at 80 DAS as influenced by the foliar application of K_2SiO_3 (0, 1 and 2 mM) under three irrigation regimes: 100% (white), 75% (green) and 50% (orange) of water requirements. CK: well-watered control, KSi 0: K_2SiO_3-untreated plants, KSi 1: K_2SiO_3 (1 mM) and KSi 2: K_2SiO_3 (2 mM). Data of the two seasons of 2018 and 2019 were subjected to combined analysis. Means were presented ± SD. Different letters are significant differences, according to Duncan's multiple range tests ($p < 0.05$).

3.5. Changes in Nonenzymatic Antioxidants

Nonenzymatic antioxidant capacity of plants was investigated by the determination of carotenoids, ascorbic acid and soluble phenols (Figure 4). Plants that were not applied by K_2SiO_3 and exposed to continuous deficit irrigation demonstrated a significant ($p \leq 0.05$) increase in ascorbic acid and soluble phenols compared to the well-watered conditions, whereas carotenoids did not show any significant differences in this respect. Applied K_2SiO_3 (1 or 2 mM) significantly enhanced carotenoids and ascorbic acid, while soluble phenols were not changed under well-irrigated conditions. Similarly, K_2SiO_3 applications, in particular at the highest concentration (2 mM), exhibited the highest significant increases in carotenoids and soluble phenols under both investigated levels of deficit irrigation (75 and 50%). On the other hand, ascorbic acid revealed an opposite trend by the treatment of 2 mM K_2SiO_3 under the moderate (75%) and lower (50%) levels of irrigation.

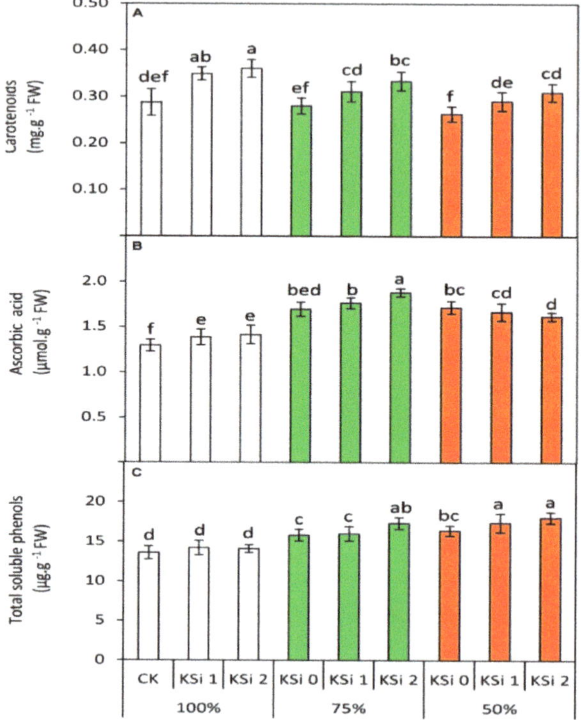

Figure 4. Nonenzymatic antioxidants including carotenoids (**A**), ascorbic acid (AsA) (**B**) and total soluble phenols (**C**) in the leaves of the maize plants at 80 DAS as influenced by the foliar application of K_2SiO_3 (0, 1 and 2 mM) under three irrigation regimes: 100% (white), 75% (green) and 50% (orange) of water requirements. CK: well-watered control, KSi 0: K_2SiO_3-untreated plants, KSi 1: K_2SiO_3 (1 mM) and KSi 2: K_2SiO_3 (2 mM). Data of the two seasons of 2018 and 2019 were subjected to combined analysis. Means were presented ± SD. Different letters are significant differences, according to Duncan's multiple range tests ($p < 0.05$).

3.6. Changes in Antioxidant Enzymes

The activities of antioxidant enzymes (POD, CAT, PPO, and APX) were determined in this study under deficit irrigation conditions and the exogenous application of K_2SiO_3 (Figure 5). No significant differences were observed between K_2SiO_3-treated, and nontreated plants in the activity of all studied antioxidant enzymes under well-irrigated conditions. Reducing irrigation levels significantly ($p \leq 0.05$) increased the activity of these enzymes compared to the well-irrigated conditions. Applied-K_2SiO_3 significantly enhanced the activity of CAT, PPO and APX under the moderate level of irrigation (75%), whereas POD was not affected. When plants were exposed to severe deficit irrigation (50%), POD exhibited a significant increase by the treatment of 1 mM K_2SiO_3. At the same time, the highest activity of CAT and PPO were obtained by the treatment of 2 mM K_2SiO_3. On the contrary, APX did not reveal any significant differences between K_2SiO_3-treated and nontreated plants under the lower level of irrigation.

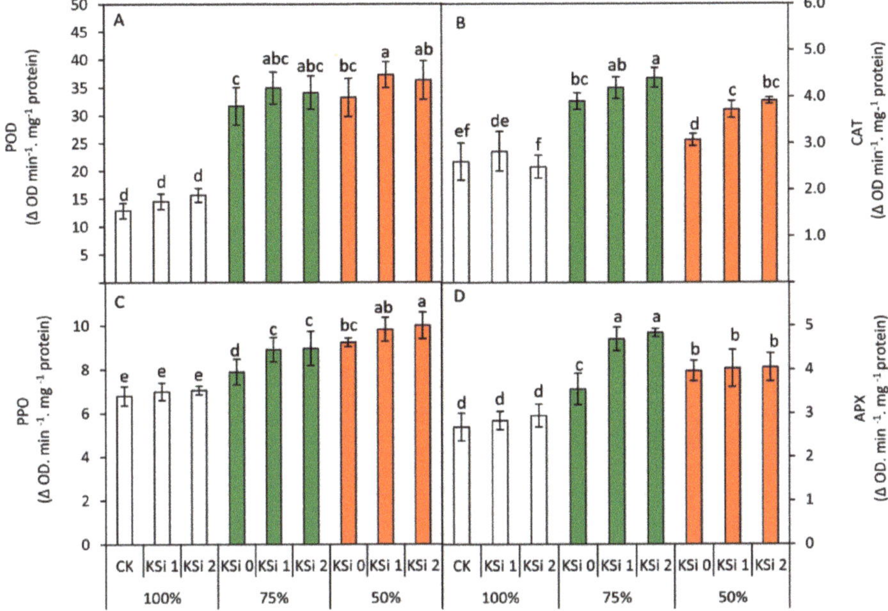

Figure 5. Activities of antioxidant enzymes including POD (**A**), CAT (**B**), PPO (**C**) and APX (**D**) in the leaves of maize plants at 80 DAS as influenced by the foliar application of K_2SiO_3 (0, 1 and 2 mM) under three irrigation regimes: 100% (white), 75% (green) and 50% (orange) of water requirements. CK: well-watered control, KSi 0: K_2SiO_3-untreated plants, KSi 1: K_2SiO_3 (1 mM) and KSi 2: K_2SiO_3 (2 mM). Data of the two seasons of 2018 and 2019 were subjected to combined analysis. Means were presented $\pm$ SD. Different letters are significant differences, according to Duncan's multiple range tests ($p < 0.05$).

3.7. Changes in Yield Parameters

Grain yield and its parameters, including ear length, ear diameter, number of grains·ear^{-1}, weight of grains·ear^{-1}, weight of grains·plant^{-1} and total grain yield (t ha^{-1}) were estimated in this investigation (Figure 6). Concerning K_2SiO_3-untreated plants, reducing irrigation level led to significant ($p \leq 0.05$), and gradual decreases in all yield parameters studied in parallel with the severity of deficit irrigation. Generally, except for the number of grains·ear^{-1} under the lower level of irrigation, applied-K_2SiO_3, specifically at 2 mM, significantly improved all studied traits regardless of the level of irrigation.

3.8. Relationships between Grain Yield and RWC, MDA, Osmolytes, Nutrients and Antioxidants

To elucidate the relationships between the grain yield of maize plants as influenced by the foliar applications of K_2SiO_3 under different irrigation regimes and RWC, MDA, osmolytes, nutrients and antioxidants, the correlation coefficient was analyzed (Figure 7). We observed that grain yield (t ha^{-1}) was significantly and positively correlated with leaf relative water content (RWC), carotenoids, N, P, K, Ca and Fe. Meanwhile, MDA, soluble sugars, soluble phenols, POD and PPO demonstrated a negative correlation. On the other hand, proline, ascorbic acid (AsA), CAT, APX and Zn did not reveal any significant correlation in this respect.

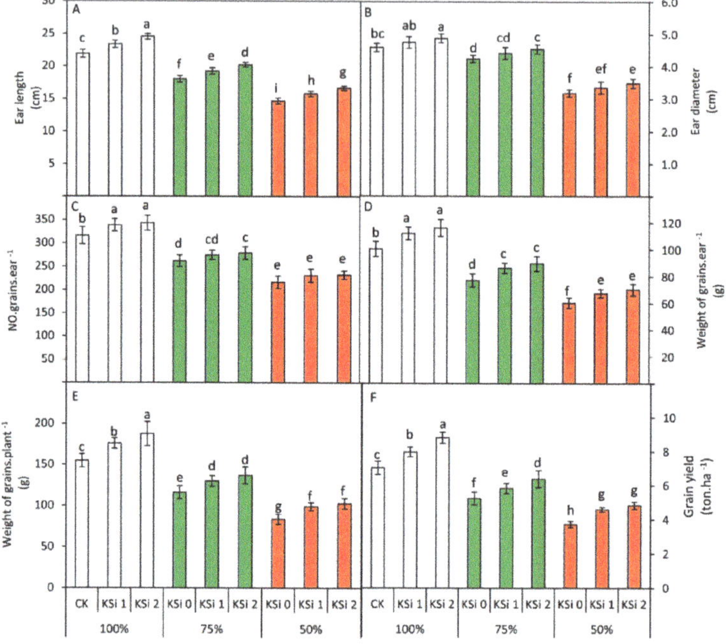

Figure 6. Yield and its parameters including averages of ear length (**A**), ear diameter (**B**), number of grains/ear (**C**) weight of grains/ear (**D**), weight of grains/plant (**E**) and grain yield (t ha^{-1}) (**F**) of the maize plants at 80 DAS as influenced by the foliar application of K$_2$SiO$_3$ (0, 1 and 2 mM) under three irrigation regimes: 100% (white), 75% (green) and 50% (orange) of water requirements. CK: well-watered control, KSi 0: K$_2$SiO$_3$-untreated plants, KSi 1: K$_2$SiO$_3$ (1 mM) and KSi 2: K$_2$SiO$_3$ (2 mM). Data of the two seasons of 2018 and 2019 were subjected to combined analysis. Means were presented ± SD. Different letters are significant differences, according to Duncan's multiple range tests ($p < 0.05$).

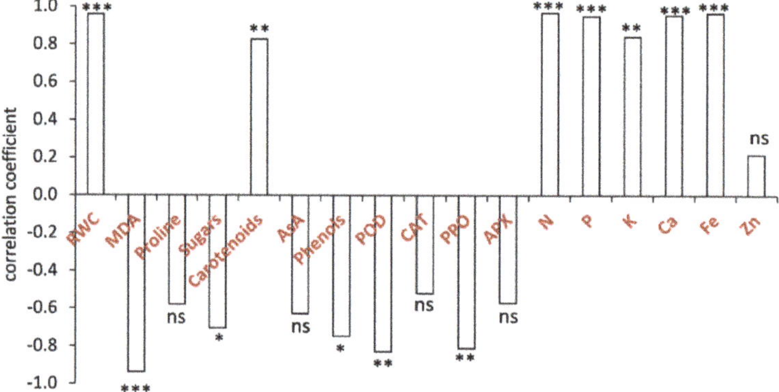

Figure 7. Relationship between the grain yield of maize crop and RWC, MDA, osmolytes, nonenzymatic antioxidants, antioxidant enzymes and mineral nutrients as influenced by the foliar application of K$_2$SiO$_3$ (0, 1 and 2 mM) under three different irrigation regimes (100, 75 and 50% of water requirements). Data of the two seasons of 2018 and 2019 were subjected to combined analysis. ns: not significant, * $p \leq 0.05$, ** $p \leq 0.01$ and *** $p \leq 0.001$.

4. Discussion

Under semiarid conditions, deficit irrigation is thought to be one of the most limiting factors that can restrict plant growth and productivity. In this study, soil analysis and climatic data showed that maize was exposed to high solar radiation and air temperatures with no precipitation during the cultivation periods in the two seasons. All of these factors exhibited drought stress on the maize plants during this study. It is well documented that drought stress reduces the growth of many plant species due to the restriction of cell division and differentiation [44]. In this study, reducing irrigation level exhibited significant decreases in shoot dry weight, root dry weight and LAI of the water-stressed plants (Figure 1). These reductions could be attributed to the disruption that occurred in the photosynthetic process through the degradation of pigments, limitation of stomatal conductance and decreasing the photochemical quantum yield [8,45]. On the other hand, root/shoot ratio as dry weight was unaffected under deficit irrigation (Figure 1). These results were in agreement with those obtained by Ma et al. [46], and may imply that phloem transport and leaf carbon exportation were less sensitive to water deficit under the circumstances of this study.

The positive effect of K_2SiO_3 on the shoot, root dry weight and LAI in the water-deficit stressed plants could be attributed to the synergistic effect of both K and Si on photosynthesis and production of assimilates [20,47]. In the present study, deficit irrigation negatively affected RWC while the K_2SiO_3 applications significantly mitigated this effect (Figure 2). Applied K positively affected leaf water content under stress conditions by maintenance of turgor potential and enhancing the integrity of cell membranes [48]. Additionally, Si could improve RWC by decreasing the rate of transpiration [49]. Lipid peroxidation is considered a pervasive biochemical response to stress in plant species due to the uncontrolled release of ROS [8]. Applied K and/or Si can promote the antioxidant capacity of the stressed plants [20,21]. This response may explain the significant decrease of MDA in the treated plants with K_2SiO_3 under water-stressed conditions (Figure 2). Proline is considered a compatible osmolyte and one of the most contributing factors that maintain intracellular redox homeostasis under stress conditions [50]. Moreover, under drought stress, proline has a crucial role in protecting the integrity of cell membranes and osmotic adjustments that allow the plant to uptake water [50,51]. Soluble sugars are the second compatible osmolytes that were determined in this investigation. The accumulation of soluble sugars during water deficit irrigation could be due to the up-regulation of genes involved in the starch-sucrose pathway [52,53]. All of the above-mentioned responses may explain the dramatic accumulation of proline and soluble sugars in the water-stressed plants under the circumstances of this study (Figure 2). The exogenous application of K_2SiO_3 resulted in a notable decrease in proline and a visible increase in soluble sugars. These effects indicate that K and/or Si may enhance the osmotic potential of leaves by stimulating the conversion of starch into soluble sugars, particularly up to the moderate level of irrigation [54,55]. Furthermore, the decrease of proline in the K_2SiO_3-treated plants may highlight the significance of K and/or Si in the protection of cell membranes and maintenance of RWC under deficit irrigation conditions (Figure 2).

Drought stress strongly affects the uptake of nutrients and it can restrict the translocation of some nutrients acropetally between plant organs [56]. Furthermore, it negatively affects active transport, permeability, and leaf transpiration [25,57]. In our study, plants exposed to moderate or severe stress exhibited a significant decline in N uptake (Figure 3A). This could be due to decreases in the activity of the N-uptake proteins (NRT1, NRT2) for inorganic nitrate (NO_3^-) and (AMT1) or ammonium (NH_4^+) [58]. Additionally, the availability of N could be reduced under the inadequate water supply [59]. The foliar application of K_2SiO_3 improved N-uptake under stressed and normal conditions. Applied-K can ameliorate the deleterious effects of drought through the regulation of stomatal movement, increasing root cell elongation, osmotic adjustment and detoxification of ROS [48]. Furthermore, silicon improves photosynthesis, antioxidant activities, and absorption of mineral nutrients of many crops [21,47]. These effects could explain the positive influence of K_2SiO_3 on N-uptake in our study. Concerning phosphorus (P), it was decreased under severe level of deficit irrigation (50%) (Figure 3B). This decrease may be attributed to reducing the concentration and/or activity of the P-uptake protein (PHT1) [58]. Moreover, under drought stress, P may be quickly converted into an immobile or insoluble form [60]. On the other hand, the increase in P-uptake prior to K_2SiO_3 application was significantly under severe and moderate stress. These effects imply that P-uptake in maize is highly dependent on the intensity of drought stress. Similarly, water stress markedly exhibited K deficiency compared to all K_2SiO_3 untreated plants (Figure 3C). This effect could be due to reduction in absorption by the roots and transpiration rate, which consequently reduced water and nutrient transport via xylem [60]. Applied-K_2SiO_3 significantly increases K content compared to the untreated plants. These results are in agreement with Jiang et al. [61], who found that the application of K can significantly increase its concentration in the different parts of maize plants such as grains and straw. In this study, Ca uptake was also inhibited by reducing water supply (Figure 3D). This impact was clear under the lower level of stress. Furthermore, the foliar application of K_2SiO_3 had no significant effect on Ca uptake under both examined treatments. Maize plants could be severely affected by Ca deficiency under drought condition because Ca is relatively an immobile nutrient and its uptake may require sufficient water supply [62].

Deficit irrigation manifestly suppressed the uptake of Fe in K_2SiO_3-untreated plants (Figure 3E). In contrast, applied K_2SiO_3, specifically at the highest concentration (2 mM), improved Fe-uptake under both treatments. Silicon (Si) can mitigate the symptoms of Fe deficiency in different plant species including soybean, cucumber and rice [63,64]. It could play a crucial role in Fe uptake and its translocation from roots to the aerial parts of the plant [65,66]. This impact could be attributed to the fact that applied-Si can enhance citrate concentration, which acts as an Fe chelator and facilitates its movement through the xylem [67]. The translocation of Zn from roots to leaves may be inhibited by Si application. This effect may be due to the fact that Si precipitates with Zn as zinc silicate around the root epidermis [68], which may reduce Zn translocation via xylem [69].

Under drought stress, plants develop a wide array of complex antioxidant systems that integrated with each other simultaneously to reduce the accumulation of ROS and oxidative damages [70]. The foliar application of K_2SiO_3 induced dramatic improvement in the concentration of carotenoids under different investigated levels of irrigation. The increase of carotenoids under water stress due to K or Si supplementation could foster the antioxidant capacity of plants under deficit irrigation [71–74]. Under stress conditions, ascorbate (ASA) could be increased through the overexpression of its synthesis related-genes such as GMP, GME, GalUR, DHAR, and MDHAR [75]. In this study, AsA was substantially increased by reducing the irrigation level (Figure 4B). This response could help in scavenging ROS and inducing the ascorbate–glutathione cycle [8,76]. Phenolic compounds could also be involved in plant tolerance to drought stress and play a significant role as a sink for carbon under stress conditions [77,78]. These effects could explain the improvement in total soluble phenols by reducing the irrigation level in this study (Figure 4C). The increase in total soluble phenols by the treatments of K_2SiO_3 could be due to the effect of Si, which may induce several changes in the phenolic compounds under abiotic and biotic stresses [79,80].

In the present study, our results showed that deficit irrigation increased the activities of POD, CAT, PPO and APX in the leaves of maize plants (Figure 5). These findings could reflect the integrated regulation between these enzymes in the tolerance of maize plants to water stress. Exogenous applications of K_2SiO_3 induced a synergistic effect leading to an increase in the activities of all studied antioxidant enzymes under deficit irrigation levels. Previous reports showed that K and/or Si could enhance the antioxidant capacity of plants under stress conditions [20,21]. In this study, these effects were confirmed by the enhancement of RWC and reduction of MDA.

It is well documented that water stress has several deleterious influences on the productivity of maize plants [17,81]. It can affect different metabolic pathways, photosynthesis and translocation of many metabolites required for grain filling [81]. Furthermore, water stress can increase the potential for unsuccessful pollination and poor kernel setting of maize by affecting the anthesis and silking stages [82]. In this study, reducing irrigation levels reduced the yield of grains (Figure 6). Applications of K_2SiO_3 not only relatively reversed these adverse effects but also increased the ultimate yield of grains under water stress. These findings could be correlated with corresponding changes in several biochemical and physiological aspects that were found during this work (Figure 7).

5. Conclusions

In this study, it was found that applied K_2SiO_3, particularly at 2 mM as a foliar spray, may have several benefits on maize crops under limited irrigation supply. These effects were associated with several changes at physiological and biochemical levels, including adjustment of RWC and osmolytes, alleviation of oxidative damage and reduction of cell membrane dysfunction, as well as enhancement of nutrient uptake of and regulation of several nonenzymatic and enzymatic antioxidant systems. These results could provide a link between the protective role of K_2SiO_3 against drought stress and the eventual yield of grains, especially under semiarid conditions.

Author Contributions: Conceptualization, M.F.M.I., H.A., R.F., G.A.E.-S., A.M.E.-S., M.H. and A.E.; methodology, H.A.E.-G., M.F.M.I., H.A., W.N.H., R.F. and G.A.E.-S.; software, A.M.E.-S., H.A.E.-G., M.H., W.H. and A.E.; validation, M.F.M.I., H.A., H.A.E.-G., R.F., G.A.E.-S., A.M.E.-S., M.H., W.N.H. and A.E.; formal analysis, M.F.M.I., H.A., R.F. and A.E.; investigation, A.M.E.-S., M.H. and W.N.H.; resources, M.F.M.I., H.A., R.F., G.A.E.-S., A.M.E.-S. and A.E.; data curation, M.F.M.I., H.A., R.F., G.A.-E.S., A.M.E.-S., M.H. and W.N.H.; writing—original draft preparation, M.F.M.I., H.A., R.F., G.A.-E.S. and A.E.; writing—review and editing, M.F.M.I., H.A.E.-G., H.A., A.A.E.-Y., R.F., G.A.E.-S., A.M.E.-S., M.H., W.N.H. and A.E. All authors have read and agreed to the published version of the manuscript.

Funding: The authors are grateful to the Researchers Supporting Project number (RSP-2020/53), King Saud University, Riyadh, Saudi Arabia.

Conflicts of Interest: The authors declare no conflict of interest.

References

1. Gallart, F.; Solé, A.; Puigdefábregas, J.; Lázaro, R. *Badland Systems in the Mediterranean*; John Wiley & Sons, Ltd.: Chichester, UK, 2002; pp. 299–326.

2. Lal, R. Soil carbon sequestration impacts on global climate change and food security. *Science* **2004**, *304*, 1623–1627. [CrossRef] [PubMed]

3. Elkeilsh, A.; Awad, Y.M.; Soliman, M.H.; Abu-Elsaoud, A.; Abdelhamid, M.T.; El-Metwally, I.M. Exogenous application of β-sitosterol mediated growth and yield improvement in water-stressed wheat (Triticum aestivum) involves up-regulated antioxidant system. *J. Plant Res.* **2019**, *132*, 881–901. [CrossRef]

4. Alhaithloul, H.A.; Soliman, M.H.; Ameta, K.L.; El-Esawi, M.A.; Elkelish, A. Changes in Ecophysiology, Osmolytes, and Secondary Metabolites of the Medicinal Plants of Mentha piperita and Catharanthus roseus Subjected to Drought and Heat Stress. *Biomolecules* **2020**, *10*, 43. [CrossRef] [PubMed]

5. Habib, N.; Ali, Q.; Ali, S.; Javed, M.T.; Zulqurnain Haider, M.; Perveen, R.; Shahid, M.R.; Rizwan, M.; Abdel-Daim, M.M.; Elkelish, A. Use of Nitric Oxide and Hydrogen Peroxide for Better Yield of Wheat (*Triticum aestivum* L.) under Water Deficit Conditions: Growth, Osmoregulation, and Antioxidative Defense Mechanism. *Plants* **2020**, *9*, 285. [CrossRef] [PubMed]

6. Ibrahim, M.F.M.; Bondok, A.M.; Al-Senosy, N.K.; Younis, R.A. Stimulation Some of Defense Mechanisms in Tomato Plants under Water Deficit and Tobacco mosaic virus (TMV). *World J. Agric. Sci.* **2015**, *11*, 289–302.

7. Badr, A.M.; Shabana, E.F.; Senousy, H.H.; Mohammad, H.Y. Anti-inflammatory and anti-cancer effects of β-carotene, extracted from Dunaliella bardawil by milking. *J. Food Agric. Environ.* **2014**, *12*, 24–31.

8. Abd Ellatif, S.; El-Sheekh, M.M.; Senousy, H.H. Role of microalgal ligninolytic enzymes in industrial dye decolorization. *Int. J. Phytoremediat.* **2020**, 1–12. [CrossRef]

9. Abdel-Daim, M.M.; Dawood, M.A.O.; AlKahtane, A.A.; Abdeen, A.; Abdel-Latif, H.M.R.; Senousy, H.H.; Aleya, L.; Alkahtani, S. Spirulina platensis mediated the biochemical indices and antioxidative function of Nile tilapia (Oreochromis niloticus) intoxicated with aflatoxin B1. *Toxicon* **2020**. [CrossRef]

10. Senousy, H.H.; Abd Ellatif, S.; Ali, S. Assessment of the antioxidant and anticancer potential of different isolated strains of cyanobacteria and microalgae from soil and agriculture drain water. *Environ. Sci. Pollut. Res.* **2020**, *27*, 18463–18474. [CrossRef]

11. Ahmad, I.; Ahmad, B.; Boote, K.; Hoogenboom, G. Adaptation strategies for maize production under climate change for semi-arid environments. *Eur. J. Agron.* **2020**, *115*, 126040. [CrossRef]

12. Ranum, P.; Peña-Rosas, J.P.; Garcia-Casal, M.N. Global maize production, utilization, and consumption. *Ann. N. Y. Acad. Sci. USA* **2014**, *1312*, 105–112. [CrossRef]
13. Dhugga, K.S. Maize biomass yield and composition for biofuels. *Crop Sci.* **2007**, *47*, 2211–2227. [CrossRef]
14. Kleinmans, J.; Densley, R.; Hurley, T.; Williams, I. BRIEF COMMUNICATION: Feed value of maize silage in New Zealand—A review. In Proceedings of the New Zealand Society of Animal Production, Adelaide, Australia, 4–7 July 2016; pp. 100–102.
15. Ostrander, B.M. Maize Starch for Industrial Applications. In *Industrial Crops: Breeding for BioEnergy and Bioproducts*; Cruz, V.M.V., Dierig, D.A., Eds.; Springer: New York, NY, USA, 2015; pp. 171–189. [CrossRef]
16. de Araujo Rufino, C.; Fernandes-Vieira, J.; Martín-Gil, J.; Abreu Junior, J.D.S.; Tavares, L.C.; Fernandes-Correa, M.; Martín-Ramos, P. Water stress influence on the vegetative period yield components of different maize genotypes. *Agronomy* **2018**, *8*, 151. [CrossRef]
17. Mi, N.; Cai, F.; Zhang, Y.; Ji, R.; Zhang, S.; Wang, Y. Differential responses of maize yield to drought at vegetative and reproductive stages. *Plant Soil Environ.* **2018**, *64*, 260–267.
18. Martineau, E.; Domec, J.-C.; Bosc, A.; Dannoura, M.; Gibon, Y.; Bénard, C.; Jordan-Meille, L. The role of potassium on maize leaf carbon exportation under drought condition. *Acta Physiol. Plant.* **2017**, *39*, 219. [CrossRef]
19. Ul-Allah, S.; Ijaz, M.; Nawaz, A.; Sattar, A.; Sher, A.; Naeem, M.; Shahzad, U.; Farooq, U.; Nawaz, F.; Mahmood, K. Potassium Application Improves Grain Yield and Alleviates Drought Susceptibility in Diverse Maize Hybrids. *Plants* **2020**, *9*, 75. [CrossRef]
20. Du, Q.; Zhao, X.; Xia, L.; Jiang, C.; Wang, X.; Han, Y.; WANG, J.; YU, H.-Q. Effects of potassium deficiency on photosynthesis, chloroplast ultrastructure, ROS, and antioxidant activities in maize (Zea mays L.). *J. Integr. Agric.* **2019**, *18*, 395–406. [CrossRef]
21. Kim, Y.-H.; Khan, A.L.; Waqas, M.; Lee, I.-J. Silicon regulates antioxidant activities of crop plants under abiotic-induced oxidative stress: A review. *Front. Plant Sci.* **2017**, *8*, 510. [CrossRef]
22. Cooke, J.; Leishman, M.R. Consistent alleviation of abiotic stress with silicon addition: A meta-analysis. *Funct. Ecol.* **2016**, *30*, 1340–1357. [CrossRef]
23. dos Santos, M.S.; Sanglard, L.M.P.V.; Martins, S.C.V.; Barbosa, M.L.; de Melo, D.C.; Gonzaga, W.F.; DaMatta, F.M. Silicon alleviates the impairments of iron toxicity on the rice photosynthetic performance via alterations in leaf diffusive conductance with minimal impacts on carbon metabolism. *Plant Physiol. Biochem.* **2019**, *143*, 275–285. [CrossRef]
24. Snyder, G.H.; Matichenkov, V.V.; Datnoff, L.E. Silicon. In *Handbook of Plant Nutrition*; CRC Press: Boca Raton, FL, USA, 2007; pp. 551–568.
25. Cao, B.-L.; Wang, L.; Gao, S.; Xia, J.; Xu, K. Silicon-mediated changes in radial hydraulic conductivity and cell wall stability are involved in silicon-induced drought resistance in tomato. *Protoplasma* **2017**, *254*, 2295–2304. [CrossRef]
26. Eneji, A.E.; Inanaga, S.; Muranaka, S.; Li, J.; Hattori, T.; An, P.; Tsuji, W. Growth and nutrient use in four grasses under drought stress as mediated by silicon fertilizers. *J. Plant Nutr.* **2008**, *31*, 355–365. [CrossRef]
27. Yin, L.; Wang, S.; Liu, P.; Wang, W.; Cao, D.; Deng, X.; Zhang, S. Silicon-mediated changes in polyamine and 1-aminocyclopropane-1-carboxylic acid are involved in silicon-induced drought resistance in *Sorghum bicolor* L. *Plant Physiol. Biochem.* **2014**, *80*, 268–277. [CrossRef] [PubMed]
28. Doorenbos, J. *Guidelines for Predicting Crop Water Requirements*; Food and Agriculture organization: Rome, Italy, 1975; Volume 24.
29. Iqbal, A.; Hidayat, Z. Potassium management for improving growth and grain yield of maize (*zea mays* L.) under moisture stress condition. *Sci. Rep.* **2016**, *6*, 34627.
30. Ünyayar, S.; Keleþ, Y.; Ünal, E. Proline and ABA levels in two sunflower genotypes subjected to water stress. *Bulg. J. Plant Physiol.* **2004**, *30*, 34–47.
31. Heath, R.L.; Packer, L. Photoperoxidation in isolated chloroplasts: I. Kinetics and stoichiometry of fatty acid peroxidation. *Arch. Biochem. Biophys.* **1968**, *125*, 189–198. [CrossRef]
32. Bates, L.; Waldren, R.; Teare, I. Rapid determination of free proline for water-stress studies. *Plant Soil* **1973**, *39*, 205–207. [CrossRef]
33. Plummer, D.T. *Practical Biochemistry*; McGraw-Hill Book Company: New York, NY, USA, 1987; pp. 179–180.
34. Cottenie, A.; Verloo, M.; Kiekens, L.; Velghe, G.; Camerlynck, R. *Chemical Analysis of Plants and Soils*; RUG. Laboratory of Analytical and Agrochemistry: Gent, Belgium, 1982; 63p.

35. de Carvalho, L.M.J.; Gomes, P.B.; de Oliveira Godoy, R.L.; Pacheco, S.; do Monte, P.H.F.; de Carvalho, J.L.V.; Nutti, M.R.; Neves, A.C.L.; Vieira, A.C.R.A.; Ramos, S.R.R. Total carotenoid content, α-carotene and β-carotene, of landrace pumpkins (Cucurbita moschata Duch): A preliminary study. *Food Res. Int.* **2012**, *47*, 337–340. [CrossRef]

36. A.O.A.C International. *Official Methods of Analysis. Association of Official Analytical Chemists. Official Method 985.33. Vitamin C, (Reduced Ascorbic Acid) in Ready-to-Feed Milk Based Infant Formula 2, 6-Dichloroindophenol Titrimetric Method*; A.O.A.C International: Washington, DC, USA, 1990; pp. 1108–1109.

37. Skalindi, F.; Naczk, M. Food phenolics: Sources, chemistry, effects, applications. *Trends Food Sci. Technol.* **1995**, *7*, 235–277.

38. Bradford, M.M. A rapid and sensitive method for the quantitation of microgram quantities of protein utilizing the principle of protein-dye binding. *Anal. Biochem.* **1976**, *72*, 248–254. [CrossRef]

39. Hammerschmidt, R.; Nuckles, E.; Kuć, J. Association of enhanced peroxidase activity with induced systemic resistance of cucumber to Colletotrichum lagenarium. *Physiol. Plant Pathol.* **1982**, *20*, 73–82. [CrossRef]

40. Cakmak, I.; Strbac, D.; Marschner, H. Activities of hydrogen peroxide-scavenging enzymes in germinating wheat seeds. *J. Exp. Bot.* **1993**, *44*, 127–132. [CrossRef]

41. Oktay, M.; Küfreviolu, I.; Kocaçalişkan, I.; Şaklrolu, H. Polyphenoloxidase from Amasya apple. *J. Food Sci.* **1995**, *60*, 494–496. [CrossRef]

42. Nakano, Y.; Asada, K. Hydrogen peroxide is scavenged by ascorbate-specific peroxidase in spinach chloroplasts. *Plant Cell Physiol.* **1981**, *22*, 867–880.

43. Snedecor, G.; Cochran, W. *Statistical Methods*, 7th ed.; Lowa: Ames, IA, USA, 1982.

44. Sacks, M.M.; Silk, W.K.; Burman, P. Effect of water stress on cortical cell division rates within the apical meristem of primary roots of maize. *Plant Physiol.* **1997**, *114*, 519–527. [CrossRef]

45. Zhou, R.; Kan, X.; Chen, J.; Hua, H.; Li, Y.; Ren, J.; Feng, K.; Liu, H.; Deng, D.; Yin, Z. Drought-induced changes in photosynthetic electron transport in maize probed by prompt fluorescence, delayed fluorescence, P700 and cyclic electron flow signals. *Environ. Exp. Bot.* **2019**, *158*, 51–62. [CrossRef]

46. Ma, X.; He, Q.; Zhou, G. Sequence of Changes in Maize Responding to Soil Water Deficit and Related Critical Thresholds. *Front. Plant Sci.* **2018**, *9*, 511. [CrossRef]

47. Chen, W.; Yao, X.; Cai, K.; Chen, J. Silicon alleviates drought stress of rice plants by improving plant water status, photosynthesis and mineral nutrient absorption. *Biol. Trace Elem. Res.* **2011**, *142*, 67–76. [CrossRef]

48. Wang, M.; Zheng, Q.; Shen, Q.; Guo, S. The critical role of potassium in plant stress response. *Inter. J. Mol. Sci.* **2013**, *14*, 7370–7390. [CrossRef]

49. Amin, M.; Ahmad, R.; Ali, A.; Hussain, I.; Mahmood, R.; Aslam, M.; Lee, D.J. Influence of silicon fertilization on maize performance under limited water supply. *Silicon* **2018**, *10*, 177–183. [CrossRef]

50. Abdelaal, K.A.; EL-Maghraby, L.M.; Elansary, H.; Hafez, Y.M.; Ibrahim, E.I.; El-Banna, M.; El-Esawi, M.; Elkelish, A. Treatment of Sweet Pepper with Stress Tolerance-Inducing Compounds Alleviates Salinity Stress Oxidative Damage by Mediating the Physio-Biochemical Activities and Antioxidant Systems. *Agronomy* **2019**, *10*, 26. [CrossRef]

51. Szepesi, Á.; Szőllősi, R. Mechanism of proline biosynthesis and role of proline metabolism enzymes under environmental stress in plants. In *Plant Metabolites and Regulation Under Environmental Stress*; Elsevier: Amsterdam, The Netherlands, 2018; pp. 337–353.

52. Das, A.; Rushton, P.J.; Rohila, J.S. Metabolomic profiling of soybeans (*Glycine max* L.) reveals the importance of sugar and nitrogen metabolism under drought and heat stress. *Plants* **2017**, *6*, 21. [CrossRef] [PubMed]

53. Kim, S.-W.; Lee, S.-K.; Jeong, H.-J.; An, G.; Jeon, J.-S.; Jung, K.-H. Crosstalk between diurnal rhythm and water stress reveals an altered primary carbon flux into soluble sugars in drought-treated rice leaves. *Sci. Rep.* **2017**, *7*, 1–18. [CrossRef]

54. Hajiboland, R.; Cheraghvareh, L.; Poschenrieder, C. Improvement of drought tolerance in tobacco (*Nicotiana rustica* L.) plants by silicon. *J. Plant Nutr.* **2017**, *40*, 1661–1676. [CrossRef]

55. Zahoor, R.; Zhao, W.; Abid, M.; Dong, H.; Zhou, Z. Potassium application regulates nitrogen metabolism and osmotic adjustment in cotton (*Gossypium hirsutum* L.) functional leaf under drought stress. *J. Plant Physiol.* **2017**, *215*, 30–38. [CrossRef] [PubMed]

56. Hu, Y.; Schmidhalter, U. Drought and salinity: A comparison of their effects on mineral nutrition of plants. *J. Plant Nutr. Soil Sci.* **2005**, *168*, 541–549. [CrossRef]

57. Arve, L.; Torre, S.; Olsen, J.; Tanino, K. Stomatal responses to drought stress and air humidity. In *Abiotic Stress in Plants-Mechanisms and Adaptations*; IntechOpen: London, UK, 2011. [CrossRef]

58. Bista, D.R.; Heckathorn, S.A.; Jayawardena, D.M.; Mishra, S.; Boldt, J.K. Effects of drought on nutrient uptake and the levels of nutrient-uptake proteins in roots of drought-sensitive and-tolerant grasses. *Plants* **2018**, *7*, 28. [CrossRef] [PubMed]

59. Bloem, J.; de Ruiter, P.C.; Koopman, G.J.; Lebbink, G.; Brussaard, L. Microbial numbers and activity in dried and rewetted arable soil under integrated and conventional management. *Soil Biol. Biochem.* **1992**, *24*, 655–665. [CrossRef]

60. Ge, T.-D.; Sun, N.-B.; Bai, L.-P.; Tong, C.-L.; Sui, F.-G. Effects of drought stress on phosphorus and potassium uptake dynamics in summer maize (*zea mays*) throughout the growth cycle. *Acta Physiol. Plant.* **2012**, *34*, 2179–2186. [CrossRef]

61. Jiang, W.; Liu, X.; Wang, Y.; Zhang, Y.; Qi, W. Responses to Potassium Application and Economic Optimum K Rate of Maize under Different Soil Indigenous K Supply. *Sustainability* **2018**, *10*, 2267. [CrossRef]

62. Naeem, M.; Naeem, M.S.; Ahmad, R.; Ihsan, M.Z.; Ashraf, M.Y.; Hussain, Y.; Fahad, S. Foliar calcium spray confers drought stress tolerance in maize via modulation of plant growth, water relations, proline content and hydrogen peroxide activity. *Arch. Agron. Soil Sci.* **2018**, *64*, 116–131. [CrossRef]

63. Gonzalo, M.J.; Lucena, J.J.; Hernández-Apaolaza, L. Effect of silicon addition on soybean (*Glycine max*) and cucumber (*Cucumis sativus*) plants grown under iron deficiency. *Plant Physiol. Biochem.* **2013**, *70*, 455–461. [CrossRef]

64. Rinny, S.; Rout, G. Effect of silicon interaction with nutrients in rice. *J. Exp. Biol. Agric. Sci.* **2018**, *6*, 717–731.

65. Carrasco-Gil, S.; Rodríguez-Menéndez, S.; Fernández, B.; Pereiro, R.; de la Fuente, V.; Hernandez-Apaolaza, L. Silicon induced Fe deficiency affects Fe, Mn, Cu and Zn distribution in rice (*Oryza sativa* L.) growth in calcareous conditions. *Plant Physiol. Biochem.* **2018**, *125*, 153–163. [CrossRef]

66. Hernandez-Apaolaza, L. Can silicon partially alleviate micronutrient deficiency in plants? A review. *Planta* **2014**, *240*, 447–458. [CrossRef]

67. Rellán-Álvarez, R.; Giner-Martínez-Sierra, J.; Orduna, J.; Orera, I.; Rodríguez-Castrillón, J.Á.; García-Alonso, J.I.; Abadía, J.; Álvarez-Fernández, A. Identification of a tri-iron (III), tri-citrate complex in the xylem sap of iron-deficient tomato resupplied with iron: New insights into plant iron long-distance transport. *Plant Cell Physiol.* **2010**, *51*, 91–102. [CrossRef] [PubMed]

68. Gu, H.-H.; Qiu, H.; Tian, T.; Zhan, S.-S.; Chaney, R.L.; Wang, S.-Z.; Tang, Y.-T.; Morel, J.-L.; Qiu, R.-L. Mitigation effects of silicon rich amendments on heavy metal accumulation in rice (*Oryza sativa* L.) planted on multi-metal contaminated acidic soil. *Chemosphere* **2011**, *83*, 1234–1240. [CrossRef] [PubMed]

69. da Cunha, K.P.V.; do Nascimento, C.W.A. Silicon effects on metal tolerance and structural changes in maize (*zea mays* L.) grown on a cadmium and zinc enriched soil. *Water Air Soil Pollut.* **2009**, *197*, 323. [CrossRef]

70. Laxa, M.; Liebthal, M.; Telman, W.; Chibani, K.; Dietz, K.-J. The role of the plant antioxidant system in drought tolerance. *Antioxidants* **2019**, *8*, 94. [CrossRef] [PubMed]

71. Kanai, S.; Moghaieb, R.E.; El-Shemy, H.A.; Panigrahi, R.; Mohapatra, P.K.; Ito, J.; Nguyen, N.T.; Saneoka, H.; Fujita, K. Potassium deficiency affects water status and photosynthetic rate of the vegetative sink in green house tomato prior to its effects on source activity. *Plant Sci.* **2011**, *180*, 368–374. [CrossRef]

72. Shen, X.; Zhou, Y.; Duan, L.; Li, Z.; Eneji, A.E.; Li, J. Silicon effects on photosynthesis and antioxidant parameters of soybean seedlings under drought and ultraviolet-B radiation. *J. Plant Physiol.* **2010**, *167*, 1248–1252. [CrossRef] [PubMed]

73. Hassanein, R.A.; El Khawas, S.A.; Khafaga, H.S.; Abd El-Nabe, A.S.; Abd Elrady, A.S. Amelioration of Drought Stress on Physiological Performance of Pearl Millet (*Pennisetum americanum*) Plant Grown Under Saline Condition Using Potassium Humate and Silicon Source. *Egypt. J. Exp. Biol.* **2017**, *13*, 57–68. [CrossRef]

74. de Camargo, M.; Bezerra, B.; Holanda, L.; Oliveira, A.; Vitti, A.; Silva, M. Silicon fertilization improves physiological responses in sugarcane cultivars grown under water deficit. *J. Soil Sci. Plant Nutr.* **2019**, *19*, 81–91. [CrossRef]

75. Zhang, Y. *Ascorbic Acid in Plants: Biosynthesis, Regulation and Enhancement*; Springer Science & Business Media: New York, NY, USA; Heidelberg, Germany; Dordrecht, The Netherlands; London, UK, 2013; pp. 111–114. [CrossRef]

76. Bartoli, C.G.; Buet, A.; Grozeff, G.G.; Galatro, A.; Simontacchi, M. Ascorbate-glutathione cycle and abiotic stress tolerance in plants. In *Ascorbic Acid in Plant Growth, Development and Stress Tolerance*; Springer: Berlin/Heidelberg, Germany, 2017; pp. 177–200. [CrossRef]

77. Hashim, A.M.; Alharbi, B.M.; Abdulmajeed, A.M.; Elkelish, A.; Hozzein, W.N.; Hassan, H.M. Oxidative Stress Responses of Some Endemic Plants to High Altitudes by Intensifying Antioxidants and Secondary Metabolites Content. *Plants* **2020**, *9*, 869. [CrossRef] [PubMed]

78. Weidner, S.; Karolak, M.; Karamac, M.; Kosinska, A.; Amarowicz, R. Phenolic compounds and properties of antioxidants in grapevine roots [*Vitis vinifera* L.] under drought stress followed by recovery. *Acta Soc. Bot. Pol.* **2009**, *78*, 97–103. [CrossRef]

79. Malčovská, S.M.; Dučaiová, Z.; Maslaňáková, I.; Bačkor, M. Effect of silicon on growth, photosynthesis, oxidative status and phenolic compounds of maize (*Zea mays* L.) grown in cadmium excess. *Water Air Soil Pollut.* **2014**, *225*, 2056. [CrossRef]

80. Shetty, R.; Fretté, X.; Jensen, B.; Shetty, N.P.; Jensen, J.D.; Jørgensen, H.J.L.; Newman, M.-A.; Christensen, L.P. Silicon-induced changes in antifungal phenolic acids, flavonoids, and key phenylpropanoid pathway genes during the interaction between miniature roses and the biotrophic pathogen Podosphaera pannosa. *Plant Physiol.* **2011**, *157*, 2194–2205. [CrossRef]

81. Li, G.; Gao, H.; Zhao, B.; Dong, S.; Zhang, J.; Yang, J.; Wang, J.; Liu, P. Effects of drought stress on activity of photosystems in leaves of maize at grain filling stage. *Acta Agron. Sin.* **2009**, *35*, 1916–1922. [CrossRef]

82. Aslam, M.; Zamir, M.; Afzal, I.; Yaseen, M.; Mubeen, M.; Shoaib, A. Drought stress, its effect on maize production and development of drought tolerance through potassium application. *Cercetari Agronomice în Moldova* **2013**, *46*, 154.

Communication

Initial Experimental Experience with a Sprayable Biodegradable Polymer Membrane (SBPM) Technology in Cotton

Michael V. Braunack [1], Raju Adhikari [2], George Freischmidt [2], Priscilla Johnston [2], Philip S. Casey [2], Yusong Wang [3], Keith L. Bristow [3,*], Lana Filipović [4] and Vilim Filipović [3,4,*]

1   CSIRO, Agriculture & Food, LB 59 Narrabri, NSW 2390, Australia; Michael.Braunack@csiro.au
2   CSIRO, Manufacturing, Private Bag 33, Clayton, VIC 3169, Australia; adhikari2428@gmail.com (R.A.); george.freischmidt@gmail.com (G.F.); priscilla.johnston82@gmail.com (P.J.); philcasey54@gmail.com (P.S.C.)
3   CSIRO, Agriculture & Food, PMB, Aitkenvale, Townsville, QLD 4814, Australia; Yusong.Wang@csiro.au
4   Department of Soil Amelioration, University of Zagreb Faculty of Agriculture, Svetošimunska 25, 10000 Zagreb, Croatia; lfilipovic@agr.hr
*   Correspondence: Keith.Bristow@csiro.au (K.L.B.); vfilipovic@agr.hr (V.F.); Tel.: +61-408-468-941 (K.L.B.); +385-1-2393-711 (V.F.)

Received: 18 March 2020; Accepted: 16 April 2020; Published: 19 April 2020

Abstract: Preformed biodegradable and next generation sprayable biodegradable polymer membrane (SBPM) formulations, which biodegrade to non-harmful products (water, carbon dioxide and microbial biomass), have been introduced as an alternative to plastic mulch films in order to mitigate plastic pollution of the environment. In this preliminary field study on cotton (*Gossypium hirsutum* L.), a novel SBPM technology was compared to preformed slotted oxo-degradable plastic (ODP) mulch film and no mulch control (CON) in terms of yield, crop water productivity (CWP), and soil temperature. The first results showed higher CWP and crop yield, and increased soil water content under the SBPM cover. This study indicates that SBPM technology could perform at similar level as ODP or comparable films under field conditions and, at the same time, provide environmentally sustainable agricultural cropping practices. Additionally, the fully treated, non-replicated SBPM plot had a wetter soil profile throughout the entire crop season. This innovative technology has shown a high potential even at this early stage of development, indicating that advances in formulation and further testing can lead to significant improvements and thus increased use in crop production systems.

Keywords: preformed plastic mulch film; crop water productivity; biodegradation; crop productivity; spray-on mulch; water use efficiency

---

1. Introduction

As the world population continues to increase, the production of food and fibre will need to increase using the same (or reduced) area of agricultural land, but with less water [1]. One of the ways to increase crop productivity is to minimise water losses by soil evaporation and crop transpiration [2], which can be achieved in the field by using various mulch surface covers and/or with improved irrigation technologies. Plastic mulch films used in crop production help to control pests, increase soil and air temperature, reduce soil evaporation, minimize soil erosion, and prevent soil particles attaching to fruits or vegetables [3]. However, a major and continuing problem with the use of plastic mulch films is disposal and non-biodegradability issues [4]. Consequently, the expanding use of plastics has led to environmental pollution, which will have long-term consequences on soil quality, the environment, and possibly human health [5]. The newer oxo-degradable mulch films reduce the amount of plastic mulch currently being disposed in landfills, but are still essentially non-biodegradable [6]. The result is that they break down to smaller fragments, causing even more environmental damage [7]. Biodegradable plastics are a promising solution, but the release of micro- and nanoparticles from biodegradable plastic upon degradation requires long term field trials in order to confirm that either complete biodegradation occurs, or that no long-term harm to the environment is caused [8].

Recently, a biodegradable spray-on mulch based on sodium alginate was developed as a potential alternative to the thin plastic mulch films currently used in horticulture [9]. Polymeric protein-based biodegradable spray coatings were tested in greenhouse trials [10] and exhibited agronomic performances comparable to commercial low-density polyethylene mulch film, indicating a similar rate of plant growth and dry matter accumulation and a complete biodegradation (<5% residues after 2 months). Similar sprayable polymer formulations (for small scale handheld sprayers and large-scale mechanised boom sprays) designed to biodegrade have been reported previously [11,12]. An increase in crop water productivity of 20%–30% for rockmelons under drip irrigation was demonstrated using a sprayable biodegradable polymer membrane (SBPM) compared to the bare soil [13]. The use of an additional viscosity modifier reduced soil wicking (polymer adsorption into the soil) by 10%–90% without compromising the system's sprayability or the general mechanical properties of the membrane, which were similar to those of the unmodified SBPM. Soil evaporation was reduced by more than 60% at a low SBPM application rate [14]. The beneficial effect of SBPM on the soil water regime in terms of the restriction of soil evaporation was even more pronounced when the polymer application rate increased (up to 1 kg m^{-2}) [15].

Using the same SBPM formulation [15], a field study was conducted to determine whether SBPM applied to irrigated cotton would be effective in conserving soil profile water and improving crop water productivity and yield. This research was conducted to give important small-scale insights, before expanding and upscaling to full field scale research. The aim was to compare SBPM with oxo-degradable plastic (ODP) mulch film and with no mulch control (CON) in terms of the soil water content, soil temperature, cotton yield, and cotton quality.

2. Materials and Methods

2.1. Experimental Field Set-Up

Field experiments were conducted during the 2014/2015 season at the Australian Cotton Research Institute (ACRI), Narrabri (149° 40′ E, 30° 10′ S), New South Wales, Australia. The soil type was classified as a grey self-mulching Vertosol with 59% clay, 28% silt, a pH of 7.27 (0.01 M CaCl$_2$) and 0.77% Organic Carbon on average [16]. Nitrogen fertiliser was applied as urea at the rate of 180 kg N ha^{-1} prior to planting. Weeds and insects were managed as per Bollgard® II protocol and plots were irrigated according to the station's schedule when an approximate soil profile water deficit of 70–80 mm was detected. Cotton (*Gossypium hirsutum* L. cultivar Sicot 74 BRF, germination percentage 96%) was sown (10 seeds per m of row) on 26 October 2015.

Three treatments, laid out as a completely randomised block design with three replicates, were tested: a thin slotted oxo-degradable polyethylene clear plastic (ODP) film, a new spray-on biodegradable polymer membrane (SBPM_gap) was applied on both sides of the plant row using a handheld pressure sprayer (Figure 1), and no mulch control (CON). The sprayable polymer formulation contained 20 wt% polymer content with 100–500 μm particle size and viscosity in the range of 50–100 mPa s. The polymer formulation that showed the best combination of film formation, water barrier and mechanical properties in the previous pot trial [13] was selected for the field trial. An unsprayed area (12 cm) between the spray lines was left to ensure that cotton seedlings would emerge. In one additional plot, SBPM was diluted by half to 0.5 L m^{-2} and applied over the plant line to determine whether cotton would emerge through the polymer membrane (SBPM_full). An additional plot with spray-on SBPM applied over the plant line (SBPM_full) was not replicated (due to the limited amount of available SBPM) and therefore was not included in the statistical analysis.

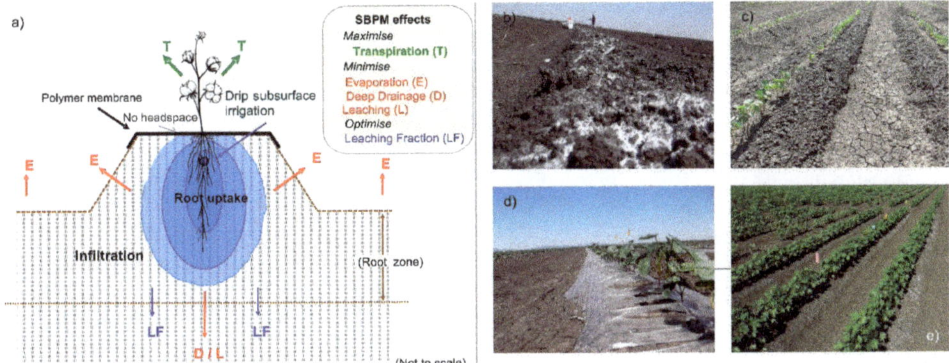

Figure 1. (**a**) Conceptual sketch of a sprayable biodegradable polymer membrane (SBPM)'s effect on relatively small-scale soil hydrology. (**b**) Soil after the application of SBPM_full spray-on, (**c**) SBPM_gap plot, (**d**) oxo-degradable plastic (ODP) slotted mulch film plot, (**e**) experimental layout in the field during October/November 2015 at the Australian Cotton Research Institute (ACRI), Narrabri.

2.2. Field Sensor Installation and Crop Data Collection

Sensors were installed prior to applying the preformed film and the spray-on polymer. The soil profile water content and seedbed soil temperature were monitored below the plant line in each plot, respectively, using multi-depth soil capacitance sensors (at 10, 20, 40, 60 and 80 cm soil depth, Odyssey, http://odysseydatarecording.com/), and with thermistor probes (at 10 cm depth, Tinytag provided by Hastings Data Loggers, www.hdl.com.au). Data logging was commenced two days after the installation to allow soil moisture to re-equilibrate. Crop water consumption (CWC, mm) was determined using a simple mass balance expression: initial profile water (IPW) + irrigation (IR) + in-crop rainfall (R) − final profile water (FPW) (i.e., CWC = (IPW + IR + R) − FPW) and crop water productivity (CWP) (kg lint mm^{-1} ha^{-1}) = lint yield (kg ha^{-1}) / (CWC − 100 mm). The threshold of 100 mm was used to account for runoff, evaporation, and plant growth at the early growing stage, as proposed in previous studies under the same climatic conditions [17,18]. This approach was recommended given the local experience in New South Wales and Queensland cultivation systems, where 60%–80% of rainfall is lost as runoff and evaporation during the initial crop growing stages.

Crop establishment was determined by counting the number of plants when no further seedlings had emerged at 51 days after sowing (DAS), and crop height was determined by measuring the height of all plants 122 DAS, when no further increase in height was observed. Crop yield was determined by picking all cotton bolls at maturity and the fibre quality was assessed using a Hi Value Instrument (HVI - Uster Technologies and https://www.cottoninfo.com.au/fibre-quality). Field crop quality parameters (CWP, yield, plant establishment, and fibre quality) were analysed by the ANOVA procedure at a level of significance of $p < 0.05$ using Genstat 16 software [19].

3. Results and Discussion

The preliminary results of this SBPM study indicate that there was no statistically significant difference between the SBPM_gap treatment and ODP or CON for almost all the tested parameters. However, with respect to water consumption and crop water productivity, the plot with the SBPM_gap had a tendency for higher crop water productivity (+12% and +8%) compared to the ODP film treatment and CON (Table 1). Cotton grown on the SBPM_gap treatment plot used 13 and 23 mm less water and produced 0.4 and 0.6 kg mm^{-1} ha^{-1} more lint compared with the CON and ODP film, respectively; a similar result to that achieved was reported in other studies using thin polyethylene (PE) film [20]. Using the same SBPM product, researchers reported a reduced soil evaporation rate of 10%–50% (in pot experiments) and a larger crop productivity (in rockmelons) of up to 30% [13]. Plant establishment (51 days) was within the target range of 8–12 plants m^{-1}, with a significantly lower establishment on the CON (Table 2). This might be the result of slightly elevated soil temperature in treated plots compared to the no mulch plot at the initial growing stages (e.g., up to 51 DAS; Figure 2). In this study, we did not find significant differences among the treatments in lint yield or fibre quality (Table 2). However, the lint yield was higher by 5%–7% in the SBPM_gap treatment, showing that this new product may have potential benefits for cotton yield. Similar novel polymeric protein-based biocomposites applied using the spray technique resulted in adequate agronomic performance (plant growth and dry matter content) [10].

Table 1. The comparison of crop water consumption and crop water productivity of cotton grown on plots treated with sprayable biodegradable polymer membrane (SBPM), oxo-degradable plastic (ODP) mulch film, and a control plot with no mulch (CON) during the 2015/16 season at the ACRI field site ($n = 9$ plots).

Treatment	Crop Water Consumption (mm)	Crop Water Productivity (kg ha^{-1} mm^{-1})
CON	667 [a]	5.1 [a]
ODP film	677 [a]	4.9 [a]
SBPM_gap	654 [a]	5.5 [a]

Means with the same letter are not significantly different at $p < 0.05$.

Table 2. The comparison of final cotton establishment in 2015/16 (51 DAS), plant height (122 DAS), lint yield and fibre quality [1] parameters of cotton grown on plots treated with sprayable biodegradable polymer membrane (SBPM), oxo-degradable plastic (ODP) mulch film, and a control plot with no mulch (CON) at the ACRI field site ($n = 9$ plots).

Treatment	Establishment (plants m^{-1})	Height (cm)	Lint yield (kg ha^{-1})	Fibre Strength (g tex^{-1})	Fibre Length (dec. inch)	Micronaire (-)
CON	7.8 [b]	98.1 [a]	2884 [a]	27.8 [a]	1.16 [a]	4.69 [a]
ODP film	10.2 [a]	100.0 [a]	2839 [a]	26.8 [a]	1.18 [a]	4.69 [a]
SBPM_gap	10.3 [a]	97.2 [a]	3032 [a]	26.9 [a]	1.18 [a]	4.69 [a]

Means with the same letter are not significantly different at $p < 0.05$. [1] Fibre quality parameter base grades: Strength > 29 g tex^{-1}, Length >1.125 dec. inch., Miconaire 3.8–4.5.

For the seedbed temperature measurements (at 10 cm), an additional plot with spray-on SBPM applied over the plant line (SBPM_full) was also included. Seedbed temperatures were slightly elevated under the ODP film and SBPM early in the season, a similar response to that observed in previous studies using thin plastic mulch film, which enables earlier planting or planting in cooler regions in Australia [21]. The SBPM_gap increased soil temperature compared to the control, but only between 21 and 38 days. It was observed that the soil temperature was lower in the treatments that took place later in the season, which was probably the result of increased soil moisture during the treatments. Similar results have been reported for cotton grown under PE film when compared with planting on non-mulched soil [6]. These studies used either non-degradable plastic or oxo-degradable plastic films which, contrary to the SBPM technology, could pose a threat to the environment in terms of the retrieval of the plastic film and soil and water pollution [22]. This preliminary study looked at the potential to incorporate SBPM technology into the existing Australian cotton (and similar) production systems where cotton is mechanically planted on raised beds and irrigated with flood–furrow irrigation. Implementing SBPM to conserve soil moisture may improve crop water productivity, with the ease of spray application and no environmental pollution or disposal costs [11,12]. The SBPM_gap treatment showed that it may have the potential to improve CWP, but the application rate and coverage width still need to be optimized to minimize production costs, while enhancing growing benefits. Research is ongoing to achieve this result. As the plant line had to be left uncovered for the seeds to emerge, the SBPM_gap application technique probably limited the full agronomic potential of SBPM cover.

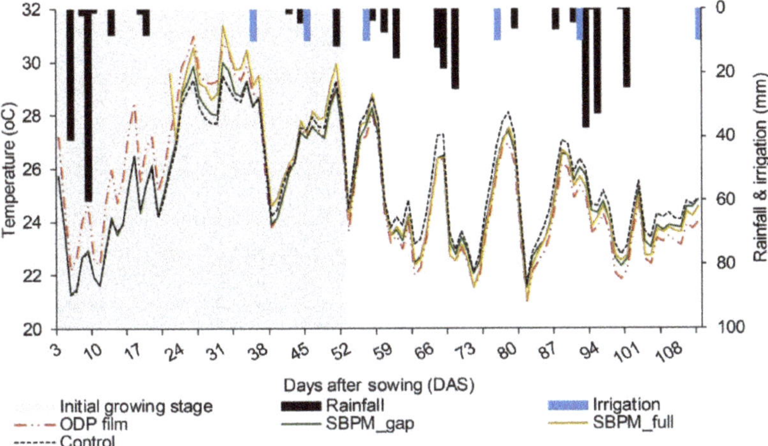

Figure 2. The effects of oxo-degradable plastic (ODP) mulch film and sprayable biodegradable polymer membrane (SBPM_gap and SBPM_full) treatments on soil temperature (°C) measured at 10 cm soil depth, during 2015/16 at the ACRI field site (*n* = 10 plots; SBPM_full not replicated).

Additionally, it was noted that in the (non-replicated) SBPM_full treatment, where the SBPM covered the plant line, the soil profile remained wetter (15–20 mm) throughout the season (Figure 3). This result is highlighted in Figure 3b, where water content in the soil profile under the SBPM_full treatment was wetter than the control plot. It is important to note that the soil water content in the SBPM_full treatment was higher from the start of the experiment, which also may indicate different soil hydraulic properties (water retention capacity) or different initial soil water status. However, both SBPM treatments (_gap and _full) indicated higher moisture content at various soil depths (not shown) and during multiple measurements taken throughout the growing season, which was probably associated with the presence of the SBPM layer at the soil–atmosphere interface (which limited soil evaporation). It was also evident that the gaps above the plant line in the slotted ODP film contributed to water loss from the soil. Both the slotted film and SBPM treatments did not impede infiltration of rainfall or irrigation water, which suggests that the technology could be used to harvest water under rain-fed conditions in a similar manner to the plastic mulch [23]. Previous research on using thin plastic mulch film in cotton was largely focussed on developing agricultural farming systems to extend cotton production to drier regions where mulch is used to harvest and conserve meagre rainfall and water under drip or saline water irrigation [6,24]. Four types of biodegradable plastic were tested as alternatives to polyethylene mulch and they all showed similar influence on soil temperature and water content, indicating that biodegradable products present viable replacement options [8]. However, this information should be treated with caution, as some authors [3] have raised concerns due to possible micro- and nano particle residues in the soil. The new SBPM technology was declared to be completely in line with the strict biodegradability and nontoxicity standards [12,13]. However, it needs to be stressed that the expectations of creating a biodegradable product which has similar properties and as low production costs as plastic mulch may be somewhat challenging (at least in its initial development phase). Nevertheless, significant improvements are currently being made with SBPMs and similar technologies in order to overcome these and any other issues [25].

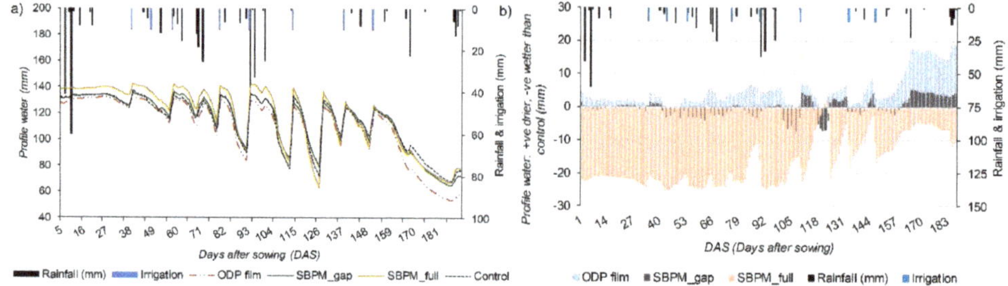

Figure 3. (**a**) The effect of oxo-degradable plastic (ODP) mulch film and sprayable biodegradable polymer membrane (SBPM_gap and SBPM_full) treatments on cumulative soil profile water content (mm) during the cotton growing season 2015/2016 at the ACRI field site (n = 10 plots; SBPM_full not replicated); (**b**) the comparison of soil profile water under the ODP, SBPM_gap and SBPM_full treatments, shown in relation to the control: positive is drier and negative is wetter than the control.

4. Conclusions

This preliminary field study shows that a novel sprayable biodegradable polymer membrane (SBPM) technology has the potential to replace preformed slotted oxo-degradable plastic (ODP) mulch films by providing a similar performance in terms of yield, crop water productivity (CWP), and soil temperature. The field testing of the new SBPM technology was limited by the availability of the product, which is still under development. A small-scale field-testing phase was undertaken to validate the agronomic performance of the new technology in order to move on to a larger-scale experimental trial. Nonetheless, results highlighted a tendency to achieve higher CWP and crop yield, increased soil temperature at the initial cotton growing stage, and increased soil water content with the SBPM cover. New large-scale field trials should focus on the application of improved SBPM formulations and their management. The SBPM technology is a promising solution with which to replace the vast amount of plastics that originate from crop production, due to its agronomic performance in comparison with plastic mulch films, simplicity of application and biodegradability into non-toxic compounds.

Author Contributions: Conceptualization, M.V.B. and P.S.C.; methodology, M.V.B., P.J., K.L.B.; software, M.V.B.; formal analysis, R.A., G.F., L.F., Y.W., V.F.; investigation, M.V.B., K.L.B.; writing—original draft preparation, M.V.B., R.A., G.F., P.J., P.S.C.; writing—review and editing, V.F. and L.F.; visualization, V.F. and Y.W. All authors have read and agreed to the published version of the manuscript.

Funding: This research was funded by CSIRO, while UKF/CSF agencies (grant number No. 3/19), Croatia, provided travel funding for Dr. Vilim Filipović and Dr. Lana Filipović.

Acknowledgments: The authors acknowledge Darin Hodgson and Shanna Smith for assisting in the establishment of the field experiment and CSIRO for funding the field experiment. The UKF visiting grant (assigned to Vilim Filipović) for bilateral research project EnviroSBPM (https://sites.google.com/view/envirosbpm/) between the University of Zagreb Faculty of Agriculture, Croatia and CSIRO Agriculture & Food is acknowledged as well. We acknowledge support by the Open Access Publication Fund, University of Zagreb Faculty of Agriculture.

Conflicts of Interest: The authors declare no conflict of interest.

References

1. Alexandratos, N.; Bruinsma, J. *World Agriculture towards 2030/2050: The 2012 Revision*; ESA Working paper No. 12-03; FAO: Rome, Italy, 2012.
2. Jabran, K.; Ullah, E.; Hussain, M.; Farooq, M.; Zaman, U.; Yaseen, M.; Chauhan, B.S. Mulching improves water productivity, yield and quality of fine rice under water-saving rice production systems. *J. Agron. Crop. Sci.* **2014**, *201*, 389–400. [CrossRef]
3. Sintim, H.Y.; Flury, M. Is biodegradable plastic mulch the solution to agriculture's plastic problem? *Environ. Sci. Technol.* **2017**, *51*, 1068–1069. [CrossRef] [PubMed]

4.	Shogren, R.L. Biodegradable mulches from renewable resources. *J. Sustain. Agric.* **2000**, *16*, 33–47. [CrossRef]

5.	Chen, N.; Li, X.; Šimůnek, J.; Shi, H.; Ding, Z.; Peng, Z. Evaluating the effects of biodegradable film mulching on soil water dynamics in a drip-irrigated field. *Agric. Water Manag.* **2019**, *226*, 105788. [CrossRef]

6.	Dai, J.; Dong, H. Intensive cotton farming technologies in China: Achievements, challenges and counter measures. *Field Crop. Res.* **2014**, *155*, 99–110. [CrossRef]

7.	European Bioplastics e.V. EUBP. EU Takes Action Against Oxo-Degradable Plastics. Available online: https://www.european-bioplastics.org/eu-takes-action-against-oxo-degradable-plastics/ (accessed on 17 May 2019).

8.	Sintim, H.Y.; Bandopadhyay, S.; English, M.E.; Bary, A.I.; DeBruyn, J.M.; Schaeffer, S.M.; Miles, C.A.; Reganold, J.P.; Flury, M. Impacts of biodegradable plastic mulches on soil health. *Agric. Ecosyst. Environ.* **2019**, *273*, 36–49. [CrossRef]

9.	Vaicekauskaite, J.; Ostrauskaite, J.; Treinyte, J.; Grazuleviciene, V.; Bridziuviene, D.; Rainosalo, E. Biodegradable linseed oil-based cross-linked polymer composites filled with industrial waste materials for mulching coatings. *J. Polym. Environ.* **2019**, *27*, 395–404. [CrossRef]

10.	Sartore, L.; Schettini, E.; Palma, L.; Brunetti, G.; Cocozza, C.; Vox, G. Effect of hydrolyzed protein-based mulching coatings on the soil properties and productivity in a tunnel greenhouse crop system. *Sci. Total. Environ.* **2018**, *645*, 1221–1229. [CrossRef] [PubMed]

11.	Adhikari, R.; Bristow, K.L.; Hornbuckle, J.; Freischmidt, G.; Casey, P. Novel sprayable biodegradable polymer membrane to minimize soil evaporation. In Proceedings of the International Conference on Technologies for Sustainable Development (ICTSD 2015), Mumbai, India, 4–6 February 2015; pp. 1–4.

12.	Adhikari, R.; Casey, P.; Bristow, K.L.; Hornbuckle, J.; Freischmidt, G. Sprayable Polymer Membrane for Agriculture. U.S. Patent WO/2015/184490, 10 December 2015.

13.	Adhikari, R.; Bristow, K.L.; Casey, P.S.; Freischmidt, G.; Hornbuckle, J.W.; Adhikari, B. Preformed and sprayable polymeric mulch film to improve agricultural crop water productivity. *Agric. Water Manag.* **2016**, *169*, 1–13. [CrossRef]

14.	Adhikari, R.; Mingtarja, H.; Freischmidt, G.; Bristow, K.L.; Casey, P.S.; Johnston, P.; Sangwan, P. Effect of viscosity modifiers on soil wicking and physico-mechanical properties of a polyurethane based sprayable biodegradable polymer membrane. *Agric. Water Manag.* **2019**, *222*, 346–353. [CrossRef]

15.	Braunack, M.V.; Zaja, A.; Tam, K.; Filipović, L.; Filipović, V.; Wang, Y.; Bristow, K.L. A Sprayable Biodegradable Polymer Membrane (SBPM) technology: Effect of band width and application rate on water conservation and seedling emergence. *Agric. Water Manag.* **2020**, *230*, 105900. [CrossRef]

16.	Hulugalle, N.R.; Weaver, T.B.; Finlay, L.A.; Lonergan, P. Soil properties, black root-rot incidence, yield and greenhouse gas emissions in irrigated cotton cropping systems sown in a Vertosol with subsoil sodicity. *Soil Res.* **2012**, *50*, 278–292. [CrossRef]

17.	French, R.; Schultz, J. Crop water productivity of wheat in a Mediterranean-type environment. I. The relation between yield, water use and climate. *Aust. J. Agric. Res.* **1984**, *35*, 743–764. [CrossRef]

18.	Holzworth, D.P.; Huth, N.I.; deVoil, P.G.; Zurcher, E.J.; Herrmann, N.I.; McLean, G.; Chenu, K.; van Oosterom, E.J.; Snow, V.; Murphy, C.; et al. APSIM—Evolution towards a new generation of agricultural systems simulation. *Environ. Modell. Softw.* **2014**, *62*, 327–350. [CrossRef]

19.	VSN International. *Genstat for Windows 18th Edition*; VSN International: Hemel Hempstead, UK, 2015.

20.	Zhang, S.; Sadras, V.; Chen, X.; Zhang, F. Crop water productivity of dryland wheat in the Loess Plateau in response to soil and crop management. *Field Crop. Res.* **2013**, *151*, 9–18. [CrossRef]

21.	Braunack, M.V.; Johnston, D.B.; Price, J.; Gauthier, E. Soil temperature and soil water potential under thin oxodegradable plastic film impact on cotton crop establishment and yield. *Field Crop. Res.* **2015**, *184*, 91–103. [CrossRef]

22.	Chen, Y.; Wu, C.; Zhang, H.; Lin, Q.; Hong, Y.; Lou, Y. Empirical estimation of pollution and contamination levels of phthalate esters in agricultural soils from plastic film mulching in China. *Environ. Earth Sci.* **2013**, *70*, 239–247. [CrossRef]

23.	Bu, L.; Liu, J.; Zhu, L.; Lou, S.; Chen, X.; Li, S.; Hill, R.L.; Zhao, Y. The effects of mulching on maize growth, yield and water use in a semi-arid region. *Agric. Water Manag.* **2013**, *123*, 71–78. [CrossRef]

24. Dong, H.; Li, W.; Tang, W.; Zhang, D. Early plastic mulching increases stand establishment and lint yield of cotton in saline fields. *Field Crop. Res.* **2009**, *111*, 269–275. [CrossRef]

25. Filipović, V.; Bristow, K.L.; Filipović, L.; Wang, Y.; Sintim, H.Y.; Flury, M.; Šimůnek, J. Sprayable biodegradable polymer membrane technology for cropping systems: Challenges and opportunities. *Environ. Sci. Technol.* **2020**. [CrossRef]

 *agronomy*

Article

Determination of Adequate Substrate Water Content for Mass Production of a High Value-Added Medicinal Plant, *Crepidiastrum denticulatum* (Houtt.) Pak & Kawano

Song-Yi Park [1,2,†], **Jongyun Kim** [3,†] **and Myung-Min Oh** [1,2,*]

[1] Division of Animal, Horticultural and Food Sciences, Chungbuk National University, Cheongju 28644, Korea; 1songyi1@gmail.com
[2] Brain Korea 21 Center for Bio-Resource Development, Chungbuk National University, Cheongju 28644, Korea
[3] Division of Biotechnology, Korea University, Seoul 02841, Korea; jongkim@korea.ac.kr
* Correspondence: moh@cbnu.ac.kr; Tel.: +82-43-261-2530
† These authors contributed equally to this work.

Received: 19 February 2020; Accepted: 10 March 2020; Published: 12 March 2020

Abstract: The effects of substrate water content on the growth and content of bioactive compounds in *Crepidiastrum denticulatum* were evaluated. Three-week-old seedlings were subjected to four levels of substrate water content (20%, 30%, 45% and 60%) and maintained for 5 weeks. Growth parameters at 5 weeks of transplanting were significantly higher with the 45% substrate water content treatment than with the other treatments. In addition, photosynthetic rate, stomatal conductance and transpiration rate increased significantly and the highest sap flow rate during the day was observed in 45% substrate water content. Total phenolic content and antioxidant capacity per shoot increased significantly with substrate water content, increasing from 20% to 45% and decreased again at 60%. Antioxidant capacity and total hydroxycinnamic acids (HCAs) content per unit dry weight of plants under the 60% treatment were significantly higher than those under the 45% treatment; however, their content per shoot was the highest under the 45% treatment. Thus, 45% substrate water content is a suitable condition for the growth of *C. denticulatum* and had positive effects on phenolic content, antioxidant capacity, and HCAs content. These results could be useful for the mass production of high-quality *C. denticulatum* in greenhouses or plant factories capable of controlling the water content of the root zone.

Keywords: antioxidant capacity; bioactive compounds; growth; hydroxycinnamic acids; hydroponics

1. Introduction

Water is one of the crucial factors for plant growth and development accounting for 80%–90% and over 50% of the fresh weight of herbaceous and woody plants, respectively [1]. Temporary water deficit in plants causes turgor loss and stomatal closure that inhibits basic metabolic processes, including photosynthesis. Excessive or constant water deficit generates reactive oxygen species (ROS) in plants, causing oxidative stress and photoinhibitory damage that eventually result in necrosis and programmed cell death [2–4]. In general, plants adapt to certain levels of water stress by promoting the biosynthesis of secondary metabolites with antioxidant properties. Water availability around the root zone directly affects the physiological and biochemical responses of plants. Therefore, controlling water content of the substrate is an important cultural practice that directly influences crop yield and quality in horticultural plant-production.

Crepidiastrum denticulatum H. is an annual or biennial species (family: Compositae) and grows naturally in East Asia and South Korea. It contains a large amount of bioactive compounds including various hydroxycinnamic acids (HCAs) such as chlorogenic acid, 3,5-di-O-caffeoylquinic acid (3,5-DCQA), chicoric acid and caftaric acid. Several previous studies have reported that *C. denticulatum* extracts have high anti-oxidative, anti-fatty liver and anti-obesity properties [5–8], and health functional foods for improving liver function have also been produced using the extract of *C. denticulatum*. One of the ecological characteristics of *C. denticulatum* is its sensitivity to the water condition around the root zone with soft rot diseases in the leaf often occurring under conditions of excessive water with frequent rainfall during summer. The occurrence of soft rot diseases was observed more often in wet soil compared to deeper and well-drained soil, resulting in the poor growth and quality of *C. denticulatum* [9].

The appropriate limitation of water supply to the root zone can be used as a cultivation technique to produce high-quality crops by promoting the biosynthesis of bioactive compounds without the inhibition of growth. In previous studies, temporary mild water stress did not inhibit the growth of lettuce, water dropwort and tomato, with a simultaneous increase in the content of polyphenolic compounds such as chicoric acid (in lettuce), anthocyanin (in water dropwort) and hydroxycinnamic acid and flavonoids (in tomato), which have antioxidant properties [10–12]. In addition, the biosynthesis of secondary metabolites in medicinal plants was also promoted by water deficit stress. For example, the content of hyperforin, the major bioactive compound in St. John's wort, was increased by 200% at 12 days of water stress compared to that in the control [13]. Water deficit treatment with a 50% reduction in irrigation increased the silymarin content of milk thistle by 170% compared to that in the control [4,13]. Recently, greenhouses or plant factories capable of controlling the root environment using various water-related sensors have been established, and the demand for high value-added crops such as medicinal plants is increasing. However, little research has been reported on the proper conditions for such cultivation. In this respect, research on the favorable water content of the substrate is needed for the stable mass production of high-quality *C. denticulatum* as a raw material for pharmaceutical products or functional foods.

Therefore, the objective of this study was to evaluate the growth and content of bioactive compounds in *C. denticulatum* according to different water content levels of the substrate. Through this study, we determined the favorable water content level of the substrate for the stable mass production of high-quality *C. denticulatum*.

2. Materials and Methods

2.1. Plant Materials and Experimental Conditions

C. denticulatum seeds collected from Pyeongchang in Korea were sown according to the method described in Park et al. [14]. Seedlings were grown for 3 weeks in a growth chamber with the following conditions: air temperature, 20 °C; relative humidity, 60%; white LEDs, PPFD 200 μmol m^{-2} s^{-1}; and light period; 16 h. Seedlings were transferred to a greenhouse for acclimation 3 days before transplanting. A total of 48 seedlings with 2–3 true leaves were transplanted into individual square plastic pots (10 × 10 × 11 cm; L × W × H) filled with commercial horticultural substrate (Myung-Moon, Dongbu Hannong Co., Seoul, Korea). The average air temperature and relative humidity of the greenhouse were 19.7 ± 0.1 °C; and 48.5 ± 0.5% (± S.E.), respectively, and average daily light integral was 9.1 ± 1.6 mol m^{-2} d^{-1} during the entire experimental period.

2.2. Treatments of Substrate Water Content

Four substrate water content levels of 20, 30, 45 and 60% were used and each level was maintained for 5 weeks starting 1 week after transplanting. Twenty-four soil water sensors (EC-5, METER group, Pullman, WA, USA) were inserted individually in 24 pots (six plants per treatment), and real-time data of volumetric water content (v/v) were collected via a data logger (CR 1000, Campbell Scientific, Logan, UT, USA) connected to the sensors. To measure the volumetric water content of the substrate, the soil water sensors were calibrated using a formula obtained from a preliminary study. The relay driver (SDM-16AC/DC, Campbell Scientific, Logan, UT, USA) connected with the data logger opened the solenoid valves of the irrigation line to supply the nutrient solution (nutrient solution for *C. denticulatum*, EC 2.0 dS m^{-1}, pH 5.5) [14] to the pots when the volumetric water content measured by the soil water sensor was lower than the set value. Two drip pins connected with a pressure compensated emitter (2L/H, Netafim, Tel Aviv, Israel) were inserted into both sides of a pot to supply the nutrient solution evenly. Figure 1 shows the changes in the volumetric water content of the substrates with the different treatments during the entire experimental period.

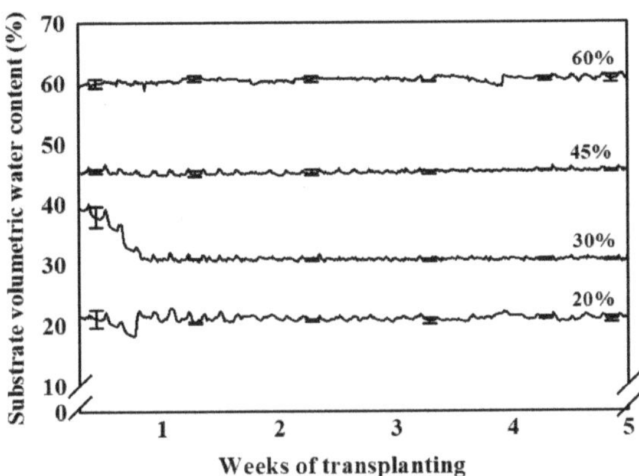

Figure 1. Average volumetric water content of the substrate at 20%, 30%, 45% and 60% water content treatments for 5 weeks. Lines and bars indicate the means and standard errors, respectively ($n = 6$).

2.3. Plant Growth Parameters

Plant growth parameters were investigated at 5 weeks after transplanting. The shoot and root were separated at the basal end and the substrate of the roots was removed by washing under running water. The remaining water was blotted using paper towels. Fresh weights of the shoot and root were measured using an electronic scale (Si-234, Denver Instrument, Bohemia, NY, USA). Shoot dry weight was measured after freeze-drying at −75 °C; for over 72 h using a lyophilizer (Alpha 24 LSCplus, CHRIST, Osterode am Harz, Germany) and root dry weight was measured after hot-air drying at 70 °C; for over 72 h. Leaf length and leaf width of the largest leaf of the plants were measured using a ruler. The leaf shape index was calculated as leaf length/leaf width. The total leaf area was measured using a leaf area meter (LI-2050A, Li-Cor, Lincoln, NE, USA).

2.4. Photosynthetic Parameters

Photosynthetic rate, stomatal conductance and transpiration rate of *C. denticulatum* were measured using a portable photosynthesis system (LI-6400, LI-COR, Lincoln, NE, USA) for 2 h, starting at 10 a.m. (3 h after sunrise) 4 weeks after transplanting. The leaf chamber conditions were set at 24 °C; block temperature, 500 μmol mol^{-1} reference CO_2, 400 μmol s^{-1} air flow and 308 μmol m^{-2} s^{-1} PPFD (average PPFD in the morning during the experiment). Six plants with fully expanded leaves per treatment were measured.

Shoots were freeze-dried after harvest and used for chlorophyll content analysis. Shoots were pulverized using a grinder (Tube Mill control, IKA, Wilmington, NC, USA). A sample of powder (40 mg) and 4 mL acetone (80%, v/v) was mixed, and then the mixture was sonicated for 15 min. The supernatant obtained by centrifugation at 15,000 × g for 2 min was diluted four times with acetone (80%, v/v). The absorbance of the solution was measured with a spectrophotometer (UV-1800, Shimadzu, Kyoto, Japan) at 663.6, 646.6 and 750 nm and chlorophyll a, chlorophyll b and chlorophyll a + b were calculated using the following equation [15].

$$\text{Chlorophyll a} = 12.25 \text{Absorbance}^{(663.6-750)} - 2.55 \text{A}^{(646.6-750)} \tag{1}$$

$$\text{Chlorophyll b} = 20.31 \text{A}^{(646.6-750)} - 4.91 \text{A}^{(663.6-750)} \tag{2}$$

$$\text{Chlorophyll a} + \text{b} = 17.76 \text{A}^{(646.6-750)} + 7.34 \text{A}^{(663.6-750)} \tag{3}$$

2.5. Sap Flow

The sap flow rate was measured to determine the effect of substrate water content on water absorption and transpiration. A micro sap flow sensor (MSF_UM, Telofarm, Seoul, Korea) was inserted into the stem of a fully expanded leaf at 4 weeks after treatment and data were continuously collected for 5 days. Sap flow values were recorded every 2 min by the data logger.

2.6. Total Phenolic Content and Antioxidant Capacity

To investigate the effects of various levels of substrate water content on the biosynthesis of secondary metabolites in *C. denticulatum*, samples were collected immediately after harvest to analyze total phenolic content and antioxidant capacity. Total phenolic content and antioxidant capacity were analyzed using a powdered sample (40 mg) obtained by grinding the freeze-dried whole shoot. Total phenols were extracted, and antioxidant capacity analyzed, as previously described in Park et al. [14]. Total phenolic content was expressed as the content of gallic acid (mg) either per unit dry weight or per shoot. Antioxidant capacity was expressed as trolox (6-Hydroxy-2,5,7,8-tetramethylchromane-2-carboxyl acid) (mM) either per unit dry weight or per shoot.

2.7. Hydroxycinnamic Acids

Caftaric acid, chlorogenic acid, caffeic acid, chicoric acid and 3,5-DCQA were extracted from a freeze-dried powder sample (100 mg) using an ultrasonicator (SK5210HP, Young Jin Corporation, Gunpo, Korea) with 70% aqueous ethanol for 90 min. Individual hydroxycinnamic acids were analyzed using a high-performance liquid chromatograph 185 (YL9100, Young Lin Instrument Co., Ltd., Anyang, Korea) according to the method previously described in Park et al. [14]. Standard curves were obtained using caftaric acid (ChemFaces, Hubei, China), chlorogenic acid, caffeic acid, 3,5-DCQA (Sigma-Aldrich, St. Louis, MO) and chicoric acid (Avention, Incheon, Korea) and the content of each compound was expressed as mg per unit dry weight of the shoot.

2.8. Statistical Analysis

In this experiment, we used a randomized complete block design with three blocks and four plants were randomly arranged in each block for each treatment. Twelve plants per treatment were used for the analysis of growth parameters, chlorophyll content, total phenolic content, antioxidant capacity and HCAs content. Photosynthetic rate, stomatal conductance and transpiration rate were measured using six plants per treatment. Statistical analysis of the results was conducted using Statistical Analysis System, 9.2 Version, SAS Institute, Cary, NC, USA (SAS). Analysis of variance (ANOVA) and Tukey's Studentized Range Test (HSD) were used to determine the statistical significance among treatments.

3. Results

3.1. Plant Growth Parameters

Different levels of substrate water content affected the growth of the shoot and root of *C. denticulatum* significantly (Figure 2; Figure 3). As substrate water content increased from 20% to 45%, fresh and dry weights of the shoot and root increased significantly, but decreased at 60%. In particular, the shoot and root biomass were higher by 2.4 and 1.8 times, respectively, with the 45% substrate water content treatment than with the 20% treatment which showed the lowest growth performance. Changes in leaf length and leaf area also showed a similar pattern to that observed in shoot growth; the highest value was recorded with the 45% substrate water content treatment. In the case of leaf width, significantly higher values were observed with the 30% and 45% substrate water content treatments than with the 20% treatment. Leaf shape index, an indicator of leaf shape, was also influenced by substrate water content and had the lowest value in plants treated with the 20% of substrate water content.

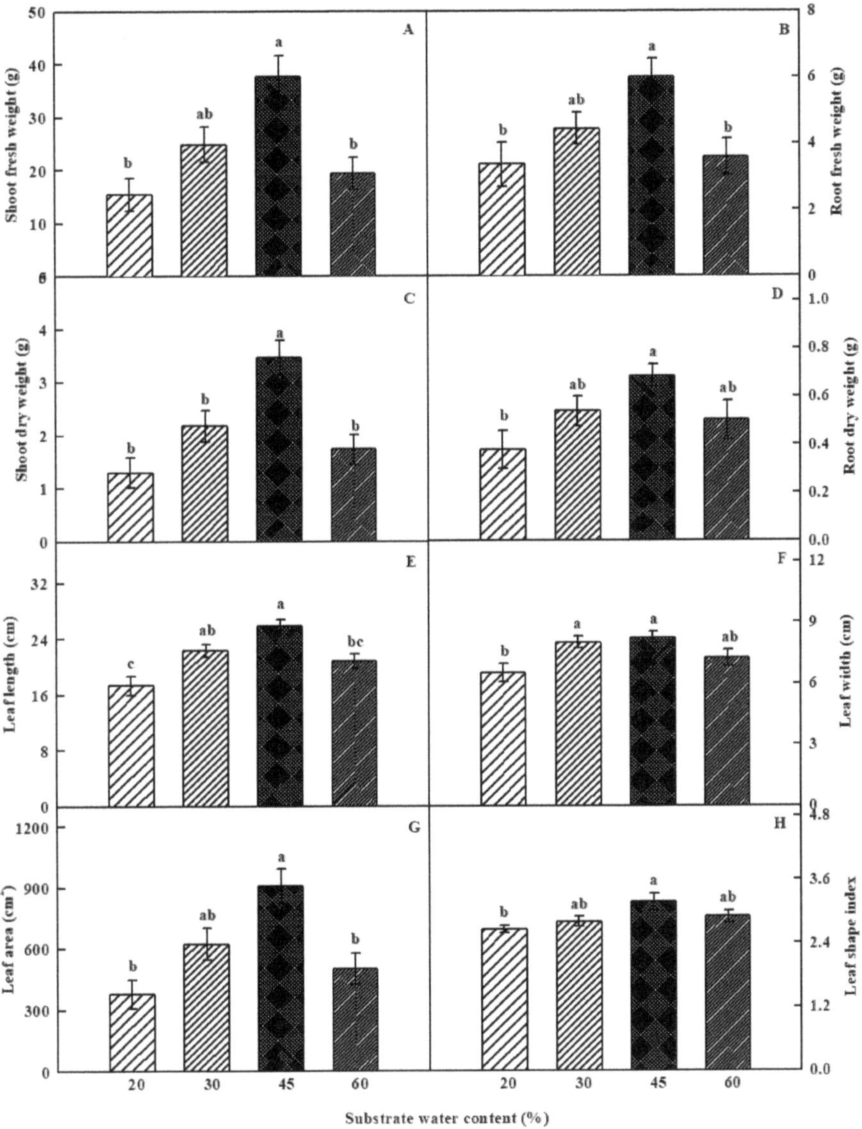

Figure 2. Effect of substrate water content on growth parameters: Fresh and dry weighs of shoot (**A** and **C**) and root (**B** and **D**), leaf length (**E**), leaf width (**F**), leaf area (**G**) and leaf shape index (**H**) of under four different substrate water content levels for 5 weeks. The data indicate the means ± S.E. ($n = 12$). Different letters above the bars indicate statistical difference by Tukey's Studentized Range Test at $p < 0.05$.

Figure 3. *Crepidiastrum denticulatum* grown under different substrate water content levels at 5 weeks after transplanting.

3.2. Photosynthetic Parameters

Photosynthetic parameters that support the increase in biomass were measured at 4 weeks after transplanting in *C. denticulatum* grown with different substrate water content (Figure 4A–C). The photosynthetic rate of plants grown under the 45% substrate water content treatment, which exhibited excellent shoot biomass, was higher than that of plants grown under the other treatments, and the photosynthetic rate of plants under the 60% substrate water content treatment was not different from that of plants under the 20% and 30% treatments. As substrate water content increased from 20% to 45%, stomatal conductance gradually increased, and then decreased again at 60%. The change in transpiration rate was similar to that observed for stomatal conductance and was significantly higher in plants grown under the 30% and 45% substrate water content treatments than in plants under the 20% and 60% treatments. The chlorophyll content of *C. denticulatum* per unit dry weight was not affected by the water content of the substrate (Figure 4D).

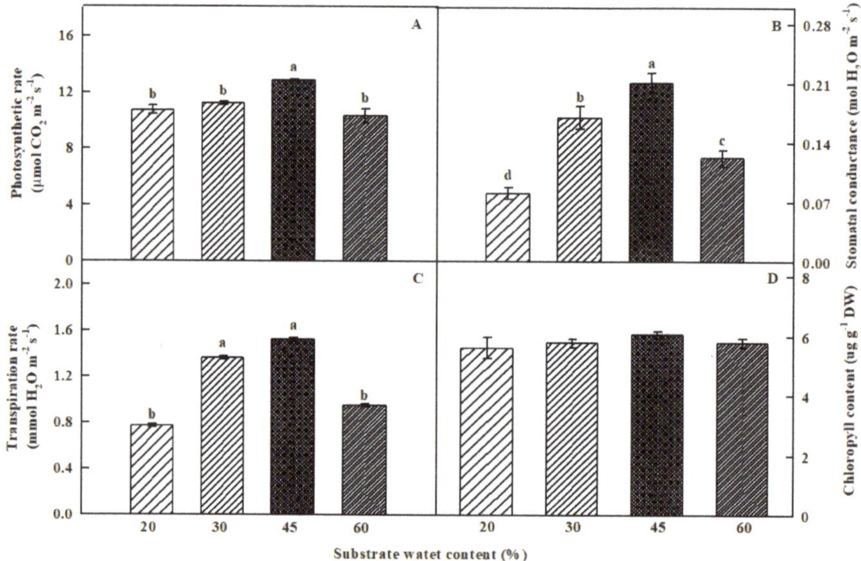

Figure 4. Effect of substrate water content on photosynthetic parameters: Photosynthetic rate (**A**), stomatal conductance (**B**), transpiration rate (**C**) and chlorophyll content (**D**) of *Crepidiastrum denticulatum* grown under four different substrate water content levels at 4 weeks after transplanting. The data indicate the means ± S.E. (photosynthetic parameters; n = 6 and chlorophyll content; n = 12). Different letters above the bars indicate statistical difference by Tukey's Studentized Range Test at $p < 0.05$.

3.3. Sap Flow

After sunrise, at around 8 a.m., sap flow values of plants in all treatments began to increase rapidly, and all values slowly increased with repeated increases and decreases until around 2 p.m. (Figure 5). All sap flow values gradually declined from after 3 p.m. until sunset. Similar to the results observed for the photosynthetic parameters, the sap flow value in plants grown under the 45% substrate water content treatment was the highest during the day. In particular, in plants grown under the 45% treatment, the sap flow value measured from 3 p.m. until sunset was clearly distinguished from that in plants grown under other treatments and remained high.

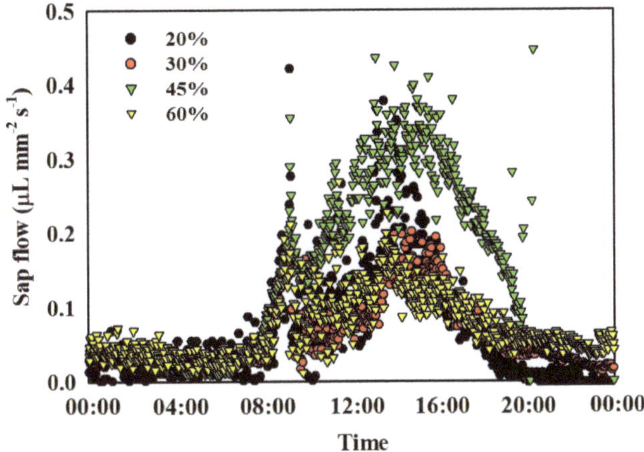

Figure 5. Sap flow values of *Crepidiastrum denticulatum* grown under four different substrate water content levels at 4 weeks after transplanting. A sap flow sensor was inserted into the stem of one plant per treatment. Sap flow values were recorded every 2 min by the data logger.

3.4. Total Phenolic Content and Antioxidant Capacity

The four levels of substrate water content showed no significant difference in total phenolic content per unit dry weight. However, the antioxidant capacity of plants under the 60% treatment was higher than that of plants under the 45% treatment and there was no difference in plants under the 20% to 45% treatments (Figure 6A,B). Total phenolic content and antioxidant capacity per shoot increased significantly from 20% to 45% of substrate water content treatments and decreased again under the 60% treatment, similar to the results observed for the shoot growth. In particular, the highest phenolic content and antioxidant capacity were also found in plants under the 45% treatment, which had the best shoot growth.

3.5. Hydroxycinnamic Acids

After harvest, four types of hydroxycinnamic acids (HCAs) were analyzed (Figure 7). Total HCAs content per unit dry weight was significantly higher in plants grown under the 60% substrate water content treatment, whereas chicoric acid content was significantly the lowest in plants grown with the 45% treatment. There was no difference between plants grown under the other three treatments. Individual HCAs and total HCAs content per shoot were significantly the highest in plants grown under the 45% substrate water content treatment that had the superior shoot biomass. In particular, the total HCAs content of plants grown under the 45% treatment was approximately three times higher than that of plants grown under the 20% treatment.

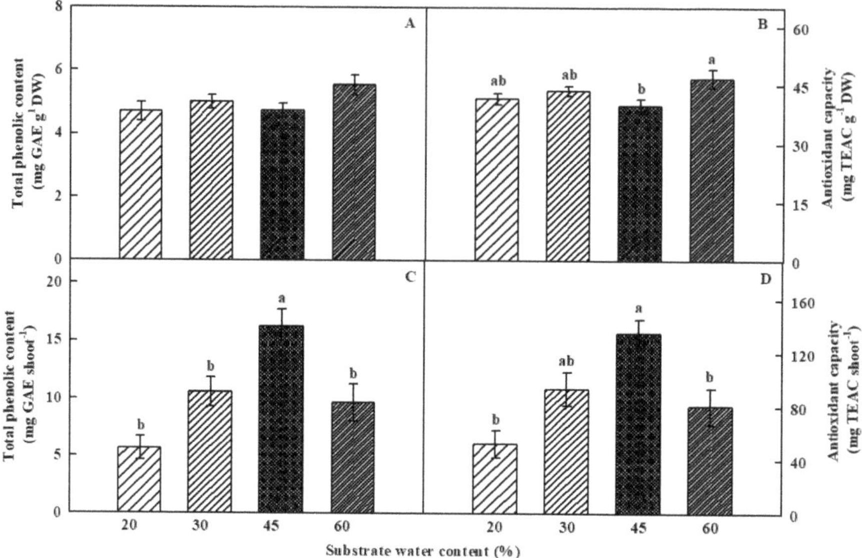

Figure 6. Effect of substrate water content on phenolic content and antioxidant capacity: Total phenolic content and antioxidant capacity per unit dry weight (**A** and **C**) and per shoot (**B** and **D**) of *Crepidiastrum denticulatum* grown under four different substrate water content levels for 5 weeks. The data indicate the means ± S.E. (n = 12). Different letters above the bars indicate statistical difference by Tukey's Studentized Range Test at $p < 0.05$.

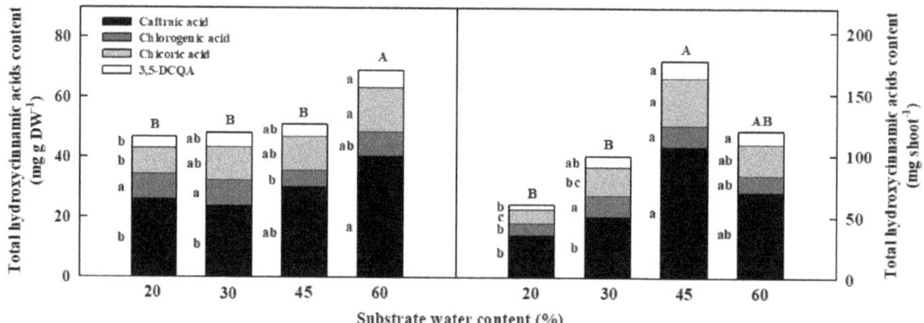

Figure 7. Effect of substrate water content on hydroxycinnamic acids' (HCAs) content: Total HCAs' content per unit dry weight (**A**) and per shoot (**B**) of Crepidiastrum denticulatum grown under four different substrate water content levels for 5 weeks. The data indicate the means ± S.E. (n = 12). Different lowercase letters indicate statistical difference in each individual compound by Tukey's Studentized Range Test at $p < 0.05$. Different uppercase letters indicate statistical difference in total HCAs by Tukey's Studentized Range Test at $p < 0.05$.

4. Discussion

In the early stage of soil drought and flooding stresses, root signals limit the movement of water and nutrients to the growing zones; these changes result in the collapse of the water potential gradient between the xylem and growing cells [16,17]. A continuous drought condition eventually causes dehydration of the shoot and stimulates the biosynthesis of abscisic acid (ABA) around the root meristem. ABA is transported to the shoot via the xylem stream and causes stomatal closure, thereby preventing water loss from leaves. This results in a restriction in the transport of large amounts of water, minerals and various chemical compounds from the roots to the shoots. Drought stress increases the content of ABA in leaves, not only by promoting the biosynthesis of ABA but also by accelerating the ABA catabolic pathway [18–20]. In contrast, excessive water content in soil or the substrate causes an oxygen-poor environment (hypoxia), which becomes a major factor in inhibiting root respiration. This inhibits metabolic activities and ATP production in plants. These plant responses limit the supply of energy for the growth of the root and eventually result in poor plant growth [2,19]. In addition, flooding stress induces the accumulation of toxic compounds such as ethanol, lactic acid and acetaldehyde, and these compounds not only suppress plant growth but also cause adverse effects such as root dysfunction and low soil redox potential [21,22]. In this experiment, the inhibition of shoot and root growth under the 20% and 60% substrate water content conditions could be explained as being caused by drought and flooding stresses, respectively (Figure 2). These results imply that the substrate water content between these values, 45%, in which plants showed the most effective growth, may be adequate for the growth of *C. denticulatum*.

Stomatal closing due to drought stress as described above prevents water loss in the leaves, while it hinders transpiration and the allocation of photosynthetic assimilates, resulting in a reduction in photosynthetic rate [21,23]. Hypoxia conditions in soil caused by flooding restrict water absorption by roots and root hydraulic conductance, leading to stomatal closure [24,25]. In addition, Ahsan et al. [3] reported that soil flooding increases photorespiration and/or decreases the activities of RuBP and RuBP activase, which is one of the main reasons for the reduction in photosynthetic rate. In our study, substrate water content of 20% (drought) and 60% (flooding) decreased stomatal conductance and transpiration rate in the leaves of *C. denticulatum* (Figure 4B,C). The observed sap flow values also supported the results of stomatal conductance and transpiration rate. During the day, plants under the 20% and 60% substrate water content treatments had lower sap flow values than plants under the 45% treatment (Figure 5). In this experiment, the decrease in transpiration rate owing to soil drought and flooding seemed to have a direct effect on photosynthetic inhibition (Figure 4A). Mutava et al. [26] reported that drought-tolerant genotypes accumulate more ABA than drought-susceptible genotypes, leading to an increased stomatal closure in dry soil. In this experiment, the stomatal conductance of plants under the 20% treatment was significantly lower than that of plants under the 60% treatment (Figure 4B), suggesting that *C. denticulatum*, which naturally grows at the foot of a mountain in dry conditions, may be more sensitive to excessive soil water content than to dry soil. In this study, chlorophyll content was not significantly affected by substrate water content (Figure 4D). In general, constant or excessive water stress decreases chlorophyll content and ultimately accelerates leaf senescence [3]. However, it should be considered that the water level and duration conditions used in this study were not severe enough to affect a change in chlorophyll content. Plants have complex defensive mechanisms to survive in tough external environments, among which the antioxidant system plays an important role in overcoming environmental stresses. Soil-water-related stresses promote the generation of reactive oxygen species (ROS) owing to the limitation of photosynthetic processes, which can activate the antioxidant system [27–29]. The typical antioxidant bioactive compounds of *C. denticulatum* are types of HCAs including caftaric acid, chicoric acid, chlorogenic acid and 3,5-DCQA [5–7,14]. The results of this experiment showed that antioxidant capacity and total HCAs per dry weight of plants under the soil flooding treatment (60%) were higher than those of plants under other treatments (Figures 6B and 7A). This result can also be explained by the concentration effect because of growth inhibition. However, considering that shoot dry weights of 20% and 30% substrate water content treatments were similar to 60% treatment, not only the inhibition of photosynthesis by flooding stress but also hypoxia-produced ROS and toxic substances are thought to activate the biosynthesis pathway of secondary metabolites [21]. At 20% water content of the substrate, total phenolic content and HCAs content did not increase, probably owing to a lack of photosynthetic assimilates. However, although antioxidant capacity and total HCAs content per dry weight of plants grown under the 60% substrate water content treatment were significantly higher than in plants under the 45% treatment, their content per shoot were the highest under the 45% substrate water content treatment because of a remarkable increase in the shoot dry biomass. Another study of the cultivation of *C. denticulatum* used the capillary wick culture system described in a previous study [14,30]. The results also showed that shoot fresh weight, photosynthetic rate, total phenolic content and antioxidant capacity per shoot of plants grown under 45% substrate water content were significantly the highest among treatments (Figure S1). These results showed that shoot biomass of *C. denticulatum* is directly related to the content of bioactive compounds, and that 45% water content of the substrate can increase not only the growth, but also the antioxidant capacity and phenolic content.

In conclusion, we confirmed that the growth and bioactive compounds of *C. denticulatum*, which is used as a plant-derived raw material for functional food, can be influenced by the water content of the substrate. The water content of 45% in the substrate increased the biomass of the shoot and root and increased phenolic content, antioxidant capacity and HCAs content per shoot. The possibility of using hydroponics for native plants was also verified by another experiment using the capillary wick culture system. The results of the current study are expected to be useful for the stable mass production of high-quality *C. denticulatum* in greenhouses or plant factories capable of controlling water content in the root zone.

Supplementary Materials: The following are available online at http://www.mdpi.com/2073-4395/10/3/388/s1, Figure S1: effects of capillary wick culture system.

Author Contributions: Methodology, investigation, resources, formal analysis, data curation, software, writing—original draft preparation, visualization, S.-Y.P.; methodology, conceptualization, formal analysis, software, writing—original draft preparation, J.K.; supervision, funding acquisition, validation, writing—review and editing, M.-M.O. All authors have read and agreed to the published version of the manuscript.

Funding: This work was supported by Korea Institute of Planning and Evaluation for Technology in Food, Agriculture, Forestry and Fisheries (IPET) through the Agriculture, Food and Rural Affairs Research Center Support Program, funded by the Ministry of Agriculture, Food and Rural Affairs (MAFRA) (717001-07-02-HD240).

Conflicts of Interest: The authors declare no conflict of interest. The funders had no role in the design of the study; in the collection, analyses, or interpretation of data; in the writing of the manuscript, or in the decision to publish the results.

References

1. Akıncı, Ş.; Lösel, D.M. Plant water-stress response mechanisms. In *Water Stress*; Rahman, I.M.M., Hasegawa, H., Eds.; InTechOpen: Rijeka, Croatia, 2012; pp. 15–42.

2. Liao, C.; Lin, C. Physiological adaptation of crop plants to flooding stress. In Proceedings of the National Science Council, Republic of China, Part B, Life Sciences, Taipei, Taiwan, 1 July 2001; Volume 25, pp. 148–157.

3. Ahsan, N.; Lee, D.; Lee, S.; Kang, K.Y.; Bahk, J.D.; Choi, M.S.; Lee, I.; Renaut, J.; Lee, B. A comparative proteomic analysis of tomato leaves in response to waterlogging stress. *Physiol. Plant.* **2007**, *131*, 555–570. [CrossRef] [PubMed]

4. Zobayed, S.M.A.; Afreen, F.; Kozai, T. Phytochemical and physiological changes in the leaves of St. John's wort plants under a water stress condition. *Environ. Exp. Bot.* **2007**, *59*, 109–116. [CrossRef]

5. Lee, H.J.; Cha, K.H.; Kim, C.Y.; Nho, C.W.; Pan, C. Bioavailability of hydroxycinnamic acids from *Crepidiastrum denticulatum* using simulated digestion and Caco-2 intestinal cells. *J. Agric. Food Chem.* **2014**, *62*, 5290–5295. [CrossRef] [PubMed]

6. Yoo, J.; Kang, K.; Yun, J.H.; Kim, M.A.; Nho, C.W. *Crepidiastrum denticulatum* extract protects the liver against chronic alcohol-induced damage and fat accumulation in rats. *J. Med. Food* **2014**, *17*, 432–438. [CrossRef]

7. Ahn, H.R.; Lee, H.J.; Kim, K.; Kim, C.Y.; Nho, C.W.; Jang, H.; Pan, C.; Lee, C.Y.; Jung, S.H. Hydroxycinnamic acids in *Crepidiastrum denticulatum* protect oxidative stress-induced retinal damage. *J. Agric. Food Chem.* **2014**, *62*, 1310–1323. [CrossRef]

8. Kim, M.; Yoo, G.; Randy, A.; Kim, H.S.; Nho, C.W. Chicoric acid attenuate a nonalcoholic steatohepatitis by inhibiting key regulators of lipid metabolism, fibrosis, oxidation, and inflammation in mice with methionine and choline deficiency. *Mol. Nutr. Food Res.* **2017**, *61*, 1600632. [CrossRef]

9. Rural Development Administration (RDA) Agriculture Science Library. Available online: http://lib.rda.go.kr/search/searchDetailART.do?ctrl=000001165804 (accessed on 15 July 2010).

10. Oh, M.M.; Carey, E.E.; Rajashekar, C.B. Regulated water deficits improve phytochemical concentration in lettuce. *J. Am. Soc. Hort. Sci.* **2010**, *135*, 223–229. [CrossRef]

11. Lee, J.Y.; Oh, M.M. Mild water deficit increases the contents of bioactive compounds in dropwort. *Hortic. Environ. Biotechnol.* **2017**, *58*, 458–466. [CrossRef]

12. Sánchez-Rodríguez, E.; Ruiz, J.M.; Ferreres, F.; Moreno, D.A. Phenolic metabolism in grafted versus nongrafted cherry tomatoes under the influence of water stress. *J. Agric. Food Chem.* **2011**, *59*, 8839–8846. [CrossRef]

13. Afshar, R.K.; Chaichi, M.R.; Jovini, M.A.; Jahanzad, E.; Hashemi, M. Accumulation of silymarin in milk thistle seeds under drought stress. *Planta* **2015**, *242*, 539–543.

14. Park, S.Y.; Oh, S.B.; Kim, S.M.; Cho, Y.Y.; Oh, M.M. Evaluating the effects of a newly developed nutrient solution on growth, antioxidants, and chicoric acid contents in *Crepidiastrum denticulatum*. *Hortic. Environ. Biotechnol.* **2016**, *57*, 478–486. [CrossRef]

15. Porra, R.J.; Thompson, W.A.; Kriedemann, P.E. Determination of accurate extinction coefficients and simultaneous equations for assaying chlorophylls a and b extracted with four different solvents: Verification of the concentration of chlorophyll standards by atomic absorption spectroscopy. *Biochim. Biophys. Acta-Bioenerg.* **1989**, *975*, 384–394. [CrossRef]

16. Mullet, J.E.; Whitsitt, M.S. Plant cellular responses to water deficit. In *Drought Tolerance in Higher Plants: Genetical, Physiological and Molecular Biological Analysis*; Belhassen, E., Ed.; Kluwer Academic Publishers: Dordrecht, The Netherlands, 1997; pp. 41–46.

17. Nonami, H.; Boyer, J.S. Wall extensibility and cell hydraulic conductivity decrease in enlarging stem tissues at low water potentials. *Plant Physiol.* **1990**, *93*, 1610–1619. [CrossRef]

18. Watanabe, S.; Sato, M.; Sawada, Y.; Tanaka, M.; Matsui, A.; Kanno, Y.; Hirai, M.Y.; Seki, M.; Sakamoto, A.; Seo, M. Arabidopsis molybdenum cofactor sulfurase ABA3 contributes to anthocyanin accumulation and oxidative stress tolerance in ABA-dependent and independent ways. *Sci. Rep.* **2018**, *8*, 16592. [CrossRef]

19. Else, M.A.; Coupland, D.; Dutton, L.; Jackson, M.B. Decreased root hydraulic conductivity reduces leaf water potential, initiates stomatal closure and slows leaf expansion in flooded plants of castor oil (*Ricinus communis*) despite diminished delivery of ABA from the roots to shoots in xylem sap. *Physiol. Plant.* **2001**, *111*, 46–54. [CrossRef]

20. Kim, J.; Malladi, A.; Van Iersel, M.W. Physiological and molecular responses to drought in *Petunia*: The importance of stress severity. *J. Exp. Bot.* **2012**, *63*, 6335–6345. [CrossRef] [PubMed]

21. Pezeshki, S.R. Wetland plant responses to soil flooding. *Environ. Exp. Bot.* **2001**, *46*, 299–312. [CrossRef]

22. Akhtar, I.; Nazir, N. Effect of waterlogging and drought stress in plants. *Int. J. Water Resour. Environ. Sci.* **2013**, *2*, 34–40.

23. Rodríguez-Gamir, J.; Ancillo, G.; González-Mas, M.C.; Primo-Millo, E.; Iglesias, D.J.; Forner-Giner, M.A. Root signalling and modulation of stomatal closure in flooded citrus seedlings. *Plant Physiol. Biochem.* **2011**, *49*, 636–645. [CrossRef]

24. Currey, C.J.; Flax, N.J.; Litvin, A.G.; Metz, V.C. Substrate volumetric water content controls growth and development of containerized culinary herbs. *Agronomy* **2019**, *9*, 667. [CrossRef]

25. Rood, S.B.; Nielsen, J.L.; Shenton, L.; Gill, K.M.; Letts, M.G. Effects of flooding on leaf development, transpiration, and photosynthesis in narrowleaf cottonwood, a willow-like poplar. *Photosynth. Res.* **2010**, *104*, 31–39. [CrossRef] [PubMed]

26. Mutava, R.N.; Prince, S.J.K.; Syed, N.H.; Song, L.; Valliyodan, B.; Chen, W.; Nguyen, H.T. Understanding abiotic stress tolerance mechanisms in soybean: A comparative evaluation of soybean response to drought and flooding stress. *Plant Physiol. Biochem.* **2015**, *86*, 109–120. [CrossRef]

27. Arbona, V.; Manzi, M.; Ollas, C.; Gómez-Cadenas, A. Metabolomics as a tool to investigate abiotic stress tolerance in plants. *Int. J. Mol. Sci.* **2013**, *14*, 4885–4911. [CrossRef]

28. Bettaieb, I.; Hamrouni-Sellami, I.; Bourgou, S.; Limam, F.; Marzouk, B. Drought effects on polyphenol composition and antioxidant activities in aerial parts of *Salvia officinalis* L. *Acta Physiol. Plant.* **2011**, *33*, 1103–1111. [CrossRef]

29. Khan, M.; Ulrichs, C.; Mewis, I. Effect of water stress and aphid herbivory on flavonoids in broccoli (*Brassica oleracea* var. *italica* Plenck). *J. Appl. Bot. Food Qual.* **2011**, *84*, 178–182.

30. Son, J.E.; Oh, M.M.; Lu, Y.J.; Kim, K.S.; Giacomelli, G.A. Nutrient-flow wick culture system for potted plant production: System characteristics and plant growth. *Sci. Hortic.* **2006**, *107*, 392–398. [CrossRef]

Article

Growth Response of Cassava to Deficit Irrigation and Potassium Fertigation during the Early Growth Phase

Daniel O. Wasonga *, Jouko Kleemola, Laura Alakukku and Pirjo S.A. Mäkelä

Department of Agricultural Sciences, University of Helsinki, P.O. Box 27, FIN-00014 Helsinki, Finland; jouko.kleemola@helsinki.fi (J.K.); laura.alakukku@helsinki.fi (L.A.); pirjo.makela@helsinki.fi (P.S.A.M.)
* Correspondence: daniel.wasonga@helsinki.fi; Tel.: +358-(0)466-170-988

Received: 27 January 2020; Accepted: 24 February 2020; Published: 26 February 2020

Abstract: Cassava (*Manihot esculenta* Crantz) experiences intermittent water deficit and suffers from potassium (K) deficiency that seriously constrains its yield in the tropics. Currently, the interaction effect between deficit irrigation and K fertigation on growth and yield of cassava is unknown, especially during the early growth phase. Therefore, pot experiments were conducted under controlled greenhouse conditions using cassava cuttings. Treatments initiated at 30 days after planting included three irrigation doses (30%, 60%, 100% pot capacity) and five K (0.01, 1, 4, 16, and 32 mM) concentrations. The plants were harvested 90 days after planting. Decreasing irrigation dose to 30% together with 16 mM K lowered the leaf water potential by 69%, leaf osmotic potential by 41%, photosynthesis by 35%, stomatal conductance by 41%, water usage by 50%, leaf area by 17%, and whole-plant dry mass by 41%, compared with full-irrigated plants. Lowering the K concentration below 16 mM reduced the values further. Notably, growth and yield were decreased the least compared with optimal, when irrigation dose was decreased to 60% together with 16 mM K. The results demonstrate that deficit irrigation strategies could be utilized to develop management practices to improve cassava productivity by means of K fertigation under low moisture conditions.

Keywords: leaf area; *Manihot esculenta*; photosynthesis; tuber; water status

---

1. Introduction

A global challenge for the agricultural sector is to produce more food with less water [1]. Developing new scientific strategies that allow crops to use water efficiently could be crucial in a world with a growing population [2]. Thus, water application strategies focused on increased agricultural water productivity, such as deficit irrigation coupled with potassium (K) fertigation to investigate multiple alternatives, have a pivotal role to play in sustainable crop production. Cassava (*Manihot esculenta* Crantz) is a major food crop for more than 800 million people in the tropics, providing more than 60% of daily calorific needs [3]. Cassava leaves are widely consumed, due to the high contents of protein, minerals, vitamins, lipids, and fiber, compared with roots that are mainly rich in carbohydrates [4,5]. Cassava is alternatively used as a processed food, animal feed, starch for pharmaceutical industries, and bioethanol for vehicles [6,7]. Even though cassava is considered drought-tolerant [8,9], the crop experiences intermittent water deficit [10] and suffers from K deficiency [11,12], which seriously constrains its yield in the tropics.

Water deficit restricts cassava growth and yield by decreasing the soil-water potential, which in turn limits stomatal conductance [13,14], resulting in reduced photosynthesis [15], number of leaves, and the individual leaf size [8,16]. It also leads to a reduction in shoot growth [17] and subsequent reduction in fresh and dry biomass [18,19]. Cassava shoot and root biomass can decrease by 70% under conditions of water deficit [20,21] but are more pronounced if the water deficit occurs during the first

1–5 months after planting [18,22] since plant leaf expansion and tuberous root development initiate during this period.

Cakmak and Engels [23] showed that water deficit increases the plant K requirement. K alleviates water deficit in plants by regulating the cell osmotic potential (Ψ_s) to maintain the turgor pressure (Ψ_p) required for cell functioning [24] and regulating the stomatal movement that aids in minimizing water losses during drought, leading to maintain the carbon dioxide (CO_2) fixation [25]. K also plays a key role in partitioning photosynthates to storage roots [26,27] and activation of enzymes [28]. It also increases heat tolerance in plants [26]. Byju et al. [29] estimated that cassava requires a total uptake of 15.6 K kg ha^{-1} to produce a single metric ton of dry root yield. Thus, improving K nutrition in drought areas with low inherent soil fertility could increase cassava productivity. Lately, cassava yields have been improved through irrigation [30,31] and nutrient application [29,32]. Studies of K nutrition in cassava have also been performed [33,34] with granular K application. However, no studies have combined water deficit strategies and K fertigation in cassava to elaborate its effect on growth and yield. Moreover, understanding how young cassava plants respond to deficit irrigation and K fertigation during the early growth phase is essential for holistic agronomic management to ensure improvements in terms of growth and marketable yield under drought conditions. Therefore, the objective of this work is to assess the effects of interaction between deficit irrigation and K fertigation on growth and yield response of biofortified cassava during the early growth phase.

2. Materials and Methods

2.1. Plant Material and Growth Conditions

Four pot experiments were conducted with single-stem cuttings (25 cm) of yellow cassava "Mutura" cultivar (Kenya Agricultural and Livestock Research Organization (KALRO), Nairobi, Kenya). Cuttings were planted in 5-L pots containing 1.7 kg of pre-fertilized potting mix (pH 5.5, N–P–K: 17–4–25, organic matter: 25–40%, Kekkilä Karkea ruukutusseos, W R8014; Kekkilä Oy, Vantaa, Finland). The potting mix was saturated with water and allowed to drain overnight, and the maximum soil water holding capacity (1600 g pot^{-1}) was calculated as the difference between water applied and water drained. The pots were placed in the greenhouse at the University of Helsinki, Finland, under controlled conditions with day/night temperatures of 28/20 °C and relative humidity of 55% ± 5%. High-pressure sodium lamps (Master son-t; Philips Lighting N.V., Eindhoven, The Netherlands) provided a 12-h photoperiod with photosynthetic photon flux density (PPFD) of 600 μmol photosynthetically active radiation (PAR) m^{-2} s^{-1} at the top of the canopy. The plants were watered every second day on the soil surface until drainage for 30 days, and the side shoots were trimmed to maintain single-stemmed plants. At 30 days after planting (DAP), treatments lasting 60 days were initiated. There were three irrigation doses (30%, 60%, 100% pot capacity) in all four experiments that were further split into a range of K (potassium chloride (KCl); Sigma-Aldrich Chemie GmbH, Munich, Germany) concentrations of 0.01 mM (EXP. I, III, IV), 1 mM (EXP. I, II, III, IV), 4 mM (EXP. II, IV), 16 mM (EXP. I, II, III), and 32 mM (EXP. III, IV) of irrigation water. Plants were watered every second day with full-strength Hoagland solution [35] in which the K concentration was modified. The experiments were arranged in a completely randomized design with four (EXP. I, in total 36 pots) to eight (EXP. II, III, IV, in total 72 pots each) replicates.

2.2. Measurements

Morpho-physiological traits were measured at 15-day intervals, beginning 30 DAP between 11:00 and 13:00 h from the three uppermost fully expanded leaves of each plant. The leaf temperature was measured, using an infrared thermometer (Fluke 574; Fluke Corporation, Everett, WA, USA). The chlorophyll content was measured with an Apogee MC-100 meter (Apogee Instruments, Logan, UT, USA). Net photosynthesis and stomatal conductance were measured with a portable photosynthesis meter (LI-6400; LI-COR, Lincoln, NE, USA). The plant height was measured from the soil level to the tip of the plant. The leaf water potential (Ψ_w) was measured according to McCutchan and Shackel [36] by first covering the leaflets with bags made of black plastic on the inside and aluminum foil on the outside to prevent leaf transpiration. After 1 h, the leaflets were detached and leaf Ψ_w was measured, using a pressure chamber (Soilmoisture Equipment Corp; Goleta, CA, USA). The osmolality was analyzed from the leaves used for the Ψ_w measurements, using a freezing-point depression osmometer (Micro-Osmometer 3300 M; Advanced Instruments, Norwood, MA, USA) as described by Mäkelä et al. [37]. The osmotic potential (Ψ_s) was calculated from the osmolality values as π (osmotic pressure, MPa) $= cRT$, where RT (R is the gas constant ($J\ mol^{-1}\ K^{-1}$), T is absolute temperature (K)) is 2.48 and c the osmolality (osmol kg^{-1}). Leaf turgor (Ψ_p) was estimated as Ψ_w–Ψ_s. The pots were weighed every second day before irrigation and K treatment application to monitor water usage.

The plants were harvested at 90 DAP by cutting the stems at the soil surface. Leaves were detached from the plants, and the roots were carefully washed with water to remove soil. The fresh weight of the leaves, stems, and roots was recorded. The green and senescent leaves were then separated, and the green leaf area was measured with a portable leaf area meter (LI-3000; LI-COR, Lincoln, NE, USA). The green leaves and tuberous roots were divided into two subsamples. One subsample of each was snap-frozen in liquid N_2 and stored at −20 °C until further analysis. The other subsample was dried in a forced-air chamber at 70 °C for 72 h, weighed, ground to pass a 0.5-mm sieve using a centrifugal mill (ZM200; Retsch, Haan, Germany) and stored at room temperature until further analysis.

2.3. Potassium Analysis

The K content was analyzed from the ground leaf and root subsamples. Plant material (250 mg) was weighed into polytetrafluoroethylene (PTFE) Teflon tubes (CEM Corp; Matthews, NC, USA) and 6 mL of 15.2 M nitric acid (68% *w/v*; VWR International BVBA, Leuven, Belgium) and 1 mL of 9.8 M hydrogen peroxide (30% *w/v*; Merck KGaA, Darmstadt, Germany) were added for microwave digestion (MARS 240/50; MARSXpress, CEM). The digested samples were filtered through paper (Whatman grade no. 42, pore size 2.5 µm; GE Healthcare, Gloucester, Cheltenham, UK), diluted in purified water and stored at −20 °C. Elemental analysis was conducted with an inductively coupled plasma-optical emission spectrometer (iCAP 6200; Thermo Fisher Scientific, Cambridge, UK) with every 20th sample as standard.

2.4. Statistical Analyses

Data of the four experiments were combined and analyzed as one experiment, after subjecting to contrast analysis for experimental differences. To show the effects of irrigation doses, K concentrations, and their interactions as fixed effects on traits measured, a two-way ANOVA was carried out. Differences were considered significant when the *p*-values were <0.05, and means were compared using Tukey's multiple range test. In addition, a two-tailed Pearson correlation was calculated to measure the patterns of relationship among the traits measured. All statistical analyses were carried out using R program (version 3.5.1; R Development Core Team, Vienna, Austria) [38].

3. Results

3.1. Physiological Parameters

The irrigation and K doses showed significant interactive effects on leaf Ψ_w and Ψ_s (Figure 1). When irrigation dose was decreased to 30% together with 0.01 mM K, leaf Ψ_w and Ψ_s were each lowered to −2.7 MPa by 90 DAP but increasing the K concentration to 32 mM increased leaf Ψ_w to −1.6 MPa and leaf Ψ_s to −1.8 MPa. Conversely, when irrigation dose was decreased to 60% together with 0.01 mM K, leaf Ψ_w and Ψ_s were each lowered to −2.3 MPa; however, increasing the K concentration to 32 mM increased leaf Ψ_w to −0.8 MPa and leaf Ψ_s to −1.3 MPa by 90 DAP. In general, leaf Ψ_w and Ψ_s values remained maximum in full irrigated plants together with 32 mM K.

The different irrigation and K doses significantly affected leaf Ψ_p but showed no interactive effects (Table 1). Leaf Ψ_p greatly reduced when irrigation dose was decreased to 30% rather than to 60%, compared with full-irrigated plants (100%). Notably, increasing the K concentration to either 16 or 32 mM resulted in increased Ψ_p by 90 DAP, regardless of the irrigation doses. In addition, the Ψ_p loss point or the critical potential was observed at 90 DAP, where leaf Ψ_p was greatly reduced when irrigation dose was decreased to 30% and K concentration was 0.01 mM.

Table 1. Leaf turgor (Ψ_p) of young cassava plants in response to deficit irrigation and K fertigation. The treatments were initiated 30 days after planting and lasted 60 days. The data from four separate experiments were combined and shown as the means ± standard error of 4–16 replicate plants.

Treatment		Leaf Turgor (MPa)				
		30 DAP	**45 DAP**	**60 DAP**	**75 DAP**	**90 DAP**
Irrigation	30%	0.46	0.39	0.29 [a]	0.18 [a]	0.11 [a]
	60%	0.48	0.43	0.37 [b]	0.36 [b]	0.30 [b]
	100%	0.48	0.49	0.50 [c]	0.51 [c]	0.55 [c]
	S.E.M (df = 3–15)	0.017	0.036	0.039	0.034	0.038
Potassium	0.01 mM	0.47	0.4	0.29	0.25 [a]	0.10 [a]
	1 mM	0.48	0.42	0.35	0.30 [ab]	0.24 [b]
	4 mM	0.46	0.43	0.38	0.34 [b]	0.32 [c]
	16 mM	0.48	0.46	0.45	0.43 [c]	0.45 [d]
	32 mM	0.48	0.48	0.47	0.46 [c]	0.49 [d]
	S.E.M (df = 3–15)	0.022	0.046	0.05	0.044	0.05
p-value (<0.05)	I	0.738	0.134	0.001	<0.001	<0.001
	K	0.964	0.710	0.068	0.004	<0.001
	I × K	0.896	0.998	1.000	0.997	0.288

Means followed by different letters in the same column are different (Tukey's test $p < 0.05$). DAP = days after planting; I = irrigation; K = potassium; S.E.M. = standard error of the mean; df = degrees of freedom.

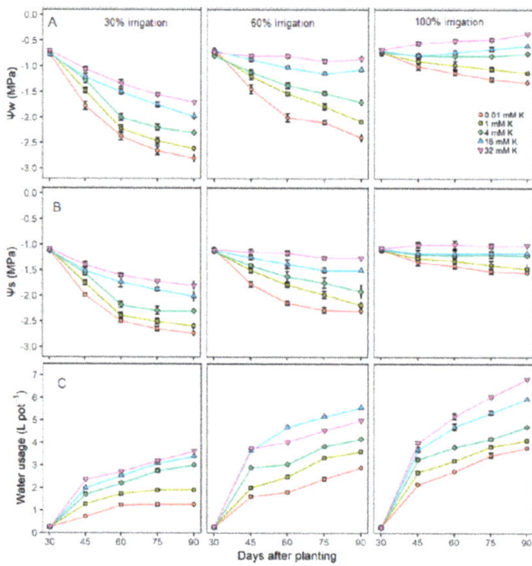

Figure 1. Leaf water potential Ψ_w (**A**), leaf osmotic potential Ψ_s (**B**), and water usage (**C**) of young cassava plants in response to deficit irrigation and K fertigation. The treatments were initiated 30 days after planting and lasted 60 days. The data from four separate experiments were combined and shown as the means ± standard error of 4 to 16 replicate plants.

Plant water usage followed a trend similar to that observed in leaf Ψ_w (Figure 1). Decreasing irrigation dose to 30% together with 0.01 mM K caused an 80% reduction in water usage between 30 and 90 DAP. However, when the K concentration was increased to 32 mM, water usage reduced by 45% compared with full-irrigated plants. Notably, decreasing irrigation dose to 60% together with 16 mM K only caused a 13% reduction in water usage compared with full-irrigated plants.

Moreover, the irrigation and K doses showed significant interactive effects on net photosynthesis, stomatal conductance, leaf temperature, and leaf chlorophyll (Figures 2 and 3). Decreasing irrigation dose to 30% together with 0.01 mM K reduced the net photosynthesis and stomatal conductance by 70% and leaf chlorophyll by 51% between 30 and 90 DAP. However, when the K concentration was increased to 16 mM, the net photosynthesis reduced by 35%, stomatal conductance by 41%, and leaf chlorophyll by 34% between 30 and 90 DAP, compared with full-irrigated plants. In contrast, decreasing irrigation dose to 60% together with 0.01 mM K reduced net photosynthesis by 56%, stomatal conductance by 42%, and leaf chlorophyll by 38% between 30 and 90 DAP. Moreover, increasing the K concentration to 16 mM reduced the net photosynthesis by 21%, stomatal conductance by 19%, and leaf chlorophyll by 18%. In all, decreasing irrigation dose to 60% together with 16 mM K resulted in the least reduction in net photosynthesis and stomatal conductance, although the least reduction in leaf chlorophyll was observed when irrigation dose was decreased to 60% together with 32 mM K.

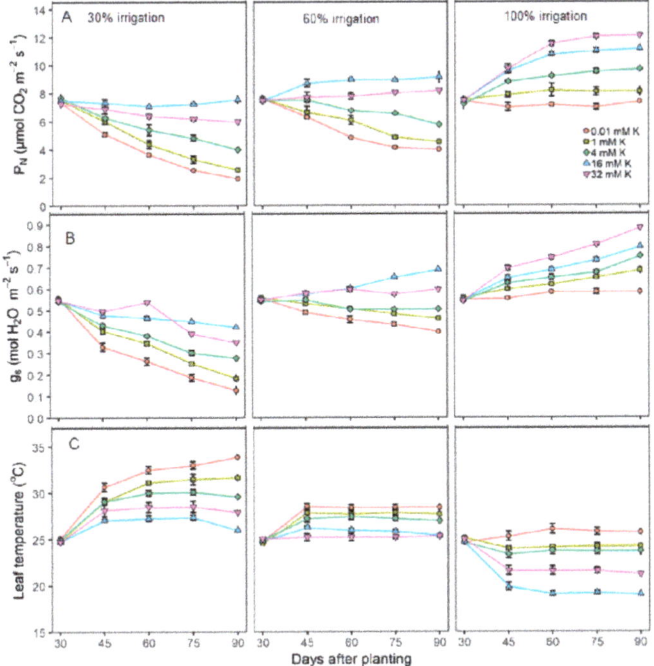

Figure 2. Net photosynthesis P_N (**A**), stomatal conductance g_s (**B**), and leaf temperature (**C**) of young cassava plants in response to deficit irrigation and K fertigation. The treatments were initiated 30 days after planting and lasted 60 days. The data from four separate experiments were combined and shown as the means ± standard error of 4 to 16 replicate plants.

The leaf temperature increased when the irrigation dose was decreased, but the leaf temperature declined when the K concentration was increased (Figure 2). At 90 DAP, the leaf temperature was highest (34 °C) when irrigation dose was decreased to 30% together with 0.01 mM K, but when the K was increased to 16 mM, the leaf temperature was low (26 °C). Likewise, when irrigation dose was decreased to 60% together with 0.01 mM K, the leaf temperature was 28 °C, but when K was increased to either 16 or 32 mM, the leaf temperature was 26 °C. In all, decreasing irrigation dose to 60% together with 32 mM K showed the lowest leaf temperature (25 °C) between 30 and 90 DAP compared with full-irrigated plants.

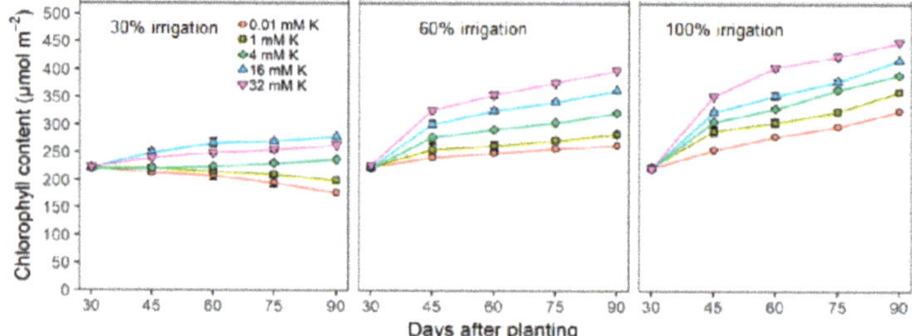

Figure 3. Leaf chlorophyll content of young cassava plants in response to deficit irrigation and K fertigation. Treatments were initiated 30 days after planting and lasted 60 days. The data from four separate experiments were combined and shown as the means ± standard error of 4 to 16 replicate plants.

3.2. Growth Parameters

The irrigation and K doses showed interactive effects on plant height (Figure 4). The plants were 49% shorter by 90 DAP when irrigation dose was decreased to 30% but 27% shorter when irrigation dose was decreased to 60%, compared with full-irrigated plants. Notably, lowering the K concentration below 16 mM resulted in shorter plants, regardless of the irrigation doses. The plants were 72% shorter when irrigation dose was decreased to 30% together with 0.01 mM K but were 52% shorter when K was increased to 16 mM K by 90 DAP, compared with full-irrigated plants. The smallest difference in heights was obtained when irrigation dose was decreased to 60% together with 32 mM K, compared with full-irrigated plants.

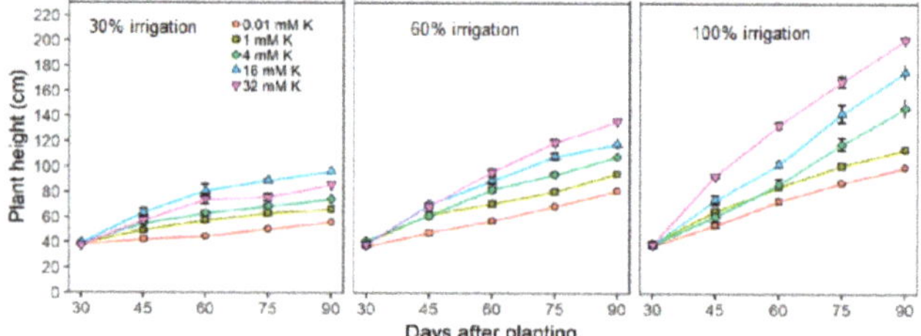

Figure 4. Plant height of young cassava plants in response to deficit irrigation and K fertigation. The treatments were initiated 30 days after planting and lasted 60 days. The data from four separate experiments were combined and shown as the means ± standard error of 4 to 16 replicate plants.

The plant leaf area, tuber number, leaf dry mass, stem dry mass, root dry mass, and whole-plant dry mass differed significantly with the irrigation and K doses and their interactions (Table 2). Decreasing irrigation doses substantially reduced these parameter values, while increasing the K increased their values. The plant leaf areas were 17% smaller and whole-plant dry mass was 41% lower when irrigation dose was decreased to 30% together with 16 mM K, compared with full-irrigated plants. In contrast, the plant leaf areas were 8% smaller and whole-plant dry mass was 13% lower when irrigation dose was decreased to 60% together with 16 mM K. This treatment combination showed the least decrease compared with full-irrigated plants. Otherwise, the largest leaf areas were obtained in full-irrigated plants together with 16 or 32 mM K, which were similar, whereas the highest whole-plant dry mass was obtained in full-irrigated plants with 32 mM K.

Table 2. Plant leaf area, leaf dry mass, shoot dry mass, root dry mass, and whole-plant dry mass of 90-day-old cassava plants in four pot experiments. Deficit irrigation and K fertigation were initiated 30 days after planting and lasted 60 days. The data from separate experiments were combined and shown as the means of 4 to 16 replicate plants.

Treatment		Plant Leaf Area (m^2)	Dry Mass (g)			
			Leaves	Shoot	Roots	Whole-Plant
Irrigation	Potassium					
30%	0.01 mM	0.37 [a]	16.9 [a]	13.6 [a]	11.0 [a]	41.4 [a]
	1 mM	0.38 [a]	20.0 [ab]	24.6 [bcd]	13.5 [ab]	58.1 [b]
	4 mM	0.44 [ab]	20.6 [ab]	24.4 [bcd]	17.7 [bc]	62.7 [b]
	16 mM	0.57 [d]	29.7 [c]	29.4 [cde]	25.5 [d]	84.6 [c]
	32 mM	0.48 [bc]	20.6 [ab]	27.7 [bcd]	27.5 [d]	75.8 [c]
60%	0.01 mM	0.38 [a]	23.3 [b]	19.0 [ab]	22.8 [d]	65.0 [b]
	1 mM	0.48 [b]	33.6 [cd]	33.3 [de]	26.8 [d]	93.6 [d]
	4 mM	0.59 [d]	33.7 [cd]	33.0 [de]	34.7 [e]	101.4 [de]
	16 mM	0.63 [de]	39.9 [e]	37.5 [ef]	46.8 [gh]	124.2 [h]
	32 mM	0.62 [de]	33.6 [de]	33.6 [de]	41.3 [fg]	108.6 [ef]
100%	0.01 mM	0.42 [ab]	34.8 [d]	20.8 [abc]	24.7 [d]	80.3 [c]
	1 mM	0.56 [cd]	32.1 [cd]	44.5 [fg]	38.5 [ef]	115.2 [fg]
	4 mM	0.62 [de]	35.3 [d]	47.2 [g]	39.1 [ef]	121.6 [gh]
	16 mM	0.68 [e]	40.5 [d]	44.8 [fg]	55.0 [i]	140.3 [i]
	32 mM	0.68 [e]	40.2 [d]	50.9 [g]	51.8 [hi]	142.9 [i]
S.E.M (df = 3–15)		0.016	0.86	1.92	1.12	1.82
p-value (<0.05)	I	<0.001	<0.001	<0.001	<0.001	<0.001
	K	<0.001	<0.001	<0.001	<0.001	<0.001
	I × K	<0.001	<0.001	0.001	<0.001	<0.001

Means followed by different letters in the same column are different (Tukey's test $p < 0.05$). S.E.M. = standard error of the mean; df = degrees of freedom; I = irrigation; K = potassium.

The leaf, stem, and root dry mass followed trends similar to that observed with whole-plant dry mass. Decreasing irrigation dose to 30% together with 16 mM K resulted in fewer tubers. Conversely, the tubers were much fewer when the K concentration was lowered below 16 mM. Substantially high numbers of tubers were obtained when irrigation dose was decreased to 60% together with 16 or 32 mM K, which did not vary.

3.3. Correlation of Physiological and Growth Traits

Statistically significant relationships were observed among the traits measured (Table 3). Net photosynthesis was highly correlated ($0.906 \leq r \leq 0.980$; $p < 0.05$) with stomatal conductance, leaf Ψ_w, leaf Ψ_s, leaf chlorophyll, water usage, and whole-plant dry mass. Positive associations ($0.719 \leq r \leq 0.897$; $p < 0.05$) were also found between plant leaf area, plant height, tuber number, and water usage. However, the leaf temperature correlated negatively ($-0.923 \leq r \leq -0.772$; $p < 0.05$) with all traits measured.

Table 3. Pearson correlation matrix for morpho-physiological traits of young cassava plants in response to deficit irrigation and K fertigation.

	LDM	RDM	WPD	TBN	LA	PLH	Chl	LT	P_N	g_s	Ψ_w	Ψ_s	Ψ_p	WUS
LDM	1													
RDM	0.826 **	1												
WPD	0.875 **	0.951 **	1											
TBN	0.719 **	0.825 **	0.834 **	1										
LA	0.736 **	0.832 **	0.887 **	0.743 **	1									
PLH	0.792 **	0.897 **	0.911 **	0.812 **	0.823 **	1								
Chl	0.857 **	0.928 **	0.936 **	0.777 **	0.832 **	0.930 **	1							
LT	−0.807 **	−0.870 **	−0.894 **	−0.809 **	−0.772 **	−0.869 **	−0.889 **	1						
P_N	0.838 **	0.915 **	0.933 **	0.823 **	0.835 **	0.928 **	0.939 **	−0.912 **	1					
g_s	0.881 **	0.906 **	0.940 **	0.781 **	0.786 **	0.892 **	0.951 **	−0.923 **	0.939 **	1				
Ψ_w	0.825 **	0.896 **	0.915 **	0.759 **	0.819 **	0.890 **	0.953 **	−0.898 **	0.941 **	0.956 **	1			
Ψ_s	0.802 **	0.869 **	0.886 **	0.725 **	0.792 **	0.865 **	0.929 **	−0.880 **	0.921 **	0.936 **	0.975 **	1		
Ψ_p	0.682 **	0.746 **	0.763 **	0.662 **	0.690 **	0.737 **	0.781 **	−0.725 **	0.763 **	0.773 **	0.820 **	0.671 **	1	
WUS	0.831 **	0.935 **	0.933 **	0.775 **	0.864 **	0.931 **	0.964 **	−0.868 **	0.931 **	0.923 **	0.946 **	0.928 **	0.761 **	1

**: differences between traits are significant at $p < 0.01$. LDM = leaf dry mass; RDM = root dry mass; WPD = whole-plant dry mass; TBN = tuber number; LA = leaf area; PLH = plant height; Chl = leaf chlorophyll; LT = leaf temperature; P_N = net photosynthesis; g_s = stomatal conductance; Ψ_w = leaf water potential; Ψ_s = leaf osmotic potential; Ψ_p = leaf turgor; WUS = water usage.

3.4. Potassium Content

The irrigation and K doses significantly affected the K content but showed no interactive effects (Table 4). When irrigation dose was decreased to 30%, the K content in the roots was 14% lower, but when irrigation dose was decreased to 60%, the K content in the roots was 9% lower. Increasing the K concentration to 32 mM resulted in the highest K content in both the leaves and roots. In general, K contents were all higher in the leaves than in the roots.

Table 4. Potassium (K) content in the leaves and roots of 90-day-old cassava plants in four pot experiments. Deficit irrigation and K fertigation were initiated 30 days after planting and lasted 60 days. The data from separate experiments were combined and shown as the means of 4 to 16 replicate plants.

Treatment		K Content (g kg^{-1} Dry Matter)	
		Leaves	**Roots**
Irrigation	30%	12.6 [a]	9.9
	60%	14.6 [b]	10.5
	100%	14.7 [b]	11.5
	S.E.M (df = 3–15)	0.89	0.73
	p–value (<0.05)	0.01	0.061
Potassium	0.01 mM	11.1 [a]	3.6 [a]
	1 mM	12.2 [a]	10.0 [b]
	4 mM	14.5 [ab]	11.3 [b]
	16 mM	14.8 [ab]	13.7 [c]
	32 mM	17.4 [b]	14.7 [c]
	S.E.M (df = 3–15)	1.08	0.49
	p–value (<0.05)	<0.001	<0.001

Means followed by different letters in the same column are different (Tukey's test $p < 0.05$). S.E.M. = standard error of the mean; df = degrees of freedom.

4. Discussion

Our findings show that water deficit and the K concentrations influence the water status of young cassava plants, thereby affecting the leaf gas exchange and causing a decline in growth and yield. Decreasing irrigation dose to 30% together with 0.01 mM K lowered leaf Ψ_w and Ψ_s greatly, while leaf Ψ_p was five times lower at the end of our experimental period (90 DAP) compared with full-irrigated plants (100%). The decrease in Ψ_s in response to the water stress imposed is an osmotic adjustment mechanism used by most plants to adjust to water-limited environments [39] and contributes to Ψ_p maintenance at low Ψ_w [40,41]. Osmotic adjustment enables leaf Ψ_p maintenance for the same leaf Ψ_w, thus supporting stomatal conductance [42], and improves root capacity for water uptake [43]. Moreover, our results showed that increasing the K concentration to 32 mM increased both leaf Ψ_w and Ψ_s. This effect could have been related to the high K$^+$ solute concentration in the leaves, which seemed to increase the Ψ_s, as evident from elemental K content analyzed in the leaves. Furthermore, K is one of the primary osmotic solutes that contribute to osmotic adjustment in plants by altering the Ψ_s and enabling plant cells to retain water and maintain Ψ_p [44,45].

Reduction in water usage parallelled the decrease in leaf Ψ_w. Significant positive correlations ($r = 0.923$) observed between water usage and stomatal conductance suggest that the reduction in water usage was probably induced by stomatal closure. Cassava responds to initial water deficit by partial stomatal closure and nearly complete stomatal closure during extreme water deficit [8,17]. Duque and Setter [46] suggested that stomatal closure protects the leaf from severe water loss and protects photosynthetic systems and cellular structures from irreversible damage. Our results showed that effective water usage was greatest when irrigation dose was decreased to 60% together with 32 mM K, given the high amount of dry mass observed relative to full-irrigated plants. High dry mass production under water deficit is achieved when plants divert a large portion of available soil moisture towards stomatal conductance [47]. The water usage and stomatal conductance data further suggest that partial stomatal closure occurred at about 60 DAP when the water level was decreased to 30% or 60% of pot capacity. The marked increase in water usage, regardless of the irrigation doses when the K concentration was increased to 32 mM, suggests that high levels of K increased the leaf Ψ_s and consequently increased Ψ_w, resulting in Ψ_p maintenance and increased water usage.

The observed depletions in net photosynthesis was significantly associated with a significant reduction in stomatal conductance. Our results showed that the decline in net photosynthesis was more pronounced when irrigation dose was decreased to 30% together with 0.01 mM K, but less pronounced when irrigation dose was decreased to 60% of pot capacity. Even more, the strong positive correlation observed between net photosynthesis and stomatal conductance ($r = 0.939$) and leaf Ψ_w ($r = 0.941$) suggests that photosynthesis was limited by stomatal closure in response to a limited water supply. The decline in leaf chlorophyll due to decreased irrigation doses additionally limited photosynthesis. This decline in leaf chlorophyll is considered a non-stomatal limiting factor [48] and is prevalent under water-deficit conditions, causing decreased photosynthetic activity [49]. Chlorophyll degradation also occurs in K-deficient plants [25], which further inhibits photosynthesis. Conversely, low-K nutrition diminishes Hill reaction activity [50] and the rate of production of adenosine triphosphate (ATP) and reduces nicotinamide adenine dinucleotide phosphate (NADP) in chloroplasts [51], resulting in reduced net photosynthesis. In contrast, increasing the K concentration to 16 mM increased net photosynthesis by increasing leaf Ψ_w and leaf Ψ_p, which in turn increased stomatal conductance, resulting in increased net photosynthesis. Moreover, Ψ_p determines stomatal aperture and closure [52,53], and the extent of stomatal closure in cassava during water deficit levels corresponds to the decline in net photosynthesis [8,54]. The enhanced need for K by plants under water deficit appears to be related to the fact that K is required for the maintenance of photosynthetic CO_2 fixation [25].

The leaf temperatures increased above the ambient greenhouse temperature (27 °C) when irrigation doses were decreased. Our results showed that increased leaf temperature corresponded to decreased net photosynthesis as indicated by the significant negative correlations between leaf temperature and net photosynthesis ($r = -0.912$), and between leaf temperature and leaf Ψ_w ($r = -0.898$). This observation implies that the rise in leaf temperature was occasioned by stomata closure in response to a decline in Ψ_w. High leaf temperature leads to heat stress, and photosynthesis is more sensitive to heat stress under water deficit [55]. Thus, C_3–C_4 plants such as cassava rely on evaporative cooling to lower leaf temperature [56]. Despite decreased irrigation doses, increasing the K concentration to 16 mM and above appeared to lower leaf temperature. This was attributed to the role of high K concentration, which increased stomatal conductance and, thus, prevented leaf temperature from reaching harmful levels [57].

Plant leaf area and plant height were greatly reduced when irrigation dose was decreased to 30% together with 0.01 mM K. These reductions could be linked to the decrease observed in leaf Ψ_w, low Ψ_p, decreased stomatal conductance, and declined net photosynthesis. Raza et al. [58] showed that water deficit initiates a series of biochemical and physiological processes that result in a reduction in crop growth and yield. The small leaf areas (0.37 m^2) and short plants (56 cm) observed at the end of the experiment when irrigation dose was decreased to 30% together with 0.01 mM K were probably due to the effects of low leaf Ψ_p. In comparison, plant leaf areas and plant height increased due to increased Ψ_p when irrigation dose was decreased to 60% of pot capacity. These observations concur with Alves and Setter [9], who found decreased leaf area expansion in cassava 8 days after initiation of water deficit. Nesreen et al. [59] reported reduced leaf area, plant height, and stem diameter in cassava subjected to water deficit under greenhouse conditions. Moreover, optimal leaf area development is important to photosynthesis and dry mass yield [60].

Leaf dry mass, root dry mass, whole-plant dry mass, and tuber number were equally reduced when irrigation dose was decreased to 30% together with 0.01 mM K. These reductions can be largely attributed to the decline in net photosynthesis and reduced leaf areas that were influenced by the low Ψ_w, Ψ_s, and Ψ_p. Duque and Setter [46] reported a 78% loss in total plant dry weight in cassava after 31-days of water deficit treatment. El-Shakwy and Cadavid [61] observed reductions in leaf area index and shoot and root biomass when they initiated water stress in three cassava cultivars for 2 to 8 months after planting. Nevertheless, increasing the K concentration to 16 mM improved the growth and yield of young cassava plants, regardless of decreased irrigation doses. This could be ascribed to the high levels of K that reduced the negative effects on plant-water relationships and improved net photosynthesis. This observation agrees with the findings of Mengel and Arneke [44], who reported improved water status and high dry mass in cassava supplied with high levels of K (4 mM) in comparison to low-K (0.1 mM) treatments. There were no tubers when irrigation dose was decreased to 30% together with 0.01 mM K, whereas four tubers were observed when K was increased to 16 mM. High levels of K increase the translocation of photosynthates to the storage roots [62], and the photosynthates initiate cassava tuberous root growth during the early growth phase [8,21].

The critical nutrient contents of K observed in this investigation fall within the sufficient range as those obtained by Nguyeh et al. [63] in field-grown cassava at four months after planting. The decreased K contents observed when irrigation doses were decreased were probably due to the limited water supply, which possibly limited the K uptake from the rhizosphere. Plants experiencing both water and K deficiency tended to show decreases in K accumulation rates, while those grown in K-rich soil maintained high rates of K accumulation during most of the season [64]. Moreover, our findings show that increasing K from 16 to 32 mM did not result in additional growth and yield benefits, regardless of the irrigation doses. Thus, it appears that 32 mM K was excessive and could have limited the uptake of other cations, creating a nutrient imbalance [65].

5. Conclusions

Our investigation showed significant interactions between deficit irrigation and K, whereby decreasing irrigation dose to 60% together with 16 mM K resulted in the least reduction in growth and yield. Thus, it seems that deficit irrigation strategies could be used as a tool to develop management practices to improve cassava productivity by means of K fertigation under low moisture field conditions. The experiments allowed analytical investigation of the effects of the irrigation and K doses without interference from underlying abiotic and biotic factors. Nevertheless, the present investigation had certain limitations such as the use of pots, which could have constrained plant growth, and the use of a controlled greenhouse environment, which varies from field environments. Therefore, further tests with several cassava cultivars under field conditions are warranted to compare these findings.

Author Contributions: Conceived and designed the experiment D.O.W., P.S.A.M., and J.K.; contributed the reagents/materials P.S.A.M.; conducted the experiments D.O.W.; supervised the research P.S.A.M., J.K., and L.A.; performed the statistical analysis D.O.W., and P.S.A.M.; wrote the first draft of the manuscript D.O.W. All authors have read and agreed to the published version of the manuscript.

Funding: This work was supported by CIMO Finland (TM-17-10552) and the Ella and Georg Ehrnrooth Foundation. The open access publication fee was covered by the University of Helsinki.

Acknowledgments: We gratefully acknowledge Markku Tykkyläinen, Afrane Yaw, and Marjo Kilpinen for their technical assistance with the greenhouse and laboratory activities. We thank the Kenya Agricultural and Livestock Research Organization (KALRO) for providing the cassava cuttings used in this investigation.

Conflicts of Interest: The authors declare no conflict of interest. The funders played no role in the design of the study; in the collection, analyses, or interpretation of data; in the writing of the manuscript or in the decision to publish the results.

References

1. Toumi, J.; Er-Raki, S.; Ezzahar, J.; Khabba, S.; Jarlan, L.; Chehbouni, A. Performance assessment of AquaCrop model for estimating evapotranspiration, soil water content and grain yield of winter wheat in Tensift Al Haouz (Morocco): application to irrigation management. *Agric. Water Manag.* **2016**, *163*, 219–235. [CrossRef]
2. Cantero-Navarro, E.; Romero-Aranda, R.; Fernández-Muñoz, R.; Martínez-Andújar, C.; Pérez-Alfocea, F.; Albacete, A. Improving agronomic water use efficiency in tomato by rootstock-mediated hormonal regulation of leaf biomass. *J. Plant Sci.* **2016**, *251*, 90–100. [CrossRef] [PubMed]

3.   Burns, A.; Gleadow, R.; Cliff, J.; Zacarias, A.; Cavagnaro, T. Cassava: the drought, war and famine crop in a changing world. *Sustainability* **2010**, *2*, 3572–3607. [CrossRef]

4.   Wobeto, C.; Corrêa, A.D.; Abreu, C.M.P.; Santos, C.D.; Abreu, J.R. Nutrients in the cassava (*Manihot esculenta* Crantz) leaf meal at three ages of the plant. *Food Sci. Technol.* **2006**, *26*, 865–869. [CrossRef]

5.   Montagnac, J.A.; Davis, C.R.; Tanumihardjo, S.A. Nutritional value of cassava for use as a staple food and recent advances for improvement. *Compr. Rev. Food Sci. Food Saf.* **2009**, *8*, 181–194. [CrossRef]

6.   Balagopalan, C. Cassava utilization in food, feed and industry. In *Cassava: Biology, Production and Utilization*; Hillock, R.J., Thresh, J.M., Bellotti, A.C., Eds.; CABI: Wallingford, Oxfordshire, UK, 2002; pp. 301–318.

7.   Tonukari, N.J.; Ezedom, T.; Enuma, C.C.; Sakpa, S.O.; Avwioroko, O.J.; Eraga, L.; Odiyoma, E. White gold: cassava as an industrial base. *Am. J. Plant Sci.* **2015**, *6*, 972–979. [CrossRef]

8.   El-Sharkawy, M.A. Cassava biology and physiology. *Plant Mol. Biol.* **2004**, *56*, 481–501. [CrossRef]

9.   Alves, A.A.C.; Setter, T.L. Response of cassava leaf area-expansion to water deficit: Cell proliferation, cell expansion and delayed development. *Ann. Bot.* **2004**, *94*, 605–613. [CrossRef]

10.   Oguntunde, P.G. Whole-plant water use and canopy conductance of cassava under limited available soil water and varying evaporative demand. *Plant Soil.* **2005**, *278*, 371–383. [CrossRef]

11.   Howeler, R.H. Long-term effect of cassava cultivation on soil productivity. *J. Field Crops Res.* **1991**, *26*, 1–18. [CrossRef]

12.   El-Sharkawy, M.A.; De Tafur, S.M. Comparative photosynthesis, growth, productivity, and nutrient use efficiency among tall and short-stemmed rain-fed cassava cultivars. *Photosynthetica.* **2010**, *48*, 173–188. [CrossRef]

13.   Ike, I.F.; Thurtell, G.W.; Thurlell, G.W. Osmotic adjustment in indoor grown cassava in response to water stress. *Physiol. Plant.* **1981**, *52*, 257–262. [CrossRef]

14.   El-Sharkawy, M.A.; Cock, J.H. Water use efficiency of cassava. I. Effects of air humidity and water stress on stomatal conductance and gas exchange. *Crop Sci.* **1984**, *24*, 497–502. [CrossRef]

15.   El-Sharkawy, M.A. Stress-tolerant cassava: the role of integrative ecophysiology-breeding research in crop improvement. *J. Soil Sci.* **2012**, *2*, 162–186. [CrossRef]

16.   Alves, A.A.C.; Setter, T.L. Response of cassava to water deficit: leaf area growth and abscisic acid. *J. Crop Sci.* **2000**, *40*, 131–137. [CrossRef]

17.   Calatayud, P.A.; Llovera, E.; Bois, J.F.; Lamaze, T. Photosynthesis in drought-adapted cassava. *Photosynthetica* **2000**, *38*, 97–104. [CrossRef]

18.   Bakayoko, S.; Tschannen, A.; Nindjin, C.; Dao, D.; Girardin, O.; Assa, A. Impact of water stress on fresh tuber yield and dry matter content of cassava (*Manihot esculenta* Crantz) in Côte d'Ivoire. *Afr. J. Agric. Res.* **2009**, *4*, 21–27.

19.   Vandegeer, R.; Miller, R.E.; Bain, M.; Gleadow, R.M.; Cavagnaro, T.R. Drought adversely affects tuber development and nutritional quality of the staple crop cassava (*Manihot esculenta* Crantz). *Funct. Plant Biol.* **2013**, *40*, 195–200. [CrossRef]

20.   El-Sharkawy, M.A. Drought-tolerant cassava for Africa, Asia, and Latin America. *BioScience* **1993**, *43*, 441–451. [CrossRef]

21.   Alves, A.A.C. Cassava botany and physiology. In *Cassava: Biology, Production and Utilization*; Hillocks, R.J., Tres, J.M., Bellotti, A.C., Eds.; CABI: Wallingford, Oxfordshire, UK, 2002; pp. 67–89.

22.   Baker, G.R.; Fukai, S.; Wilson, G.L. Response of cassava to water deficits at various stages of growth in the subtropics. *Aust. J. Agric. Res.* **1989**, *40*, 517–528. [CrossRef]

23.   Cakmak, I.; Engels, C. Role of mineral nutrients in photosynthesis and yield formation. In *Mineral Nutrition of Crops*; Rengel, Z., Ed.; Haworth Press: New York, NY, USA, 1999; pp. 141–168.

24.   Ahmad, I.; Maathuis, F.J.M. Cellular and tissue distribution of potassium; physiological relevance, mechanisms and regulation. *J. Plant Physiol.* **2014**, *171*, 708–714. [CrossRef] [PubMed]

25.   Cakmak, I. The role of potassium in alleviating detrimental effects of abiotic stresses in plants. *J. Plant Nutr. Soil Sci.* **2005**, *168*, 521–530. [CrossRef]

26.   Oosterhuis, D.M.; Loka, D.A.; Kawakami, E.M.; Pettigrew, W.T. The physiology of potassium in crop production. *Adv. Agron.* **2014**, *126*, 203–233.

27.   Susan John, K.; Suja, G.; Sheela, M.N.; Ravindran, C.S. Potassium: the key nutrient for cassava production, tuber quality and soil productivity—An overview. *J. Root Crops* **2010**, *36*, 132–144.

28. Zörb, C.; Senbayram, M.; Peiter, E. Potassium in agriculture–Status and perspectives. *J. Plant Physiol.* **2014**, *171*, 656–669.

29. Byju, G.; Nedunchezhiyan, M.; Ravindran, C.S.; Mithra, V.S.S.; Ravi, V.; Naskar, S.K. Modelling the response of cassava to fertilizers: a site-specific nutrient management approach for greater tuberous root yield. *Commun. Soil Sci. Plant Anal.* **2012**, *43*, 1149–1162. [CrossRef]

30. Odubanjo, O.O.; Olufayo, A.A.; Oguntunde, P.G. Water use, growth and yield of drip irrigated cassava in a humid tropical environment. *Soil Water Res.* **2011**, *6*, 10–20. [CrossRef]

31. Santanoo, S.; Vongcharoen, K.; Banterng, P.; Vorasoot, N.; Jogloy, S.; Roytrakul, S.; Theerakulpisut, P. Seasonal variation in diurnal photosynthesis and chlorophyll fluorescence of four genotypes of cassava (*Manihot esculenta* Crantz) under irrigation conditions in a tropical savanna climate. *Agronomy* **2019**, *9*, 206. [CrossRef]

32. Sopheap, U.; Patanothai, A.; Aye, T.M. Nutrient balances for cassava cultivation in Kampong Cham province in Northeast Cambodia. *Int. J. Plant Prod.* **2012**, *6*, 37–58.

33. Imas, P.; John, S.K. *Potassium Nutrition of Cassava*; e-ifc No. 34; International Potash Institute: Zug, Switzerland, 2013; Volume 34, pp. 13–18.

34. Ezui, K.S.; Franke, A.C.; Mando, A.; Ahiabor, B.D.K.; Tetteh, F.M.; Sogbedji, J.; Janssen, B.H.; Giller, K.E. Fertiliser requirements for balanced nutrition of cassava across eight locations in West Africa. *J. Field Crops Res.* **2016**, *185*, 69–78. [CrossRef]

35. Hoagland, D.R.; Arnon, D.I. The water-culture method for growing plants without soil. *Calif. Agric. Exp. Stat.* **1950**, *Cir. 347*, 1–39.

36. McCutchan, H.; Shackel, K.A. Stem-water potential as a sensitive indicator of water stress in prune trees (*Prunus domestica* L. cv. French). *J. Am. Soc. Hortic. Res.* **1992**, *117*, 606–611. [CrossRef]

37. Mäkelä, P.; Munns, R.; Colmer, T.D.; Condon, A.G.; Peltonen-Sainio, P. Effect of foliar applications of glycinebetaine on stomatal conductance, abscisic acid and solute concentrations in leaves of salt- or drought-stressed tomato. *Aust. J. Plant Physiol.* **1998**, *25*, 655–663. [CrossRef]

38. R Core Team. *A Language and Environment for Statistical Computing*; R Foundation for Statistical Computing: Vienna, Austria, 2018.

39. Morgan, J.M. Osmoregulation and water stress in higher plants. *Ann. Rev. Plant Physiol.* **1984**, *35*, 299–319. [CrossRef]

40. Radin, J.W. Physiological consequences of cellular water deficits: osmotic adjustment. In *Limitations to Efficient Water Use in Crop Production*; Taylor, H.M., Jordan, W.R., Sinclair, T.R., Eds.; American Society of Agronomy, Inc, Crop Science Society of America, Inc., Soil Science Society of America, Inc.: Madison, WI, USA, 1983; pp. 267–276.

41. Blum, A. Drought resistance, water-use efficiency, and yield potential–are they compatible, dissonant, or mutually exclusive? *Aust. J. Agric. Res.* **2005**, *56*, 1159–1168. [CrossRef]

42. Sellin, A. Hydraulic and stomatal adjustment of Norway spruce trees to environmental stress. *Tree Physiol.* **2001**, *21*, 879–888. [CrossRef]

43. Chimenti, C.A.; Marcantonio, M.; Hall, A.J. Divergent selection for osmotic adjustment results in improved drought tolerance in maize (*Zea mays* L.) in both early growth and flowering phases. *Field Crops Res.* **2006**, *95*, 305–315. [CrossRef]

44. Mengel, K.; Arneke, W.W. Effect of potassium on the water potential, the pressure potential, the osmotic potential and cell elongation in leaves of Phaseolus vulgaris. *J. Plant Physiol.* **1982**, *54*, 402–408. [CrossRef]

45. Wang, M.; Zheng, Q.; Shen, Q.; Guo, S. The critical role of potassium in plant stress response. *Int. J. Mol. Sci.* **2013**, *14*, 7370–7390. [CrossRef]

46. Duque, L.O.; Setter, T. Cassava response to water deficit in deep pots: root and shoot growth, ABA, and carbohydrate reserves in stems, leaves and storage roots. *Trop. Plant Biol.* **2013**, *6*, 199–209. [CrossRef]

47. Blum, A. Effective use of water (EUW) and not water-use efficiency (WUE) is the target of crop yield improvement under drought stress. *Field Crops Res.* **2009**, *112*, 119–123. [CrossRef]

48. Anjum, S.A.; Xie, X.; Wang, L.; Saleem, M.F.; Man, C.; Lei, W. Morphological, physiological and biochemical responses of plants to drought stress. *Afr. J. Agric. Res.* **2016**, *6*, 2026–2032.

49. Tuna, A.L.; Kaya, C.; Muhammad, A. Potassium sulfate improves water deficit tolerance in melon plants grown under glasshouse conditions. *J. Plant Nutr.* **2010**, *33*, 1276–1286. [CrossRef]

50. Spencer, D.; Possinghami, J.V. The effect of nutrient deficiences on the Hill reaction of isolated chloroplasts from tomato. *Aust. J. Biol. Sci.* **1960**, *13*, 441–455. [CrossRef]

51. Terry, N.; Ulrich, A. Effects of Potassium Deficiency on the Photosynthesis and Respiration of Leaves of Sugar Beet. *J. Plant Physiol.* **1973**, *51*, 783–786. [CrossRef]
52. Boyer, J.S. Leaf enlargement and metabolic rates in corn, soybean, and sunflower at various leaf water potentials. *J. Plant Physiol.* **1970**, *46*, 233–235. [CrossRef]
53. Franks, P.J.; Cowan, I.R.; Tyerman, S.D.; Cleary, A.I.; Lloyd, J.; Farquhar, G.D. Guard-cell pressure aperture characteristics measured with the pressure probe. *Plant Cell Environ.* **1995**, *18*, 795–800. [CrossRef]
54. Itani, J.; Oda, T.; Numao, T. Studies on mechanisms of dehydration postponement in cassava leaves under short-term soil water deficits. *J. Plant Prod Sci.* **1999**, *2*, 184–189. [CrossRef]
55. Björkman, O.; Badger, M.R.; Armond, P.A. Response and adaptation of photosynthesis to high temperatures. In *Adaptation of Plants to Water and High Temperatures Stress*; Turner, N.C., Kramer, P.J., Eds.; Wiley: New York, NY, USA, 1980; pp. 233–249.
56. El-Sharkawy, M.A. International research on cassava photosynthesis, productivity, eco-physiology and responses to environmental stresses in the tropics. *Photosynthetica.* **2006**, *44*, 481–512. [CrossRef]
57. Crawford, A.J.; McLachlan, D.H.; Hetherington, A.M.; Franklin, K.A. High temperature exposure increases plant cooling capacity. *Curr. Biol.* **2012**, *22*, 386–397. [CrossRef] [PubMed]
58. Raza, M.A.S.; Saleem, M.F.; Shah, G.M.; Khan, I.H.; Raza, A. Exogenous application of glycinebetaine and potassium for improving water relations and grain yield of wheat under drought. *J. Soil Sci. Plant Nutr.* **2014**, *14*, 348–364. [CrossRef]
59. Nesreen, A.S.H.; Eisa, S.S.; Amany, A. Morphological and Chemical Studies on Influence of Water Deficit on Cassava. *World J. Agric. Res.* **2013**, *9*, 369–376.
60. Farooq, M.; Wahid, A.; Kobayashi, N.; Fujita, D.; Basra, S.M.A. Plant drought stress: effects, mechanisms and management. *Agron. Sustain. Dev.* **2009**, *29*, 185–212. [CrossRef]
61. El-Sharkawy, M.A.; Cadavid, L.F. Response of cassava to prolonged water stress imposed at different stages of growth. *J. Exp. Agric.* **2002**, *38*, 253–264. [CrossRef]
62. Vreugdenhil, D. Source-to-sink gradient of potassium in the phloem. *Planta* **1985**, *163*, 238–240. [CrossRef]
63. Nguyen, H.; Schoenau, J.J.; Nguyen, D.; Van Rees, K.; Boehm, M. Effects of long-term nitrogen, phosphorus, and potassium fertilization on cassava yield and plant nutrient composition in North Vietnam. *Plant Nutr.* **2007**, *25*, 425–442. [CrossRef]
64. Grzebisz, W.; Gransee, A.; Szczepaniak, W.; Diatta, J. The effects of potassium fertilization on water-use efficiency in crop plants. *J. Plant Nutr. Soil Sci.* **2013**, *176*, 355–374. [CrossRef]
65. Howeler, R. *Sustainable Soil and Crop Management of Cassava in Asia*; Centro Internacional de Agricultura Tropical (CIAT): Cali, Colombia, 2014; Volume 389, p. 280.

MDPI

St. Alban-Anlage 66

4052 Basel

Switzerland

Tel. +41 61 683 77 34

Fax +41 61 302 89 18

www.mdpi.com

Agronomy Editorial Office

E-mail: agronomy@mdpi.com

www.mdpi.com/journal/agronomy